# STANDARDS FOR OPERATION OF NUTRIENT REMOVAL FACILITIES

## WEF Standard 37-24

2024

Water Environment Federation
601 Wythe Street
Alexandria, VA 22314–1994 USA
https://www.wef.org

ISBN 978-1-57278-466-6



## *About WEF*

The Water Environment Federation (WEF) is a not-for-profit technical and educational organization of 30,000 individual members and 75 affiliated Member Associations representing water quality professionals around the world. Since 1928, WEF and its members have protected public health and the environment. As a global water sector leader, our mission is to connect water professionals; enrich the expertise of water professionals; increase the awareness of the impact and value of water; and provide a platform for water sector innovation. To learn more, visit www.wef.org.

This edition edited by William C. McConnell, PE, BCEE

**With contributions by:**

Mohamad Baydoun, PhD
Jeanette Brown, PhD, PE, BCEE, F.WEF
Raj Chavan, PhD, PE, PMP
Alexandra T. Doody, PE
Georgine Grissop, PE, BCEE
Jennifer E. Lawrence, PhD, PE
Laura R. Marcolini, PE
Arifur Rahman, PhD, PE
David L. Wagoner, PE, NC IV WW Operator

**Review provided by:**

Sara Arabi, PhD, PE, BCEE
Raj Chavan, PhD, PE, PMP
Xiaohong He, PhD, PE
Paul Jue
Xiaojin Li, PhD
Yanjin Liu, PhD, PE
Layne McWilliams, PE, JD
Faisal Mullalayam BTech, MSc-Ing
Steven Reusser, PE
Keval Satra, PE
Scott Weirich
Jianfeng Wen, PhD, PE
David A. Winters, PE

Text originally prepared by the **Operation of Nutrient Removal Facilities** Task Force

William C. McConnell, PE, *Chair*

Nancy Afonso
Archis R. Ambulkar, MS
Vincent Apa, PE, BCEE
Sara Arabi, PhD, PEng
Richard G. Atoulikian, PMP, BCEE, PE
Vinod Barot, PhD
Somnath Basu, PhD, PE, BCEE
Jason Beck, PE
Katherine Bell, PhD, PE, BCEE
Mario Benisch, PE
Joshua P. Boltz, PhD, PE, BCEE
Lucas Botero, PE, BCEE
Bob Bower, BS, WWTPO 4
Jeanette Brown, PE, BCEE, D.WRE, F.WEF
Marie S. Burbano, PhD, PE, BCEE
James Clifton
John T. Cookson Jr., PhD
Rhodes R. Copithorn, PE, BCEE
John B. Copp, PhD
Catherine Crabb
Allegra K. da Silva, PhD
Bob Dabkowski
Sarah V. Dailey, PE
Tania Datta, PhD
Michael John Dempsey, PhD, C.Biol.
Timur Deniz, PhD, PE, BCEE
Leon S. Downing, PhD, PE
Ufuk Erdal, PhD, PE

Richard Finger
M. Kim Fries, M.A.Sc., PEng
Michael H. Gerardi
Mikel E. Goldblatt, PE
Gary M. Grey
Georgine Grissop, PE, BCEE
John Hahn
Susan Hansler, M.A.Sc., PEng
Ladan Holakoo, PhD, PE, PEng
Sarah Hubbell
Joseph A. Husband, PE, BCEE
Maria Inneo
Samuel S. Jeyanayagam, PhD, PE, BCEE
Gary R. Johnson, PE, BCEE
Ishin Kaya, PEng
Michael P. Keleman, MSEV, REHS
Frank Kulick
Robert Lagrange, PhD
Cory Lancaster
Dennis Lindeke
Yanjin Liu, PhD, PE
Jorj Long
Huijie Lu
Ting Lu, PhD
Nehreen Majed, PhD
Randall B. Marx, PhD, PE
Neil Massart, PE
Samir Mathur, PE, BCEE
Michael McGhee, PE
Jun Meng, MS, EIT
Indra N. Mitra, PhD, PE, MBA, BCEE
George Nakhla, PhD, PE, PEng
Robert C. O'Day, CHMM, QEP, ASP
Annalisa Onnis-Hayden, PhD
Jurek Patoczka, PhD, PE
Marie-Laure Pellegrin, PhD, PE
Peter Petersen
Heather M. Phillips, PE, BCEE
Scott Phipps, PE
Alvin Pilobello
Barry Rabinowitz, PhD, PEng, BCEE
Steven Reusser, PE

**Manuals of Practice and Standards of the Water Environment Federation**

The WEF Technical Practice and Publications Advisory Panel (formerly the Technical Practice Committee and the Committee on Sewage and Industrial Wastes Practice of the Federation of Sewage and Industrial Wastes Associations) was created by the Federation Board of Control on October 11, 1941. The primary function of the Advisory Panel is to originate and produce, through appropriate subcommittees, publications dealing with technical aspects of the broad interests of the Federation. These publications are intended to provide background information through a review of technical practices and detailed procedures that research and experience have shown to be functional and practical.

# Contents

## Chapter 4 Nitrification in Biofilm Reactors 35

## Chapter 10 Chemical Precipitation of Phosphorus 213

# List of Tables

# List of Figures

# Preface

An increasing number of water resource recovery facilities (WRRFs) are required to remove nutrients, and many facilities that have accomplished nutrient removal for some time are facing increasingly stringent performance requirements. At the same time, WRRF owners and operators are under pressure to meet these treatment requirements while simultaneously minimizing operations and maintenance costs—to do more with less. Successfully operating a stable, reliable, and efficient facility to remove nitrogen and/or phosphorus requires a high level of operator attention, skill, and knowledge. This updated edition provides guidance regarding the successful, efficient operation of nutrient removal facilities.

Water Environment Federation Standard 37-24 (WEF 37) is written primarily for facility managers and operators but will also be useful for design engineers and regulators and provides an understanding of the theory and typical design approaches for processes currently used for nitrogen and phosphorus removal to optimize facilities' ability to consistently meet regulatory requirements and provides general operational guidance on process control and troubleshooting methodologies. Early chapters of these standards provide technical information regarding fundamental biological and chemical processes that are in use at nutrient removal facilities, namely, nitrification, denitrification, enhanced biological phosphorus removal, and chemical phosphorus removal. Subsequent chapters illustrate configurations that use these processes at WRRFs to remove nitrogen, phosphorus, or both. Later chapters describe ways that operators may use, monitor, and control these processes to meet their facilities' treatment goals.

**Authors' and reviewers' efforts were supported by the following organizations:**

AECOM
AtkinsRealis
CDM Smith
City of Atlanta, Georgia
Elixir Enviro Systems Pvt Ltd, India
HR Green, Inc.
Jacobs
Madison Metropolitan Sewerage District
Manhattan College
Marcolini & Associates, Inc.
Parametrix
Stantec

# Chapter 1

# Introduction



## 1.0 INTRODUCTION

A *nutrient* is defined as a substance used in an organism's metabolism that must be taken in from its environment. Nutrients are essential to all forms of life, including those found within any natural waterbody. Natural levels of nutrients in the environment help in the growth of plants and animals. Aquatic life is supported and maintained through a healthy balance of the right amount and form of nutrients.

However, elevated levels of nutrients, particularly nitrogen and phosphorus, above critical levels specific to each waterbody, can result in excess growth of aquatic plant life, especially algae. The largest contributors of nitrogen and phosphorus to freshwater and marine ecosystems are local runoff (nonpoint sources), discharges from local industrial facilities, and municipal water resource recovery facilities (WRRFs) (point sources). A point-source pollutant can be traced to the specific location where the contribution enters the waterbody, whereas nonpoint-source pollutants enter a body of water from many diffuse sources.

There are numerous WRRFs worldwide that are required by local regulations to reduce nitrogen and phosphorus concentrations in their effluent. Effective nitrogen removal from a wastewater stream relies heavily on biological processes, wheras phosphorus removal may require a biological process, chemical precipitation, or a combination of both. Successful operation of a biological nutrient removal (BNR) facility requires both operator knowledge and involvement in the daily operations of the facility. To successfully achieve the required level of nutrient reduction within the treatment process, the operator must perform additional testing, sampling, analysis, decision making, and monitoring beyond that required for a standard secondary process, which is primarily focused on biochemical oxygen demand and suspended solids removal. In addition, successful operation of a BNR facility requires a thorough understanding of process control and troubleshooting.

In 1972, the U.S. Environmental Protection Agency (U.S. EPA) enacted the Clean Water Act (CWA), creating a federal program designed to achieve the goals of protecting and restoring waters of the United States (U.S. EPA, 1972). Since that time, many additional regulations and technological advances have been made to further protect the nation's natural waterbodies. The Water Quality Act of 1987 (U.S. EPA, 1987) authorized individual states to establish their own water quality standards. A provision of the CWA mandated that each state prepare a water quality inventory that lists the state's impaired waters. The total maximum daily load (TMDL) process was established based on a combination of these rules to specifically meet the overall goals on a local level. A TMDL is the maximum daily amount of any pollutant that can enter a specific waterbody before the quality of the water is deemed unfit for its designated uses. Thus, the TMDL of a pollutant is the threshold or upper limit loading for each specific body of water. Total maximum daily loads must be established for both point-source and nonpoint-source pollutants.

As part of the TMDL program, each state is responsible for allocating loads among point and nonpoint sources identified under each TMDL. The U.S. EPA encourages states to consider a range of options that both are technically feasible and demonstrate consistency and are based on cost effectiveness and equity. Final allocation determinations are policy decisions and should reflect public perceptions about acceptable trade-offs between these measures.

## 2.0 OBJECTIVE

With the global objective of reducing point-source contributions of nutrient loadings to local waterbodies, states have placed regulatory limits on numerous WRRFs specifically related to nitrogen and phosphorus. The purpose of WEF 37 is to provide facility managers and operators with an understanding of the theory and design approaches for processes currently used for nitrogen and phosphorus removal to optimize their facilities' ability to consistently meet regulatory requirements and to provide general operational guidance on process control and troubleshooting methodologies. WEF 37 also benefits design engineers and other technical professionals by broadening their understanding of BNR processes and enhancing their ability to relate to the facility's daily operations following construction and implementation of their design.

## 3.0 FORMAT

The chapters of these standards are organized such that the reader develops an understanding of the theory before reading about process control parameters and requirements and troubleshooting methodologies. Chapter 2 explains the pertinent characteristics of raw wastewater and its overall effect on nutrient removal processes within a WRRF. This chapter provides the distinction between typical influent characteristics of a raw wastewater flow stream and typical effluent permit limits often required of a BNR facility. The differential between influent and effluent concentrations sets the stage for the remainder of this standards document to describe successful configurations of both attached growth (biofilm) and suspended growth processes for removing nitrogen and phosphorus.

The theory and basics of nitrification are introduced in Chapter 3. Nitrification is the process by which ammonia is converted to nitrite and, subsequently, nitrate using two specific groups of autotrophic bacteria. The text in this chapter provides a stepwise process for both operators and design engineers to best understand and optimize nitrification.

Although use of biofilm process in municipal wastewater treatment has perhaps been reduced over time, biofilm technologies remain common. Chapter 4 presents a brief history of these processes and presents theory and examples of the nitrifying process in biofilm reactors. Biofilm processes have long been used in both rotating biological contactors and trickling filters. Use of biofilm reactors has been reduced greatly over the years, resulting in a large increase in the use of suspended growth in the activated sludge process. Nutrient removal requirements are causing biofilm processes to be reconsidered in new configurations, particularly in combination with the suspended growth process to enhance current facilities.

The overall process of nitrogen removal involves both nitrification and denitrification. Chapter 5 begins the discussion on denitrification and presents stoichiometry, optimal conditions, and their overall application in variations to the standard activated sludge process.

Chapter 6 includes an overview of enhanced biological phosphorus removal (EBPR). This chapter focuses on process fundamentals, the microbiology of the process, and the biochemistry of EBPR. Additionally, the chapter provides facility operators with key operational considerations and effects on sludge handling with biological phosphorus removal processes in addition to a summary of ongoing research related to EBPR.

The discussion in Chapter 7 begins to integrate concepts of Chapters 3, 4, and 5 and describes design details and benefits of combined nitrification and denitrification treatment systems. Such systems are used to achieve carbonaceous biochemical oxygen demand removal, nitrification, and denitrification in a single sludge configuration. Building on previous discussions, Chapter 7 also provides an in-depth discussion of design and operational criteria and the expected performance of various process configurations for nitrogen removal.

Chapter 8 provides a narrative that builds on the fundamentals of EBPR presented in Chapter 6 and describes the components of a functional EBPR system. Alternate configurations of EBPR process and operating strategies are presented.

Chapter 9 describes the interaction between nitrogen and phosphorus removal in combined systems. Removal of phosphorus in a combined nitrogen and phosphorus removal facility can be achieved both chemically and biologically; however, the biological alternative has several significant advantages. This chapter focuses on combined nitrogen and phosphorus removal systems and their operational conditions, process control strategies, and troubleshooting.

The discussion in Chapter 10 continues with phosphorus removal and presents concepts and processes related to the chemical precipitation of phosphorus in a WRRF. Removal of phosphorus using chemical precipitation is common and has been in practice for many decades, either as a primary removal process in conjunction with a biological removal process or as a supplemental, polishing, or backup process.

With tightening budgets and increasingly stringent regulations, operators are under constant pressure to optimize the performance of nutrient removal processes in terms of effluent quality and/or reducing ongoing operations costs associated with power and chemical use. Chapter 11 provides a summary of cost-effective operational considerations for various nutrient removal processes. This chapter describes the details of BNR processes as related to the various parameters for operators to monitor and control within their WRRFs.

Chapter 12 addresses concerns and management strategies of recycling nutrients from solids management processes. These recycle streams can represent a significant nutrient load on the nutrient removal process and must be managed to ensure compliance with effluent quality goals and to improve treatment efficiency. A discussion on emerging processes that bypass the conversion of nitrite to nitrate and convert nitrite directly to nitrogen gas (anammox) is also included.

Chapter 13 provides a summary of process control, instrumentation, and automation related to operation of BNR WRRFs. Nutrient removal facilities can be complex, requiring significant monitoring and control of processes to meet effluent requirements.

Chapter 14 describes in detail laboratory analysis performed as part of the operation and monitoring of BNR processes.

Chapter 15 presents several case studies covering full-scale facilities that are designed and operated for BNR to meet effluent total nitrogen limits. Each facility covered in this chapter has a unique process modification to the activated sludge process specifically designed to reduce nitrogen levels in the effluent. The chapter includes a case study of four-stage Bardenpho, modified Ludzack–Ettinger, a sequencing batch reactor, and integrated fixed-film activated sludge, among others.

Similar to Chapter 15, Chapter 16 includes several case studies, designed and operated to meet effluent total phosphorus limits between 0.05 and 1.0 mg/L for optimal EBPR. All the facilities presented in this chapter have been in operation for a minimum of 2 years. The case studies provide a unique summary of the challenges encountered by facility operators during the startup and commissioning phases of the project.

Finally, a BNR troubleshooting guide is provided as an appendix to theses standards. Operators can use this guide to quickly identify potential problem sources that may be encountered during the operation of a BNR facility and possible solutions.

## 4.0 REFERENCES

U.S. Environmental Protection Agency. (1972). *Federal Water Pollution Control Amendments of 1972.*

U.S. Environmental Protection Agency. (1987). *Water Quality Act of 1987.*

# Chapter 2

# Wastewater Constituents That Affect Nutrient Removal



## 1.0 INTRODUCTION

Wastewater characteristics, influent loadings, and effluent limits dictate treatment needs and processes involving the removal of organics, solids, and nutrients in a water resource recovery facility (WRRF). This chapter focuses on wastewater characteristics that affect nitrogen and phosphorus removal.

Excessive nitrogen and phosphorus discharges from WRRFs to receiving waters have environmental consequences, including eutrophication processes that result in excessive growth of plant and algae, ammonia toxicity, and nitrate contamination in groundwater. Ammonia toxicity causes fish mortality and health issues in aquatic reproduction. Nitrate contamination in groundwater can result in public health issues such as blood disorders affecting infants (e.g., methemoglobinemia, which is commonly referred to as "blue baby" syndrome).

## 1.1 Wastewater Characteristics and Nutrients

Wastewater characteristics and influent nutrient loadings to WRRFs depend on factors such as generation sources, collection system size and operation, and inflow/infiltration. These characteristics are typically measured as biochemical oxygen demand (BOD), total suspended solids (TSS), pH, and macronutrients (e.g., nitrogen and phosphorus). Table 2.1 summarizes these parameters, and Chapter 14 discusses procedures for their measurement.

## 1.2 Nutrient Sources for Wastewater

### *1.2.1 Nitrogen Sources*

Ammonia ($NH_3$), ammonium ion ($NH_4^+$), nitrite ($NO_2^-$), nitrate ($NO_3^-$), and organic nitrogen are common forms of nitrogen found in wastewater. Decaying plants, animals, and their wastes; human wastes; and fertilizers are significant sources of nitrogen, as can be certain industrial contributors. Influent nitrogen loading to WRRFs is commonly measured as total Kjeldahl nitrogen (TKN), which is the sum of ammonia (as nitrogen) and organic nitrogen.

### *1.2.2 Phosphorus Sources*

Fertilizers, manure, detergents, household cleaning products, and human and animal waste are typical sources of phosphorus. Common forms of phosphorus are orthophosphate, polyphosphates (condensed phosphates), and organic phosphates (phospholipids and nucleotides). Phosphorus can be present in wastewater in both soluble and particulate forms.

# 2.0 WASTEWATER CHARACTERISTICS THAT AFFECT NUTRIENT REMOVAL

Raw wastewater characteristics are typically characterized based on physical, chemical, and biological properties, as shown in Table 2.1.

## 2.1 Physical Characteristics

### *2.1.1 Temperature*

Wastewater temperature affects the solubility of gases (oxygen being of particular importance) and reaction rate kinetics for biological and chemical processes. Microorganisms are sensitive to temperature and exhibit slower growth in colder temperatures and faster growth in warmer temperatures. Refer to Chapters 3 and 5 for details of the effects of temperature on microbial activity on nitrification and denitrification process rates.

### *2.1.2 Total Solids*

Wastewater solids consist of TSS and total dissolved solids, with each portion further categorized as volatile solids and fixed solids. Volatile solids are an indicator of biodegradability. The biodegradable fraction of volatile suspended solids (VSS) contributes to BOD, nitrogen, and phosphorus load, and is typically 70% to 80% of TSS for domestic wastewater. Higher VSS content is typical of domestic wastes; however, combined sewers typically lower this ratio because of the contribution of inert solids.

## 2.2 Chemical Characteristics

This section presents common chemical characteristics of wastewater (Table 2.1).

**TABLE 2.1** Typical Raw Wastewater Characteristics (Adapted from Tchobanoglous et al., 2014)

| Characteristic | Units | Concentration[a] Low Strength | Medium Strength | High Strength |
|---|---|---|---|---|
| pH | | | Typical range—6 to 9 | |
| Alkalinity, as $CaCO_3$ | mg/L | 150 | 200 | 250 |
| Solids, total (TS) | mg/L | 537 | 806 | 1612 |
| Dissolved, total (TDS) | mg/L | 374 | 560 | 1121 |
| Fixed | mg/L | 224 | 336 | 672 |
| Volatile | mg/L | 150 | 225 | 449 |
| Suspended solids, total | mg/L | 130 | 225 | 449 |
| Fixed | mg/L | 29 | 43 | 86 |
| Volatile (VSS) | mg/L | 101 | 152 | 304 |
| Settleable solids | mg/L | 8 | 12 | 23 |
| Biochemical oxygen demand, 5-day, 20 °C ($BOD_5$, 20 °C) | mg/L | 133 | 200 | 400 |
| Total organic carbon (TOC) | mg/L | 109 | 164 | 328 |
| Chemical oxygen demand (COD) | mg/L | 339 | 508 | 1016 |
| Nitrogen (total as N) | mg/L | 23 | 35 | 69 |
| Organic | mg/L | 10 | 14 | 29 |
| Free ammonia | mg/L | 14 | 20 | 41 |
| Nitrite | mg/L | 0 | 0 | 0 |
| Nitrate | mg/L | 0 | 0 | 0 |
| Phosphorus (total as P) | mg/L | 3.7 | 5.6 | 11 |
| Organic | mg/L | 2.1 | 3.2 | 6.3 |
| Inorganic | mg/L | 1.6 | 2.4 | 4.7 |
| Potassium | mg/L | 11 | 16 | 32 |
| Chlorides | mg/L | 39 | 59 | 118 |
| Sulfate | mg/L | 24 | 36 | 72 |
| Oil and grease | mg/L | 51 | 76 | 153 |
| Volatile organic compounds (VOCs) | mg/L | <100 | 100–400 | >400 |
| Total coliform | No./100 mL | $10^6$–$10^8$ | $10^7$–$10^9$ | $10^7$–$10^{10}$ |
| Fecal coliform | No./100 mL | $10^3$–$10^5$ | $10^4$–$10^6$ | $10^5$–$10^8$ |
| *Cryptosporidum* oocysts | No./100 mL | $10^{-1}$–$10^1$ | $10^{-1}$–$10^2$ | $10^{-1}$–$10^3$ |
| *Giardia lamblia* cysts | No./100 mL | $10^{-1}$–$10^1$ | $10^{-1}$–$10^3$ | $10^{-1}$–$10^4$ |

*Notes.* mg/L = g/m$^3$. Values should be increased by the amount of constituent present in the domestic water supply.
[a]Low strength is based on an approximate wastewater flowrate of 570 L/cap·d (150 gpd/cap), medium strength is based on an approximate wastewater flowrate of 380 L/cap·d (100 gpd/cap), and high strength is based on an approximate wastewater flowrate of 190 L/cap·d (50 gpd/cap).

### 2.2.1 Dissolved Oxygen

Aeration equipment provides the dissolved oxygen needed for the nitrification process. Because nitrification involves the oxidation of ammonia and organic nitrogen to nitrite and nitrate, the presence of dissolved oxygen is necessary. Because the denitrification process requires anoxic conditions (i.e., no dissolved oxygen), lower dissolved oxygen is required for efficient conversion of nitrite and nitrate to nitrogen gas.

### 2.2.2 pH

*pH* is defined as the negative log of hydrogen ion concentration represented by a numerical scale from 1 to 14 standard units, with lower than pH 7 (neutral condition) indicating acidic conditions and higher than pH 7 indicating basic conditions. Biological processes are sensitive to the pH of wastewater, with the best results observed close to neutral values (6 to 9 pH range). Domestic wastewater typically falls within this range, whereas industrial wastewater pH may vary significantly.

### 2.2.3 Alkalinity

Alkalinity is the buffering capacity of a solution to resist change in pH. The ions present in wastewater that generally contribute most of the alkalinity are carbonate ($CO_3^{2-}$), bicarbonate ($HCO_3^-$), and hydroxyl ($OH^-$) ions. Total alkalinity is the sum of all these alkalinities. Alkalinity is required for nitrification processes and is partly gained back through the denitrification process. Chapter 11 presents an in-depth discussion of alkalinity.

### 2.2.4 Biochemical Oxygen Demand

Wastewater BOD is a measure of oxygen consumed during biochemical oxidation of organic matter by microorganisms. Products of this oxidation process include carbon dioxide ($CO_2$), nitrite-nitrogen and nitrate-nitrogen, water, new cells, and debris, which is used to create new cell mass and to maintain cells. Oxygen required to complete these reactions is referred to as *BOD*. Commonly reported BOD is based on a 5-day test and is reported as $BOD_5$. Refer to Chapter 14 for more detailed information.

### 2.2.5 Chemical Oxygen Demand

Chemical oxygen demand (COD) represents oxygen consumed during a laboratory procedure that chemically oxidizes organic matter in wastewater. The COD value is always higher than the BOD value because it also takes into account reducing chemicals that are not available for biological uptake. The typical raw domestic wastewater COD-to-BOD ratio ranges between 1.9 and 2.2 to 1. A higher ratio indicates the presence of a significant amount of nonbiodegradable or refractory materials, which is an indication of industrial wastewater or inhibitory substances.

### 2.2.6 Total Nitrogen and Related Fractions

The typical nitrogen concentration range in domestic wastewater includes organic and inorganic compounds (ammonia, nitrite, and nitrate). Nitrogen load in untreated domestic wastewater is typically composed of TKN, with negligible nitrite and nitrate, whereas industrial wastewater could include nitrite and nitrate. Figure 2.1 shows a simplified version of the nitrogen cycle.

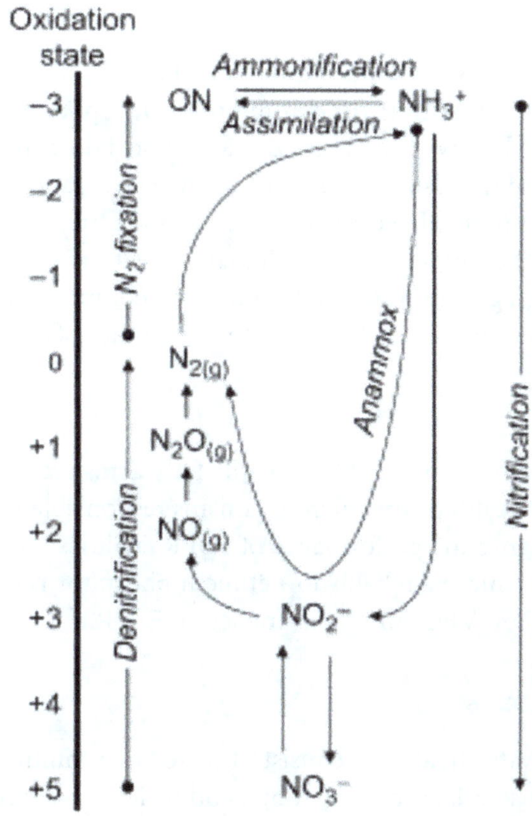


**FIGURE 2.1** Simplified Version of Nitrogen Cycle (Water Environment Federation, 2011)

### 2.2.7 *Total Phosphorus and Related Fractions*

As Table 2.1 indicates, domestic wastewater total phosphorus concentrations typically range between 3.7 and 11 mg/L. Sidestreams within biological nutrient removal (BNR) facilities, especially those that incorporate sludge digestion, can return significant phosphorus loads back from sidestreams to the process. The presence of industrial discharges (such as detergents, etc.) and the nature of drinking water treatment processes that may contain corrosion inhibitors rich in phosphorus also indirectly affect wastewater total phosphorus concentrations. Phosphorus in wastewater is composed of both inorganic and organic forms. The inorganic forms, which are soluble, include orthophosphate and polyphosphates. The orthophosphate form ($PO_4^{3-}$) is the simpler form of phosphorus and typically accounts for 70% to 90% of the total phosphorus. It is also the form that is available for biological metabolism without further breakdown and is precipitated by metal salts during chemical phosphorus removal. Related fractions of orthophosphate are associated with pH. Between pH 6.8 and 6.5, an inversion of proportion occurs between hydrogen phosphate ions and dihydrogen phosphate ions (61% to 39% to 38% to 62%, respectively) (Campbell & Reece, 2005). Therefore, chemical addition to further lower phosphorus would be most effectively introduced to react in an environment having pH near 6.8 (or slightly below) to achieve the greatest effect.

Polyphosphates are complex in nature and are broken down to orthophosphate (or derivations thereof) during the treatment process. Organically bound phosphorus includes a variety of more complex forms of phosphorus that derive from proteins, amino acids, and nucleic acids that, to some extent, are degraded and are present as waste products.

### 2.2.8 Trace Metals

Trace amounts of metals help support microbial growth and metabolism during biological treatment processes. Trace metals such as cadmium (Cd), chromium (Cr), copper (Cu), iron (Fe), lead (Pb), manganese (Mn), mercury (Hg), nickel (Ni), and zinc (Zn) are important constituents in water (Tchobanoglous et al., 2014). Most of these metals help with biological growth, and requirements for the presence of trace metals have been recognized. However, if trace metals are present in excess quantity, they may be toxic to microorganisms and eventually inhibit biological treatment processes. Sources of these metals include residential dwellings, groundwater infiltration, and commercial and industrial dischargers. Analyzing trace metals in treated effluent helps assess the suitability of water for reuse.

### 2.2.9 Volatile Fatty Acids—Carbon

Volatile fatty acids (VFAs) are a fraction of COD. Volatile fatty acids are readily biodegradable and are vital to the predominance and health of the microorganisms responsible for enhanced biological phosphorus removal (EBPR) and hence the performance of EBPR facilities. Insufficient VFAs will reduce the ability of the EPBR process to achieve a reliably low effluent phosphorus concentration. Chapters 6 and 8 provide detailed information on VFAs and their impact on EBPR.

## 2.3 Biological Characteristics

Biological processes for wastewater treatment consist of mixed communities containing a wide variety of microorganisms including bacteria, protozoa, fungi, and rotifers. Bacteria can be classified based on their oxygen requirements (aerobic, anaerobic, facultative) as well as their carbon requirements (heterotrophic, autotrophic). Aerobic bacteria need an oxygen-rich environment, whereas anaerobic bacteria require no oxygen. Facultative bacteria are capable of living in the presence and absence of oxygen. Heterotrophic bacteria need organic carbon for growth, whereas autotrophic bacteria are able to rely solely on carbon dioxide ($CO_2$) for growth. The microbial mass initiates various biochemical reactions that cycle the available forms of nitrogen and phosphorus.

Nitrogen removal in WRRF processes generally requires an established population of aerobic autotrophic organisms to achieve nitrification—the two-stage conversion of ammonia first to nitrite and then to nitrate. Anaerobic autotrophic organisms are required to bypass the nitrite-to-nitrate conversion and convert ammonia and nitrite to nitrogen gas. Facultative heterotrophic microorganisms, when in an anoxic environment, will convert nitrate to nitrogen gas.

Populations of polyphosphate-accumulating organisms (PAOs) and glycogen-accumulating organisms (GAOs) are found in EBPR facilities. Glycogen-accumulating organisms directly compete with PAOs, particularly at lower pH levels (i.e., below 7.0 standard units) (see Chapter 6 for more detailed information on GAOs). Table 2.2 classifies bacteria related to nutrient removal processes by electron donor, electron acceptor, sources of cell carbon, and end products.

# 3.0 EFFECT OF WATER RESOURCE RECOVERY FACILITY FLOW AND LOADING

Water resource recovery facility design and operation should address hydraulic and mass loading conditions to achieve effluent limits. Depending on the time of day, month, season, and so on, hydraulic and mass loads can vary significantly. To dampen the effect of peak flows and loads on treatment processes, some WRRFs provide equalization units for temporary storage of excess wastewater. In addition,

TABLE 2.2 Classification of Nutrient-Cycling Microorganisms (Adapted from Tchobanoglous et al., 2014)

| Type of Bacteria | Common Reaction Name | Carbon Source | Electron Donor Substrate Oxidized | Electron Acceptor | Products |
|---|---|---|---|---|---|
| Aerobic heterotrophic | Aerobic oxidation | Organic compound | Organic compound | $O_2$ | $CO_2$, $H_2O$ |
| Aerobic autotrophic | Nitrification | $CO_2$ | $NH_4^+$, $NO_2^-$ | $O_2$ | $NO_2^-$, $NO_3^-$ |
| Facultative heterotrophic | Denitrification | Organic compound | Organic compound | $NO_2^-$, $NO_3^-$ | $N_2$, $CO_2$, $H_2O$ |
| Anaerobic heterotrophic | Acid formation | Organic compound | Organic compounds | Organic compounds | Volatile fatty acids (VFAs) (acetate, propionate, butyrate) |
| | Dissimilatory nitrate reduction to ammonium | Organic compound | Organic compound | $NO_3^-$ | $NH_4^+$ |
| Anaerobic autotrophic | Anammox | $CO_2$ | $NH_4^+$ | $NO_2^-$ | $N_2$, $NO_3^-$ |

appropriate design criteria for various pieces of process equipment may differ based on average, maximum, or peak flow conditions. It is important to understand these different flow and mass loading conditions for WRRF design and process optimization. Figure 2.2 illustrates a typical diurnal flow and loading profile. Flowrates and corresponding loadings significantly affect system performance. Various common flow-related terminologies are described in the following sections.

## 3.1 Average Daily Flows

- Average daily flow and monthly average flow—total wastewater inflow volume during the specific time period, divided by the number of days in that period, and total wastewater inflow volume during a 1-month period, divided by the number of days in that month, respectively
- Annual average daily flow—total wastewater inflow volume during any consecutive 365 days, divided by 365
- Three-month average daily flow—total wastewater inflow volume during 3 consecutive months, divided by the number of days in this period

## 3.2 Maximum Flows

- Maximum daily flow—largest volume of wastewater inflow during any consecutive 24-hour period
- Maximum monthly average flow—the highest monthly average flow during any one calendar year

## 3.3 Peak Flows

- Peak 2-hourly flow—maximum flow averaged during a 2-hour period
- Peak hourly flow—maximum flow averaged during 1 hour
- Peak instantaneous flow—maximum instantaneous flow at any given time

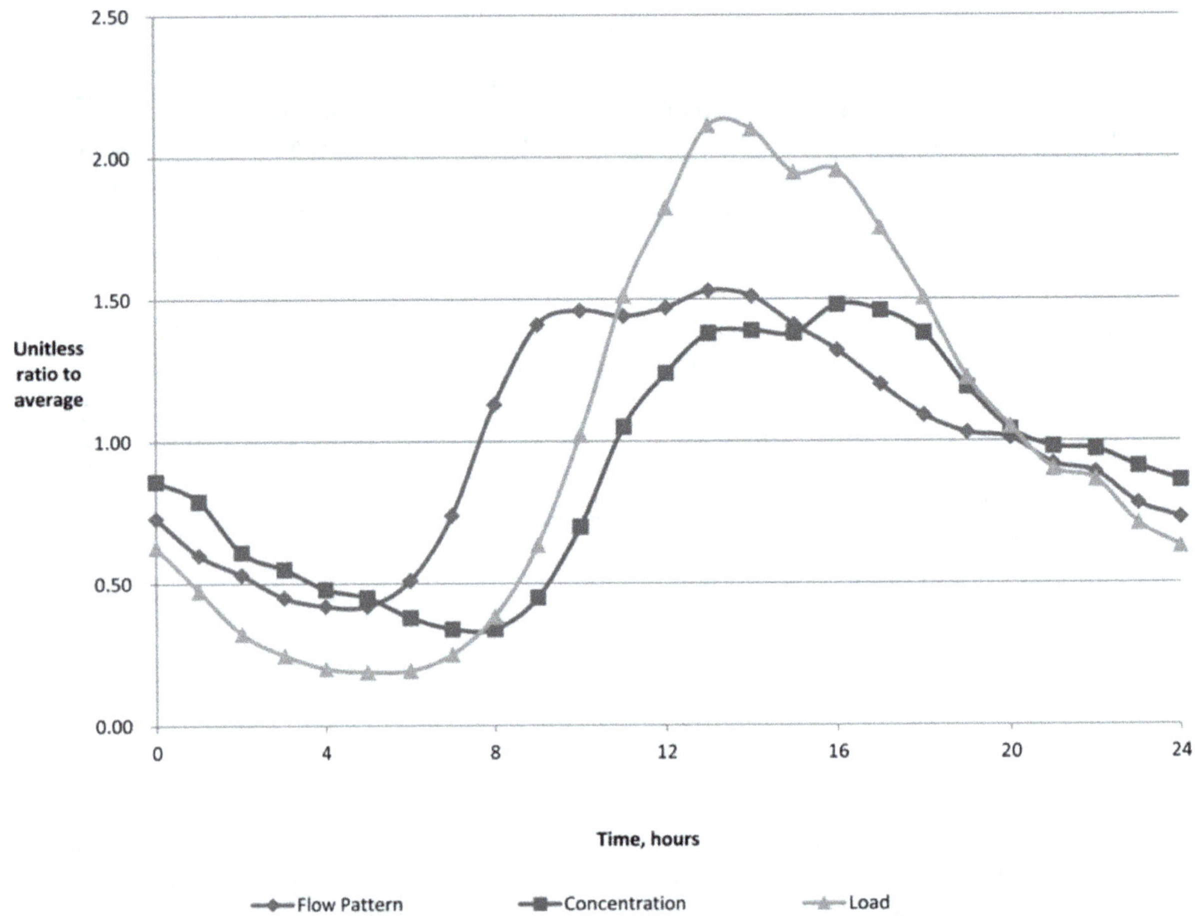


FIGURE 2.2 Typical Hourly Variation in Flow and Load of Municipal Wastewater (Data from Grady et al., 2011)

## 3.4 Internal Sidestream Flows

Sidestreams represent flows generated within a WRRF that are recycled back to be treated. These streams can include filtrate or centrate from thickening and dewatering processes or digester supernatant and are typically sent back to the head of the WRRF. In general, they are low in volumes (approximately 5% to 10% compared to influent flows) and may be rich in soluble organics, solids, and nutrients. Sidestreams can affect aeration demands and BNR significantly if they are intermittent (e.g., if dewatering only takes place 8 hours per day, 2 or 3 days per week). The effects of sidestream flows and loads are addressed in detail in Chapter 12.

## 3.5 Constituent Ratios

The effectiveness and efficiency of nitrogen and phosphorus removal is greatly influenced by the amount of carbon available, often expressed as process influent COD : nitrogen and COD : phosphorus ratios, respectively. This is described in detail in subsequent chapters. Sufficient biodegradable COD is required to efficiently drive the denitrification reaction in nitrogen-removal processes, often at a COD : N ratio in the range of 3 to 7. Enhanced biological phosphorus removal also requires enough COD, with a sufficient fraction of that COD being either VFAs or readily biodegradable that can ferment to VFAs. For good EBPR, the COD : P ratio may need to be 60 : 1.

# 4.0 PRETREATMENT REQUIREMENTS FOR INDUSTRIAL DISCHARGERS

The U.S. Environmental Protection Agency's (U.S. EPA's) National Pretreatment Program identifies specific requirements that apply to all industrial users or nondomestic sources of wastewater before entering the sewer system (U.S. EPA, 2023). These objectives are achieved by enforcing the discharge standards outlined in the following sections.

## 4.1 Prohibited Discharge Standards

Prohibited discharge standards forbid the discharge of any pollutant(s) to a WRRF that causes pass through or interference. These national standards apply to all industrial users discharging to a WRRF, regardless of whether or not the WRRF has an approved pretreatment program or whether the industrial user has been issued a control mechanism or permit. These standards are intended to provide protection for publicly owned treatment works. The prohibited discharge standards are listed in 40 CFR 403.5 ("National Pretreatment Standards: Prohibited Discharges").

## 4.2 Categorical Pretreatment Standards

Categorical pretreatment standards limit pollutant discharges to WRRFs from specific process wastewaters of particular industrial categories. These national technology-based standards apply regardless of whether or not the WRRF has an approved pretreatment program or the industrial user has been issued a control mechanism or permit. Such industries are called *categorical industrial users*. The standards are promulgated by the U.S. EPA in accordance with Section 307 of the Clean Water Act and are designated in "Effluent Guidelines & Limitations" (Parts 405 to 471) as "Pretreatment Standards for Existing Sources" and "Pretreatment Standards for New Sources."

## 4.3 Local Limits

Local limits reflect specific needs and capabilities at individual WRRFs and are designed to protect the WRRF and its receiving waters and sludge disposal practices. Regulations of 40 CFR 403.8(f)(4) state that WRRF pretreatment programs must develop local limits or demonstrate that they are not necessary; in addition, 40 CFR 403.5 states that local limits are needed when pollutants are received that could result in pass through or interference at the WRRF. Essentially, local limits translate the General Prohibited Discharge Standards of 40 CFR 403.5 to site-specific needs.

# 5.0 NUTRIENT EFFLUENT PERMIT REQUIREMENTS

To preserve and improve the quality of receiving waters, WRRFs are required to comply with effluent discharge permit requirements. Types of permits or concepts used for developing discharge limits are discussed in the following sections.

## 5.1 Technology-Based Permit Limits

For technology-based permits, effluent requirements are based on specific treatment levels achievable by the available technology, that is, considering the "best available technology" or "limit of technology." This treatment level might not be related to the assimilative capacity of receiving streams or water quality standards.

#### 5.1.1 Monthly Average-Based Permit Limits

Monthly average-based permit limits are based on 30-day average effluent limits. This reporting period requires the WRRF to operate at optimum efficiency at all times. If few samples indicate high levels of total nitrogen, for example, then the facility must operate at a higher level of performance for the rest of the month to maintain compliance with the monthly average. Typically, there are allowances for higher daily or weekly results, providing flexibility for peak daily or weekly influent loads.

#### 5.1.2 Annual Average-Based Permit Limits

An effluent permit that is based on an annual average during a 1-year period allows more flexibility to operate at or near the permitted level. If there is an excursion above the annual average limit, then there is more time to operate at a higher level of treatment to maintain compliance with the annual average limit.

#### 5.1.3 Seasonal-Based Permit Limits

Seasonal permits require a different permit level in the summer versus the winter; this may allow the facility to be operated in a "less aggressive" manner during the "off-season." For example, a facility with a seasonal nitrification permit may be able to operate at a lower solids retention time in the winter than otherwise would be required for a year-round permit.

### 5.2 Water-Quality-Based Permit Limits

Water-quality-based permits are based on assimilative capacity of the receiving stream and, in general, are based on an annual average load allocation from the facility. As flows increase, the level of treatment required to meet the permit increases. At lower flows below design capacity, the allowable effluent concentration is greater. An annual load allocation allows some flexibility in seasonal operation if the WRRF is capable of operating below load allocation on an average monthly basis or if it is not yet at design capacity. In some states, water-quality-based limits are enforced in terms of a concentration limit, which governs how the WRRF must be operated.

## 6.0 REFERENCES

Campbell, N. A., & Reece, J. B. (2005). *Biology* (7th ed.). Benjamin Cummings.

Grady, Jr., C. P. L., Daigger, G. T., Love, N. D., & Filipe, C. D. M. (2011). *Biological wastewater treatment* (3rd ed.). Marcel Dekker.

National pretreatment standards: Prohibited discharges. 40 C.F.R. § 403.5 (2023). https://www.ecfr.gov/current/title-40/section-403.5

Tchobanoglous, G. Stensel, H. D., Tsuchihashi, R., Burton, F. L., Abu-Orf, M., Bowden, G., Pfrang, W., & Metcalf & Eddy. (2014). *Wastewater engineering: Treatment and resource recovery* (5th ed.). McGraw Hill-Education.

U.S. Environmental Protection Agency. (2023). *National pretreatment program overview.* https://www.epa.gov/npdes/national-pretreatment-program-overview

Water Environment Federation. (2011). *Nutrient removal* (WEF Manual of Practice No. 34).

## 7.0 SUGGESTED READINGS

U.S. Environmental Protection Agency. (1987). *Guidance manual for preventing interference at POTWs.*

Water Environment Federation. (2015). *The nutrient roadmap.*

# Chapter 3

# Nitrification



## 1.0 INTRODUCTION

Nitrification is the process of biological oxidation of ammonia (which exists mostly as ammonium nitrogen [$NH_4$-N] in typical wastewater) to nitrite-nitrogen ($NO_2$-N) and further oxidation to nitrate-nitrogen ($NO_3$-N). Nitrification is important because it converts ammonia to nitrate and is the first step in biological nitrogen removal (denitrification), which will be discussed in Chapter 5. Conversions to nitrite and nitrate involve two specific groups of autotrophic bacteria: ammonia-oxidizing bacteria (AOB) and nitrite-oxidizing bacteria (NOB). Ammonia-oxidizing bacteria carry out the oxidation of ammonia to nitrite, and NOB carry out nitrite conversion to nitrate. Both these reactions should operate at optimal rates for nitrate production. Ammonia-oxidizing bacteria and NOB are often referred to as *nitrifiers* and are autotrophic bacteria. Ammonia-oxidizing bacteria and NOB both belong to the phyla *Proteobacteria*. Nitrite-oxidizing bacteria need nitrite as their electron donor, so they depend on the presence of AOB for their substrate production. Autotrophic bacteria, specifically chemoautotrophic bacteria, differ from heterotrophic bacteria that consume organic material (biochemical oxygen demand [BOD]) in that chemoautotrophic bacteria synthesize cellular material from inorganic carbon (carbon dioxide) under typical operating conditions. Oxidation of ammonia or nitrite provides the energy needed for cell synthesis. These bacteria are obligate aerobes because they grow only when dissolved oxygen is available; however, the lack of oxygen is not lethal because nitrifiers periodically withstand both anaerobic and anoxic conditions. There are emerging processes that bypass the conversion of nitrite to nitrate and convert nitrite directly to nitrogen gas (denitritation), thus completing the nitrogen removal process. These processes are discussed in Chapter 12.

## 1.1 Stoichiometry

The stoichiometry of biochemical reactions associated with nitrification defines the proportion of reactants and products involved in this process. Understanding stoichiometry is important because it defines the basic inputs and outputs for each of the steps in the process and can determine which of these inputs will limit the reaction. The stoichiometric equation that defines the molar ratios for the oxidation of ammonia to nitrite by AOB is

$$NH_4^+ + 1.5\ O_2 \Rightarrow NO_2^- + 2\ H_2O + 2H^+ \tag{3.1}$$

Similarly, the stoichiometric equation that describes the oxidation of nitrite to nitrate by NOB is

$$NO_2^- + 0.5\ O_2 \Rightarrow NO_3^- \tag{3.2}$$

These reactions also generate biomass associated with the growth of AOB and NOB. Unlike heterotrophs, AOB and NOB obtain carbon for cell growth from an inorganic source, that is, carbon dioxide. The total yield of these nitrifying organisms for the combination of both steps in the process is substantially lower than for heterotrophs, generally ranging from 0.06 to 0.20 g volatile suspended solids (VSS)/g $NH_4$-N oxidized. The stoichiometric equations do not reflect that a portion of nitrogen is used in assimilation (bacterial growth). However, because the quantity of nitrogen removed through this assimilation process into nitrifier biomass is typically less than 2% of the ammonia nitrified, it can generally be ignored. On the other hand, nitrogen assimilation by heterotrophs can be quite high in the range of 10% to 35%, depending on influent BOD/N (U.S. Environmental Protection Agency [U.S. EPA], 2010).

The overall energy reaction for the two-step nitrification process combines Equations 3.1 and 3.2, as follows:

$$NH_4^+ + 2\ O_2 \Rightarrow NO_3^- + 2\ H^+ + H_2O \tag{3.3}$$

The overall stoichiometric equation that accommodates both ammonia oxidation and cell synthesis reactions is given by

$$\begin{aligned} 0.13\ NH_4^+ + 0.225\ O_2 + 0.02\ CO_2 + 0.005\ HCO_3^- \rightarrow 0.005\ C_5H_7O_2N\ + \\ 0.125\ NO_3^- + 0.25\ H^+ + 0.12\ H_2O \end{aligned} \tag{3.4}$$

Based on the stoichiometry of the reaction shown in Equation 3.3, 2 mol of oxygen is required to oxidize 1 mol of ammonia to nitrate, which is equivalent to 4.57 g $O_2$/g $NH_4^+$-N oxidized.

During nitrification, 2 mol of hydrogen ion are also formed. The formation of 1 mol of hydrogen is equivalent to the consumption of 1 mol of alkalinity or 0.5 mol of $CaCO_3$ (50 g $CaCO_3$). The consumption of alkalinity during nitrification may decrease the pH of the wastewater if there is insufficient alkalinity available and can adversely affect the nitrification process, as discussed later in this chapter. It is important to understand the differences between heterotrophic bacteria, which synthesize cellular material from organic carbon (BOD), and autotrophic bacteria (AOB and NOB), which synthesize cellular material from carbon dioxide. Table 3.1 presents typical oxygen and alkalinity relationships of the nitrification process.

TABLE 3.1 Key Nitrification Relationships

| Parameter | Coefficient Units |
|---|---|
| Oxygen use | 4.57 g $O_2$/g $NH_4$-N |
| Alkalinity consumption | 7.14 g alkalinity (as $CaCO_3$)/g $NH_4$-N |

## 1.2 Kinetics

### 1.2.1 *Biomass Growth and Ammonia Utilization*

The first step in nitrification (i.e., conversion of ammonia to nitrite) is the rate-limiting step because AOB typically have lower specific growth rates than NOB. Growth of AOB and NOB is the result of the oxidation of ammonia and nitrite, respectively. The Monod specific growth rate equation is used to describe the effect of limiting substrates on microbial growth (Monod, 1949). Both ammonia and dissolved oxygen are substrates for AOB growth. The concentrations of either or both of these substances could be low enough to limit the specific AOB growth rate in wastewater treatment systems.

It is important to understand that nitrifiers cannot store food (ammonia). Therefore, they must metabolize it before the mixed liquor discharges to the clarifier and enters an anaerobic environment. Thus, for a given solids retention time (SRT), the process must provide a hydraulic retention time large enough for the slowly metabolizing nitrifiers to complete nitrification of all the ammonia.

Assuming no alkalinity limitation, the growth rate of AOB can be expressed as

$$\mu_{AOB} = \mu_{max,AOB}\,[S_{nh}/(S_{nh} + K_{nh,AOB})][S_o/(S_o + K_{o,AOB})] \qquad (3.5)$$

Where

$\mu_{AOB}$ = specific growth rate of AOB biomass (g biomass formed/g biomass present per day in system), $d^{-1}$;
$\mu_{max,AOB}$ = maximum specific growth rate of AOB (g biomass formed/g biomass present per day in system), $d^{-1}$;
$K_{nh,AOB}$ = half-saturation coefficient for AOB (mg N/L);
$S_{nh}$ = growth-limiting substrate ($NH_4$-N) concentration (mg N/L);
$S_o$ = dissolved oxygen concentration of bulk mixed liquor or wastewater (mg $O_2$/L); and
$K_{o,AOB}$ = oxygen half-saturation coefficient for AOB (mg $O_2$/L).

Equation 3.5 is a simplification because it does not explicitly consider free ammonia, the substrate for AOB. Figure 3.1 shows specific growth as a function of the $NH_4$-N concentration when dissolved oxygen is not limiting. The specific growth rate increases almost linearly (approximating first-order kinetics) with respect to $NH_4$-N concentration near the origin of the plot where the $NH_4$-N concentration is low. At higher $NH_4$-N concentrations, the specific growth rate approaches an asymptotic value (the maximum specific growth rate), thus exhibiting zero-order behavior. The $NH_4$-N half-saturation coefficient is the $NH_4$-N concentration at which the specific growth rate of the AOB is one-half of its maximum value. A similar plot could be drawn to depict the relationship between the specific growth rate and dissolved oxygen concentration when $NH_4$-N is not limiting. Under this condition, the oxygen half-saturation coefficient would be the dissolved oxygen concentration, in which the specific growth rate is one-half of its maximum value.

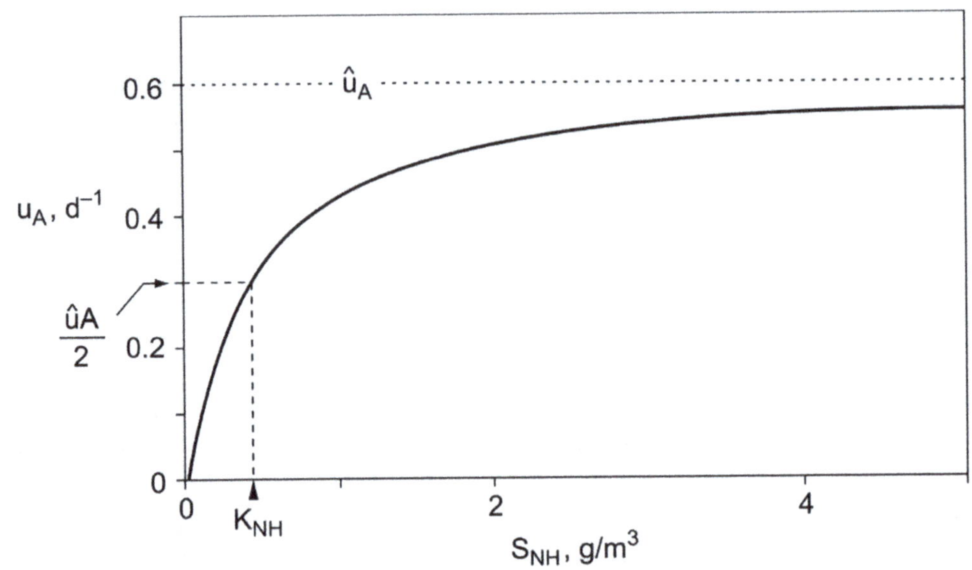


FIGURE 3.1 Relationship Between Specific Growth Rate of AOB and Ammonia-Nitrogen Concentration as Predicted by the Monod Equation (Dissolved Oxygen Is Assumed to Be Nonlimiting)

Nitrate-oxidizing bacteria can use both ammonia and nitrite (nitrous acid) as nitrogen sources. Considering $NH_4$-N and $NO_2$-N as nitrogen sources, the specific growth rate equation can be written as

$$\mu_{NOB} = \mu_{max,NOB}\,[S_{nh}/(S_{nh} + K_{nh,NOB})]\,[S_o/(S_o + K_{o,NOB})][S_{NO2}/(S_{NO2} + K_{NO2,NOB})] \quad (3.6)$$

Where

$\mu_{NOB}$ = specific growth rate of NOB biomass (g biomass formed/g biomass present per day in system), $d^{-1}$;
$\mu_{max,NOB}$ = maximum specific growth rate of NOB (g biomass formed/g biomass present per day in system), $d^{-1}$;
$S_{nh}$ = growth-limiting substrate ($NH_4$-N) concentration (mg N/L);
$K_{nh,NOB}$ = half-saturation coefficient for NOB (mg N/L);
$S_o$ = dissolved oxygen concentration of bulk mixed liquor or wastewater (mg $O_2$/L);
$K_{o,NOB}$ = oxygen half-saturation coefficient for NOB (mg $O_2$/L);
$S_{NO2}$ = growth limiting substrate ($NO_2$-N) concentration (mg N/L); and
$K_{NO2,NOB}$ = $NO_2$-N half-saturation coefficient for NOB (mg N/L).

Where nitrous acid ($NO_2$-N) is the only source of nitrogen, the expression can be written as

$$\mu_{NOB} = \mu_{max,NOB}\,[S_o/S_o + K_{o,NOB}][S_{NO2}/S_{NO2} + K_{NO2,NOB}] \quad (3.7)$$

Nitrous acid typically does not accumulate in large concentrations in biological treatment systems under stable operation because the maximum growth rate of NOB is significantly higher than that of AOB. As a result, the growth rate of AOB generally controls the overall rate of nitrification (i.e., as soon as nitrite is produced by AOB, it is used by NOB). However, facilities that transition into and out of nitrification or have dissolved oxygen limitations or toxic substances entering the facility may experience conditions in which the NOB growth rate is unable to keep up with the AOB growth rate, resulting in elevated effluent nitrite concentrations. This condition is known as *nitrite lock* and results in a significant increase in effluent chlorine demand caused by high levels of nitrite in the effluent (5.1 mg of chlorine is consumed per milligram of nitrite). If the effluent has a high nitrate concentration but low ammonia concentration (typical of water resource recovery facilities [WRRFs] that are nitrifying), chlorine reacts as free chlorine and is removed by the reaction with nitrite, thus increasing the amount of chlorine needed to meet the required effluent coliform concentration by 5 : 1. When significant ammonia is present, chlorine reacts preferentially with ammonia (as opposed to nitrite) to produce mostly mono-chloramine, which is an effective disinfectant.

In addition, accumulation of nitrite in a nitrification process can indicate the presence of an inhibitory or toxic substance in the wastewater, which may require further evaluation, as discussed later in this chapter. In situations in which effluent permits contain both seasonal nitrification and seasonal disinfection requirements (using chlorine), it is advisable to establish complete nitrification before the start of chlorination.

Bacterial growth rate of AOB, as defined by the time required to double their population, is 10 to 20 times slower than that of aerobic heterotrophic bacteria. Because the SRT of an activated sludge nitrification system is inversely proportional to the net growth rate of the slowest growing nitrifiers (AOB), the aerobic SRT needed to effectively nitrify must be significantly longer than the typical SRT needed to maintain a stable heterotrophic bacterial population (i.e., secondary treatment). Typically, the SRT is between 4 and 6 days for a nitrifying facility, which correlates to a food-to-microorganism ratio of approximately 0.15. The SRT in a biological system is typically defined as

$$\text{SRT} = \Theta = (\text{Mass of biological solids in the reactor}/\text{Mass of biological solids leaving the system/day}). \quad (3.8)$$

Because determination of the mass of active biological solids can be difficult, the overall solids maintained in, and wasted from, the reactor can be used to determine SRT. At steady state, the solids leaving the system will equal the solids produced within the system plus any inert solids that enter the system and pass through the process. Therefore, if the unit solids production rate of the biological treatment system (i.e., the net yield) is known, the required mass of reactor solids can be determined to achieve the target SRT. Typically, net yield values are reported as the mass of solids produced per each unit mass of carbonaceous oxygen demand or 5-day BOD ($BOD_5$) removed.

The growth rate of nitrifying microorganisms in the system is related to SRT, as follows:

$$\Theta = 1/(\mu_n - b_n) \quad (3.9)$$

Where

$\mu_n$ = specific growth rate of nitrifiers (g VSS added·d/g VSS in system) and
$b_n$ = endogenous decay coefficient for nitrifiers (g VSS destroyed·d/g VSS in the system).

Several environmental factors significantly affect the nitrifier growth rate, which, in turn, influences the minimum SRT required to allow for accumulation and maintenance of a sufficient population of nitrifying organisms. The effects of significant environmental factors that influence nitrification are presented in the following sections.

## 1.2.2 Environmental Conditions

### 1.2.2.1 Wastewater Temperature

Growth rates of AOB and NOB are particularly sensitive to the liquid temperature in which they live. Nitrification occurs in wastewater temperatures from 4 to 45 °C (39 to 113 °F), with an optimum rate occurring at approximately 30 °C (86 °F) (U.S. EPA, 1993). However, most WRRFs operate with a liquid temperature between 10 and 25 °C (50 and 77 °F). It is generally recognized that the nitrification rate doubles for every 8 to 10 °C rise in temperature. A number of relationships between maximum nitrifier growth rate ($\mu_{max}$) and wastewater temperature have been developed. The most commonly accepted expression for this relationship for wastewater over a temperature range from 5 to 30 °C (41 to 86 °F) is the following (U.S. EPA, 1993):

$$\mu_{max} = 0.47\ e^{0.098(T-15)} \tag{3.10}$$

Where

$\mu_{max}$ = maximum specific growth rate of microorganisms (g nitrifiers/g nitrifiers in system•d) and
$T$ = wastewater temperature (°C).

This relationship between maximum nitrifier growth rate and wastewater temperature is illustrated graphically in Figure 3.2. It is important to note that the endogenous decay rate ($b_n$) for nitrifying organisms is also temperature dependent and that the coefficients must be corrected for temperature.

### 1.2.2.2 Dissolved Oxygen Concentration

Nitrifiers are obligate aerobes, meaning they are only able to grow under aerobic conditions, although they can survive in anaerobic or anoxic conditions. Consequently, the dissolved oxygen concentration in the bulk liquid can have a significant effect on the nitrifier growth rate. The value at which dissolved oxygen concentration reduces the rate of nitrification varies on a site-specific basis, depending on temperature, organic loading rate, SRT, and diffusional limitations (i.e., floc size and density). It is generally

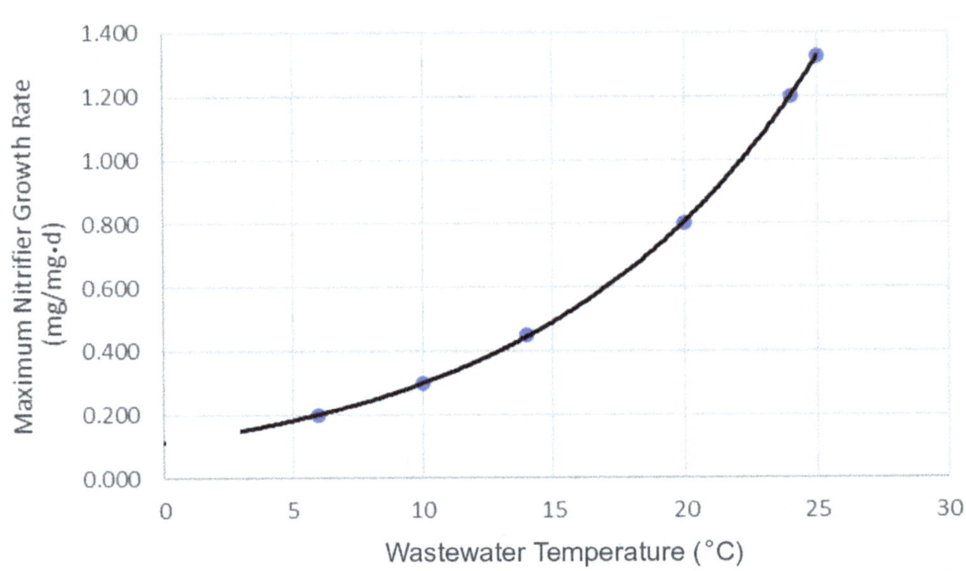


FIGURE 3.2 Influence of Temperature on the Nitrification Rate

accepted that nitrification is not limited at dissolved oxygen concentrations greater than 2.0 mg/L. However, this may not be true for systems using high biomass concentrations or biofilm systems or in cyclical systems or transition zones between unaerated and aerated reactors, where dissolved oxygen concentrations less than 2.0 mg/L may affect nitrification. In addition, some activated sludge systems may have large, dense floc (possibly indicated by very low sludge volume indexes [<65]), which will require higher dissolved oxygen levels to achieve consistent and effective nitrification because of poor oxygen mass-transfer efficiencies within the dense floc. Operators should check the diurnal variation of dissolved oxygen concentration, if possible, because there may be a significant difference between peak and low loading periods.

Although currently most nitrifying processes are operated at a dissolved oxygen (DO) concentration of approximately 2 mg/L, there have been recent studies that have shown nitrification can occur at very low dissolved oxygen concentrations. Kisling et al. (2022) demonstrated that nitrification occurred at dissolved oxygen concentrations as low as 0.2 mg/L, achieving the same effluent results as they did at the typical dissolved oxygen concentrations of 2 mg/L. They demonstrated that a gradual decrease in dissolved oxygen concentration over time allowed the AOB and NOB to adapt to these low dissolved oxygen concentrations. Furthermore, they showed that other organisms, such as phosphate accumulating organisms, also were responsible for nitrification. Another study by Miller et al. (2022) also showed that nitrification was occurring at low dissolved oxygen concentrations. They described a WRRF in Pueblo, Colorado, that achieved effluent total inorganic nitrogen (TIN) concentrations of 11 mg/L while operating at a dissolved oxygen between 0.2 and 0.5 mg/L. They also described a WRRF in Rochester, Minnesota, that achieved an effluent TIN concentration of approximately 10 mg/L while operating at a dissolved oxygen concentration of 0.2 to 0.3 mg/L. Although this is still in the development stage, it is important to understand that low dissolved oxygen nitrification is a real possibility that can save a considerable amount of aeration energy and costs.

#### *1.2.2.3 pH and Alkalinity*

Equations 3.3 and 3.4 describe the effect that nitrification has on alkalinity levels within the treatment reactors. These equations illustrate that, in situations in which low-alkalinity wastewater is being treated, the nitrification process can have a substantial effect on alkalinity levels and, ultimately, pH. Reactor pH levels have been shown to have a significant effect on the rate of nitrification (U.S. EPA, 1993). An important factor to consider regarding the effects of pH on the nitrification process is the degree of acclimation that the particular system has achieved. Wide swings in pH have been demonstrated to be detrimental to nitrification performance, although acclimation generally allows satisfactorily performance with consistent pH control within the range of 6.5 to 8.0 standard pH units. It is generally recommended that sufficient alkalinity be present throughout the nitrifying reactors by maintaining a minimum effluent alkalinity of at least 50 and preferably 100 mg/L (as $CaCO_3$). Low pH shifts the ammonia and ammonium equilibrium more toward ammonium, reducing the actual concentration of ammonia, which is the organisms' food. Low pH slows the nitrifier metabolic rate because the food concentration is reduced and requires higher detention time to compensate. Because the acid dissociation constant ($pK_a$) of ammonia is 9.25, a pH of 7.25 allows only 1% of the total ammonia to be in ammonia form; a pH of 6.75 reduces the concentration to less than 0.5%.

#### *1.2.2.4 Inhibition*

Nitrifiers are particularly susceptible to inhibition from a variety of organic and inorganic substances. In addition, nitrifiers are particularly susceptible to wide fluctuations in the concentration of inhibitory substances but may exhibit only minor effects if these substances are in low concentrations and are consistently applied to the system. Toxic organic compounds include organic solvents, amines, phenolic compounds, alcohols, cyanates, ethers, carbamates, and benzenes (U.S. EPA, 2010). A full list of known organic chemicals that have been identified as inhibitory are identified in the *Nutrient Control Design Manual* (U.S. EPA, 2010). Nitrifying bacteria do not seem to be overly affected by salinity. A study by Campos et al. (2002) found 100% nitrification at concentrations of 13.7 g/L NaCL but did show a decline at 20 g/L.

Nitrifier performance can also be significantly affected by heavy metals, including nickel (at or above 0.25 mg/L), chromium (at or above 0.25 mg/L), and copper (at or above 0.10 mg/L) (Metcalf & Eddy, Inc./AECOM, 2014). Un-ionized or free ammonia can be inhibitory to AOB and NOB, depending on temperature and pH conditions in the reactor. At a pH of 7.0 and a temperature of 20 °C, inhibition is expected to begin at an ammonia-plus-ammonia ($NH_3$-N + $NH_4^+$-N) concentration of 1000 and 20 mg/L for AOB and NOB, respectively (U.S. EPA, 1993). These concentrations can occur in centrate from the anaerobic digestion process, the discharge from animal feedlots, and industrial wastewater (U.S. EPA, 2010).

#### 1.2.2.5 Flow and Load Variations

Provided that environmental conditions do not limit growth of nitrifying organisms, the quantity or mass of AOB and NOB that grow in the system will be a function of the applied ammonia load. As such, variations in flow and nitrogen load to the system that result in either a significantly reduced hydraulic retention time (HRT) or increased ammonia load may result in an increase in effluent ammonia. Intermittent loads, such as the sidestream from anaerobic digestion centrate, or industrial discharges may have a very high ammonia concentration; it is important to distribute those loads to prevent shock loading on the process. Short-HRT systems, such as many biofilm processes, are more likely to experience this reduction in process efficiency, even during normal diurnal variations. Longer-HRT systems, such as extended aeration, sequencing batch reactor, and oxidation ditch activated sludge systems, are less likely to experience increased effluent ammonia levels because of variations in flow and ammonia load.

## 2.0 SUSPENDED GROWTH SYSTEMS

### 2.1 General

Suspended growth systems are those in which the bacteria and microorganisms are suspended in the wastewater (typical activated sludge process) as opposed to attached growth in which the bacteria and microorganisms are attached to substrate such as plastic or rock. Suspended growth biological treatment systems operate in a fashion that allows control of the amount of biomass in the process and, therefore, the net growth rate of the biomass. The operating range available is limited by the physical capacities of the system, most notably, aeration tank volume, aeration capacity, clarifier surface area, and return activated sludge pump capacity. The ability to control the overall biomass growth rate within the system provides the opportunity to manipulate the system to achieve the concurrent growth of heterotrophic (carbonaceous BOD [CBOD]-consuming) and autotrophic (nitrifier) biomass populations.

Suspended growth nitrification can be achieved in various reactor configurations; however, these configurations must be designed and operated to meet the following two overriding criteria:

1. The biomass inventory must be retained in the system for a sufficient period to allow a stable population of nitrifiers to develop and be maintained in the process (i.e., there must be sufficient SRT); and

2. The HRT of the system must be long enough to allow the biomass ample time to remove the pollutant load (BOD and ammonia) entering the system to the extent necessary to maintain permit compliance.

The $SRT_{aerobic}$ is the average period that any solid or particle is retained in the aerated portion of a suspended growth process reactor. It is important to note that if the WRRF is only designed for nitrification, then the overall SRT is identical to $SRT_{aerobic}$ because all the bioreactors are aerated. In that case, the term should be SRT and not $SRT_{aerobic}$. The use of $SRT_{aerobic}$ is only relevant if the facility is removing nitrogen, phosphorus, or both when there are different environmental conditions in the reactors, in which case you might have an $SRT_{anoxic}$ and an $SRT_{anaerobic}$. For biological nutrient removal (BNR) operations, typically the overall SRT ($SRT_{aerobic} + SRT_{anoxic} + SRT_{anaerobic}$) is used for all process control calculations. Because this chapter is only considering nitrification, the term *SRT* will be used. The SRT, also referred to as *mean cell residence time* or *aerobic sludge age*, is most commonly defined as the mass of solids in the aerated portion of the reactor tanks divided by the mass of solids removed from the system (wasted) per day, as shown in the following equation:

$$\text{SRT} = \text{Mass of MLSS in aerobic portion of reactor/} \\ \text{Mass of MLSS wasted per day} \tag{3.11}$$

This equation has been simplified based on the following assumptions:

- Only the mixed liquor suspended solids (MLSS) inventory in the aerated portion of the system is considered to contribute to nitrification because the nitrifier growth rate is assumed to be negligible at low or zero dissolved oxygen conditions (i.e., anoxic zones); and
- The mass of effluent total suspended solids (TSS) is ignored because it is significantly lower than the mass of MLSS in the waste sludge and typically makes up a small fraction of the total sludge wasted. (If it is necessary to calculate SRT during process upsets or wet weather conditions, then effluent TSS should be included in the calculation.)

These two assumptions are generally true for most full-scale applications. However, it is important to note that the mass of effluent TSS may be significant in situations in which the effluent contains high concentrations of effluent TSS, such as during a secondary clarifier washout event or in activated sludge systems where the quantity of waste sludge generated is small as a result of the low organic loading (e.g., separate sludge nitrification systems). Separate activated sludge systems are discussed in Section 2.3 of this chapter.

The overall approach for operating a suspended growth nitrification system presented here is to determine the target SRT using the equations presented and adjust sludge wasting rates to maintain the actual SRT of the system at or above the target SRT. Because nitrification is sensitive to wastewater temperature, dissolved oxygen concentration, pH, and ammonia concentrations, each of these factors may have a significant effect on the target SRT determination, and each often requires a different target SRT during specific operating conditions (i.e., colder temperatures will often result in a higher target SRT).

### 2.1.1 Determining the Target Aerobic Solids Retention Time

The following steps can be used to determine the target SRT:

1. Calculate the maximum nitrifier growth rate ($\mu_{max}$) based on wastewater temperature (Equation 3.10).

2. Calculate the specific nitrifier growth rate (μn) based on the reactor ammonia concentration. Assuming pH values are stable and approximately neutral ($6.5 < pH < 8.0$) and dissolved oxygen concentrations of greater than 2.0 mg/L are maintained, proceed to the next step.
3. Calculate the minimum aerobic SRT, which equates to the SRT when nitrifiers are reproducing as fast as they are being wasted out of the system.
4. Select the process design factor (PDF) to account for the variability in influent wastewater characteristics and the safety factor. The PDF is the product of a peaking factor and a safety factor and typically ranges between 1.5 and 3.0. The peaking factor can be determined to account for the variability in process influent wastewater characteristics, internal facility recycles, facility operation (return activated sludge [RAS], waste activated sludge [WAS], and dissolved oxygen control), effluent permit requirements, and the physical process configuration. Highly variable influent wastewater characteristics, intermittent or poorly controlled internal facility recycles, stringent permit limits, intermittent or manually controlled operation, and smaller process tanks warrant larger peaking factor values. Consistent influent and internal facility recycles; less frequent and stringent permit requirements (effluent quality and/or permit compliance period); well-automated processes; and facilities with larger, more forgiving process tanks or equalization allow smaller peaking factor values. The safety factor is generally a means of reflecting the level of uncertainty in design or process performance. Use of a safety factor may be warranted in situations with newer, less proven technologies and may be omitted if all variations to the process influent are well defined and accounted for in the design (Grady & Lim, 1980).
5. Calculate the target SRT by multiplying the minimum SRT times the PDF.

Use of daily values for wastewater temperature and SRT should be avoided. Instead, moving average values are recommended to dampen intermittent operations and sampling variability. Generally, a 7-day moving average is satisfactory to minimize the effect of variability without creating a significant "lag" effect on calculated values. Figure 3.3 illustrates daily versus 7-day moving average SRT for an operating facility. Note how SRT increases during colder months and decreases during warmer months.

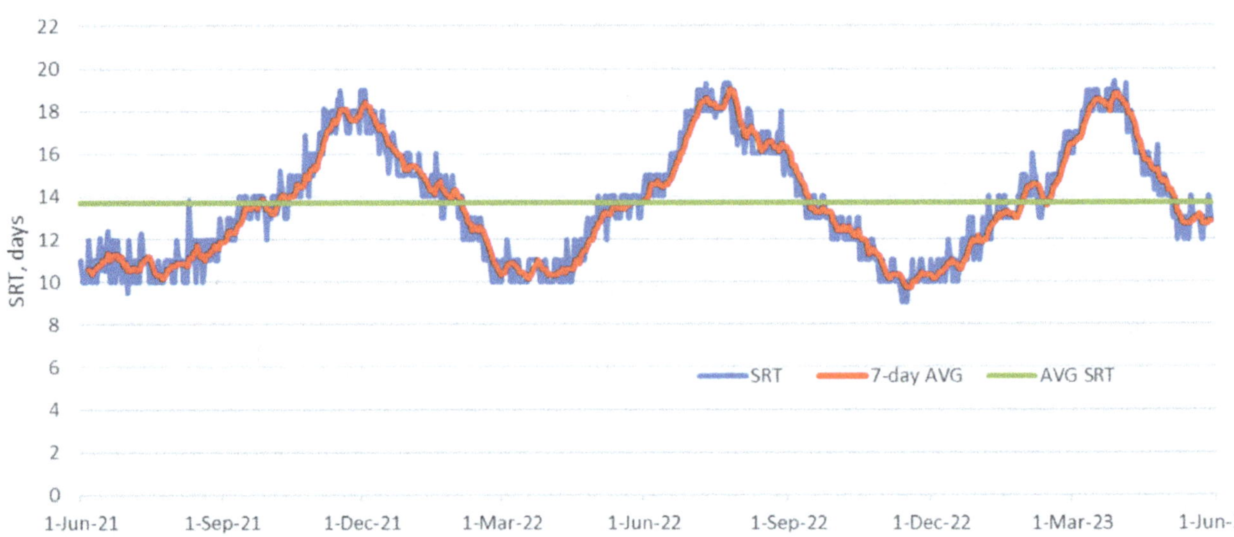


FIGURE 3.3 Daily Versus 7-Day Moving Average SRT

### 2.1.2 Example 3.1—Single Sludge Suspended Growth Nitrification

The following is an example procedure for determining the design SRT for a single sludge suspended growth nitrification system.

The following secondary influent conditions and secondary effluent requirements are given:

- Average daily flow = 4000 $m^3$/d (1.06 mgd);
- Secondary influent $BOD_5$ = 150 mg/L;
- Secondary influent total Kjeldahl nitrogen (TKN) = 35 mg/L;
- Secondary influent ammonia-nitrogen ($NH_4$-N) = 21 mg/L;
- Secondary influent wastewater temperature = 15 °C;
- Maximum monthly and average flow ratio = 1.3;
- Diurnal peak and average flow ratio = 1.2; and
- Secondary effluent $NH_4$-N = 1.0 mg/L (note that secondary influent flow, BOD, and TKN loading should be based on the most stringent permit criteria [monthly, weekly, etc.] anticipated for the time period).

The design SRT for a single sludge suspended growth nitrification system can be determined using the following steps:

1. Calculate the nitrifier maximum growth rate, endogenous decay coefficient, and half-saturation constant as follows:

   The value of $\mu_{maxAOB}$ = 0.9 at 20 °C, b = 0.17 at 20 °C, $K_o$ = 0.5, and $K_{nh}$ = 0.5 at 20 °C, but the temperature for this wastewater is 15 °C. To correct for temperature, the following formula is used:

   $$\mu_{maxAOB@15} = \mu_{maxAOB@20}\,\Theta^{(15\text{-}20)} = 0.9 \times 1.072^{(15\text{-}20)} = 0.636\ \text{g/g}\cdot\text{d}$$
   $$b_{15} = 0.17 \times 1.029^{15\text{-}20} = 0.147$$
   $$K_{nh15} = 0.5 \times 1.0^{15\text{-}20} = 0.5$$

2. Calculate the specific growth rate as follows:

   $$\mu_{AOB} = \mu_{max-AOB}\frac{SNH}{SNH + K_{NH}} \times \frac{S_o}{S_o + K_{o,AOB}} - b_{AOB\mu}$$

   $$\mu_{AOB} = \frac{0.636\ \text{g}}{\text{g}\cdot\text{d}}\left\{\frac{0.5}{0.5+0.5} \times \frac{2}{2+0.5}\right\} - \frac{0.147}{\text{d}} = \frac{0.107}{\text{d}}$$

3. Calculate the theoretical SRT as follows:

   $$\text{SRT} = 1/\mu_{AOB} = 1/(0.107/\text{d}) = 10\ \text{days}$$

4. Determine the PDF as follows:

   PDF = Peaking factor × safety factor
   Peaking factor = Maximum month flow factor × diurnal flow factor
   Peaking factor = 1.3 × 1.2 = 1.56

The safety factor is intended to account for variability in secondary influent wastewater characteristics and process operation (dissolved oxygen control, sludge wasting, etc.) and process performance uncertainty. For example,

Safety factor = 1.5 (selected)
PDF = 1.56 × 1.5 = 2.34

5. Calculate the target SRT as follows:

Design SRT = Theoretical SRT × PDF
Design SRT = 10 days × 2.34 = 23.4 days

## 2.2 Single Sludge Systems

Single sludge suspended growth nitrification systems are defined as a process configuration including only one set of clarifiers per train of reactors for oxidation of both CBOD and nitrogenous (TKN and ammonia) pollutants. Because of this configuration, these systems generally operate at higher CBOD-to-TKN ratios than separate sludge systems. Because of the common practice of incorporating denitrification and/or biological phosphorus removal with nitrifying systems, single sludge BNR systems have become the predominate approach when nitrification is required. Figure 3.4 presents a schematic of a single sludge nitrification system.

## 2.3 Separate Sludge Systems

Separate sludge suspended growth systems, as shown in Figure 3.5, are defined as a process configuration that includes a separate set of reactors and clarifiers for each step of the process, typically for CBOD removal and nitrification. Separate sludge nitrification is characterized by low CBOD-to-TKN ratios (typically less than 5:1 and often less than 2:1) in the second stage. This influent loading to the nitrification system typically results in particularly low WAS generation rates in the nitrification stage. In some situations, a portion of the primary effluent is bypassed to the separate sludge nitrification system to provide sufficient loading to generate more biomass than is lost in the effluent TSS.

Separate sludge nitrification systems have been implemented for a variety of reasons, including the following:

- to provide tertiary treatment following an upstream biofilm (trickling filter) process;
- to protect the nitrification system from an influent wastewater that contains materials toxic to nitrifiers (i.e., metals); and
- to allow optimized operation of the separate carbonaceous and nitrifying activated sludge systems to meet specific effluent requirements or to provide a fully nitrified wastewater to a downstream denitrification process.

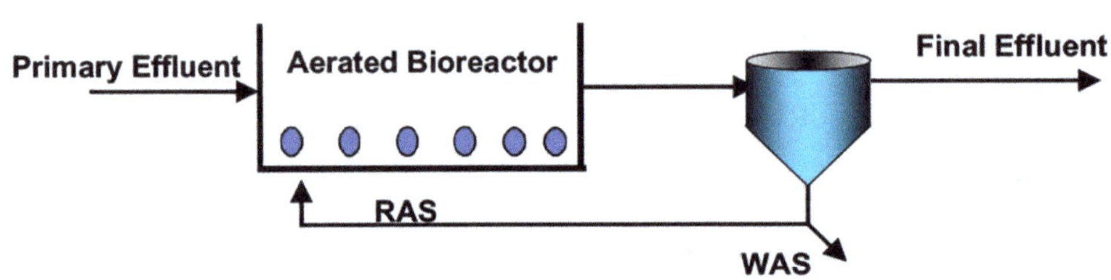


FIGURE 3.4 Single Sludge Nitrification System for BOD Removal and Nitrification

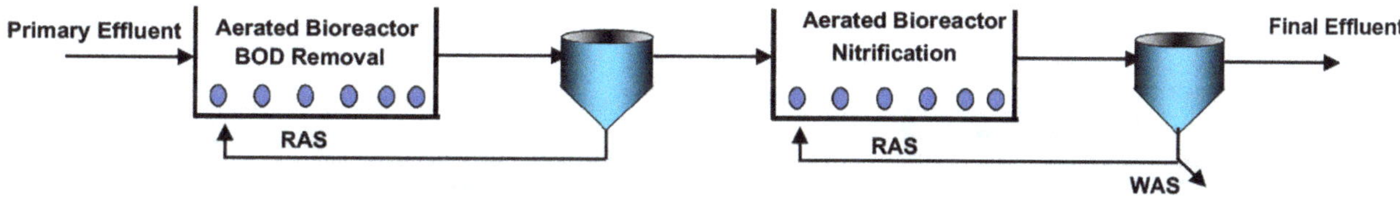


FIGURE 3.5 Separate Sludge Nitrification System

In general, separate sludge suspended growth nitrifying systems have become less common recently, primarily because of increased costs associated with the construction of separate clarifiers for each stage and the tendency to integrate BNR (nitrification and denitrification) systems together. Separate stage systems have experienced numerous operational challenges mainly related to low biomass growth rates in the nitrification stage and "weak" floc structure that is susceptible to shearing, which can result in effluent solids. However, there are some large separate sludge WRRFs (e.g., Blue Plains Wastewater Treatment Plant in Washington, DC) that have a long history of successful operation and performance.

## 3.0 BIOFILM SYSTEMS

Biological wastewater treatment processes with biomass attached to some type of inert media are termed *biofilm reactors* (aka fixed film, attached growth, or fixed growth). True biofilm systems are differentiated from coupled treatment systems that incorporate a separate biofilm reactor followed by an activated sludge process without intermediate clarification and hybrid systems that include suspended and biofilm processes within the same reactor. A complete description of biofilm processes used in wastewater treatment is presented in Chapter 4.

## 4.0 COUPLED SYSTEMS

A *coupled process* can be defined as a process configuration that combines two different treatment processes. Most coupled processes consist of a biofilm reactor followed by a suspended growth reactor. The most common coupled processes currently in use are the trickling filter and solids contact and the roughing filter and activated sludge (RF/AS) processes. Figure 3.6 shows one configuration of the RF/AS process. Combining the two processes is often a means of cost-effectively upgrading an existing facility and may also be used for new designs of secondary treatment facilities. Using dual processes offers the opportunity to use the best attributes of each process while limiting some of their individual weaknesses. Because most coupled-process configurations include a trickling or roughing filter as the initial process, they are not necessarily an optimum configuration for total nitrogen removal. However, upgrading a facility using a coupled process is often a cost-effective alternative to implementing nitrification or a modest level of total nitrogen removal, especially if the trickling filters already exist and the activated sludge process represents the upgrade. Some operational issues associated with coupled systems include the accumulation of filter snails in the activated sludge basin, which can require additional operation and maintenance attention, and the presence of nuisance filter flies.

Integrated fixed-film activated sludge (IFAS) and moving bed bioreactor (MBBR) processes have become increasingly popular technologies for WRRFs trying to increase nitrification and total nitrogen removal capacity. Both processes use biofilms, with IFAS incorporating the biofilm process to an aeration tank, and MBBRs relying solely on biofilm processes in a flow-through tank configuration. The

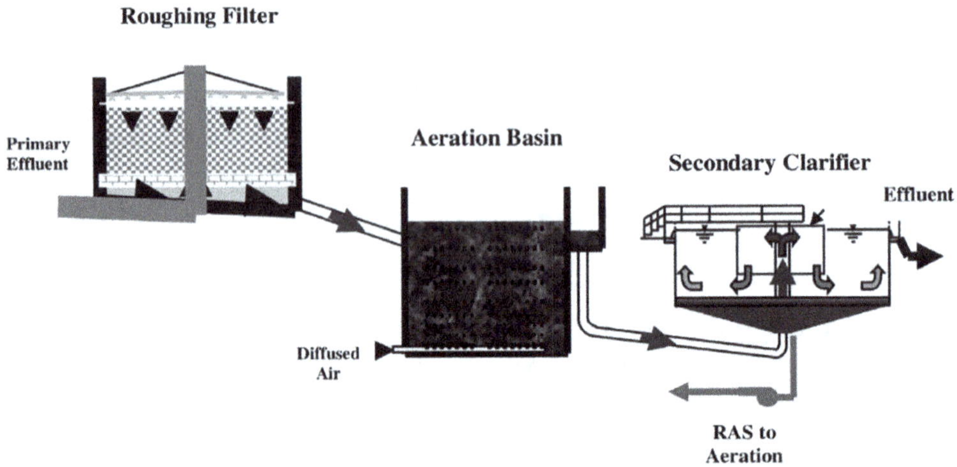


FIGURE 3.6 The RF/AS Process

IFAS process uses RAS (Figure 3.7a), whereas the MBBR does not use RAS recycle (Figure 3.7b). Both processes can be used to increase nitrification capacity, and although they can be designed and built as new processes, they are often implemented as upgrades to existing activated sludge facilities.

The IFAS process uses either fixed or free-floating media that can be made of plastic, foam, or fabric. The fixed media are anchored within the reactor and typically resemble a structured plastic filter media (net or mesh) or bundles of ropes. Media are installed in an activated sludge basin to encourage the establishment of biofilm in the tank, thus increasing the solids inventory in the basin. Table 3.2 presents a brief description of different types of fixed and free-floating media that are currently in use in IFAS

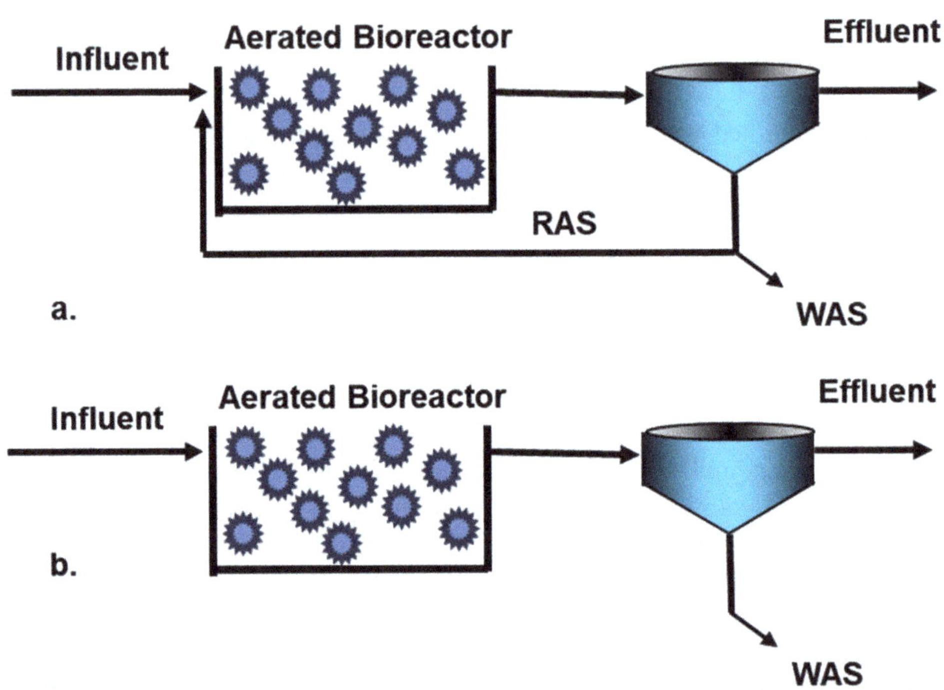


FIGURE 3.7 (a) Schematic of a Simple IFAS Application Showing RAS Recycle From a Secondary Clarifier and (b) Schematic of a Simple MBBR Application

TABLE 3.2 Types of Media Used in IFAS Systems

| Media type | Advantages | Disadvantages |
|---|---|---|
| **Fixed Media** | | |
| Fabric web type | Simple and quick to install, low initial cost, low maintenance, no risk of media loss | Prone to fouling (rags and red worm infestations), accumulation and trapping of debris, may require primary treatment upgrade |
| Rope type | Simple and quick to install, low initial cost, low maintenance, no risk of media loss | Prone to fouling (rags and red worm infestations), media breakage and entanglement, field assembly required |
| Poly vinyl chloride sheet (trickling filter media) | Simple and quick to install, low initial cost, low maintenance, no risk of media loss | Structural media may impede mixing and air distribution, excess biomass accumulation may cause clogging, rags can cause clogging |
| **Free-Floating Media** | | |
| Polypropolyene finned cylinders | Excellent mixing, staged implementation, may convert to MBBR | Potential for media loss (abrasion and washout), fouling of aeration system and screens, difficult to maintain the aeration system, need to account for media distribution |

systems. Integrated fixed-film activated sludge systems are often implemented to improve nitrification capacity at existing facilities without increasing the solids load rate on secondary clarifiers. The IFAS systems have been shown to achieve 100% nitrification with 50% or less of aeration volume compared to conventional activated sludge systems.

The key to the IFAS process is the increased total biomass inventory, which can be achieved by combining the attached biomass with the suspended MLSS. The amount of biomass that can be retained on the different types of media depends on multiple factors, including loading rate, dissolved oxygen concentration, media type, temperature, mixing, suspended activated sludge biomass concentration, and overall SRT. Because the attached biomass is retained in the activated sludge basin and not sent to the clarifiers, use of IFAS technology can increase the capacity of the activated sludge system in the same tank volume without significantly affecting the clarifiers.

To implement IFAS technology, the following design and operational issues must be considered:

- Mixing requirements—enough mixing energy is needed to ensure that free-floating media remain well mixed and uniformly distributed in the tank. The mixing requirement is also critical for sloughing of biomass and maintenance of a thin biofilm to reduce mass-transfer limitations.
- Aeration requirements for nitrification systems—a higher dissolved oxygen concentration (greater than 3.0 to 4.0 mg/L) is typically required in the suspended phase to overcome the biofilm mass-transfer limitation and ensure that the biofilm is completely aerobic. In addition, many IFAS systems require the use of course-bubble diffusers as opposed to fine-bubble diffusers, which can result in additional blower capacity.
- Effluent screens—when free-floating media are used, effluent screens or media retention sieves (typically 6-mm screens) must be installed in the bioreactors to retain the media within the reactors. The requirement that all tank effluent pass through the screens can result in a foam trap that can increase foam accumulation in the IFAS tanks, which may require special foam-control tools (i.e., defoamant surface spray and selective surface wasting). In addition, the screens may

also result in the accumulation of floatables and other debris, which can block or clog the effluent screens. For this reason, upgrades to preliminary and primary treatment may need to be considered when implementing an IFAS system.

- Media retention and distribution—when floating media are used, aeration diffuser placement and power need to be carefully considered to ensure that the media are well distributed throughout the tank and that they do not accumulate at the end of the tank because of excessive bulk flow velocities. Recirculation of the media to the head of the tank and/or implementation of a hydraulic knife at the effluent screens may be needed to avoid media buildup at the screens. Incidents of effluent screen failures resulting in total or partial loss of floating media have been reported.

Facility operators considering installing an IFAS system as an upgrade may also need to consider upgrading influent screens (e.g., 6-mm screens for facilities with primary clarifiers and 3-mm screens for facilities without primary clarifiers). Additionally, the effluent screen or sieve approach velocity should be monitored and maintained at a level (<35 mg/h) that reduces media stacking toward the effluent end of the tank. Finally, operators can control the approach velocity by reducing RAS rates. Tools to control froth and scum accumulation may also be needed for some installations.

One significant operational issue associated with IFAS systems is the need to remove media for tanks and aeration system maintenance. During such an event, the facility must also have plans for transport and storage of the media when removed. Fixed-media IFAS systems avoid many of the media retention and washout issues but will still require equipment and facilities for removal during aeration tank maintenance.

Integrated fixed-film activated sludge systems can be used to increase both nitrification and denitrification capacity but are most commonly used to increase nitrification capacity. Figure 3.8 depicts a schematic of IFAS for a four-stage BNR process.

Moving bed bioreactors are similar to free-floating media IFAS systems, except there is no RAS recycle and the process relies almost entirely on biofilm activity. Many of the same issues associated with free-floating IFAS media are also a concern in MBBR systems (i.e., media loss, foaming, etc.). The key advantage of MBBR processes is their ability to retain slow-growing organisms and to maintain a high equivalent SRT. For this reason, MBBRs are currently being implemented as separate centrate treatment technologies that use anaerobic ammonia oxidation (anammox) biomass and deammonification processes. These types of systems require specialized biomass to be retained in the reactor for long SRTs (>30 days). One such process is the MBBR nitritation and anammox process, which allows for high levels of specialized biomass to be retained for long SRTs while providing the specific biofilm niche needed to carry out this two-path BNR removal process. Additional information on anammox, IFAS,

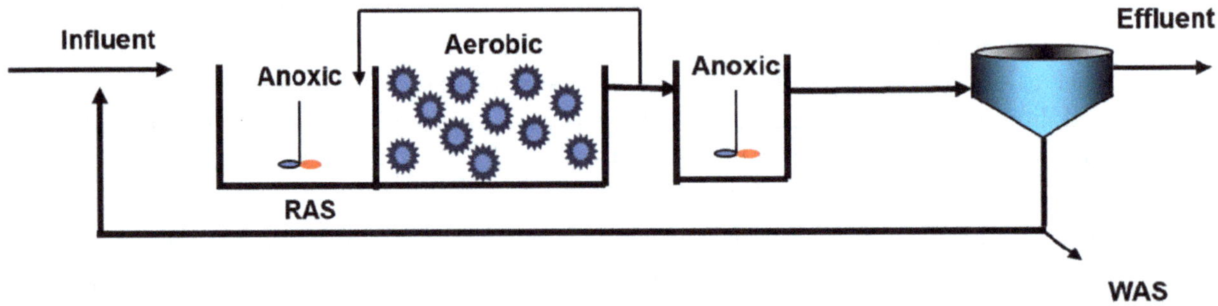


FIGURE 3.8 Schematic of IFAS Implementation in a BNR Process With Media Placed in the Aerobic Zone to Increase Nitrification Capacity

and MBBR processes can be found in the Water Environment Federation's (2010) *Biofilm Reactors* and in Chapters 4 and 7 of this book.

## 5.0 OPERATIONS SUMMARY

Operations and process control are key to ensuring a WRRF meets permit. Biological nutrient removal processes require a high level of understanding of how the process works and what is needed to ensure optimum process performance. Attention to all operating considerations is important to ensure an efficient process and one that consistently meets required permits. For nitrification, operational considerations include

- dissolved oxygen,
- alkalinity,
- inhibition,
- hydraulic retention time, and
- nitrite formation.

Dissolved oxygen: Nitrification is an aerobic process, so it is important to ensure sufficient oxygen is available for the microbial process. As a rule of thumb, maintain a dissolved oxygen concentration between 1.5 and 2 mg/L. The dissolved oxygen concentration will vary depending on the diurnal flow, seasonal effects, and wet weather events. The best way to monitor dissolved oxygen and ensure adequate dissolved oxygen is with in-line continuous dissolved oxygen instrumentation. If instruments are not available, the operator should check dissolved oxygen concentrations at a minimum of twice per day at low flow and peak flows using a calibrated handheld meter.

Alkalinity: The first step of the nitrification reaction, conversion of ammonia to nitrite by AOB, produces nitrous acid. If there is insufficient alkalinity, the pH will decrease, inhibiting the conversion of nitrite to nitrate. At low pH values, approximately 6.4, nitrification is inhibited (U.S. EPA, 1993). It is important for operators to monitor the pH in the bioreactors and effluent daily and to determine alkalinity on the influent or primary effluent and the final effluent daily. As a general rule, there should be a minimum of 70 mg/L of alkalinity in the final effluent. This ensures sufficient alkalinity to maintain a pH of 6.8 to 7 (Metcalf & Eddy Inc./AECOM, 2014).

Inhibition: Operators understand that it is impossible to know what is entering a WRRF at any given time. There are many chemicals that can cause inhibition of the nitrification process, including metals and organic compounds. In addition, sidestreams from thickening and dewatering processes with high concentrations of ammonia can also be inhibitory. It may be easy to control in-facility processes by equalizing the load prior to entering the bioreactors, but it may not be easy to control external discharges. If an operator suspects that some industrial or toxic waste has entered the WRRF, it is important to take samples to try to identify the substance. Once identified, it may be possible to prevent the discharge in the future.

Nitrite formation: Typically, the first indication that nitrite is being produced is an increase in chlorine dosage to maintain a residual. Nitrite formation can be caused by a low dissolved oxygen concentration or low pH in the reactors or inhibitory chemicals entering the facility. Instruments that measure nitrite and nitrate concentration in the reactors and effluent can help inform the operator that there is a problem. Some actions that can be taken are increasing the dissolved oxygen concentration or adding alkalinity. Nitrite production caused by inhibition is more complex. The problem might not correct itself until all the inhibitory chemical is washed out of the facility.

## 6.0 REFERENCES


Campos, J. L., Mosquera-Corral, A., Sanchez, M., Méndez, R., & Lema, J. M. (2002). Nitrification in saline wastewater with high ammonia concentration in an activated sludge unit. *Water Research, 36*(10), 2555–2560.

Grady, C. P. L., & Lim, H. C. (1980). *Biological wastewater treatment: Theory and applications.* Marcel Dekker.

Kisling, T., Khatiwada, S., McCullough, K., Klaus, S., Wilson, C., Rauch-Williams, T., & Bott, C. (2022). *Mechanistic evaluation of the adaptation of nitrifiers and PAOs to low DO operation.* WEFTEC 2022, New Orleans, LA, United States. Water Environment Federation.

Metcalf & Eddy, Inc./AECOM. (2014). *Wastewater engineering: Treatment and resource recovery.* McGraw Hill Education Publishing.

Miller, M. W., Srinivasan, V., Al-Omari, A., Esping, D., Regmi, P., Bauhs, K., Sturm, B., Downing, L., & Jimenez, J. (2022). *More than just energy savings: Understanding the benefits of low DO operation.* WEFTEC 2022, New Orleans, LA, United States. Water Environment Federation.

Monod, J. (1949). The growth of bacterial cultures. *Annual Review of Microbiology, 3*, 371–394.

U.S. Environmental Protection Agency. (1993). *Nitrogen control manual* (EPA-625/R-93-010).

U.S. Environmental Protection Agency. (2010). *Nutrient control design manual* (EPA-600/R-10/100).

Water Environment Federation. (2010). *Biofilm reactors* (Manual of Practice No. 35).

# Chapter 4

# Nitrification in Biofilm Reactors



## 1.0 INTRODUCTION

*Nitrification* is the oxidation of ammonium to nitrate using specialized bacteria that occur naturally in soil and water. Whereas Chapter 3 discusses nitrification in general, this chapter focuses on nitrification control in biofilm systems. The term *biofilm* is typically used for a community of microorganisms growing on a surface or support material, sometimes called media, biofilm carrier, or in the case of a membrane aerated bioreactor (MABR), the membrane. Historically, bacteria growing on an attached surface in a wastewater system were referred to as *fixed film reactors*. Water Environment Federation's (WEF's) publication *Biofilm Reactors* (WEF, 2011) provides design and operational considerations for these systems. Biofilms can be problematic when they occur inside pipes, valves, and so on, but advantageous when they grow on media in a biological treatment process. Often, the same microorganisms occur in both situations; the only difference is where they occur and how they are described (e.g., *fouling* when microorganisms are not wanted and *biofilms* when they are wanted).

Biofilm reactors are often implemented for increased nitrification in wastewater applications because the fixed nature of bacteria in biofilms allows for retention and proliferation of the "slow"-growing nitrifying bacteria, typically and predominantly *Nitrosomonas* and *Nitrobacter*. These organisms are highly susceptible to process upset resulting from alkalinity limitation, lack of dissolved oxygen, seasonal temperature swings, toxic shock, and suspended growth processes and are susceptible to nitrifier washout. Fixed film processes have been coupled to or incorporated within suspended growth systems, taking advantage of the merits of both systems, and are generally referred to as integrated fixed-film and activated sludge (IFAS) processes. Given the increasing importance of nutrient removal using smaller footprint systems, nitrifying biofilm reactors constitute an important technology for water resource recovery operators. The fundamentals of reactions in biofilms and how to control these reactions in full-scale reactors are critical pieces of knowledge for effectively operating nitrifying biofilm reactors.

## 2.0 BASIC BIOFILM STRUCTURE AND FUNCTION

Physical transport of nutrients and oxygen into the biofilm layer must occur for biological activity to take place. The rate at which materials from wastewater are transported to the biofilm is dominated by diffusion. Diffusion is directly related to liquid velocity in the reactor. Liquid velocity is increased by increasing the flow of the liquid, either through pumping or mixing. As mixing intensity is increased, the rate of diffusion into the biofilm increases. This increased diffusion rate results in a higher rate of transport of "food" to bacteria in the biofilm and higher biological activity.

In addition to understanding the rate of food transport into the biofilm, the ecology of the biofilm is also important to understand. Each layer of biofilm consumes substrates, nutrients, and oxygen, resulting in a lower concentration of each of these compounds in every subsequent layer. The different concentrations of substrate, oxygen, and nutrients in each layer result in varying selective pressures on the ecology, which then results in a stratification of bacteria. When multiple biofilm reactors are operated in series (stages), specific bacterial populations may be maintained in each stage. When a single reactor or single stage is used or when biofilm thickness goes uncontrolled, the bacterial population can be expected to vary through the layers of biofilm.

### 2.1 Systems With Inward Dissolved Oxygen Diffusion

For systems including trickling filters, rotating biological contactors (RBC), moving bed biofilm reactors (MBBR), biologically active filters (BAF), and submerged aerated filters (SAF), in which oxygen is provided from the bulk fluid and exterior to the biofilm surface, oxidation of organic (carbonaceous) matter and ammonia (nitrification) can occur in the outer aerobic zone, whereas denitrification can occur in the anoxic base of the biofilm, and the following citation holds. Typically, heterotrophs are found on the outer edge of an aerobic biofilm, followed by nitrifying bacteria, and then anoxic and anaerobic metabolism at the base of the biofilm (Okabe et al., 1996).

A simplified schematic of the stratification of a biofilm within processes such as trickling filters, RBC, MBBR, BAF, and SAF, with each subsequent layer being exposed to different compound concentrations, is shown in Figure 4.1. Figure 4.1 also shows idealized concentration profiles of substrates

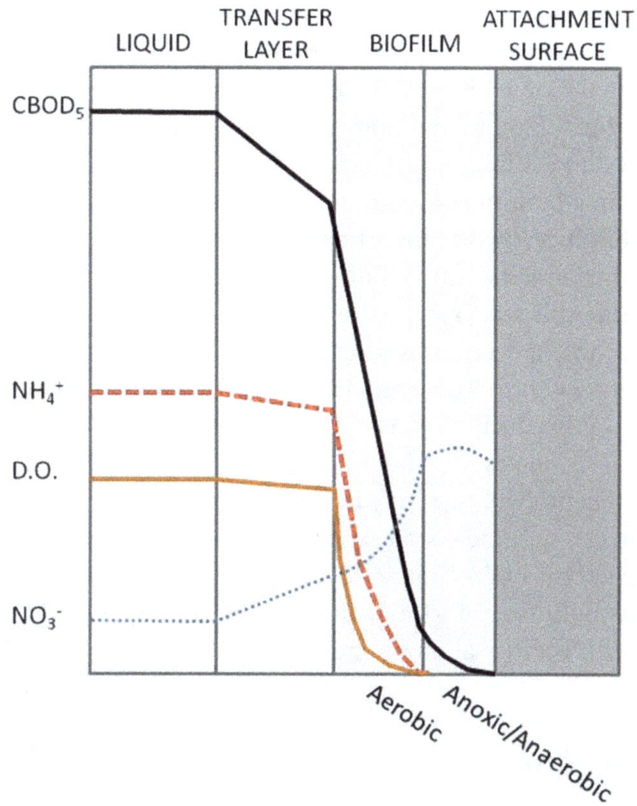


FIGURE 4.1 Idealized Concentration Profiles of Substrates Being Consumed (Dissolved Oxygen, CBOD, and $NH_4^+$-N) or Produced ($NO_3^-$-N) by Organisms in a Biofilm, Typical of Trickling Filters, RBC, MBBR, BAF, and SAF

being consumed (dissolved oxygen, carbonaceous biochemical oxygen demand [$CBOD_5$], and ammonia [$NH_4^+$-N]) or produced (nitrate-nitrogen [$NO_3^-$-N]) by the organisms in a biofilm. Substrate concentrations in the wastewater (liquid) are scaled to represent a secondary effluent concentration of 20, 30, and 10 mg/L of biochemical oxygen demand (BOD), suspended solids, and $NH_4^+$-N, respectively, which diffuses from the bulk water through the mass-transfer boundary layer (MTBL) and into the biofilm. The reduced concentration of $NO_3^-$-N compared to $NH_4^+$-N represents 10% of the nitrogen being assimilated into biomass, and the change in concentration across the boundary layer is proportional to the individual substrate diffusion coefficient, with CBOD idealized as glucose. The outer biofilm is aerobic, but the inner biofilm is anoxic or anaerobic depending on the nitrate concentrations.

Within the biofilm, as the concentrations of substrates (i.e., BOD and ammonia) change with penetration depth, the selective pressure for bacterial selection also changes. For each layer deeper in the biofilm shown in Figure 4.1, a lower BOD, ammonia, and dissolved oxygen concentration is observed. In contrast, the concentration of materials being produced (e.g., nitrate [$NO_3^-$]) rises as it diffuses out of the biofilm because it is being formed as the end product of ammonia oxidation (i.e., nitrification) by the nitrifiers.

The depth of oxygen penetration into a biofilm has been examined experimentally. Microelectrodes, which are essentially small-tipped dissolved oxygen sensors (i.e., the tip diameter is less than 10 μm), have been used to examine oxygen penetration into biofilms. In a biofilm of a trickling filter treating municipal wastewater, the dissolved oxygen concentration was found to reach zero at a depth of only 400 μm (0.016 in.) under steady-state conditions with raw wastewater. When the activity of the biofilm was increased by feeding the system an artificial nutrient broth, the dissolved oxygen concentration reached zero at a depth of only 100 μm (0.004 in.) (Kuenen et al., 1986). Using biofilm from an eel aquaculture trickling filter, Schramm et al. (1996) demonstrated that oxygen consumption and nitrate production only occurred in the outer 50 μm (0.002 in.) of the biofilm. These results highlight the rapid consumption of oxygen within a biofilm and how efficient transfer of oxygen into a biofilm is critical to optimize the amount of biofilm exhibiting aerobic activity.

Gradients develop within a biofilm as a result of different rates of diffusion from the wastewater and different rates of consumption by biofilm microorganisms. For a nitrifying biofilm, a partially penetrated biofilm exists when either ammonia or oxygen is depleted before reaching the bottom of the biofilm. Biofilms grown in full-scale reactors might be completely penetrated immediately after a hydraulically triggered detachment event or during an organic shock load, but it is likely that most biofilm reactors generally operate with partially penetrated biofilms. Once a nitrifying biofilm is partially penetrated, increasing the biofilm thickness has no nitrification benefit because the ammonia, oxygen, and/or macronutrients cannot penetrate to the full depth of the biofilm, and therefore, cells that are deeper than the penetration depth will be starved of these nutrients. However, the inner portions of a partially penetrated biofilm can become anoxic, leading to some levels of denitrification in the biofilm even if the biofilm is in an aerobic basin (Points et al., 2010). If biofilm thickness becomes excessive, biofilm zones existing near the biofilm support may be deprived of oxygen and/or macronutrients, resulting in anaerobic conditions in this ecological niche and potentially uncontrolled biofilm detachment (or sloughing) and/or odors.

### 2.2 Biofilm Systems With Outward Dissolved Oxygen Diffusion

Membrane aerated bioreactors are biofilm processes that use a gas-permeable membrane to deliver oxygen to a biofilm that populates on the membrane surface, which is in contact with the bulk fluid containing mixed liquor suspended solids (MLSS) and pollutants. The membrane delivers oxygen outward into the biofilm. This is the opposite mode of oxygen diffusion to that shown in Figure 4.1. Membrane cassettes or modules are submerged in MLSS, and oxygen transport is provided through the membrane pores or via molecular diffusion, as depicted in Figure 4.2. Simultaneous nitrification and denitrification may occur and will depend on carbon and nitrogen ratios available within anoxic layers of the biofilm. Typically, coarse-bubble diffusers are installed beneath MABR cassettes/modules and provide mixing that's required to ensure pollutant transport to the biofilm and to promote biofilm detachment (i.e., control of biofilm thickness). Use of aeration to promote pollutant transport and control of biofilm thickness is common to all biofilm processes. Similar to biofilm systems with inward dissolved oxygen diffusion, the MABR may be staged to achieve low total nitrogen concentrations. According to a study published by Severn Trent and OxyMem (2018), the MABR's apparent sludge yield (0.13 kg total suspended solids [TSS]/kg chemical oxygen demand [COD] delivered) is less than that of conventional activated sludge, typically 0.3 to 0.4 kg TSS/kg COD delivered, and the energy efficiency associated with oxygen transfer (4.5 kg$O_2$/kW·h) is than that of CAS processes, typically 2 kg $O_2$/kW·h. Greenhouse gas emissions, $N_2O$ mitigation and life-cycle cost savings may be afforded with the MABR process and may be considered by designers.

## 3.0 NITRIFYING BIOFILM OPERATION—OVERVIEW

The primary difference between a suspended growth activated sludge reactor and a biofilm reactor is that suspended growth reactors are biomass-limited, whereas biofilm reactors are controlled by resistance to mass transfer (i.e., diffusion). Mass transfer is the driving force to convey pollutants from the

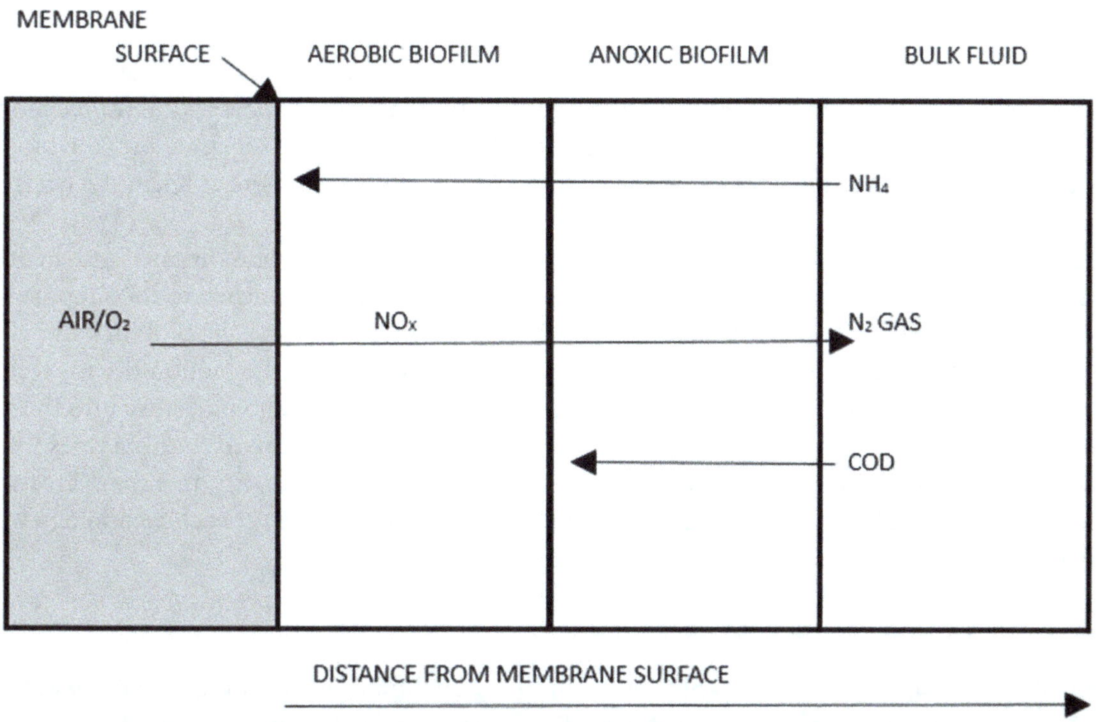


FIGURE 4.2 Diffusion Pathways Typical of MABR Systems for Dissolved Oxygen and COD Consumed, Ammonia-Nitrogen Oxidized, Nitrate and Nitrite Denitrified to $N_2$ Gas

liquid to the biofilm, in which biological and chemical reactions reduce pollutant concentrations. Whereas a wide range of nitrifying biofilm reactor technologies are currently available and implemented in practice, the overriding principles for operating biofilm reactors for improved nitrification can be summarized by the following two areas: mass-transfer control and biofilm thickness management. Although nitrification rate control through mass-transfer control and biofilm management are discussed in separate sections in this chapter, they are intimately related, and modifications to either of the parameters will have a direct effect on the other parameter.

## 3.1 Nitrification Rate Control—Mass Transfer

The nitrification rate of the biofilm is tied to the thickness of the MTBL (also known as *transfer layer* or *diffusion distance*) and concentrations in the bulk liquid. This mass transfer is described by Fick's law. The thickness of the boundary layer can be thought of as the distance pollutants must travel from the well-mixed liquid to reach the biofilm. The nitrification rate is inversely proportional to diffusion distance, meaning that smaller diffusion distances result in higher nitrification rates. The smaller this diffusion distance, the "faster" pollutants can reach the biofilm, resulting in a higher amount of "food" (or substrate) in the biofilm for bacteria to consume.

The distance pollutants have to travel to reach the biofilm is one aspect of how much food reaches the bacteria. The other factor is the concentration of substrate in the liquid. Higher liquid concentrations lead to a larger mass of pollutants reaching the biofilm, resulting in increased bacterial activity. Pollutant concentrations in the liquid are controlled by influent loadings and effluent requirements; however, the oxygen concentration can be manipulated for most biofilm reactor technologies by increasing the oxygen concentration in the wastewater, thereby increasing reaction rates in the biofilm.

### 3.1.1 Decreasing Diffusion Distance

Penetration of food into the biofilm can be directly tied to the thickness of the mass-transfer boundary layer, or diffusion distance, and to reactions within the biofilm. *Transfer layer* is a term related to mass-transfer dynamics and can be thought of as the diffusion distance for substrate into the biofilm. The faster the liquid is moving, the shorter the diffusion distance and the more efficient the transfer of compounds into the biofilm.

Whereas a high liquid velocity would theoretically lead to a small diffusion distance and, hence, high reaction rates, a balance must be achieved for the biofilm to remain attached to the support media. Achieving sufficient liquid velocity to achieve high levels of mass transfer into the biofilm and maintain adequate biofilm thickness is dependent on wastewater characteristics and the biofilm reactor technology implemented. Some operational control measures to decrease the diffusion distance into the biofilm include increased mechanical mixing, increased aeration flowrate, and increased flushing rates. For each of these techniques, more intense mixing or flow results in higher liquid velocity, thereby decreasing the diffusion distance. Details of these specific means for mass-transfer control for each technology will be discussed later in this chapter.

### 3.1.2 Liquid Concentration Management

For a nitrifying biofilm reactor, it is important to address bulk liquid BOD, ammonia, and oxygen concentrations. The higher the liquid concentration of a nutrient, the further it will penetrate a biofilm. As the concentration of BOD transferred into the biofilm increases, increased heterotrophic competition occurs for space and oxygen, resulting in decreased nitrification rates. To help limit the effect of BOD on nitrification rates, a BOD loading rate to the biofilm reactor should be maintained that minimizes diffusion of BOD in the biofilms. This loading varies for different reactor technologies. If the BOD loading rate exceeds the recommended loading for a given technology, it is critical to either (1) increase BOD removal upstream of the biofilm reactor or (2) increase the surface area within the reactor for biofilm attachment. Depending on the type of biofilm reactor, adding media to increase the biofilm attachment surface area can be difficult and/or costly. Operational control to limit the effect of excessive BOD in the reactor can be accomplished by increasing BOD removal upstream of the nitrifying biofilm reactor through methods such as chemically enhanced primary treatment (CEPT) or upstream biological processes.

Increasing the dissolved oxygen concentration in the liquid will also result in increased oxygen penetration into the biofilm and, potentially, much thicker biofilms. This results in a thicker aerobic layer, which can allow for both the higher heterotrophic population and the nitrifying population to coexist in the same biofilm. Although this can be an effective adjustment, it can also lead to excessive biofilm thickness. Excessive thickness, in turn, can lead to sloughing events and elevated effluent TSS concentrations.

Lower bulk liquid ammonia concentrations will result in slower nitrification rates in the biofilm because of the decreased transfer of ammonia into the biofilm. When stringent effluent ammonia limitations are in place (e.g., lower than 1 mg/L), a multi-staged nitrifying biofilm reactor can be implemented, with the first stage having a higher bulk liquid ammonia concentration (and thus a higher nitrification rate) and the later stage having a lower bulk liquid ammonia concentration (and thus selecting for a biofilm ecology that thrives in an environment with a low ammonia concentration).

For both bulk liquid BOD and ammonia concentrations, increasing the attachment surface area can decrease the bulk liquid concentration of these pollutants because more surface is available for the growth of active biofilm. Thus, for the same biofilm thickness, more surface area results in more biomass. The resulting increase in active biofilm volume can increase reaction rates, thereby decreasing the bulk liquid concentrations. However, for several biofilm reactor technologies, it is not possible to increase the attachment surface area without significant infrastructure changes.

Controlling the dissolved oxygen concentration in bulk liquid can be the most direct operational control parameter. Whether the dissolved oxygen concentration is controlled via mechanical mixing, diffused aeration, or water recirculation and flushing, it is critical to ensure that sufficient oxygen is present in the liquid so that oxygen can diffuse throughout the entire biofilm thickness. As discussed previously, the dissolved oxygen may only penetrate a biofilm to a depth of 400 µm (0.016 in.) under steady-state conditions in a wastewater system (Kuenen et al., 1986). In addition, increasing the bulk liquid dissolved oxygen typically results in increased fluid velocity and, hence, a shorter diffusion distance and higher transfer rates into the biofilm. A higher dissolved oxygen concentration in the bulk liquid or a smaller diffusion distance are directly tied to a higher biofilm oxygen concentration, resulting in a higher level of activity within the biofilm and, therefore, higher nitrification rates.

### 3.1.3 Relationship of Diffusion Thickness and Liquid Concentration

For nitrifying biofilm reactors, controlling the penetration of dissolved oxygen into the biofilm is directly tied to mechanisms for mixing, whether they are mechanical mixing, flushing water rates, diffused aeration, or other means of fluid movement. As the liquid velocity at the biofilm surface is increased, so, too, is the dissolved oxygen transfer rate into the biofilm. The higher liquid velocity reduces the diffusion distance, which also increases the nitrification rate. However, the liquid velocity must be monitored to ensure that velocities are not getting so high that they scour the biofilm off the attachment surface.

The point of excessive fluid velocity is highly dependent on the reactor type and specific field conditions (temperature, BOD loadings, alkalinity, etc.). From an operations standpoint, increasing the fluid velocity to decrease diffusion distance and increase the dissolved oxygen concentration should be done in a stepwise manner. These stepwise increases well help ensure that excessive biofilm detachment is not being induced from an attempt to increase the nitrification rate. Slowly increasing aeration rates, flushing rates, or backwash cycles can help prevent excessive biofilm loss. Losing the biofilm will obviously have a detrimental effect regardless of the bulk liquid concentrations because, without sufficient biofilm, nitrification will not occur at a sufficient rate.

## 3.2 Nitrification Rate Control—Biofilm Thickness Management

Mass-transfer control affects the transfer of substrates into the biofilm, but biofilm thickness and ecology affect the chemical and biological reactions that occur within the biofilm. A biofilm system will respond to varying environmental conditions (e.g., temperature and alkalinity) and electron donor, electron acceptor, and/or macronutrient availability. The physical response will be a change in biofilm mass, which is mainly observed as a change in biofilm thickness. This change in biofilm thickness results in a change to the substrate penetration depth into the biofilm, which affects the ecology, reaction rates, and overall health of the biofilm. In reactors with removable media (e.g., MBBRs and BAFs), this biofilm thickness information can be measured and used quantitatively in process models and qualitatively as an indicator of nitrification rate and the presence of anoxic and anaerobic layers within the biofilm.

### 3.2.1 Substrate Penetration

Substrate concentration gradients within a biofilm result from either partial or complete penetration by any soluble substrate diffusing into the biofilm. For a nitrifying biofilm, partial penetration exists when the ammonia, oxygen, and/or macronutrient is depleted before reaching the biofilm support. Biofilms grown in full-scale reactors might be completely penetrated immediately after a hydraulically triggered detachment event, during an organic shock load, or during certain points of the diurnal flow pattern. However, it is likely that most biofilm reactors generally operate with partially penetrated biofilms.

For facilities with low-level ammonia and/or phosphorus effluent limitations, care should be taken in the final stage of a nitrifying biofilm system to ensure that ammonia and phosphorus, in addition to oxygen, penetrate the biofilm fully. The concentration of ammonia and phosphorus affects this penetration, as does maintaining a shorter diffusion distance.

### 3.2.2 Biofilm Ecology

Biofilm ecology is controlled by substrate concentrations within the biofilm, which are directly tied to the mass-transfer controls discussed previously. For a nitrifying biofilm reactor, ecology concerns are mainly associated with the competition for oxygen between heterotrophic bacteria consuming BOD and autotrophic, nitrifying bacteria oxidizing ammonia. Substantial work has been conducted concerning the competition between these two groups of bacteria in biofilms for wastewater treatment (Fernandez-Polanco et al., 2000; Nogueira et al., 2005; Okabe et al., 1996). As the influent BOD concentration increases, heterotrophs and nitrifiers form a distinct stratification, with nitrifiers in the interior and heterotrophs at the exterior (Satoh et al., 2000). This stratification at high bulk liquid BOD has been verified in multiple reactor types (Cole et al., 2004; Downing & Nerenberg, 2008a, 2008b; Fernandez-Polanco et al., 2000; Hibiya et al., 2003; Okabe et al., 1996; Satoh et al., 2004; Terada et al., 2003).

For nitrifying biofilm reactors, the key to selecting for a biofilm dominated by nitrifying bacteria is to maintain a bulk liquid BOD concentration low enough to prevent excessive heterotrophic growth and to achieve sufficient oxygen penetration to maintain a high level of nitrification activity. The mass-transfer factors are controlled by the operational tools discussed previously, but the management of mass-transfer factors also has a positive effect on biofilm ecology (e.g., increased oxygen transfer increases aerobic biofilm volume and the volume of nitrifying bacteria), resulting in further optimization of nitrification rates.

### 3.2.3 Biofilm Detachment

Detachment by physical removal or grazing is a natural event that is necessary for successful operation of a biofilm reactor. Detachment is the only process by which biomass is wasted in a biofilm reactor. In a hybrid reactor or IFAS system, where suspended solids are present, biofilm detaches into the MLSS and wasting occurs with removal of the suspended growth typically through clarification. Bryers (1984) described four biofilm detachment processes: abrasion, erosion, sloughing, and predator grazing. *Abrasion* (initiated by particle collision) and *erosion* (initiated by fluid-induced shear in the vicinity of the biofilm surface) refer to the removal of small groups of cells and represent a relatively constant removal of biofilm biomass. *Sloughing* refers to the loss of large segments of biofilm approximately equal to the biofilm thickness (Morgenroth, 2003). Predators such as macrofauna and microfauna graze biofilms and consume biomass (Boltz et al., 2008). Depending on the reactor type, the detachment mechanism may be one or more of the aforementioned processes. Whichever method is implemented in a biofilm reactor, controlled and consistent detachment of biomass is critical to a healthy, stable, nitrifying biofilm reactor. Just as an activated sludge system needs to have consistent wasting of biomass, so does a biofilm system. If the only wasting of biofilm is via large, uncontrolled sloughing events, not only will the facility likely experience spikes in effluent TSS, but also the remaining biofilm will likely experience periods of decreased nitrification capacity.

Excess biofilm accumulation can result in increased hydraulic head loss, odors, reduced media specific surface area, increased biofilm weight (which can exceed design structural loads), and media clogging. All biofilm reactor types rely on biofilm thickness control, which is commonly achieved by increasing the liquid velocity. This control occurs either at a constant rate or during periodic increases in fluid velocity to shear excess biofilm. If excess biofilms develop, substrates and/or nutrients do not penetrate to the base of the biofilm. As noted previously, these nutrient-deficient zones at the base of the biofilm can result in large biomass inactivity at the base of the biofilm, where attachment mechanisms are important. This can result in large biomass loss events (sloughing), reducing the overall biomass in the reactor and detrimentally affecting the observed nitrification rate.

### 3.3 General Operation and Control Parameters for Nitrifying Biofilm Reactors

Effective biofilm reactor operation consists of actively controlling the reactor's liquid velocity and managing biofilm thickness. Active control allows for management of mass transfer into the biofilm, management of the diffusion distance, and management of the biofilm thickness. For many technologies, the same operational tool is used to achieve both of these control objectives. As will be discussed for each technology later in this chapter, adjusting fluid velocity is one of the main operational tools available to a biofilm reactor operator. Whether adjusting flushing rates, aeration rates, backwash cycles, or scour periods, changes in biofilm reactor operation are generally geared toward management of the fluid velocity. This requires operational observation of biofilm thickness, biofilm consistency, nitrification rates, and, potentially, oxygen concentrations within the reactor. For each technology, methods for monitoring these parameters and operational responses to changes in parameters will be discussed.

For common nitrifying biofilm reactor technologies, the main operational decisions will involve increasing or decreasing the liquid velocity. A summary of operational goals and general operation and control parameters is shown in Table 4.1; the table also includes an indication of whether the liquid velocity should be increased or decreased in response to the operational change. Like all wastewater processes, these control parameters are not independent of each other. However, an understanding of their effect on biofilm reactor performance and how to effectively control each area is important to successful operation of a biofilm reactor system.

## 4.0 BIOFILM REACTOR ANCILLARY COMPONENTS

Whereas operation of the nitrifying biofilm reactor itself is the main focus of this chapter, the effect of operations associated with several ancillary processes not directly tied to the biofilm is important for overall permit compliance and maintenance procedures.

### 4.1 Influent Screening

Several biofilm processes require ultrafine screening of influent wastewater to limit the accumulation of rags and debris within the biofilm reactor. Several biofilm reactor manufacturers recommend the use of 3-mm screens or finer for facilities without primary clarifiers and 6-mm screens for facilities with primary clarifiers.

### 4.2 Phosphorus Removal

Unless an activated sludge system is used in conjunction with a biofilm reactor, phosphorus removal is typically achieved through biological assimilation and chemical precipitation. Enhanced biological phosphorus removal is only achieved when biomass is exposed to cyclic anaerobic and aerobic conditions, which are only achievable in a system where the bacteria are exposed to both anaerobic and

**TABLE 4.1** General Operation and Control Parameters for Nitrifying Biofilm Reactors

| Operational Goal | Operation Modification | Liquid Velocity Change | Operation and Control |
|---|---|---|---|
| Improve nitrification rate—mass transfer limitation | Increase transfer into the biofilm | Increase | Increase oxygen transfer rate to increase dissolved oxygen concentration in biofilm<br>Increase upstream BOD removal to reduce heterotrophic competition in the biofilm<br>Increase fluid velocity to decrease MTBL thickness |
| Improve nitrification rate—biofilm thickness limitation | Increase biofilm thickness | Decrease | Decrease fluid velocity/mixing intensity<br>Increase dissolved oxygen concentration<br>Increase biofilm attachment area by adding additional media if sufficient reactor capacity exists (% fill < maximum fill achievable) |
| Prevent biofilm septicity—biofilm thickness limitation | Decrease biofilm thickness | Increase | Increase fluid velocity and mixing intensity to increase shear force<br>Implement cyclical backwash cycles<br>Increase biofilm attachment area by adding additional media |
| Prevent biofilm septicity—mass transfer limitation | Increase oxygen transfer | Increase | Increase oxygen transfer rate to increase dissolved oxygen concentration in biofilm |
| Prevent excessive biomass loss (sloughing)—mass transfer limitation | Increase transfer into the biofilm | Increase | Ensure sufficient phosphorus and nitrogen entering the biofilm reactor<br>Increase oxygen transfer rate to increase dissolved oxygen concentration in biofilm<br>Ensure sufficient alkalinity to buffer pH<br>Prevent oxygen and nutrient starvation |
| Prevent excessive biomass loss shear—biofilm thickness limitation | Decrease shear force on the biofilm | Decrease | Decrease fluid velocity, aeration energy, and/or mixing intensity |

aerobic conditions (see Chapter 6). This cycling between anaerobic and aerobic conditions is typically not achieved in pure biofilm reactor systems and can be accommodated in IFAS systems.

When chemical precipitation for phosphorus removal is used in conjunction with a biofilm process, it is critical that enough soluble phosphorus enters the biofilm reactor to ensure diffusion of phosphorus to the base of the biofilm. The soluble phosphorus concentration in the biofilm reactor is a critical component of phosphorus removal optimization in a biofilm system. Frequently, a multipoint addition system is used for chemical phosphorus removal in a biofilm system. Depending on site-specific hydraulic conditions and biofilm thickness, different phosphorus concentrations may be required to enter the biofilm reactor to prevent nutrient deficiency at the base of the biofilm and associated excessive biofilm sloughing events. If excessive sloughing events occur, it may be beneficial to examine the phosphorus concentration entering the biofilm process to determine if the biofilm is receiving sufficient phosphorus.

## 4.3 Solids Separation

Differences in solids generated in a biofilm system greatly affect settling performance. The ability of aerobic biofilm reactors to achieve effluent performance is dependent on the operation of the downstream settling process. Suspended solids remaining in the effluent contribute to BOD and organic nitrogen and phosphorus and have the potential to consume residual dissolved oxygen, which may result in nitrogen gas generation because of denitrification in the clarifier sludge blanket, causing sludge to float. Solids separation performance is affected by many factors, which are characterized in *Standards for Design of Water Resource Recovery Facilities* (WEF, 2023). Effluent solids particles from a biofilm reactor are denser than activated sludge flocs, but poor bioflocculation results in a large range of biofilm aggregate sizes. This, in turn, results in a wide distribution of particle sizes and a fraction of biofilm reactor effluent solids that will almost never settle in a conventional clarifier (Ødegaard et al., 2010).

Because of the production of effluent solids aggregates with poor settling characteristics, several operational modifications have been developed to improve flocculation before solids separation. Solids contact zones, where settled solids from final clarification are recycled to an aerobic tank downstream of the biofilm process, have been implemented for several biofilm reactor technologies. This has resulted in production of increased solids settleability because of bioflocculation in the aerobic contact zone. Chemical flocculation has also been implemented for improved flocculation, in which a small flocculation basin is added between the biofilm reactor and clarifier. Chemical flocculants are also effective at precipitating phosphorus, and chemical-based flocculation downstream of a biofilm process is an option for nitrifying biofilm reactor facilities where an effluent phosphorus permit is in place.

Detailed operational considerations for the solids separation process itself are the same for biofilm reactors as they are for other treatment systems. Dissolved air flotation, conventional clarification, high-rate clarification, and filtration have all been successfully applied to various biofilm reactor technologies (Ødegaard et al., 2010). Detailed discussions on solids separation technologies can be found in *Standards for Design of Water Resource Recovery Facilities* (WEF, 2023). It should be noted that despite the lower effluent suspended solids concentrations and lower yields observed from biofilm reactors, it is typically not recommended to forgo the solids separation step because of occasional sloughing events that cause a high level of variability in effluent TSS concentrations.

## 4.4 Microfauna Management

Growth of microfauna has been a significant concern for many biofilm reactor installations. Microfauna vary depending on whether a biofilm reactor is a submerged biofilm or a non-submerged biofilm. For non-submerged biofilm reactors (e.g., trickling filters and RBCs), filter flies proliferate when the biofilm reactors have semidry pockets of old biomass available for grazing by fly larvae. Increasing wetting rates of the reactor will assist in reducing the suitability of the environment for propagation of filter flies.

Snails can become a significant problem, particularly in nitrifying trickling filters. Snails consume biofilm as a source of food and grow in dark, moist environments. In some instances, snails consume nitrifying bacteria faster than they are produced, which can result in loss of ammonia removal efficiencies. Snails also excrete ammonia wastes. There are several approaches to reduce snail populations. One approach is ammonia washing, which is a process in which the biofilm reactor is isolated during off-peak hours and the influent and recirculation flow stream is charged with water containing approximately 100 mg/L of ammonia (Daigger & Boltz, 2011). Sludge centrate, digester supernatant, or other ammonia-rich water may be used for this purpose. The addition of sufficient caustic to raise the pH to 9.5 or 10 is necessary to ensure a high concentration of toxic, un-ionized ammonia. The ammonia solution should be recirculated for several hours or a day if possible. The high strength of dissolved gaseous ammonia will discourage snail growth, and the high recirculation rate will purge the snail shells and eggs from the tower. The elevated pH should be maintained by adding more caustic, if necessary. This procedure should be repeated, as necessary, to mitigate the snail population.

Ponding is a problem typically associated with rock trickling filters and occurs when the buildup of biomass, plant life, or inert solids does not permit proper flow of wastewater through the rock matrix. Increasing Spülkraft flushing intensity (the "SK rate") (see Section 5.1.1) or taking the filter out of service to "dry out" the biomass or plant life may improve flow. If excessive grit or inert material is the cause, then media replacement is likely necessary.

For submerged biofilms, the main microfauna challenge has been associated with redworm proliferation. Several biofilm reactor systems have experienced a bloom of a type of redworm population that feeds on the biomass on the media. The worms are obligate aerobes. Thus, high levels of dissolved oxygen favor their growth. Occasional, un-aerated operation (in the range of 1 to 3 hours) can help to limit the worm growth potential. Maintaining a healthy thin biomass through shearing by dedicated aeration can help keep worms at bay, although this results in increased oxygen concentration, which can select for more worm growth. Controlling the worms consists of creating an anoxic condition in the area of the worm bloom by turning the air off for several hours. Nitrifying bacteria will survive this treatment because they can tolerate periods of oxygen depletion of at least 24 hours.

### 4.5 Peak Flow Management

Unlike suspended growth reactors, biofilm reactors are not at risk of biomass "washout" during peak flow events. Fluid velocity must be maintained below the point that would create excess biofilm detachment during peak flow events, but the attached biofilm will not be carried with the fluid flow as a suspended biomass would be. In addition, because of the lower effluent suspended solids concentration from a biofilm reactor, downstream clarifiers are typically not controlled by solids loading rates as an activated sludge clarifier would be.

## 5.0 NITRIFYING BIOFILM REACTOR TECHNOLOGIES

A wide range of biofilm reactor technologies are currently available and being implemented for nitrification. Figure 4.3 is a schematic of some of the most common nitrifying biofilm reactor technologies. From an application standpoint, the trickling filter, MBBR, and IFAS are the most commonly installed reactor technologies, although full-scale examples of all of the reactor types shown in Figure 4.3 do exist. A detailed discussion of these three most common reactor technologies and a brief description of each of the additional technologies are included in this section. It is important to remember that the fundamentals of biofilm reactor operation are the same for each technology; only the mechanisms for control vary from reactor technology to reactor technology. For further information pertaining to the functions, designs, and details of each reactor technology, refer to *Biofilm Reactors* (WEF, 2011).

### 5.1 Trickling Filter

Trickling filters are aerobic biofilm reactors that receive either primary-treated or fine-screened wastewater that is distributed or trickled over the media. Nitrifying trickling filters receive secondary-treated wastewater typically from carbon-oxidizing trickling filters. As the wastewater travels down through the trickling filter contacting the biofilm, air flows either up or down through the trickling filter by natural convection or low-pressure fans. The direction of the airflow for natural convection depends on the density difference between the air within the trickling filter and exterior ambient air densities, which can vary daily with air temperature. *Diurnal stagnation* is the term used to describe the condition in which

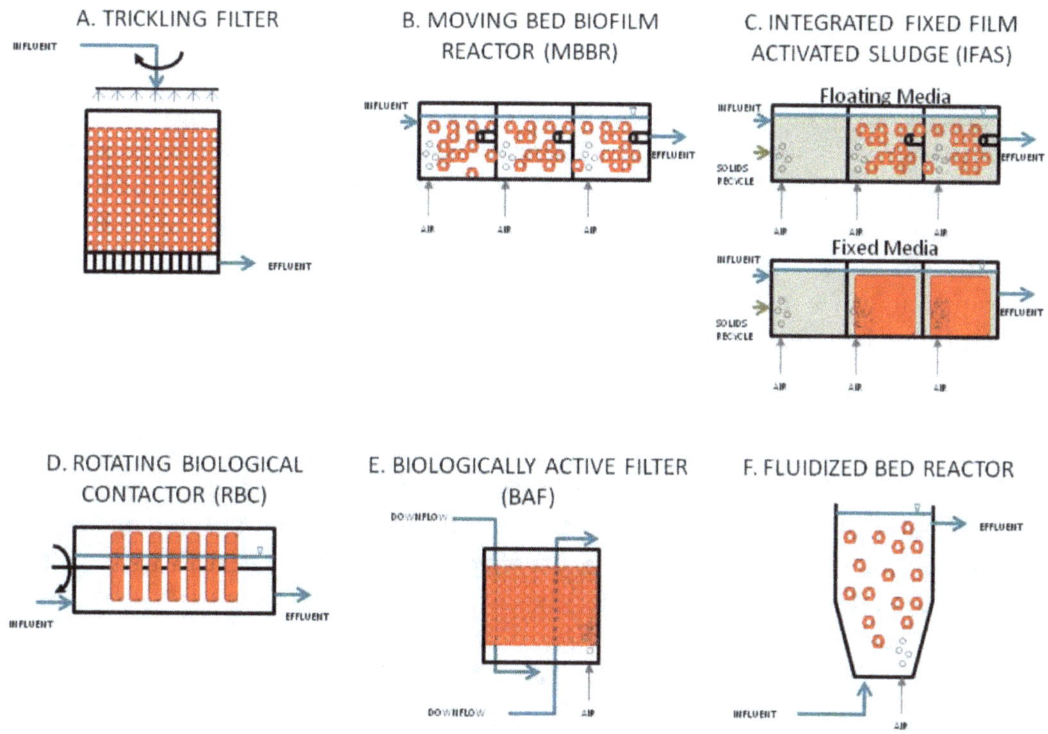


**FIGURE 4.3** Examples of Nitrifying Biofilm Reactor Technologies

air densities are equal and airflow is limited. During diurnal stagnation conditions, odors proliferate from the trickling filter. Treated wastewater leaves the media and enters the underdrains, where it is collected and either recirculated to the top of the trickling filter or exited from the trickling filter process.

Trickling filters are classified by hydraulic and organic loading rates. As defined in *Standards for Design of Water Resource Recovery Facilities* (WEF, 2023), filters are classified as either roughing, carbon-oxidizing, carbon-oxidizing and nitrifying, or nitrifying filters. The classifications are based on the intended service for BOD, BOD and ammonia, and ammonia removal and give loading and expected removal efficiencies.

The trickling filter consists of media, a containment structure, underdrains, and a distributor. If forced ventilation is used, low-pressure fans and a cover are also required. The media serve to provide surface area for biofilm growth and can be structured sheet, rock, or random dump-type media. Structured sheet media can be cross flow, vertical flow, or a combination of the two. Cross-flow media act to continually redistribute flow to ensure utilization of the entire media surface. Low-density surface area media (100 $m^2/m^3$) (31 sq ft/cu ft) are generally used for BOD removal, medium-density surface area media (160 $m^2/m^3$) (48 sq ft/cu ft) are used for nitrification and low-load BOD removal, and high-density surface area media (225 $m^2/m^3$) (68 sq ft/cu ft) are used for tertiary nitrification applications. Vertical-flow media are used when an unusually high solids concentration is applied to the trickling filter or when high-rate roughing is desired where sloughed biomass is more easily flushed from the media. Vertical-flow media are typically used in conjunction with two layers of cross-flow media above for improved distribution. Rock and random dump media are not self-supporting and require structural support to contain the media. *Standards for Design of Water Resource Recovery Facilities* (WEF, 2023) presents a comparison of the physical properties of various types of trickling filter media. Underdrains collect treated wastewater and convey it for recirculation or to downstream unit processes, creating a plenum for the transfer of air throughout the media (Grady et al., 1999).

### 5.1.1 Mass-Transfer Control

Flow distribution, application rates, and recirculation rates are the main control mechanisms for trickling filter reactors. Distribution of flow to the top of the trickling filter is accomplished through fixed nozzles or a rotating distributor. Harrison and Timpany (1988) reported fixed nozzles as giving reduced performance because of incomplete and noncyclic wetting. The rotary distributor can be operated hydraulically or motorized; however, motorized distributors offer more operational control. Reversing jets can be used to reduce the speed of hydraulic rotary distributors; however, damage to structured sheet media can occur at the radial location of the reversing jets on the distribution arm. Periodic pan testing of the trickling filter distributor is necessary to ensure proper media wetting and distribution of wastewater. Periodic cleaning and adjustment of rotary distributor orifice plates are typically required to correct poor distribution.

*Wetting rate* refers to the average flow per unit plan view surface area of the tricking filter, typically given in liters per square meters per minute (gallons per minute per square feet), and is a function of the applied flow (i.e., influent plus recirculation) and tower diameter. Spülkraft flushing intensity, or the SK rate (millimeters per pass), is the instantaneous rate of wastewater applied to the trickling filter. The SK rate is calculated using the following equation:

$$\mathrm{SK} = \mathrm{THL}/(N_{\mathrm{a}} \cdot \omega_{\mathrm{d}}) \tag{4.1}$$

Where

SK = Spülkraft flushing intensity (mm/pass);
THL = total hydraulic loading (influent plus recirculation), $m^3/d$;
$N_{\mathrm{a}}$ = number of distribution arms (typically two or four); and
$\omega_{\mathrm{d}}$ = rotational speed, rev/min.

Correct trickling filter hydraulics are critical for proper operation. Improper flushing at inadequate SK rates retains excess biomass and macrofauna (e.g., filter flies and snails). Both the wetting rate and SK rate depend on the flowrate to the distributor, which is influenced greatly by recirculation. During low-flow periods, recirculation is necessary to maintain a minimum flow for the rotation of hydraulic rotary distributors and thus keep the biomass wetted. Similarly, for the SK rate, the distributor speed changes with flowrate because of the jet effect. Therefore, the SK rate must be assessed for operation at various flowrates to determine optimum performance. Recommended operating and flushing dosing rates for rotary distributors range from 25 to 230 mm/pass (1 to 9 in./pass) (WEF, 2023).

Recirculation typically occurs from underdrains and is controlled through the placement of a weir. Clarification of the filter effluent before recirculation improves trickling filter performance as a result of TSS removal. However, improved trickling filter performance can easily be negated if the facility influent plus recirculation hydraulic loads negatively affect clarifier operation. Additionally, recirculation returned to primary clarification can produce anaerobic conditions in the feed and, subsequently, in the top of the trickling filter, generating nuisance *Beggiatoa* populations. *Beggiatoa* use sulfur compounds from anaerobic decay, which is indicated by a white "plating" or "matting" biomass. Proper recirculation returns dissolved oxygen to the top of the trickling filter, which improves performance in instances in which limited dissolved oxygen exists in the trickling filter influent, especially in shallow filters.

Trickling filters that provide both BOD and ammonia removal can benefit from increased recirculation by reducing BOD concentrations applied to the media, thereby limiting competition between heterotrophs and autotrophs. The lower liquid BOD concentration limits BOD diffusion into the biofilm. Utilization of the entire media surface area for ammonia removal can be achieved from the dilution effects. Empirical models, such as the modified Velz equation, show the effects of recirculation on BOD removal (WEF, 2023). However, excessive recirculation can bypass biomass on the media and cause excessive loss of biomass, resulting in poor performance.

Covers are used to control odors and in instances in which ambient air temperatures are relatively low to eliminate freezing of the trickling filter. When covers are used, low-pressure fans are required for air circulation. Forced ventilation provides both warmer air and increased oxygen transfer to the biofilm.

### 5.1.2 Solids Separation

Aerobic solids contact chambers and chemical flocculation have both been shown to improve solids settling. In facilities without aerobic solids contact basins between the clarifiers and trickling filter, addition of chemical coagulants for phosphorus removal downstream of the trickling filter can significantly improve solids settleability.

Macrofauna, mainly in the form of snails, are typically removed during the clarification process. Clarification before recirculation can minimize potential plugging of media from snail shells, which affects both air and water flow. Periodic removal of snail shells from underdrains may be necessary to maintain trickling filter performance and extend the life of media in configurations that recirculate directly from the trickling filter. Proper design of the media support system can also minimize the potential for plugging; in extreme instances, vertical-flow media replacement is necessary based on the improved flushing it provides.

### 5.1.3 Nitrification Rate

For nitrification to take place, the CBOD concentration of wastewater must be sufficiently low to limit competition between heterotrophic and autotrophic populations because a bulk liquid BOD concentration in excess of 30 mg/L essentially causes nitrification to cease (WEF, 2000). This is increasingly important in combined carbon- and ammonia-oxidizing trickling filters. Carbon-oxidizing trickling filters remove the more easily biodegradable soluble BOD, while the particulate BOD remains relatively unchanged. Diurnal flow variations may cause conditions in the trickling filter to cycle across this threshold value, thereby eliminating nitrification. Increased recirculation can improve nitrification performance by increasing oxygen transfer to the biofilm and decreasing the MTBL.

Ammonium-nitrogen concentration, dissolved oxygen concentration, alkalinity deficiency, and temperature factors affect the nitrification rate in trickling filters. It is generally understood that nitrification rates are reduced as a result of ammonia limitations starting at ammonium-nitrogen concentrations between 1 and 4 mg/L (WEF, 2023). Figure 4.4 provides the approximate upper and lower limits of ammonium removal rates measured at five nitrifying trickling filter facilities (Oakey & Albertson, 1989). It is important to note that the values used to develop Figure 4.4 have not been corrected for temperature or dissolved oxygen concentration. For combined carbon and ammonia oxidation, trickling filter effluent ammonia concentration is generally accepted to be limited to 2 mg/L (WEF, 2023); however, in lightly loaded trickling filters operated under proper conditions, lower ammonia concentrations may be achieved because a selection for a microbial ecology well suited for low ammonia concentrations will develop.

### 5.1.4 Trickling Filter Startup and Shutdown

After all the mechanical equipment has been checked out according to the manufacturer's recommendations, the flow of wastewater to the filter should be begun at the lowest rate possible while maintaining the highest possible recirculation rate. In addition, a pan test should be performed to ensure proper

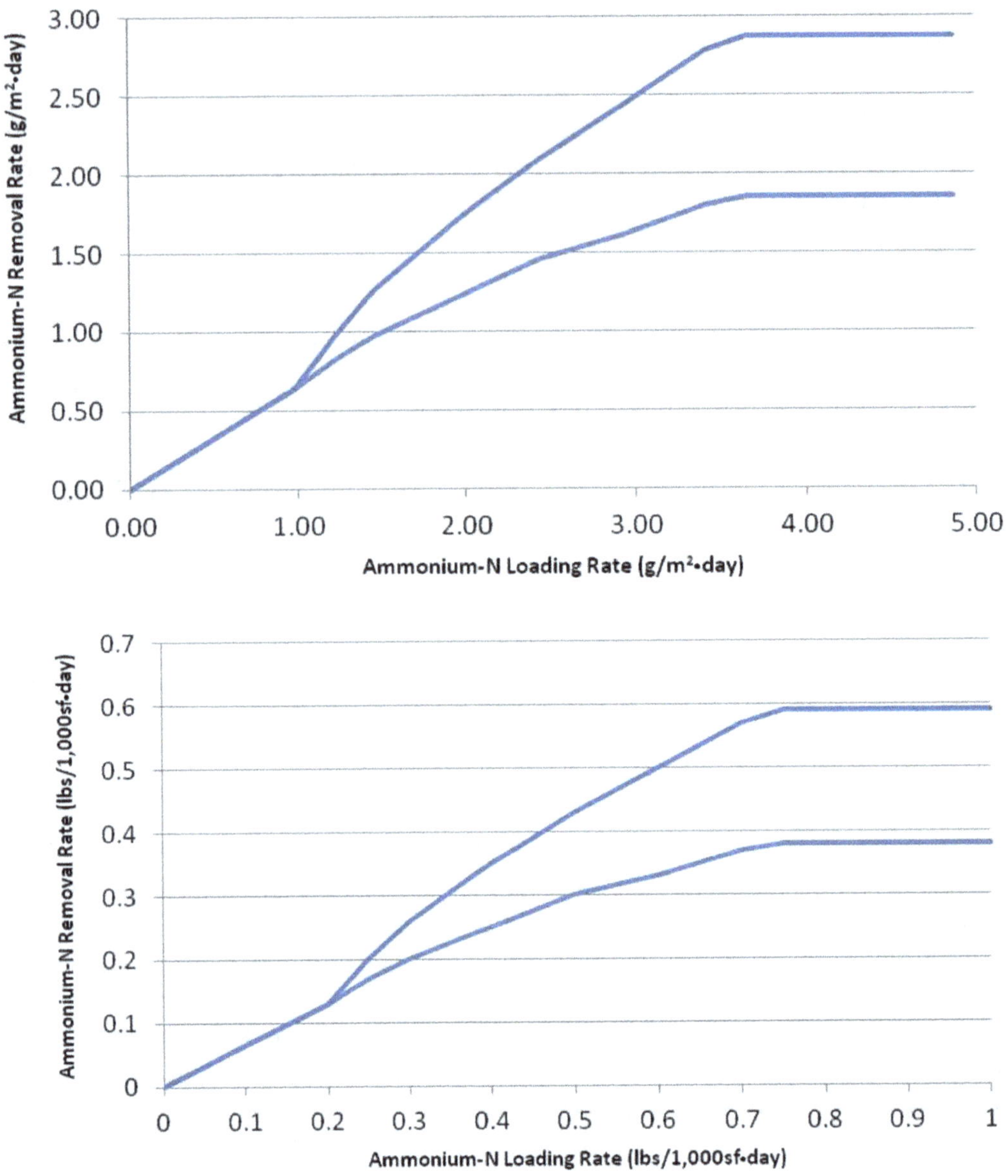


FIGURE 4.4 Trickling Filter Nitrification (Oakey & Albertson, 1989)

distribution and wetting of the media. The initial effluent may not be suitable for discharge but should improve rapidly after the first week. Recirculating effluent to other operating units may be advisable to avoid discharge violations.

A shutdown procedure is necessary to prevent the possible generation of offensive odors, accumulation of dried biomass that may clog the filter when restarted, and damage to the media from extreme heat or cold. To flush biomass from the media, the mechanical distributor should be used at the slowest speed recommended by the manufacturer at the highest possible flowrate for an extended period of time. The flushing water may be chlorinated to clean the media and kill biomass. The chlorinated water should be recirculated until the chlorine residual is safe to discharge. For domed filters, the condition of the media and the temperature in the dome should be monitored at least once per shift to avoid buildup of heat or ammonia gas from the decomposition of any remaining biomass. It is important to continue to operate ventilating and odor control fans to cool the domed space and remove accumulated gases.

## 5.2 Moving Bed Biofilm Reactors

Moving bed biofilm reactors can be designed for aerobic or anoxic conditions. This discussion focuses on the operation of nitrifying MBBR systems. The MBBR is strictly a biofilm reactor; it does not incorporate suspended growth (although some may be present because of detachment) and has no solids recycling from the final clarification step. Moving bed biofilm reactors contain a plastic biofilm carrier volume ranging from 25% to 67% of the liquid volume, and the biofilm carriers with biofilm attachment are neutrally buoyant, thereby moving freely within the aeration basin by way of diffused air mixing energy. The value of the volume of biofilm carriers (supplied in bulk) divided by the operating liquid volume of the basin is referred to as the *carrier fill fraction*. Retention sieves typically are installed on one MBBR wall and allow treated effluent to flow through to the next treatment step while retaining free-moving plastic biofilm carriers. An MBBR may be a single reactor or several reactors in series. Reactors with a length-to-width ratio greater than 2 to 1 can result in nonuniform distribution of free-moving plastic biofilm carriers. Carbon-oxidation, nitrification, or combined carbon-oxidation and nitrification MBBRs use a diffused aeration system to uniformly distribute plastic biofilm carriers and meet process oxygen requirements. The MBBR is capable of meeting similar carbon-oxidation and nitrogen removal as activated sludge (Andreottola et al., 2000).

### 5.2.1 *Mass-Transfer Control*

The main control mechanism for mass transfer in a nitrifying MBBR is the airflow rate. Air is introduced to an MBBR system through coarse-bubble diffusers mounted on the floor of the reactor. The diffusers are designed to produce a rolling pattern in the tank, effectively achieving oxygen transfer and mixing with the aeration system. Because of the presence of biofilm carriers in the reactor, coarse-bubble diffusers have been reported to have a higher oxygen-transfer efficiency than coarse-bubble diffusers in an activated sludge system, with the MBBR oxygen-transfer efficiency in the range of 2.5% to 3.5% per meter of submergence (0.8% to 1.2% per foot of submergence) (McQuarrie & Boltz, 2010).

Increasing the airflow increases both the fluid velocity over the biofilm and the dissolved oxygen concentration in the bulk liquid. For mixing requirements, an airflow rate of 5 to 10 $m^3/(m^2 \cdot h)$ (0.27 to 0.52 scfm/sq ft) is typical. These airflows ensure sufficient mixing of the floating media and sufficient fluid velocity across the biofilm growing in the carrier. To ensure sufficient oxygen supply, operating with a dissolved oxygen concentration in the range of 4 to 6 mg/L is typically recommended. The balance between biofilm thickness and airflow management is a site-specific condition.

### 5.2.2 *Solids Separation*

Retention of floating media in an MBBR is accomplished using retention sieves that are typically located in the reactor along the effluent wall of each zone. The screen design varies by manufacturer, but typically, anoxic zones incorporate flat screens mounted on the face of the cell wall, and aerobic zones incorporate cylindrical screens that are mounted horizontally facing into the flow (refer to examples in Figure 4.5). This type of design is effective in preventing clogging and allows detached bacteria and inert solids to pass freely to the downstream stage, either another biofilm reactor, a flocculation chamber or clarification process. The media itself tend to scrub the screens clean. Flat screens, where used, should be equipped with air knives to encourage media scouring and to avoid excessive head loss resulting from accumulations on the screen surface. Preventive maintenance schedules should include screen inspections and cleaning.

Moving bed biofilm reactors are low hydraulic retention time biofilm reactors that have suspended solids remaining in the effluent stream as a result of biological transformation processes. Therefore,

FIGURE 4.5 Examples of Cylindrical MBBR Retention Sieves

MBBR process performance is dependent on a downstream liquid–solid separation unit. Generally, the suspended solids concentration in the MBBR effluent stream has been reported to be 150 to 250 g suspended solids/m$^3$ when treating medium- to high-strength municipal wastewater (Ødegaard et al., 2010). Solids separation after the retention sieves has been achieved using a wide range of technologies, including dissolved air flotation, conventional clarification, high-rate clarification, and filtration (Ødegaard et al., 2010). For increased bioflocculation of MBBR effluent solids, aerobic solids contact and chemical coagulation have both been implemented with success. For chemical coagulation, it is important to achieve sufficient mixing and contact time to ensure proper flocculation before the final clarifiers. For systems required to meet an effluent phosphorus limitation, the addition of a chemical coagulant after the MBBR process can increase the flocculation of MBBR effluent TSS, reducing the final effluent TSS concentration.

### *5.2.3 Nitrification Rate*

In applications in which sufficient alkalinity and ammonia are present (at least initially), the nitrification rate will increase as organic loading is reduced until the dissolved oxygen concentration becomes rate-limiting. In well-established nitrifying MBBR biofilms, the availability of oxygen limits the rate of nitrification on the carriers as long as the ratio of oxygen to ammonium-nitrogen is below 2.0. Thus, unlike suspended growth systems, the reaction rate in a moving bed reactor exhibits a linear or near-linear dependence on the bulk dissolved oxygen concentration under oxygen-limited conditions. The increased mixing energy under higher aeration velocities also helps to improve transfer from the bulk liquid to the biofilm. If organic loading is held constant (i.e., biofilm thickness and composition), the nitrification rate can be expected to increase linearly with an increasing dissolved oxygen concentration. Until fairly low bulk liquid ammonia concentrations (1 to 2 mg/L) are observed in the reactor, an increase in the bulk dissolved oxygen level helps to increase the nitrification rate.

Although reduced nitrification rates are observed at lower temperatures, at the governing "wintertime" design temperature, this effect can be offset by the combined effect of higher attached biofilm concentrations on the carriers typically observed at lower temperatures (less than 10 °C) (Bjornberg et al., 2009) and by maintaining a higher bulk dissolved oxygen concentration in the reactor. At lower wastewater temperatures, a higher biofilm attachment (i.e., a higher value of grams of dry solids attached per square meter of fixed-film surface area) is commonly observed. Dissolved oxygen solubility is higher in wastewater with lower temperatures; therefore, a higher bulk dissolved oxygen concentration can be attained without an increase in airflow or blower energy demand. The result of this balance of oxygen concentration and biomass density is that the nitrification activity per unit surface area of carrier can be maintained effectively despite a reduction in specific biofilm activity during winter months.

### 5.2.4 Biomass Growth

The operator should monitor the accumulation of growth inside the protected surface area of plastic media. If growth appears to be excessive, the roll pattern may be too gentle, resulting in too weak an aeration-induced shear force. If the roll pattern is sufficient, and there is an excessive biofilm, organic loading may be excessive. To rectify this, the organic loading may have to be reduced by implementing CEPT or by increasing the biofilm carrier fill fraction, or increasing the number of biological treatment trains. Implementation of CEPT in primary clarifiers can lead to phosphorus limitation in the biofilm, so that should be a last resort. Any of these methods biofilm carriers, implementing CEPT, or increasing biological treatment basin volume is analogous to a capacity design change. Airflow control is typically the main control mechanism in the MBBR, with increased airflow rates used to increase biofilm shearing. Biomass growth can range from 5 to 30 mg $TSS/m^2$, depending on loading, temperature, and operating conditions (McQuarrie & Boltz, 2010).

An MBBR is typically a multizone system, with each zone representing a separate, completely mixed zone. Media located in upstream zones of a multizone MBBR typically will have a thicker biofilm, with a greater portion of the biofilm composed of heterotrophs (BOD oxidizers). Media in downstream zones will have a thinner biofilm with a higher proportion of autotrophs (nitrifiers).

Because the media are free-floating and completely mixed within a zone of the MBBR, the biofilm attachment and therefore thickness will adjust itself to conditions in the bulk liquid. During extended periods of high load, the biofilm will get thicker; conversely, during extended periods of low load, the biofilm will get thinner. Typically, the entire aeration basin will contain the media, even if they are divided into several cells. Therefore, there is no need to relocate media to other sections of the basin in response to changing load patterns.

### 5.2.5 Moving Bed Biofilm Reactor Startup Considerations

Although establishment of a heterotrophic biofilm occurs quickly in an MBBR, the nitrifying biofilm can take longer to establish. If an excessive heterotrophic biofilm develops during startup in the nitrifying portions of the reactors, the shift from this heterotrophic biofilm to a nitrifying biofilm can be prolonged. To avoid excessive heterotrophic growth during startup, it is important to maintain a low bulk liquid BOD concentration in the nitrifying portion of the MBBR by either lightly loading the system from a hydraulic loading rate at startup or by increasing BOD removal in the primary clarifiers.

Foaming episodes may occur occasionally in MBBRs, especially during startup or process upset. Foaming is prevalent in MBBRs that contain plastic biofilm carriers resulting from the difference in surface tension between wastewater and the carriers. The surface tension difference and propensity to foam is mitigated once biofilm is established. Foaming that occurs during process upset may become trapped in the MBBR reactor because the partition wall between two successive reactors extends above the water surface, so designers and operators may consider implementing foam capture methods along sidewalls. A non-silicone, water-based defoaming agent should be used when foam must be controlled in a reactor. Use of a silicone-based defoamer will coat the carrier media, impeding diffusion to the biofilm, which may affect performance. It is extremely important to use defoamers that do not use silicon-compound agents because these types of defoamers are incompatible with plastic-carrier media.

### 5.2.6 Media Transfer and Inventory Management

Although there is little to no maintenance required inside the tank of an MBBR, it is still prudent to consider how one might manage to transfer and store media on-site should a reactor ever need to be taken out of service for in-basin maintenance. The contents of a reactor, including carriers, can be transferred using a 10-cm (4-in.) recessed impeller pump. Depending on the design fill fraction, it may be possible to transfer the media to another reactor. The drawback of this is that it is difficult to reapportion the media once two reactors are combined. After pumping the media back to the reactor, the only reasonably accurate method for measuring the media fill fraction is to dewater the reactor and measure the height of the media in both reactors to check the fill fraction. Preferably, another tank or idle unit process could be used for temporary storage of the media. This way, it is easier to ensure that the proper fill fraction has been restored to the reactor.

## 5.3 Integrated Fixed-Film Activated Sludge

Integrated fixed-film activated sludge facilities can be designed by adding biofilm support media (biofilm carrier particles) to activated sludge basins. The IFAS process can use either fixed bed or moving bed biofilm carriers (i.e., media). The moving bed biofilm carrier types include plastic, sponge-type, and cellulosic materials such as Kenaf.

The principle of the IFAS system is to enhance BOD removal and nitrification above the removal that could have been achieved using MLSS in a suspended growth-only reactor of like volume. The operator must understand the IFAS system's design criteria and understand the limitations related to solids loading allowed on the secondary clarifier. The operator must then consider seasonal fluctuations in wastewater temperatures and manage the mixed liquor suspended solids concentration (and therefore the suspended growth solids retention time [SRT]) such that mixed liquor suspended solids settling is adequate and the fixed-film population optimizes nitrifier growth on the biofilm carrier surface at critical times. The decision to carry a suspended growth SRT below or above that required for nitrification should be based upon design criteria and the fixed-film system's capacity for nitrification.

### 5.3.1 Nitrification Rate

Nitrification occurs in both the suspended growth and biofilm in an IFAS system. For nitrification in the biofilm portion of IFAS, the nitrification rate is driven by the reactions occurring in the bulk liquid (nitrification and BOD oxidation) and dissolved oxygen concentration. The IFAS media are typically placed after the first one-fourth to one-third of the aeration basin to limit exposure of the biofilm to high BOD concentrations. The following steps can be taken to increase nitrification in an IFAS system:

- Decrease the mixed liquor wasting rate to increase the mixed liquor suspended solids concentration and increase the nitrifying bacteria population in the suspended growth.
- Increase the dissolved oxygen concentration in the aeration basin zones with IFAS up to 4 to 6 mg/L.
- Increase the biofilm attachment area by increasing the fill fraction of biofilm carriers (media) in the IFAS portions of the biological treatment train.
- Increase BOD removal upstream of the IFAS basins.

Increasing the dissolved oxygen concentration in the aeration basin zones with IFAS media is the most readily controlled parameter. Increasing the airflow rate increases the dissolved oxygen concentration and the fluid velocity over the biofilm, resulting in increased mass transfer. Longer-term adjustments to the typical operating mixed liquor concentration can also be addressed.

### 5.3.2 Integrated Fixed-Film Activated Sludge Startup Considerations

Development of a heterotrophic biofilm on IFAS carriers can limit nitrifying biofilm development and lead to extended startup periods. Because biofilm zones in the IFAS system are typically designed for the lower BOD regions of the aeration basin, it is important to limit the BOD concentration in these zones during startup to limit the formation of excess heterotrophic biofilm. The attachment media can also be added in stages to allow for proper wetting of the media, which, in turn, allows for staged biofilm development.

### 5.3.3 Interaction of Biofilm and Suspended Bacteria

By accomplishing a portion of treatment in a biofilm, IFAS helps reduce the overall operating mixed liquor concentration. This reduces the solids loading to the final clarifier, resulting in increased capacity for clarifiers that are not hydraulically limited. Biofilm erosion from the carriers also has a seeding effect on suspended growth because the biofilm that is sheared from the carrier can remain in the bulk liquid.

### 5.3.4 Biofilm Thickness Control

For floating media, biofilm thickness control is similar to MBBR operation. Several approaches have been developed for biofilm thickness management for fixed-media IFAS systems. Many fixed IFAS systems are installed with fine-bubble aeration, which does not provide the liquid velocity to dislodge biomass. One of the more common strategies is to use periodic, high-intensity air sparging to increase fluid velocity around the biofilm structure. This sparging can also produce large air bubbles that dislodge biomass. If an active biofilm management strategy such as air sparging is not implemented in fixed-media IFAS systems, the biofilm modules risk excess biofilm accumulation and decreased nitrification rates because of lack of oxygen transfer into the inner portions of the biofilm.

## 5.4 Additional Biofilm Reactor Technologies

A brief summary of additional biofilm reactor technologies is presented in the following subsections. Each of these reactors would be governed by the same principles as previously discussed, with specific control mechanisms differing for each technology.

### 5.4.1 Rotating Biological Contactors

As a secondary treatment process, RBCs have been applied in instances in which average effluent water-quality standards are less than or equal to 30 mg/L BOD and TSS. When the RBC is used in conjunction with effluent filtration, the process is capable of meeting more stringent effluent water-quality limits of 10 mg/L BOD and TSS. Nitrification RBCs can produce effluent having less than 1 mg/L of ammonia-nitrogen remaining in the effluent stream, although this is not typical. The RBC uses a cylindrical, synthetic media bundle that is mounted on a horizontal shaft. The media are partially submerged (typically 40%) and slowly rotate (1 to 1.6 rpm) to expose the biofilm to substrates in the bulk of the liquid (when submerged) and to air (when not submerged).

The RBC process has the following advantages: operational simplicity, low energy costs, and rapid recovery from shock loadings. However, the literature has documented several examples of RBC malfunctions resulting from shaft, media, or media support system structural failure; poor treatment performance; accumulation of nuisance macrofauna; poor biofilm thickness control; and inadequate performance of air-driven systems for shaft rotation.

Many RBC facilities have been retrofitted to include load cells to monitor RBC shaft weight and to include coarse-bubble air sparging beneath the RBC process such that biofilm thickness may be monitored and controlled.

### 5.4.2 Biological Filters

Biological wastewater treatment and suspended solids removal are carried out in BAFs under either aerobic or anoxic conditions. In a BAF, the media act simultaneously to support the growth of biomass and as a filtration medium to retain filtered solids. Accumulated solids are removed from the BAF through backwashing. There is a direct interaction between media characteristics and the process because the configuration (i.e., sunken media or floating media) and flow and backwash regimes depend on media density. Media may be natural mineral, structured plastic, or random plastic. Backwashing the filters maximizes capture and run times and guarantees proper effluent quality. Proper backwashing requires filter bed expansion and rigorous scouring, followed by efficient rinsing. Poor filter cleaning will result in shortened filter runs, accumulation of solids, and deteriorating performance. Accumulation of solids and media ("mudballing") produces short-circuiting of water flow and can result in excessive media loss.

### 5.4.3 Expanded and Fluidized Bed Reactors

Expanded and fluidized bed biofilm reactors are similar in that they contain a bed of small particles (approximately 1 mm) of a suitable biomass support material on which biofilm grows and through which wastewater flows upward. Because the upflow velocity is fixed but flow through the works is not, there is automatically a variable recirculation rate. This type of reactor can hold a high concentration of active biofilm (up to 40 g $m^{-3}$) with a large surface area (900 to 2800 $m^2$ $m^{-3}$). Because the bed is expanded, it does not trap solids and, therefore, does not need backwashing. Being pseudo plug flow, the concentration of $NH_3$-N in the final effluent can be reduced to low levels (i.e., below 1 mg/L). Originally, dissolved oxygen was supplied by introducing air through a sparger or diffuser (Figure 4.2); however, recent improvements to design use countercurrent aeration in a separate column, which results in improved oxygen transfer. Management of the system is through control of the organic loading rate, to minimize competition from heterotrophic bacteria (which depend on the effectiveness of the upstream process), and control of the dissolved oxygen concentration at the top of the bed to ensure adequate oxygen supply. Occasional blooms of red oligochaete worms can be controlled by turning off aeration for a few hours, which kills the worms but not the nitrifying bacteria.

### 5.4.4 Submerged Fixed-Media Reactors

Structured sheet media have been installed for use in both aerobic biofilm and IFAS applications. Low- and medium-density cross-flow trickling filter media have historically been used for these installations. The structured sheet media modules are fixed in place using a retention system, with diffusers placed beneath the media towers. Similar to the mixed-media configuration of trickling filters, where cross-flow and vertical-flow media are used in combination to provide improved distribution of wastewater and the ability to flush solids, the submerged structured sheet media system uses cross-flow distribution media for improved air distribution below vertical-flow media for greater surface area and air-lift pumping. The diffused air provides air-lift pumping of wastewater through the vertical-flow media and downward flow occurs through "downcomer" regions where no media are installed. This allows for repeated

**TABLE 4.2** Biofilm Reactor Technology Operational Table

| Operational Goal | Biofilm Modification | Liquid Velocity Change | Operation and Control—General | Operation and Control—Trickling Filter | Operation and Control—MBBR | Operation and Control—IFAS |
|---|---|---|---|---|---|---|
| Improve nitrification rate—mass transfer limitation | Increase mass transfer to the biofilm | Increase | Increase oxygen transfer rate to increase dissolved oxygen concentration in biofilm<br>Increase upstream BOD removal to reduce heterotrophic competition in the biofilm<br>Increase fluid velocity to decrease MTBL thickness | Increase nutrient flow technique recirculation rate to transfer more dissolved oxygen to bulk phase (i.e., mix with influent)<br>Increase primary trickling filter (i.e., BOD removal) recirculation rate to improve BOD removal rate (this will also raise inlet dissolved oxygen to NTF) | Increase aeration rate to increase oxygen in bulk liquid<br>Increase aeration rate to decrease MTBL thickness<br>Increase BOD removal in upstream MBBR zones | Increase aeration rate to increase oxygen in bulk liquid<br>Increase aeration rate to decrease MTBL thickness<br>Increase MLSS to decrease bulk liquid BOD concentration |
| Improve nitrification rate—biofilm thickness limitation | Increase biofilm thickness | Decrease | Decrease fluid velocity and mixing intensity<br>Increase biofilm attachment area by adding additional media if sufficient reactor capacity exists (% fill < maximum fill achievable)<br>Addition of nutrients, such as phosphorus and potassium, may be necessary to promote the growth of nitrifying bacteria within the biofilm. A residual of 0.15 mg/L orthophosphate (P) is generally required to support nitrification. | Reduce flushing intensity (SK rate)<br>Increase primary trickling filter (i.e., BOD removal) recirculation rate to improve BOD removal rate<br>If operated for combined BOD and ammonia removal, increased recirculation may also improve performance (applied soluble BOD concentration < ~30 mg/L) | Decrease aeration rate to decrease shear force on biofilm and decrease biofilm erosion | Decrease MLSS concentration/decrease bulk liquid SRT to increase bulk liquid concentrations<br>Decrease aeration rate to decrease shear force on biofilm and decrease biofilm erosion |

*(continued)*

**TABLE 4.2** Biofilm Reactor Technology Operational Table (*Continued*)

| Operational Goal | Biofilm Modification | Liquid Velocity Change | Operation and Control—General | Operation and Control—Trickling Filter | Operation and Control—MBBR | Operation and Control—IFAS |
|---|---|---|---|---|---|---|
| Improve nitrification rate—biofilm thickness limitation (*continued*) | | | Proper training of staff is essential to ensure that they understand how to operate and maintain the biofilm system to ensure optimal nitrification performance. | Operate trickling filters in series to reduce heterotrophic and autotrophic competition (if currently operated in parallel) | | |
| Prevent biofilm septicity—biofilm thickness limitation | Decrease biofilm thickness | Increase | Increase fluid velocity and mixing intensity to increase shear force<br>Implement cyclical backwash cycles<br>Increase biofilm attachment area by adding additional media | Increase flushing intensity (SK rate)<br>Increase recirculation rate to increase bulk dissolved oxygen concentration (increased recycle rate may potentially increase SK rate; verify through calculation because distributor speed may increase for a hydraulically propelled distributor) | Increase aeration intensity to increase shear force and resulting biofilm erosion<br>Adjust aeration roll pattern to increase shear force | Increase MLSS/increase bulk liquid SRT to decrease bulk liquid BOD and ammonia concentrations<br>Increase aeration rate to increase shear force and biofilm erosion<br>Adjust aeration roll pattern to increase shear force |
| Prevent biofilm septicity—mass transfer limitation | Increase oxygen mass transfer | Increase | Increase oxygen transfer rate to increase dissolved oxygen concentration in biofilm | Increase recirculation rate to increase bulk dissolved oxygen concentration | Increase aeration rate to increase oxygen in bulk liquid<br>Increase aeration rate to decrease MTBL thickness and increase mass transfer rate | Increase aeration rate to increase oxygen in bulk liquid<br>Increase aeration rate to decrease MTBL thickness and increase mass transfer rate<br>Increase MLSS to decrease bulk liquid BOD concentration |

| Operational Goal | Biofilm Modification | Liquid Velocity Change | Operation and Control—General | Operation and Control—Trickling Filter | Operation and Control—MBBR | Operation and Control—IFAS |
|---|---|---|---|---|---|---|
| Prevent excessive biomass loss (sloughing)—mass transfer limitation | Increase mass transfer to the biofilm | Increase | Ensure sufficient phosphorus and nitrogen entering biofilm reactor<br>Increase oxygen transfer rate to increase dissolved oxygen concentration in biofilm<br>Ensure sufficient alkalinity to buffer pH<br>Prevent oxygen and nutrient starvation | Ensure sufficient phosphorus and nitrogen entering biofilm reactor<br>Increase oxygen transfer rate by increasing recirculation rate to increase dissolved oxygen concentration in biofilm<br>Ensure sufficient alkalinity to buffer pH | Ensure sufficient phosphorus and nitrogen entering biofilm reactor<br>Increase aeration rate to decrease MTBL thickness and increase mass transfer rate<br>Ensure sufficient alkalinity to buffer pH | Ensure sufficient phosphorus and nitrogen entering biofilm reactor<br>Increase aeration rate to decrease MTBL thickness and increase mass transfer rate<br>Increase MLSS to decrease bulk liquid BOD concentration<br>Ensure sufficient alkalinity to buffer pH |
| Prevent excessive biomass loss (sloughing)—biofilm thickness limitation | Decrease shear force on biofilm | Decrease | Decrease fluid velocity and mixing intensity | Reduce flushing intensity (SK rate) while still maintaining adequate wetting | Reduce aeration rate to reduce shear force and biofilm erosion | Reduce aeration rate to reduce shear force and biofilm erosion |

recirculation within the tower, which results in contact of the wastewater with the fixed biomass and provides air scouring simply from process air requirements.

Biofilm control occurs in the vertical-flow media as air bubbles pass the attached biofilm. The high mixing rate developed from air-lift pumping minimizes the thickness of the mass-transfer boundary layer, enhancing nitrification performance. The media surface areas range from 157 $m^2/m^3$ (48 sq ft/cu ft) to 315 $m^2/m^3$ (96 sq ft/cu ft), with surface nitrification rates as high as 1.66 $g/(m^2 \cdot d)$ at 20 °C (Ye et al., 2010). The bulk liquid dissolved oxygen concentration is typically maintained between 3 and 5 mg/L. The structured sheet has a limited effect on the oxygen-transfer efficiency of fine-bubble diffusers that are used to improve oxygen-transfer efficiency for the system (Zhu et al., 2011).

### 5.5 Biofilm Reactor Technologies—General Operations Table

Operation of each of these biofilm reactor technologies and emerging and yet-to-be-developed technologies will generally follow the same principles outlined in Table 4.1. The specific mechanisms for adjustment vary by technology, and technology-specific operational tools to achieve the main operational goals are summarized in Table 4.2.

## 6.0 BIOFILM REACTOR DESIGN

An in-depth discussion of the principles and practices of biofilm reactor design is presented in *Biofilm Reactors* (WEF, 2011). This publication covers details of biofilm rector design and should be referenced for questions pertaining to specific reactor selection and evaluation. Recommended loading rates can be used as a preliminary evaluation of overall loading to an operating biofilm reactor technology and can help determine if the system is significantly overloaded or underloaded. These loadings only represent a cursory evaluation of the system; given site-specific environmental conditions, these applicable loading rates can vary significantly.

## 7.0 REFERENCES

Andreottola, G., Foladori, P., Ragazzi, M., & Tatano, F. (2000). Experimental comparison between MBBR and activated sludge system for the treatment of municipal wastewater. *Water Science and Technology, 41*(4–5), 375–382.

Bjornberg, C., Lin, W., & Zimmerman, R. (2009, October 10–14). Effect of temperature on biofilm growth dynamics and nitrification kinetics in a full-scale MBBR system. *Proceedings of the 82nd Annual Water Environment Federation Annual Technical Exhibition and Conference* [CD-ROM] (pp. 4407–4426), Orlando, FL, United States. Water Environment Federation.

Boltz, J. P., Goodwin, S. J., Rippon, D., & Daigger, G. T. (2008). A review of operational control strategies for snails and other macro-fauna infestations in trickling filters. *Water Practice and Technology, 2*(4).

Bryers, J. D. (1984). Biofilm formation and chemostat dynamics: Pure and mixed culture considerations. *Biotechnology & Bioengineering, 26*, 948–958.

Cole, A. C., Semmens, M. J., & LaPara, T. M. (2004). Stratification of activity and bacterial community structure in biofilms grown on membranes transferring oxygen. *Applied and Environmental Microbiology, 70*(4), 1982–1989.

Daigger, G. T., & Boltz, J. P. (2011). Trickling filter and trickling filter-suspended growth process design and operation: A state-of-the-art review. *Water Environment Research, 83*(5), 388–404.

Downing, L., & Nerenberg, R. (2008a). Effect of bulk liquid BOD concentration on activity and microbial community structure of a nitrifying, membrane-aerated biofilm. *Applied Microbiology and Biotechnology, 81*, 153–162.

Downing, L., & Nerenberg, R. (2008b). Effect of oxygen gradients on performance and microbial community structure of a nitrifying, membrane-aerated biofilm. *Biotechnology & Bioengineering, 101*(6), 1193–1204.

Fernandez-Polanco, F., Mendez, E., Uruena, M. A., Villaverde, S., & Garcia, P. A. (2000). Spatial distribution of heterotrophs and nitrifiers in a submerged biofilter for nitrification. *Water Research, 34*(16), 4081–4089.

Grady, L. E., Daigger, G. T., & Lim, H. (1999). *Biological wastewater treatment* (2nd ed.). Marcel Dekker.

Harrison, J. R., & Timpany, P. L. (1988). Design considerations with the trickling filter solids contact process. *Proceedings of the Joint Canadian Society of Civil Engineers, ASCE National Conference on Environmental Engineering* (pp. 753–762), Vancouver, British Columbia.

Hibiya, K., Terada, A., Tsuneda, S., & Hirata, A. (2003). Simultaneous nitrification and denitrification by controlling vertical and horizontal microenvironment in a membrane-aerated biofilm reactor. *Journal of Biotechnology, 100*(1), 23–32.

Kuenen, J. G., Jorgensen, B. B., & Revsbech, N. P. (1986). Oxygen microprofiles of trickling filter biofilms. *Water Research, 20*(12), 1589–1598.

McQuarrie, J. P., & Boltz, J. P. (2010). Moving bed biofilm reactor technology: Process applications, design, and performance. *Water Environment Research, 83*(6), 560–575.

Morgenroth, E. (2003). Detachment: An often-overlooked phenomenon in biofilm research. In S. Wuertz, P. Bishop, & P. Wilderer (Eds.), *Biofilm in wastewater treatment*. IWA Publishing.

Nogueira, R., Elenter, D., Brito, A., Melo, L. F., Wagner, M., & Morgenroth, E. (2005). Evaluating heterotrophic growth in a nitrifying biofilm reactor using fluorescence in situ hybridization and mathematical modeling. *Water Science & Technology, 52*(7), 135–141.

Oakey, R. W., & Albertson, O. E. (1989). Diffusion's role in regulating rate and masking temperature effects in fixed film nitrification. *Journal Water Pollution Control Federation, 61*, 510.

Ødegaard, H., Cimbritz, M., Christensson, M., & Dahl, C. P. (2010, August 14–18). Separation of biomass from moving bed biofilm reactors (MBBRs). *Proceedings of the Water Environment Federation and International Water Association Biofilm Reactor Technology Conference*, Portland, OR, United States. Water Environment Federation.

Okabe, S., Hiratia, K., Ozawa, Y., & Watanabe, Y. (1996). Spatial microbial distributions of nitrifiers and heterotrophs in mixed-population biofilms. *Biotechnology & Bioengineering, 50*(1), 24–35.

Points, A., Downing, L., & Yu, J. (2010, October 2–6). Nitrification and denitrification in an aerobic integrated fixed-film activated sludge system. *Proceedings of the 83rd Annual Water Environment Federation Technical Exhibition and Conference* [CD-ROM] (pp. 556–559), New Orleans, LA, United States. Water Environment Federation.

Satoh, H., Okabe, S., Norimatsu, N., & Watanabe, Y. (2000). Significance of substrate C/N ratio on the structure and activity of nitrifying biofilms determined by in situ hybridization and the use of microelectrodes. *Water Science & Technology, 41*(5), 317–321.

Satoh, H., Ono, H., Rulin, B., Kamo, J., Okabe, S., & Fukushi, K.-I. (2004). Macroscale and microscale analyses of nitrification and denitrification in biofilms attached on membrane aerated biofilm reactors. *Water Research, 38*(6), 1633–1641.

Schramm, A., Larsen, L. H., Revsbech, N. P., Ramsing, N. B., Amann, R., & Schleifer, K.-H. (1996). Structure and function of a nitrifying biofilm as determined by in situ hybridization and the use of microelectrodes. *Applied Environmental Microbiology, 62*, 4641–4647.

Severn Trent & OxyMem. (2018). *Case study: Severn Trent Water.* https://www.oxymem.com/severn-trent-casestudy

Terada, A., Hibiya, K., Nagai, J., Tsuneda, S., & Hirata, A. (2003). Nitrogen removal characteristics and biofilm analysis of a membrane-aerated biofilm reactor applicable to high-strength nitrogenous wastewater treatment. *Journal of Bioscience and Bioengineering, 95*(2), 170–178.

Water Environment Federation. (2000). *Aerobic fixed-growth reactors.*

Water Environment Federation. (2011). *Biofilm reactors* (WEF Manual of Practice No. 35).

Water Environment Federation. (2023). *Standards for design of water resource recovery facilities* (WEF 8-23).

Ye, J., Kulick, F. M., III, & McDowell, C. S. (2010, October 2–6). Biofilm performance of high surface area density vertical-flow structured sheet media for IFAS and fixed bed biofilm reactor (FBBR) applications. *Proceedings of the 83rd Annual Water Environment Federation Technical Exhibition and Conference* [CD-ROM] (p. 234), New Orleans, LA, United States. Water Environment Federation.

Zhu, J., Kulick, F. M., III, & Koch, K. (2011, October 15–19). Impact of structured sheet media on the oxygen transfer efficiency of fine and coarse bubble diffusers. *Proceedings of the 84th Annual Water Environment Federation Technical Exhibition and Conference* [CD-ROM], Los Angeles, CA, United States. Water Environment Federation.

# Chapter 5

# Denitrification



## 1.0 INTRODUCTION

Denitrification is the biological process that removes nitrogen from the wastewater. Nitrogen typically enters a water resource recovery facility (WRRF) as ammonium-nitrogen with some organic nitrogen. Nitrification (Chapter 3) converts ammonium-nitrogen to nitrate-nitrogen. The denitrification process converts the nitrate-nitrogen to nitrogen gas, which has limited solubility in water and evolves from the water to the atmosphere. Denitrification requires an anoxic environment, which is defined as having no dissolved oxygen, but the presence of nitrite or nitrate-nitrogen. Nitrification must occur for denitrification to occur. Denitrification is dependent on efficient nitrification. Nitrification uses autotrophic bacteria (Chapter 3), whereas denitrification uses heterotrophic bacteria, the same bacteria used for biochemical oxygen demand (BOD) removal. Heterotrophic bacteria use organic carbon as their carbon source. In an aerobic system, such as for BOD removal, the organic chemicals are the electron donors, and oxygen is the electron acceptor. The byproducts are carbon dioxide and water. If there is no dissolved oxygen available, the heterotrophic bacteria still use organic carbon as the electron donor but use nitrite or nitrate as the electron acceptor. The byproducts of this reaction are carbon dioxide, water, and nitrogen gas—thus the removal of nitrogen from the wastewater.

## 2.0 PROCESS FUNDAMENTALS

### 2.1 Chemical Thermodynamics

The chemical transformation or degradation of a compound by microbial action is based on chemical thermodynamics. The reaction that results is that which yields the highest positive energy. All microbial species obtain energy as a result of this favorable energy flow. If thermodynamics are not favorable, the transformation of chemicals will not occur. This energy flow is obtained from the coupling of oxidation-reduction reactions, which are frequently referred to as *redox reactions*. The amount of free energy depends on the Gibbs free energies of the substrates and the products. A useful concept for solution-based redox reactions is electron activity (pE), a measure of the available electrons in much the same manner that pH measures the availability of protons. The amount of free energy that can be obtained by microorganisms from redox reactions is directly proportional to the electron activity. Understanding if an energy yield is positive or negative is a key component of knowing how compounds can be biologically degraded or transformed to desired chemical species.

Understanding microbial metabolism modes is important to successful operation of denitrification processes and troubleshooting poor denitrification performance. Microorganisms have developed wide varieties of metabolism systems. These are characterized by the nature of the reductant and the oxidant. For aerobic respiration, the electron acceptor is molecular oxygen. Anaerobic respiration uses an oxidized inorganic or organic compound other than oxygen as the electron acceptor. It is the electron acceptor that provides the name for the metabolism mode.

Microorganisms take advantage of this by using the transfer of electrons to produce energy-rich bonds within the microorganism's cell. Microorganisms in the presence of a potential energy source produce enzymes that catalyze the natural thermodynamic reaction to yield an increase in the rate of energy flow (Cookson, 1995). For denitrification, the electron acceptor must be nitrate ($NO_3^-$). The oxidation reaction is nitrate oxidizing the organic compound or compounds measured as BOD. The organic compound loses electrons, which are transferred to nitrate. This overall reaction consists of two half reactions: (1) the oxidation of organic compounds and (2) the reduction of nitrate. The end products are carbon dioxide and nitrogen gas. Other chemicals in water can serve as electron acceptors (e.g., oxygen, sulfate, ferric iron, another organic compound, and carbon dioxide). The electron acceptor that enters this reaction with the organic compound establishes the metabolism mode.

If thermodynamics are not favorable for nitrate to be the electron acceptor, denitrification will not occur. Certain changes in water chemistry can result in unfavorable conditions for denitrification. Any compound or, in some instances, gases that can serve as an electron acceptor and have a higher pE than nitrate will block denitrification. One example is the presence of oxygen. Understanding these principles can help solve potential problems resulting from changes in water chemistry that can change chemical thermodynamics.

## 2.2 Stoichiometry and Kinetics

Heterotrophic denitrification is the dissimilatory reduction of ionic nitrogen oxides such as nitrate ($NO_3^-$) and nitrite ($NO_2^-$) to gaseous oxides such as nitric oxide (NO) and nitrous oxide ($N_2O$) and, eventually, to nitrogen gas ($N_2$) under anoxic conditions using organic electron donors and assimilative carbon sources (Knowles, 1982). Nitrogen oxides act as terminal electron acceptors in the absence of oxygen. Various organic carbons—such as methanol, acetate, glycerol, or glucose—can serve as electron donors and carbon sources for growth. The overall reaction when using methanol as a carbon source for denitrification is as follows (note that cell growth is not taken into account):

$$5CH_3OH + 6NO_3^- \rightarrow 3N_2 + 5CO_2 + 7H_2O + 6OH^- \quad (5.1)$$

The amount of nitrate that can be removed from a given wastewater is a function of organic carbon availability and environmental and operating conditions and is primarily determined as the stoichiometric ratio between the electron donor and acceptor (i.e., carbonaceous oxygen demand [COD]-to-nitrate—as nitrogen—ratio [C/N]). Many studies have reported a wide range of COD/N required for satisfactory denitrification, that is, between 3 and 7 g COD/g N (Chiu & Chung 2003; Li et al., 2008; Sobieszuk & Szewczyk, 2006). However, this ratio largely depends on the nature of the influent concentrations of biodegradable COD and nitrogen species, solids retention time, and the microbial ecology of the denitrification process itself. For instance, if the influent wastewater COD/N is not sufficient, an external carbon source is required to achieve complete nitrate reduction to $N_2$. Furthermore, denitrifying bacteria with a higher growth yield coefficient require a higher dosage of an external electron donor, as follows (Grady et al., 1999):

$$\text{Carbon dose} = \frac{2.86}{1-Y} x(NO_{3,in}^- - NO_{3,eff}^-) \quad (5.2)$$

Where

$(NO_{3,in}^- - NO_{3,eff}^-)$ = the degree of denitrification required (mg N/d, calculated by multiplying the difference between the influent and effluent concentrations of nitrate [mg N/L] by the flowrate [L/d]);

$Y$ = the anoxic growth yield of denitrifying bacteria on specific carbon sources (mg biomass COD produced/mg COD removed); and

Carbon dose = in mg COD/d.

On the other hand, the kinetics of denitrification directly influence anoxic reactor zone sizing in biological nutrient removal (BNR) reactors. Microbial kinetics are generally described by the Monod model (Equation 5.3 and Figure 5.1). Biokinetic parameters that are essential to full-scale denitrification process design and operation, such as maximum specific growth rate ($\mu_{max}$) and half-saturation coefficients (*Ks*), can be obtained by nonlinear regressions of the following equations:

$$\mu = \frac{\mu_{max} S_{COD}}{1 + K_{S,COD} S_{COD}} \quad (5.3)$$

$$\frac{dX_H}{dt} = \mu X_H = \frac{\mu_{max} S_{COD}}{1 + K_{S,COD} S_{COD}} X \quad (5.4)$$

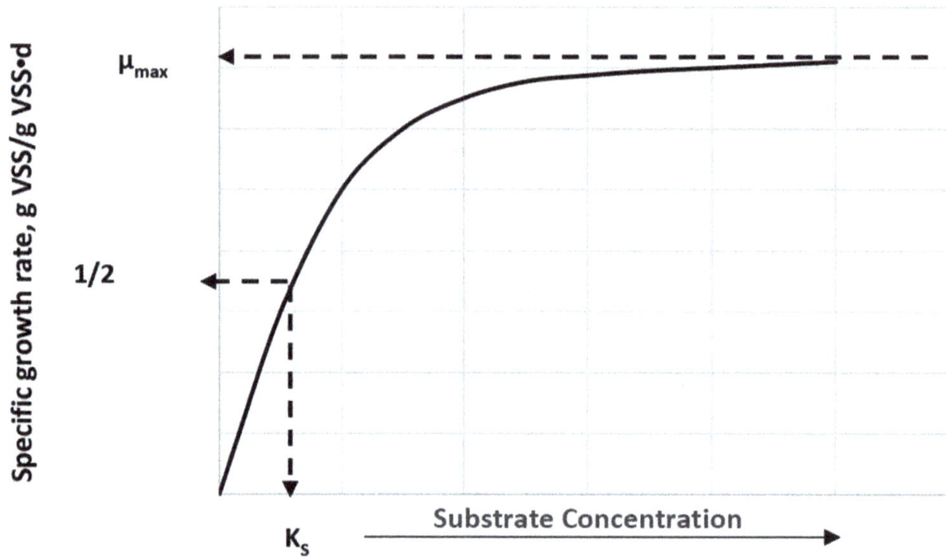


FIGURE 5.1 Monod Model That Defines the Relationship Between the Growth Rate (μ) and the Concentration of the Limiting Nutrient (*S*)

$$\frac{dS_{NO}}{dt} = \frac{1-Y_H}{2.86\,Y_H}\frac{dX_H}{dt} \tag{5.5}$$

$$\frac{dS_{COD}}{dt} = \frac{1}{Y_H}\frac{dX_H}{dt} \tag{5.6}$$

Where

$\mu$ = the specific growth rate ($t^{-1}$);
$\mu_{max}$ = the maximum specific growth rate ($t^{-1}$);
$K_{S,COD}$ = the half-saturation coefficient of organic carbon (mg/L);
$Y_H$ = the growth yield of heterotrophs (mg COD biomass formed/mg COD removed);
$X_H$ = the biomass concentration (mg/L); and
$S$ = the substrate concentration (mg/L, $S_{COD}$: organic carbon; $S_{NO}$: nitrate).

Using methanol as the carbon source, values for $\mu_{max}$ range from 0.2 to 2.0 $d^{-1}$ according to Gaudy and Gaudy (1980). The $\mu_{max}$ is also a function of temperature, as reported by Stensel et al. (1973). The specific substrate use rate ($q$) is related to μ by the growth yield, $Y$ (Equation 5.7). The specific denitrification rate (SDNR) is a direct measure of denitrification activity (Equation 5.8) and can be obtained by denitrification rate batch assays, as follows:

$$q = \mu/Y \tag{5.7}$$

$$\frac{dS_{NO}}{dt} = \mathrm{SDNR}x X_H \tag{5.8}$$

A double-substrate Monod model can be used to describe the effects of both carbon and nitrate concentrations on the specific denitrification rate (Equation 5.9). However, the half-saturation coefficient of nitrate ($K_{S,NO}$) is less than 1 mg/N and has little influence on measured SDNR values (Marazioti et al., 2003; Wild et al., 1995). The SNDR is a function of the food-to-microorganism (F/M) ratio in an anoxic reactor and the content of readily biodegradable BOD in the feed wastewater to the anoxic reactor. If the F/M is 0.5 g BOD/g mixed liquor suspended solids (MLSS)/d in the anoxic reactor and readily biodegradable BOD comprises 30% of the total BOD, SNDR becomes approximately 0.125 g $NO_3$-N/g MLSS/d, as follows:

$$SDNR = SDNR_{max} x \frac{S_{COD}}{K_{S,COD} + S_{COD}} x \frac{S_{NO}}{K_{S,NO} + S_{NO}} \tag{5.9}$$

For treatment processes with a single anoxic zone, the denitrifying reactor can be sized based on the SDNR values, the concentration of volatile suspended solids (VSS) in the mixed liquor, and the degree of denitrification required, as follows:

$$V_{ANX} = \frac{S_{NO,in} - S_{NO,eff}}{SDNR(X_{VSS})} \tag{5.10}$$

Where

$V_{ANX}$ = the calculated volume of the anoxic zone (L),
$S_{NO,in} - S_{NO,eff}$ = the degree of denitrification required (mg N/d),
SDNR = the specific denitrification rate (mg N/mg VSS/d), and
$X_{VSS}$ = the concentration of VSS in the anoxic zone (mg VSS/L).

## 2.3 Denitrifying Organisms

Unlike nitrifying bacteria, which are phylogenetically closely related (Purkhold et al., 2000), denitrifying bacteria are distributed in a large variety of physiological and taxonomic groups belonging to various subclasses of *Proteobacteria* (Gerardi, 2006). Because methanol is the most commonly used carbon source supporting wastewater denitrification, methylotrophic denitrifiers have been widely studied (Anthony, 2011; Labbe et al., 2007; Nurse, 1980; Sperl & Hoare, 1971). Methanol denitrification selects special groups of bacteria (Christensson et al., 1994; Ginige et al., 2004) that can be further divided into three subgroups: obligate, restricted facultative, and typical facultative methylotrophs (de Vries et al., 1990). The obligate methylotroph can only use C1 compounds (e.g., *Hyphomicrobium methylovorum* KM146 and *Methylococcus capsulatus*) (de Vries et al., 1990). The C1 compounds are those that contain only one carbon atom (e.g., methanol). The restricted methylotroph can grow on a limited range of complex organic compounds (e.g., *Methylophilus glucosoxydans* and *Methylophilus rhizosphaerae*) (de Vries et al., 1990). Typical facultative methylotrophs, such as *Methylobacterium extorquens* AM1, can grow on a wide range of polycarbon compounds (Chistoserdova et al., 2003; O'Keeffe & Anthony, 1980).

## 2.4 Favorable Operating Conditions

### 2.4.1 Anoxic

Oxygen inhibits denitrification by providing a better electron acceptor for denitrifying species to generate energy. The Gibbs standard free energy of water–oxygen is −78.73 kJ/eeq (the definition is as follows: 1 eeq = amount of substrate releasing 1 mol $e^-$ during total carbonaceous oxidation [e.g., substrate ≥ $CO_2$, $NH_4^+$]; note that 1 eeq of an organic electron donor = 8 g COD), and that for nitrate-nitrogen is −72.20 kJ/eeq (Rittmann & McCarty, 2001). The threshold oxygen inhibition concentration is approximately 0.2 mg $O_2$/L (Knowles, 1982). In practice, oxidation-reduction potential (ORP), which is a measure of the activity or strength of oxidizers and reducers in relation to their concentration in wastewater, has been used to indicate aerobic, anoxic, and anaerobic conditions. The ORP generally shows a strong response to dissolved oxygen, especially at low concentrations, and is a better monitoring and controlling parameter under anoxic and anaerobic conditions than dissolved oxygen concentrations. In general, ORP values lower than −200 mV indicate anaerobic conditions, values between −200 and +200 mV indicate anoxic conditions, and values higher than +200 mV are for aerobic conditions. The ORP in an anoxic reactor must be below +50 mV to trigger a denitrification reaction. Typically, denitrification occurs between −50 and +50 mV. It is important to note that ORP values would be expected using an ORP probe with a silver-to-silver chloride ratio reference electrode. However, if another reference electrode system is used, such as a hydrogen or a calomel electrode, the ORP values will be different. For instance, a hydrogen electrode registers an ORP approximately 200 mV more positive than the silver-to-silver chloride ratio reference. It is important to speak with your instrument supplier to understand the type of ORP electrode and the expected readings from your process. The inhibition of oxygen to denitrification is reversible, and the response time of the enzymatic system to the electron acceptor switch is generally in the magnitude of minutes (Kornaros & Lyberatos, 1998). An expended Monod expression of

$$\frac{dS_{NO}}{dt} = \frac{1-Y}{2.86\,Y}\mu_{max}\left(\frac{S_{COD}}{K_{S,COD}+S_{COD}}\right)\left(\frac{S_{NO}}{K_{S,NO}+S_{NO}}\right)\left(\frac{K_{I,O}}{K_{I,O}+S_{O}}\right)\eta X_{bH} \tag{5.11}$$

Where

$S_S$, $S_{NO}$, and $S_O$ = concentrations of the organic carbon source, nitrate-nitrogen, and oxygen (in COD mg/L, mg N/L, and mg $O_2$/L, respectively);

$K_{S,COD}$ and $K_{S,NO}$ = the half-saturation coefficients of the organic carbon source (mg COD/L) and nitrate (mg N/L), respectively;

$K_{I,O}$ = the inhibition coefficient of oxygen on nitrate reduction, which was found to be around 0.1 mg $O_2$/L using both mixed and pure cultures (Grady et al., 1999; Kornaros & Lyberatos, 1998); and

$\eta$ = the correlation factor for growth under anoxic conditions (dimensionless, with the typical value of 0.8).

### 2.4.2 Carbonaceous Oxygen Demand–to-Nitrate Ratio

The performance of single sludge predenitrification processes depends on the composition of influent wastewater, and low ratios of COD/N can lead to poor denitrification, whereas relatively high COD/N can detrimentally affect nitrification (Sharma & Gupta, 2004). The most critical form of COD is readily biodegradable, soluble COD that can be metabolized quickly and efficiently by organisms.

For postdenitrification systems with external carbon augmentation, the optimal C/N will vary depending on the forms of carbon source. Tam et al. (1994) reported an optimal 5-day BOD ($BOD_5$)/$NO_x$-N ($NO_x$ is either nitrate-nitrogen [$NO_3$-N] or nitrite-nitrogen [$NO_2$-N]) of 2.48 for a denitrification process with methanol and acetic acid as the carbon source. A theoretical optimal carbon-to-nitrogen ratio (C/N) may be calculated using the following stoichiometric relationship (using methanol as an example):

$$6\ NO_3^- + 5\ CH_3OH + H_2CO_3 \rightarrow 3\ N_2 + 8\ H_2O + 6\ HCO_3^- \quad (5.12)$$

Converting this to a mass basis reduces the equation (in grams) to the following:

$$84g\ NO_3^-\text{-}N + 160g\ CH_3OH + 62g\ H_2CO_3 \rightarrow 84g\ N_2 + 144g\ H_2O + 366g\ HCO_3^- \quad (5.13)$$

One gram of methanol has the theoretical oxygen equivalent of 1.5 g of oxygen. Therefore, 1.91 g methanol/g $NO_3^-$-N is required to reduce nitrate completely. In addition to providing energy for the reduction of nitrate, the methanol or carbonaceous BOD (CBOD) consumed also creates new biomass. As a result, more methanol or CBOD are required than are presented in Equations 5.12 and 5.13 to reduce each unit of nitrate. The amount of new biomass generated and the portion used for denitrification are specific to each compound. Equations for calculating methanol consumption using nitrate, nitrite, and oxygen are as follows (U.S. Environmental Protection Agency, 1975):

$$NO_3^- + 1.08\ CH_3OH + 0.24\ H_2CO_3 \rightarrow 0.04\ C_5H_7NO_2 + 0.48\ N_2 + 1.23\ H_2O + HCO_3^- \quad (5.14)$$

$$NO_2^- + 0.67\ CH_3OH + 0.53\ H_2CO_3 \rightarrow 0.056\ C_5H_7NO_2 + 0.47\ N_2 + 1.68\ H_2O + HCO_3^- \quad (5.15)$$

$$O_2^- + 0.93\ CH_3OH + 0.056\ NO_3^- \rightarrow 0.056\ C_5H_7NO_2 + 1.04\ N_2 + 0.59\ H_2CO_3 + 0.056\ HCO_3^- \quad (5.16)$$

These three equations can be converted into a methanol dose based on nitrate, nitrite, and dissolved oxygen concentrations, as follows:

$$\text{Methanol dose (mg/L)} = 2.47\ NO_3^-\text{-N} + 1.53\ NO_2^-\text{-N} + 0.87\ \text{Dissolved oxygen} \quad (5.17)$$

Provided that dissolved oxygen discharged or recycled to a denitrification reactor is minimal, the general rule that 3 mg/L $BOD_5$ is removed for every 1 mg/L $NO_3^-$-N denitrified is a reasonable approximation.

### 2.4.3 Temperature

Similar to other heterotrophs, the effect of temperature on the kinetics of denitrifying bacteria is mathematically approximated based on the following Arrhenius equation:

$$k = Ae^{-Ea/RT}$$
$$k_1 = k_2\theta^{T1-T2} \quad (5.18)$$

Where

$k$ = specific denitrification rate at temperature $T$ (mg $NO_3$-N/mg VSS/d),
$A$ = the prefactor,
$Ea$ = activation energy in J/mol,
$R$ = ideal gas constant (8.314 J/mol/K),
$T$ = absolute temperature, and
$\theta$ = temperature coefficient (dimensionless).

Denitrification rate versus temperature is, therefore, a bell-shaped curve, and the maximum denitrification rate is obtained at the optimal temperature (ranging from 20 to 60 °C [Dawson & Murphy, 1972;

Knowles, 1982]), and then declines rapidly above this temperature. Various values have been reported based on a number of site-specific studies, yielding a range from 1.03 to 1.20 (Dawson & Murphy, 1972).

### 2.4.4 External Carbon Source

Nearly all separate-stage denitrification reactors and even some denitrification reactors within single sludge systems use some form of supplemental carbon or substrate augmentation. Most separate-stage denitrification processes follow CBOD and nitrification processes. Therefore, there is a limited quantity of biodegradable organic carbon available to serve as the substrate for denitrification. The form of substrate and, in particular, the ease or difficulty of biodegradation of that substance is a key parameter related to the rate and efficiency of the denitrification process. Several organic carbon sources have historically been used to augment the rate of the denitrification process, including methanol, ethanol, acetate, molasses, and brewery wastes. Several industrial wastes were also tested as organic carbon sources. Some organic wastes, such as formaldehyde and dextrose waste, were less efficiently degraded than distillery oils or methanol, but the majority of food industry wastes exhibit high denitrification rates and carbon-to-nitrogen ratios from 2 to 6. The addition of residue from food industrial processes also shows improved nitrogen removal performance at a full-scale municipal WRRF in Finland. The use of anaerobic fermentation products from the organic fraction of municipal solid waste has been suggested for enhanced denitrification (Bolzonella et al., 2001) in an integrated BNR and solid waste management system, which has already been applied in full-scale size at the Treviso (Italy) WRRF.

Soluble and readily degradable substrates support the highest rates of denitrification. Although methanol is the most widely used exogenous carbon source, it is not as efficient as other carbon substrates (e.g., ethanol and acetate on the kinetic basis) (Onnis-Hayden & Gu, 2008). Denitrifying bacteria enriched by ethanol and acetate also grow at higher yields. The stoichiometric and kinetic coefficients of denitrification with various carbon sources are summarized in Table 5.1.

TABLE 5.1 Representative Stoichiometric and Kinetic Parameters at 20 °C for Denitrifying Biomass Grown on Different Carbon Sources

| Carbon Source | COD/N | Y (g VSS/g COD) | $\mu_{max}(d^{-1})$ | SDNR (mg N/g VSS·h) | Reference |
|---|---|---|---|---|---|
| Methanol | 4.1–4.5 | 0.23–0.25 | 0.77–2 | 32–91 | Christensson et al. (1994) |
| | 4.7 | | 0.4–1.0 | 6.07 | Mokhayeri et al. (2006) |
| | | 0.4 | | 3.2 | Peng et al. (2007) |
| Ethanol | | 0.42 | | 9.6 | Peng et al. (2007) |
| | | 0.22 | | 27 | Akunna et al. (1993) |
| | 5.877 | 0.36 | | | McCarty et al. (1969) |
| Acetate | | 0.65 | | 12 | Peng et al. (2007) |
| | 3.5 | | 1.2–3.5 | | Mokhayeri et al. (2006) |
| | | 0.3 | | 3.6 | Kujawa & Klapwijk, 1999 |
| MicroC1000™ | 7.87 | 0.44 | 1.99 | 3.31 | Akunna et al. (1993) |
| MicroC2000™ | 6.46 | 0.39 | 2.05 | 9.34 | Du et al. (2010) |
| Glucose | | | | 2.7 | Du et al. (2010) |
| | | | | | Akunna et al. (1993) |
| | | 0.38 | | | Muller et al. (2003) |
| Biodiesel | 4.8 | | | 1.8 | Ramalingam et al. (2007) |

Adaptation of denitrifying communities to change external carbon sources may involve both enrichment of new populations and the regulation of enzyme expression in existing populations. Methanol addition often requires an adaptation period of up to several months before *denitrification* rates significantly increase (Nyberg et al., 1992). This may be attributed to necessary population shifts in the microbial community (Ginige et al., 2004) to enrich methylotrophs. In contrast, activated sludge responds immediately to acetate and ethanol (Hallin et al., 1996; Hallin & Pell, 1998; Isaacs et al., 1994) because these two carbons select *Azoarcus*, *Dechloromonas*, *Thauera*, and *Acidovorax*-like denitrifiers, which have broad substrate specificities (Hallin & Pell, 1998; Hwang et al., 2006). Denitrifying species identified in activated sludge processes with different carbon sources are summarized in Table 5.2.

## 2.5 Benefits of Denitrification

### 2.5.1 Permit Compliance

Denitrification reduces nitrite and nitrate ions that are discharged, thereby protecting the quality of the receiving water and preventing eutrophication. Activated sludge processes that have a total nitrogen discharge limit must denitrify.

### 2.5.2 Control of Undesired Filamentous Organism Growth

Denitrification promotes the formation of firm and dense floc particles. Firm and dense floc particles are resistant to shearing action and exhibit desired settling characteristics. An anoxic environment produced through denitrification favors the growth of facultative anaerobic, floc-forming bacteria and discourages the growth of strict aerobic, filamentous organisms and weak facultative anaerobic, filamentous organisms. Strict aerobic, filamentous organisms can only use free molecular oxygen. Therefore, their

**TABLE 5.2** Denitrifying Populations Grown on Different Types of Carbon Sources

| Carbon Source | Dominant Functional Groups | Techniques | References |
|---|---|---|---|
| Methanol | *Methylophilales, Methyloversatilis, Hyphomicrobium, Paracoccus* | Culture isolation, SIP[a], 16S rRNA gene sequencing, FISH[b] | Ginige et al. (2004), Blaszczyk et al. (1985), Lemmer et al. (1997), Baytshtok et al. (2008), Baytshtok et al. (2009) |
| Ethanol | *Methyloversatilis, Azoarcus Dechloromonas, Pseudomonas, Hydrogenophaga* | Stable isotope probing, 16S rRNA sequencing | Hwang et al. (2006), Baysthtok et al. (2008) |
| Acetate | *Comamonas, Acidovorax, Thauera, Dechloromonas, Paracoccus, Rhodobacter* | Stable isotope probing, 16S rRNA gene sequencing, FISH | Osaka et al. (2006), Ginige et al. (2005) |
| Glycerol | *Comamonas, Bradyrhizobium, Diaphorobacter, Tessaracoccus* | Stable isotope probing | Lu & Chandran (2010) |
| Glucose | *Arthrobacter, Phenylobacterium, Pseudomonas, Proteobacterium* | Denaturing gradient gel electrophoresis | Nakatsu et al. (2005) |
| Landfill leachate | *Thauera, Acidovorax, Alcaligenes* | 16S rRNA gene sequencing | Etchebehere et al. (2001) |

[a]SIP = stable isotope probing.
[b]FISH = fluorescent in situ hybridization.

growth is stopped or the organisms die in an anoxic environment. Although weak facultative, anaerobic, filamentous organisms can use free molecular oxygen and nitrite ions or nitrate ions, these filamentous organisms cannot compete successfully for nitrite ions or nitrate ions with facultative anaerobic, floc-forming bacteria.

### 2.5.3 Return of Alkalinity

Approximately 3.57 mg of alkalinity as $CaCO_3$ is produced for each milligram of nitrate converted to molecular nitrogen. Nearly all activated sludge processes that denitrify must nitrify, which results in a consumption of alkalinity for approximately 7.14 mg as $CaCO_3$/mg $NH_4^+$ oxidized. Therefore, the alkalinity that is returned during denitrification is approximately one-half the amount of alkalinity that is lost during nitrification.

### 2.5.4 Energy Savings

Each gram of nitrate consumed in respiration saves 2.86 g of oxygen. If an activated sludge facility is configured so that facultative bacteria can use the nitrate produced during nitrification while consuming CBOD, the facility can realize substantial energy cost savings as a result of the decrease in aeration.

# 3.0 DENITRIFICATION SYSTEMS

## 3.1 Suspended Growth Systems

Single-sludge, suspended growth denitrification processes accomplish COD removal, nitrification, and denitrification with the same biomass. In these systems, biomass is exposed to different environmental conditions within various zones and shares the same set of secondary clarifiers. These systems achieve COD and nitrogen removal using linked reactors in series and a common set of secondary clarifiers. This consolidation of processes offers the potential of reduced capital and operating costs and, at the same time, a greater operational challenge related to proper balancing that is sometimes required between the different process environments.

Combined nitrification and denitrification processes can be classified in various ways, such as biofilm versus suspended growth, flow regime, staging of process sequences, or method of aeration. The common aspect consistent throughout all of these processes is alternating environments that are used sequentially to nitrify and denitrify. Nitrification must be completed to a large extent before denitrification can be achieved. Various factors such as dissolved oxygen concentration (high or low) and toxic compounds can influence the efficiency of these systems. In addition, the following three parameters can limit denitrification and, ultimately, total nitrogen removal:

- Nitrate—incomplete nitrification will result in low concentrations of nitrate, reducing nitrogen removal through denitrification and resulting in high-effluent total nitrogen concentrations.
- Substrate (CBOD)—insufficient carbon in the preanoxic zone reduces denitrification efficiency.
- Denitrification capacity (detention time)—insufficient reactor volume reduces the detention time necessary for denitrification.

Another limiting factor is the capacity of secondary clarifiers. Secondary clarifiers for BNR facilities must be designed conservatively to avoid washout (Water Environment Federation [WEF], 2023).

### 3.2 Tertiary Denitrification Systems

Denitrification can also be accomplished in biofilm reactors, typically applied to tertiary treatment processes downstream of activated sludge systems. These tertiary denitrification systems are commonly configured as biologically active filters (BAF), operating under anoxic conditions. In a BAF, the filter media not only are used to trap total suspended solids (as an effluent polishing filter would) but also serve to support the growth of anoxic biofilm. Denitrification is achieved by maintaining anoxic conditions in the BAF and by provision of a supplemental carbon source to drive denitrification (see Section 4). When operated correctly per the manufacturer's recommendations, BAFs are able to achieve effluent nitrate of less than 1 mg/L.

Biologically active filters are predominantly implemented as vendor-provided systems, each with specific media type and reactor dimensions, such that loading rate criteria and operating characteristics are system-specific. Regardless of the system, however, the same fundamental considerations described above for suspended-growth systems control the ability to denitrify: avoidance of dissolved oxygen in the anoxic reactor, the need for sufficient supplemental carbon, and the denitrifying capacity of the process.

## 4.0 CARBON AUGMENTATION

A carbon source is necessary for denitrification for nitrogen removal to serve as an electron donor. The fundamentals of denitrification (stoichiometry and kinetics) are covered in Section 2.0 of this chapter. In the denitrification process, a readily biodegradable carbon source must be available. These electron donor or carbon requirements can be met by influent wastewater soluble BOD, from natural or enhanced cell decay and lysis, or by the addition of an external supplemental carbon source (Onnis-Hayden & Gu, 2008). This section will discuss when an external or internal carbon source is needed and provide information on various types of carbon sources, their applications, and handling and dosage requirements.

### 4.1 The Need for Carbon Use to Meet Effluent Nitrogen Limits

When total nitrogen, total inorganic nitrogen, and nitrate-nitrogen effluent limits are in place, it is necessary to have a sufficient supply of available carbon for denitrification. This is especially the case when effluent total nitrogen limits are 10 mg/L or less. If effluent total nitrogen limits are 5 mg/L or less, having a source of external supplemental carbon available to consistently meet the limit is typically required.

#### *4.1.1 Internal Carbon Sources*

Internal carbon sources are present in most WRRFs. Carbon is present in raw influent wastewater, and endogenous carbon is present in the postanoxic zone after aerobic treatment. Internal carbon is also available from volatile fatty acids (VFAs) from sludge fermentation or a separate septage stream received at the WRRF.

##### *4.1.1.1 Raw Wastewater and Primary Effluent*

Raw influent wastewater and primary effluent contain BOD and COD that are necessary for denitrification. The readily biodegradable fraction of COD is used for denitrification and typically represents 15% to 25% of the total COD in the influent to the anoxic reactor (Metcalf & Eddy, Inc./AECOM, 2013). Water resource recovery facilities without primary clarifiers generally have higher amounts of carbon present in the unsettled wastewater than facilities with primary clarification. The influent wastewater characteristics of each individual WRRF need to be evaluated to determine the availability of degradable carbon available for denitrification in the wastewater.

Influent or primary effluent BOD or COD will typically need to be 4 to 5 times the total Kjeldahl nitrogen (TKN) for effective denitrification in the primary anoxic zone. When available COD is less than this, denitrification will still take place until the available COD is consumed.

#### 4.1.1.2 Endogenous Carbon

Endogenous carbon results from cell mass decay and is typically used as a carbon source in postanoxic processes in which all of the available influent carbon has been consumed in the aerobic treatment process. Specific denitrification rates are low, that is, from 0.01 to 0.04 g $NO_3$-N/g mixed liquor volatile suspended solids under endogenous respiration (WEF, 2011). Slow SDNRs require a large postanoxic zone to accomplish denitrification.

#### 4.1.1.3 Sludge Fermentation

Volatile fatty acids resulting from sludge fermentation are typically used as a carbon source with biological phosphorus removal to drive biological release and subsequent uptake of soluble phosphorus. Use of VFAs for biological phosphorus is described in Chapter 6.

For nitrogen removal, VFAs can serve as an electron donor in the form of acetic and propionic acids. The amount of soluble COD in fermentate is estimated to be between 400 and 800 mg/L. This is not a high concentration for supplemental carbon addition and, therefore, will require a considerable volume of primary sludge fermentate available for denitrification. Mixed liquor or return activated sludge fermentation is also feasible at facilities that do not have primary clarifiers or facilities that use primary sludge to enhance digester biogas production versus fermentation.

#### 4.1.1.4 Cell Lysis

Cell lysis is a process in which cell walls break down and the cell contents, which contain carbon, are released to the mixed liquor. This process occurs naturally in the endogenous respiration process discussed in Section 4.1.1.2. Cell lysis can also result from an external force (electrical or mechanical) to burst cells and release carbon. Proprietary processes such as pulse electric field, ultrasound disintegration, and hydrodynamic cavitation provide a carbon source for denitrification from cell lysis (Metcalf & Eddy, Inc./AECOM, 2013).

#### 4.1.1.5 Septage

Septage can potentially be a source of carbon for denitrification; however, careful analysis of the septage is required. Along with COD, septage contains high levels of TKN, which can negate the benefit as a carbon source in the septage (WEF, 2023).

### 4.1.2 External Carbon Sources

Sections 4.1.2.1 through 4.1.2.7 describe the most commonly available external carbon sources used for denitrification. Additional information on evaluating external carbon sources is available in Water Environment Research Foundation's *Protocol to Evaluate Alternative External Carbon Sources for Denitrification at Full-Scale Wastewater Treatment Plants* (Gu & Onnis-Hayden, 2010).

#### 4.1.2.1 Methanol

Historically, methanol ($CH_3OH$) has been the most commonly used external carbon source for denitrification. Methanol use in both suspended growth and biofilm denitrification systems has been well documented in the literature. Advantages of methanol include a low COD/N requirement and a low sludge or solids yield. Methanol requires a period of acclimation in a biological process that can vary from a week to several weeks depending on the wastewater temperature. Methanol is used by a specific group of facultative methylotroph organisms, as discussed earlier in this chapter. These organisms are slow growing, and a population must be established for denitrification to occur.

On a unit-cost basis, methanol is often the least costly of the external carbon sources. However, methanol has been subject to historical pricing fluctuations as it is manufactured from natural gas and almost exclusively imported from outside of the United States. Methanol is a flammable liquid and requires specialized pumping, storage, and handling facilities that are regulated under Sections 30, Flammable and Combustible Liquids Code, and 820, Fire Protection in Wastewater Treatment and Collection Facilities, of the National Fire Protection Association codes (NFPA, 2023a, 2023b) (see Section 3.2).

#### 4.1.2.2 Ethanol

Ethanol ($CH_3CH_2OH$) is classified as a flammable liquid and requires the same specialized handling requirements as methanol. Despite the high COD concentration, ethanol is typically not cost effective compared to methanol because of a higher unit cost and higher biomass yield (COD/N). As a result, ethanol use for biological denitrification in wastewater applications is sparse at best. Pure ethanol is regulated by the Alcohol and Tobacco Tax and Trade Bureau; therefore, denatured ethanol is the most suitable option for biological denitrification.

#### 4.1.2.3 Sugar Solutions

Sugar solutions are categorized as carbohydrate-based solutions that may include compounds such as glucose, sucrose, tri and polysaccharides, molasses, and corn sweeteners, in addition to other commodity carbohydrate compounds. Sugar solutions are widely available and nonhazardous, making them suitable options for smaller facilities that are sensitive to safety. Denitrifying organisms grown on carbohydrate solutions result in a higher biomass yield that can create operational challenges, particularly with biofilm denitrification systems. Sugar solutions have a higher freezing point than methanol and are subject to biological degradation over time. To allow for a long shelf life for stability products such as MicroC 1000™, a proprietary carbohydrate solution contains a preservative for long-term stability.

Molasses has been used as a supplemental carbon source; however, it may contain nitrogen and phosphorus that must be evaluated in the product before use to avoid adding additional nutrients to the wastewater that will need to be removed. In some applications in which molasses has been used with UV disinfection, there have been issues in which coating of the surface of UV lamps has been observed, resulting in additional maintenance for cleaning the UV lamps.

#### 4.1.2.4 Acetic Acid Solutions

Acetic acid ($CH_3COOH$) has been used in wastewater treatment applications for both denitrification and biological phosphorous removal. Acetic acid has favorable kinetics and does not require an acclimation period. Commonly available acetic acid solution strengths are 20%, 56%, 80%, and 100%. More dilute solutions are used when concerns about handling hazardous materials need to be minimized. Higher-strength solutions require special handling and solutions greater than 80% may be categorized as flammable liquids. Acetic acid has a high freezing point, and heated storage may be necessary in colder climates. Acetic acid is typically more expensive than other available supplemental carbon sources.

#### 4.1.2.5 Sodium and Potassium Acetate

Acetate salts such as sodium and potassium acetate have the same functional group ($CH_3COO^-$) as acetic acid but are available in a crystalline or powder form that requires being dissolved in water before use. The solubility of these acetate salts in water results in handling and storage challenges, even at moderate temperatures. Sodium and potassium acetate are nonhazardous; however, they are not widely used because of their high cost relative to other carbon sources.

#### 4.1.2.6 Glycerol

Glycerol, or glycerin ($C_3H_8O_3$), has gained popularity as a supplemental carbon source at WRRFs in the last several years for both suspended growth and biofilm processes. Crude glycerin is a byproduct of the biodiesel production process. Unrefined crude glycerin may contain undesirable fats, soaps, biodiesel, and other intermediate and end products of the biodiesel process. Unrefined crude glycerin can also contain high concentrations of methanol, which results in a flammable classification. A number of refined glycerin products are available that have a fixed glycerin concentration (typically 70%) and COD greater than 1 kg/L and that have had methanol and other undesirable products removed from the crude glycerin.

Glycerin provides favorable kinetics, in addition to a low freezing point, and can be used in most climates without freeze protection. With the exception of methanol, glycerin-based products are often the least expensive option when evaluated on a dollar(s)-per-kilogram-of-nitrogen-removed basis. Glycerin does not require an acclimation period or specialized storage and feed equipment.

#### 4.1.2.7 Waste Products

A number of waste products such as soda bottling wastes, acetic acid solutions, brewery wastes, and cheese whey have been used as supplemental carbon sources. These waste streams can provide a local source of carbon that can be used for denitrification. Concerns with any waste product include consistency and type of COD, nitrogen and phosphorus content, availability on a continuous basis, cost, and the potential for toxic compounds in the waste product. A careful analysis should be performed on the waste product from an analytical, pilot-testing, and financial perspective before a product is used full scale in a treatment process.

## 4.2 Supplemental Carbon Handling and Storage Facilities

Supplemental carbon handling and storage facilities are designed for either flammable or nonflammable systems. Nonflammable carbon sources can potentially be used in systems originally designed for flammable liquids. Flammable liquids fall under the requirements of the National Fire Protection Association (NFPA) and cannot be used in nonflammable-rated storage and handling facilities.

### 4.2.1 Regulations

Use of certain external supplemental carbon sources may be regulated by NFPA, the Occupational Safety and Health Administration, and the U.S. Environmental Protection Agency. The regulations and their enforcement are site specific and often fall under the jurisdiction of the state and local fire marshal. In the aftermath of the 2006 methanol fire and explosion at the Bethune Point Wastewater Treatment Plant in Daytona Beach, Florida, the U.S. Chemical Safety and Hazard Investigation Board (2007) issued a report that spelled out specific design changes and training requirements for the use of flammable liquids at WRRFs (NFPA, 2023a, 2023b).

Therefore, the local fire marshal office should be consulted for specific local requirements for bulk liquid storage and handling requirements because they vary by jurisdiction.

#### 4.2.2 Storage Facilities

Bulk storage tanks can be made of carbon steel, stainless steel, fiberglass, and reinforced polyethylene. The type of external carbon or waste product will require analysis by the tank manufacturer to ensure compatibility. Carbon products are also available in intermediate bulk carrier totes and 208-L (55-gal) drums to be used for small use or temporary installations. Applications should include either a double-wall tank or a suitable containment structure in case of a spill or failure of the bulk tank.

#### 4.2.3 Pumping and Conveyance Facilities

Pumping and conveyance systems, like bulk storage tanks, are specific to the carbon source to be used. Generally, nonflammable products can be used in all systems, and flammable products require specialized NFPA-approved systems. Piping systems can range from polyvinyl chloride to stainless steel or carbon steel. Some hazardous flammable products require specialized double-wall containment piping (NFPA, 2023a, 2023b). Chemical metering pumps are used in most applications for product dosing. Common pump types include hose or peristaltic pumps, diaphragm pumps, and gear pumps.

### 4.3 Carbon Dosage Points and Mixing Requirements

Poor mixing and high dissolved oxygen in the anoxic zone or process can result in high carbon use and poor denitrification performance. For any carbon source to be effective, proper dosing and feed point specific to the process are needed. In suspended growth systems in general, the external carbon source needs to feed into a well-mixed point at the head of the process tank. Dissolved oxygen will need to be limited to less than 0.5 mg/L entering the anoxic process to not consume excess supplemental carbon before achieving anoxic conditions for denitrification.

Anoxic tank mixing can be accomplished with a number of different mixing technologies. With all mixing technologies, it is important that the entire anoxic process tank be completely mixed with no dead zones. Mixers should use the minimum mixing energy to completely mix the process tank and not add additional dissolved oxygen to the process.

With biofilm processes, it is necessary to limit influent dissolved oxygen entering the filter process, or excessive supplement carbon will need to be fed to the process to lower dissolved oxygen to anoxic conditions before denitrification can be accomplished. Minimizing dissolved oxygen can be accomplished by limiting turbulent influent pumping conditions to the filter and minimizing the drop into the filter from influent feed weirs.

## 5.0 OPERATIONS SUMMARY

Denitrification processes require good process control and monitoring. Process control and monitoring requirements include maintaining an anoxic environment, ensuring effective nitrification, ensuring a carbon to nitrogen ratio of approximately 5, having instruments and ensuring they are maintained and calibrated, and managing cold and wet weather events.

- Anoxic environment: An anoxic environment is critical. The facultative heterotrophs that are used for denitrification prefer oxygen. When oxygen is not available, they use nitrite or nitrate as their electron acceptor, which produces nitrogen gas and removes it from the water column. Even small concentrations of dissolved oxygen can reduce the efficiency of the denitrification process. Operators need to ensure that the dissolved oxygen is less than 0.3 mg/L. The best way to ensure that concentration is to have either OPR or low-level dissolved oxygen sensors in the anoxic reactors.
- Effective nitrification: To have good nitrogen removal, the operator needs to ensure that nitrification is optimized (see Chapter 3).
- Carbon-to-nitrogen ratio: The denitrifying organisms are heterotrophs and require organic carbon as the electron donor. Most influent wastewater has a sufficient carbon-to-nitrogen ratio; however, if the process has a post-anoxic zone, such as with a four-stage Bardenpho process, the operator may need to add an external source of carbon such as methanol, glycerol, or other products. Samples should be taken at least 3 days per week and tested for COD and nitrogen compounds.
- Instrumentation: Good and well-calibrated instrumentation located in critical spots in the bioreactors help operators ensure an optimized process. The instruments should include the following:
    - dissolved oxygen
    - ORP
    - nitrate
    - ammonia
    - pH
    - solids in the reactors (MLSS)
- Cold and wet weather events: Operators cannot control the weather, but they can help minimize the effects of cold and wet weather events. As the temperature decreases, the growth rate of the microorganisms also decreases. To compensate for that, the operator can increase the MLSS to ensure a higher population of organisms in the reactors. Wet weather can wash out MLSS from the reactors. That results in having a low population of organisms and a reducing in process efficiency, perhaps resulting in permit violations. This is especially a problem in cold weather when the growth rates are low. To compensate for this, operators can shut off the aeration equipment for a brief period at high flow to ensure the MLSS stays in the reactors. The air can be turned on again as soon as the flows subside.

## 6.0 REFERENCES

Akunna, J. C., Bizeau, C., & Moletta, R. (1993). Nitrate and nitrite reductions with anaerobic sludge using various carbon sources: Glucose, glycerol, acetic acid, lactic acid and methanol. *Water Research, 27*(8), 1303–1312.

Anthony, C. (2011). How half a century of research was required to understand bacterial growth on C1 and C2 compounds, the story of the serine cycle and the ethylmalonyl-CoA pathway. *Science Progress, 94*(2), 109–137.

Baytshtok, V., Kim, S., Yu, R., Park, H., & Chandran, K. (2008). Molecular and biokinetic characterization of methylotrophic denitrification using nitrate and nitrite as terminal electron acceptors. *Water Science & Technology, 58*(2), 359–365.

Baytshtok, V., Lu, H. J., Park, H., Kim, S., Yu, R., & Chandran, K. (2009). Impact of varying electron donors on the molecular microbial ecology and biokinetics of methylotrophic denitrifying bacteria. *Biotechnology & Bioengineering, 102*(6), 1527–1536.

Blaszczyk, M., Galka, E., Sakowicz, E., & Mycielski, R. (1985). Denitrification of high concentrations of nitrites and nitrates in synthetic medium with different sources of organic carbon. III. Methanol. *Acta Microbiologica Polonica, 34*(2), 195–205.

Bolzonella, D., Innocenti, L., Pavan, P., & Cecchi, F. (2001). Denitrification potential enhancement by addition of anaerobic fermentation products from the organic fraction of municipal solid waste. *Water Science & Technology, 44*(1), 187–194.

Chistoserdova, L., Chen, S.-W., Lapidus, A., & Lidstrom, M. E. (2003). Methylotrophy in *Methylobacterium extorquens* AM1 from a genomic point of view. *Journal of Bacteriology, 185*(10), 2980–2987.

Chiu, Y. C., & Chung, M. S. (2003). Determination of optimal COD/nitrate ratio for biological denitrification. *International Biodeterioration & Biodegradation, 51*, 43–49.

Christensson, M., Lie, E., & Welander, T. (1994). A comparison between ethanol and methanol as carbon sources for denitrification. *Water Science & Technology, 30*(6), 83–90.

Cookson, J. T. (1995). *Bioremediation engineering: Design and application.* McGraw-Hill.

Dawson, R. N., & Murphy, K. L. (1972). The temperature dependency of biological denitrification. *Water Research, 6*(1), 71–83.

de Vries, G. E, Kües, U., & Stahl, U. (1990). Physiology and genetics of methylotrophic bacteria. *FEMS Microbiology Letters, 75*(1), 57–101.

Du, Y. Q., Liu, S., Onnis-Hayden, A., & Gu, A. Z. (2010). *Investigation of MicroCG$^{TM}$ and Microcglycerin$^{TM}$ a alternative electron donor/carbon sources for enhancing biological nutrient removal in wastewater treatment.* Technical Report prepared for Environmental Operation Solutions, Inc.

Etchebehere, C., Errazquin, I., Barrandeguy, E., Dabert, P., Moletta, R., & Muxi, L. (2001). Evaluation of the denitrifying microbiota of anoxic reactors. *FEMS Microbiology Ecology, 35*(3), 259–265.

Gaudy, A. F. J., & Gaudy, E. T. (1980). *Microbiology for environmental scientists and engineers.* McGraw-Hill.

Gerardi, M. H. (2006). *Wastewater bacteria.* Wiley-Interscience.

Ginige, M. P., Hugenholtz, P., Damis, H., Wagner, M., Keller, J., & Blackall, L. (2004). Use of stable-isotope probing, full-cycle rRNA analysis, and fluorescence in situ hybridization-microautoradiography to study a methanol-fed denitrifying microbial community. *Applied Environmental Microbiology, 70*(1), 588–596.

Ginige, M. P., Keller, J., & Blackall, L. L. (2005). Investigation of an acetate-fed denitrifying microbial community by stable isotope probing, full-cycle rRNA analysis, and fluorescent in situ hybridization-microautoradiography. *Applied Environmental Microbiology, 71*(12), 8683–8691.

Grady, C. P. L., Jr., Daigger, G. T., & Lim, H. C. (1999). *Biological wastewater treatment* (2nd ed.). Marcel Dekker.

Gu, A., & Onnis-Hayden, A. (2010). *Protocol to evaluate alternative external carbon sources for denitrification at full-scale wastewater treatment plants* (WERF Report NUTR1R067b). Water Environment Research Foundation.

Hallin, S., & Pell, M. (1998). Metabolic properties of denitrifying bacteria adapting to methanol and ethanol in activated sludge. *Water Research, 32*(1), 13–18.

Hallin, S., Rothman, M., & Pell, M. (1996). Adaptation of denitrifying bacteria to acetate and methanol in activated sludge. *Water Research, 30*(6), 1445–1450.

Hwang, C., Wu, W. M., Gentry, T., Carley, J., Carroll, S., Schadt, C., Watson, D., Jardine, P., Zhou, J., Hickey, R., Criddle, C., & Fields, M. (2006). Changes in bacterial community structure correlate with initial operating conditions of a field-scale denitrifying fluidized bed reactor. *Applied Microbiology and Biotechnology, 71*(5), 748–760.

Isaacs, S. H., Henze, M., Søeberg, H., & Kümmel, M. (1994). External carbon source addition as a means to control an activated sludge nutrient removal process. *Water Research, 28*(3), 511–520.

Knowles, R. (1982). Denitrification, *Microbiology Review, 46*(1), 43–70.

Kornaros, M., & Lyberatos, G. (1998). Kinetic modelling of *Pseudomonas denitrificans* growth and denitrification under aerobic, anoxic and transient operating conditions. *Water Research, 32*(6), 1912–1922.

Kujawa, K., & Klapwijk, B. (1999). A method to estimate denitrification potential for predenitrification systems using NUR batch test. *Water Research, 33*(10), 2291–2300.

Labbe, N., Laurin, V., Juteau, P., Parent, S., & Villemur, R. (2007). Microbiological community structure of the biofilm of a methanol-fed, marine denitrification system, and identification of the methanol-utilizing microorganisms. *Microbial Ecology, 53*(4), 621–630.

Lemmer, H., Zaglauer, A., Neef, A., Meier, H., & Amann, R. (1997). Denitrification in a methanol-fed fixed-bed reactor. Part 2: Composition and ecology of the bacterial community in the biofilms. *Water Research, 31*(8), 1903–1908.

Li, Y.-M., Li, J., Zheng, G.-H., Luan, J.-F., Fu, Q.-S., & Gu, G.-W. (2008). Effects of the COD/$NO_3$-N ratio and pH on the accumulation of denitrification intermediates with available pyridine as a sole electron donor and carbon source. *Environmental Technology, 29*(12), 1297–1306.

Lu, H., & Chandran, K. (2010). Diagnosis and quantification of glycerol assimilating denitrifying bacteria in an integrated fixed-film activated sludge reactor via $^{13}C$ DNA stable-isotope probing. *Environmental Science & Technology, 44*(23), 8943–8949.

Marazioti, C., Kornaros, M., & Lyberatos, G. (2003). Kinetic modeling of a mixed culture of *Pseudomonas denitrificans* and *Bacillus subtilis* under aerobic and anoxic operating conditions. *Water Research, 37*(6), 1239–1251.

McCarty, P. L., Beck, L., & Amant, P. S. (1969, May 6–8). Biological denitrification of wastewaters by addition of organic materials. *Proceedings of the 24th Industrial Waste Conference (Engineering Extension Series)*, (pp. 1271–1286), Lafayette, IN, United States. Purdue University.

Metcalf & Eddy, Inc./AECOM. (2013). *Wastewater engineering: Treatment and resource recovery* (5th ed.). McGraw-Hill.

Mokhayeri, Y., Nichols, A., & Murtry, S. (2006). Examining the influence of substrates and temperature on maximum specific growth rate of denitrifiers. *Water Science & Technology, 54*(8), 155–162.

Muller, A., Wentzel, M. C., Loewenthal, R. E., & Ekama, G. A. (2003). Heterotroph anoxic yield in anoxic aerobic activated sludge systems treating municipal wastewater. *Water Research, 37*(10), 2435–2441.

Nakatsu, C. H., Carmosini, N., Baldwin, B., Beasley, F., Kourtev, P., & Konopka, A. (2005). Soil microbial community responses to additions of organic carbon substrates and heavy metals (Pb and Cr). *Applied and Environmental Microbiology, 71*(12), 7679–7689.

National Fire Protection Association. (2023a). *NFPA 30: Flammable and combustible liquids code.*

National Fire Protection Association. (2023b). *NFPA 820: Standard for fire protection in wastewater treatment and collection facilities.*

Nurse, G. R. (1980). Denitrification with methanol: Microbiology and biochemistry. *Water Research, 14*(5), 531–537.

Nyberg, U., Aspegren, H., Andersson, B., la Cour Jansen, J., & Villadsen, I. S. (1992). Full-scale application of nitrogen removal with methanol as carbon source. *Water Science & Technology, 26*(5–6), 1077–1086.

O'Keeffe, D. T., & Anthony, C. (1980). The two cytochromes in the facultative methylotroph *Pseudomonas* AM1. *The Journal of Biochemistry, 192*(2), 411–419.

Onnis-Hayden, A., & Gu, A. Z. (2008, October 18–22). Comparisons of organic sources for denitrification: Biodegradability, denitrification rates, kinetic constants and practical implication for their application in WWTPs. *Proceedings of the 81st Annual Water Environment Federation Technical Exhibition and Conference* [CD-ROM], Chicago, IL, United States. Water Environment Federation.

Osaka, T., Yoshie, S., Tsuneda, S., Hirata, A., Iwami, N., & Inamori, Y. (2006). Identification of acetate- or methanol-assimilating bacteria under nitrate-reducing conditions by stable-isotope probing. *Microbial Ecology, 52*(2), 253–266.

Peng, Y.-z., Ma, Y., & Wang, S.-y. (2007). Denitrification potential enhancement by addition of external carbon sources in a pre-denitrification process. *Journal of Environmental Sciences, 19*(3), 284–289.

Purkhold, U., Pommerening-Röser, A., Juretschko, S., Schmid, M. C., Koops, H.-P., & Wagner, M. (2000). Phylogeny of all recognized species of ammonia oxidizers based on comparative 16S rRNA and *amoA* sequence analysis: Implications for molecular diversity surveys. *Applied and Environmental Microbiology, 66*(12), 5368–5382.

Ramalingam, K., Fillos, J., Deur, A., & Beckmann, K. (2007, December 12–13). Specific denitrification rates with alternate external sources of organic carbon. *Proceedings of the 2nd External Carbon Source WERF Workshop,* Washington, DC, United States. Water Environment Research Foundation.

Rittmann, B. E., & McCarty, P. L. (2001). *Environmental biotechnology: Principles and applications.* McGraw-Hill.

Sharma, R., & Gupta, S. K. (2004). Influence of chemical oxygen demand/total Kjeldahl nitrogen ratio and sludge age on nitrification of nitrogenous wastewater. *Water Environment Research, 76*(2), 155–161.

Sobieszuk, P., & Szewczyk, K. W. (2006). Estimation of (C/N) ratio for microbial denitrification. *Environmental Technology, 27*(1), 103–108.

Sperl, G. T., & Hoare, D. S. (1971). Denitrification with methanol: A selective enrichment for *Hyphomicrobium* species. *Journal of Bacteriology, 108*(2), 733–736.

Stensel, H. D., Loehr, R. C., & Lawrence, A. W. (1973). Biological kinetics of suspended-growth denitrification. *Journal Water Pollution Control Federation, 45*(2), 249–261.

Tam, N. F. Y., Leung, G. L. W., & Wong, Y. S. (1994). The effects of external carbon loading on nitrogen removal in sequencing batch reactors. *Water Science & Technology, 30*(6), 73–81.

U.S. Chemical Safety and Hazard Investigation Board. (2007, March). *Methanol tank explosion and fire, Bethune Point Wastewater Treatment Plant, city of Daytona Beach, Florida* (Report No. 2006-03-I-FL).

U.S. Environmental Protection Agency. (1975). *Process design manual for nitrogen control.* U.S. Environmental Protection Agency and the National Environmental Research Center.

Water Environment Federation. (2011). *Nutrient removal* (WEF Manual of Practice No. 34).

Water Environment Federation. (2023). *Standards for design of water resource recovery facilities* (WEF 8-23).

Wild, D., von Schulthess, R., & Gujer, W. (1995). Structured modelling of denitrification intermediates. *Water Science & Technology, 31*(2), 45–54.

# Chapter 6

# Enhanced Biological Phosphorus Removal Fundamentals

## 1.0 INTRODUCTION

This chapter will provide the critical information and knowledge on fundamentals of enhanced biological phosphorus removal (EBPR) process. The chapter primarily covers the microbiology, influencing factors and operational parameters that can provide overall understanding of EBPR process mechanisms and operations. Many of the topics covered in this chapter are also discussed elsewhere in these standards; therefore, the relevant chapters are cross referenced.

## 2.0 PROCESS FUNDAMENTALS

Enhanced biological phosphorus removal relies on the selection and proliferation of a microbial population capable of storing orthophosphate in excess of its biological growth requirements. This group of organisms, referred to as *polyphosphate-accumulating organisms* (PAOs), relies on operational conditions that impose a selective advantage for them while putting other groups at a temporary disadvantage with respect to access to food. Once this is achieved by implementation of special design and operational conditions (e.g., alternative anaerobic and aerobic cycles), PAOs gain the selective advantage to grow and function, resulting in excessive accumulation of orthophosphate in activated sludge. With proper biomass wasting, phosphate removal can then be achieved.

Enhanced biological phosphorus removal research has clearly shown that PAOs are a subset of heterotrophs. Several mechanisms have been proposed to explain the enhanced uptake of phosphorus by microorganisms in wastewater. It has been shown that, for biological phosphorus removal to occur in water resource recovery facilities (WRRFs), biomass first needs to pass through an oxygen- and nitrate-free phase (i.e., an anaerobic phase) before entering a phase in which an electron acceptor is present (i.e., an anoxic phase in which nitrate is present or an aerobic phase in which oxygen is present). The oxygen- and nitrate-free phase can be achieved in a separate reactor, the first section of a plug flow reactor, or a part of a sequencing batch reactor cycle (for schemes for EBPR, refer to Chapter 8).

When wastewater enters the anaerobic phase, the PAOs accumulate carbon sources (volatile fatty acids [VFAs]) as an internal polymer called *polyhydroxyalkanoate* (PHA). The main forms of these PHAs are poly-beta-hydroxybutyrate (PHB) and poly-beta-hydroxyvalerate (PHV). The energy to store this polymer is obtained from the breakdown of glycogen and hydrolysis of polyphosphate bonds. Polyphosphates are formed by a series of high-energy bonds; PAOs can subsequently obtain energy by breaking these bonds. Because polyphosphate is broken down to orthophosphate for energy supply, the phosphate concentration in the anaerobic phase increases. The ability of PAOs to accomplish anaerobic VFA uptake and to store PHA polymers is the main mechanism through which they gain selective advantage in EBPR systems.

The anaerobic phase needs to be followed by an oxygen- or nitrate-rich phase, that is, an anoxic phase (anoxic phosphorus removal) or an aerobic phase (aerobic phosphorus removal). During this phase, the stored PHB is consumed, generating energy for growth and uptake of orthophosphate from the liquid phase and generating energy and carbon for replenishment of the glycogen and polyphosphate pools.

Figure 6.1 presents the process schematic and concentration profiles of the mean measurable components for EBPR operated under anaerobic-aerobic conditions, while Figure 6.2 shows a schematic representation of the significant biochemical processes involved.

Under these conditions, the orthophosphate concentration decreases. Most importantly, because the amount of biomass containing large amounts of polyphosphate is increasing under these conditions (PAOs are able to store up to 15% of their dry weight [Bond et al., 1999]), a net phosphorus removal occurs with the wasted sludge.

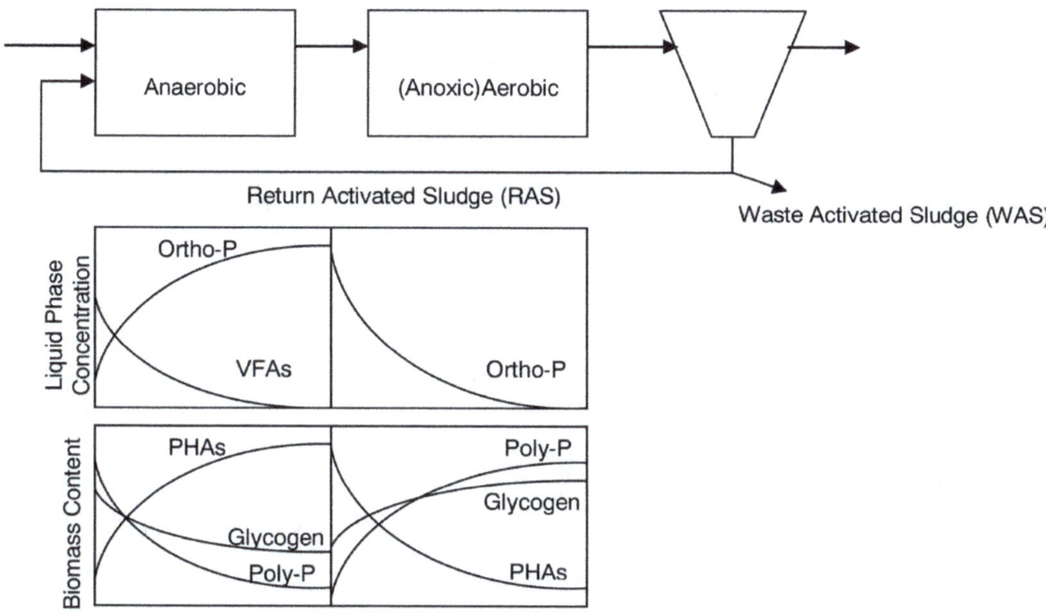


FIGURE 6.1 Typical Concentration Profiles Observed in a Generic EBPR System

Initially, little attention was paid to reactions occurring in the anaerobic phase, and aerobic uptake was thought to occur because of stress caused to the organisms during the different stages of operation. As such, nearly all attention was initially focused on aerobic reactions and optimization of that phase. Considerably more insight has now been gained in the metabolisms underlying the EBPR process, which is presented in Section 3.0 of this chapter.

The anaerobic phase was believed to provide a unique, positive environment for PAOs, enabling them to reserve the necessary amount of carbon to themselves without having to compete with other microorganisms (Matsuo et al., 1992). Later research provided evidence that other organisms, termed *glycogen-accumulating organisms* (GAOs), are also able to store carbon sources and can compete with PAOs in such systems.

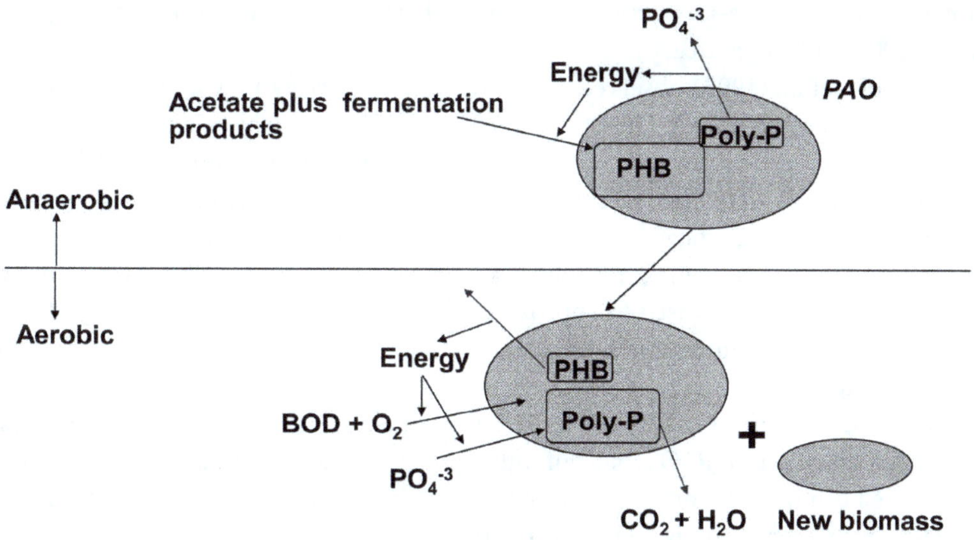


FIGURE 6.2 Schematic Representation of Biological Phosphorus Removal in an Activated Sludge System

It has been assumed that the availability of VFAs, with acetate as the main constituent, is a prerequisite for EBPR. In the absence of these components, fermentation of readily biodegradable carbon sources under anaerobic conditions is necessary. Results of later research also indicated good phosphorus removal with direct use of readily biodegradable material (see Section 4.1).

There is also evidence that EBPR can occur under continuously aerated "aerobic" conditions, in which phosphorus and the carbon source (acetate) are temporarily separated and not added simultaneously (Ahn et al., 2007). This occurs with anaerobic-aerobic EBPR systems.

## 2.1 Aerobic Phosphorus Removal

In the early days of EBPR research, nearly all attention was focused on aerobic processes. A link between anaerobic processes and aerobic processes was not recognized. Aerobic phosphorus, then called "overplus" or "luxury" uptake (Levin & Shapiro, 1965), was supposed to result from stress conditions because of the dynamic feeding of activated sludge facilities. It is now recognized that aerobic processes, use of PHA, replenishment of glycogen, and uptake of phosphorus are linked to anaerobic processes.

## 2.2 Anoxic Phosphorus Removal

Initially, it was believed that PAOs lacked the ability to denitrify and, hence, aerobic conditions were considered essential for the proliferation of these microorganisms for the effectiveness of phosphorus removal. This hypothesis was initially supported by observations that EBPR was negatively affected by nitrate entering the anaerobic phase, that a lower denitrification rate occurred compared to systems with non-PAO heterotroph organisms, and that a net phosphorus release typically occurred rather than an uptake under anoxic conditions (Barker & Dold, 1996). Later interpretations of earlier results and investigations, however, proved that at least a fraction of the PAOs can accumulate phosphate under anoxic conditions (Barker & Dold, 1996; Kuba et al., 1993, 1994). Enhanced biological phosphorus removal can, therefore, take place under anoxic conditions (in the presence of nitrate) as long as no readily biodegradable substrates are present (Oehmen et al., 2007, 2010; Seviour et al., 2003).

The use of nitrate rather than oxygen in PAO sludges is advantageous for several reasons. The availability of organic substrates in wastewater can limit both biological phosphorus removal and denitrification. With anoxic phosphorus removal, the same organics can be used for nitrate and phosphorus removal. This double use of carbon source will result in reduced sludge production; the use of nitrate as an electron acceptor for at least a portion of the phosphate uptake will also reduce oxygen demand (Carvalho et al., 2007; Kuba et al., 1996).

Additionally, Kuba et al. (1993) observed a clear difference in growth yield between anaerobic-aerobic sludges and anaerobic-anoxic ones, that is, 0.35 and 0.25 mg suspended solids/mg carbonaceous oxygen demand (COD), respectively. Using a two-sludge system, that is, denitrifying PAOs (DPAOs) and nitrifiers completely separated in two sludges with recirculation of the nitrified supernatant from the nitrifying stage to the anoxic stage, Kuba et al. (1996) observed the required COD to be up to 50% less than that for conventional aerobic phosphorus and nitrogen removal systems. Oxygen requirements and sludge production decreased to approximately 30% and 50%, respectively. Kuba et al. (1993) also observed a lower sludge volume index value, indicating better sludge settleability compared to anaerobic-aerobic systems.

Experimental evidence indicates that two different populations of PAOs exist in EBPR systems (Kern-Jespersen & Henze, 1993): PAOs that can only use oxygen as a terminal electron acceptor and DPAOs that can use both oxygen and nitrate as terminal electron acceptors. Filipe and Daigger (1999) stated that DPAOs have a disadvantage when competing with PAOs because of a lower thermodynamic efficiency of anoxic growth compared to aerobic growth.

Kuba et al. (1994) and Chuang et al. (1996) provided evidence that DPAOs can use both internally stored PHAs and external substrate for denitrification. When external substrate is used, phosphorus release is observed, whereas phosphorus uptake is expected when internally stored PHAs are degraded. Both mechanisms occur concurrently. According to Chuang et al. (1996), the kinetic competition observed is determined by the polyphosphate content of the microorganisms. With increased polyphosphate content, an increased specific phosphorus release rate is observed. More details on the microorganisms present in anoxic phosphorus removal and the metabolic pathways and mechanisms involved in anoxic phosphorus removal are presented in the following sections.

## 3.0 MICROBIOLOGY OF ENHANCED BIOLOGICAL PHOSPHORUS REMOVAL

There has been interest in the microbiology of the process, and as a result, several reviews on the microbiology of EBPR systems have been published (e.g., Blackall et al., 2002; Forbes et al., 2009; Oehmen et al., 2007; Seviour et al., 2003; and Seviour & McIlroy, 2008). Bacteria that are involved directly or indirectly in the EBPR process are fermentative (acid-producing) bacteria, PAOs, and GAOs.

### 3.1 Fermentative (Acid-Producing) Bacteria

Short-chain VFAs are absorbed by PAOs and stored as PHBs within the cell of these microorganisms. Volatile fatty acids are generated from complex particulate and soluble organic molecules during the fermentation process, which consists of the following three steps: hydrolysis, acidogenesis, and acetogenesis. These three steps are the first of four involved in the anaerobic digestion of complex organics.

Bacteria involved in hydrolysis and fermentation are not well investigated in full-scale EBPR facilities, but recent studies show that hydrolysis seems to be carried out by relatively few and specialized species (Nielsen et al., 2012). Several filamentous bacteria are involved such as *Microthrix*, which produce lipase and consume long-chain fatty acids (Nielsen et al., 2009), *Chloroflexi*, and the epiphytic bacteria *Candidatus epiflobacter*, which produce proteases and consume amino acids (Xia et al., 2007). Nielsen et al. (2012) found that fermenting bacteria constitute a large fraction (20%) of the microbial community in EBPR and are dominated by *Firmicutes* and *Actinobacteria*.

### 3.2 Polyphosphate-Accumulating Organisms

Based on cultivation experiments, *Gammaproteobacteria*, of the genus *Acinetobacter*, were traditionally believed to be the only PAOs. However, it has become clear today that *Acinetobacter* can accumulate polyphosphate but do not possess the aforementioned PAO metabolism. Furthermore, cultivation-independent methods and quantitative fluorescent in situ hybridization (FISH) have demonstrated that the relative abundance of *Acinetobacter* in EBPR systems was dramatically overestimated because of cultivation biases, further confirming that *Acinetobacter* is not important to EBPR.

Clone library analyses followed by FISH instead indicated that *Rhodocyclus*-related *Betaproteobacteria*, given the name *Candidatus Accumulibacter phosphatis*, were more likely candidates (Hesselmann et al., 1999). However, not all clone libraries from EBPR processes have detected these populations, which is likely attributable to difficulties in extracting DNA from such heavily capsulated clustered cells. *Accumulibacter* have frequently shown the previously defined PAO phenotype in both laboratory- and full-scale EBPR systems (Crocetti et al., 2000; Kong et al., 2004).

Combining FISH with 4',6-diamidino-2-phenylindole (DAPI) staining for detection of polyphosphates has revealed that *Accumulibacter* are not the only PAO populations in EBPR systems (He et al.,

2008; Onnis-Hayden et al., 2011; Wong et al., 2005). In fact, evidence shows that *Accumulibacter* and *Beta proteobacteria* often comprise only a relatively small proportion of cells accumulating polyphosphate, especially in full-scale facilities.

Many full-scale EBPR facility communities contain high numbers of polyphosphate staining and as-yet uncultured *Tetrasphaera*-related *Actinobacteria* (Kong et al., 2006); however, this class of bacteria did not assimilate acetate or synthesize PHAs anaerobically (Kong et al., 2005). According to Seviour and McIlroy (2008), any population that accumulates more phosphorus than it requires for growth and that stains positively for polyphosphate should be considered as a putative PAO, regardless of whether it synthesizes PHAs anaerobically or not.

Several studies have indicated that *Accumulibacter* are the significant PAO populations in both denitrifying and conventional EBPR processes (Carvalho et al., 2007; Kong et al., 2004; Zeng et al., 2003), emphasizing the metabolic versatility of these bacteria. However, differences in physiologies among *Accumulibacter* strains may exist. Carvalho et al. (2007), for example, suggested that rod-shaped *Accumulibacter* used nitrate, nitrite, and oxygen as electron acceptors for EBPR, whereas the more common *Accumulibacter* morphotype of large coccal cells used nitrite and oxygen. Further work determined that *Accumulibacter* populations established in a laboratory-scale bioreactor under nitrite-reducing conditions could not reduce nitrate (Guisasola et al., 2009). The differences in nitrate reduction activities suggested finer phenotypic and ecological differences among members of the *Accumulibacter* lineage (Flowers et al., 2009).

## 3.3 Glycogen-Accumulating Organisms

Enhanced biological phosphorus removal facilities often experience upsets, and chemical phosphorus removal is used as a backup to polish their effluents (Seviour et al., 2003). In some instances, external disturbances such as high rainfall, excessive nitrate loading to the anaerobic reactor, or nutrient limitation may explain these process upsets. In other instances, different studies have linked such upsets to the appearance of a bacterial population known generically as *GAOs* (Oehmen et al., 2007; Seviour et al., 2003).

Under anaerobic conditions, GAOs have the ability to assimilate substrates like acetate and, under aerobic conditions, to use these to synthesize intracellular PHAs. Like PAOs, GAOs are thought to metabolize this stored PHA; however, like PAOs, they synthesize intracellular glycogen instead of polyphosphate. Glycogen-accumulating organisms are viewed as potential competitors of PAOs for anaerobic substrate uptake and, thus, are a likely cause of EBPR failure (Cech & Hartman, 1993; Liu, Mino, Nakamura, and Matsuo, 1996; Oehmen et al., 2007; Seviour et al., 2003). Therefore, it is recognized that maintaining conditions favoring the proliferation of PAOs over GAOs is critical to the stability of the EBPR process (Gu et al., 2008).

Because GAOs always seem to be present but are suppressed as a minority in well-functioning EBPR processes (Cech & Hartman, 1993; Matsuo, 1994), GAOs may possibly function as a scavenger for soluble COD and may become the dominating population in deteriorated EBPR processes (Liu, Mino, Nakamura, and Matsuo, 1996). Indeed, as GAOs are able to use different mechanisms for anaerobic uptake of substrates, they will be in a favorable position whenever the internal polyphosphate reserve of PAOs is depleted because of external unfavorable conditions (e.g., nitrate inhibition, unfavorable growth conditions, or unknown factors).

Several factors have been identified that influence competition between PAOs and GAOs. These include influent COD-to-bioavailable-phosphorus ratio (as milligrams per liter of influent COD per milligrams per liter of influent phosphorus), solids retention time (SRT), substrate type, hydraulic retention time (HRT), temperature, pH, dissolved oxygen, and feeding strategy (Filipe et al., 2001a, 2001b; Oehmen, Vives, et al., 2005; Oehmen, Yuan, et al., 2005; Rodrigo et al., 1996; Whang & Park, 2006).

Finally, the precise identity of these GAOs is still largely unknown. This is a significant reason for the difficulty in fully comprehending the role of different GAOs in EBPR system failure.

Polyphosphate-accumulating organisms (Accumilibacter and *Tetrasphaera*), and GAOs (Defluvicoccus, Competibacter, and Propionivibrio) are typically quantified by the FISH imaging method. A mix of general bacterial probes (EUB mix) quantify the total viable cells of which the fraction of PAOs/GAOs are quantified. Figure 6.3 highlights PAOs and GAOs found in EBPR facilities (Water Research Foundation [WRF], 2014).

## 4.0 FACTORS INFLUENCING ENHANCED BIOLOGICAL PHOSPHORUS REMOVAL

In this section, external factors influencing the EBPR process will be discussed. Although initial research mainly focused on aerobic processes, it has gradually become clear that good phosphorus removal activity can only be obtained when anaerobic microorganisms are subjected to conditions favoring storage of

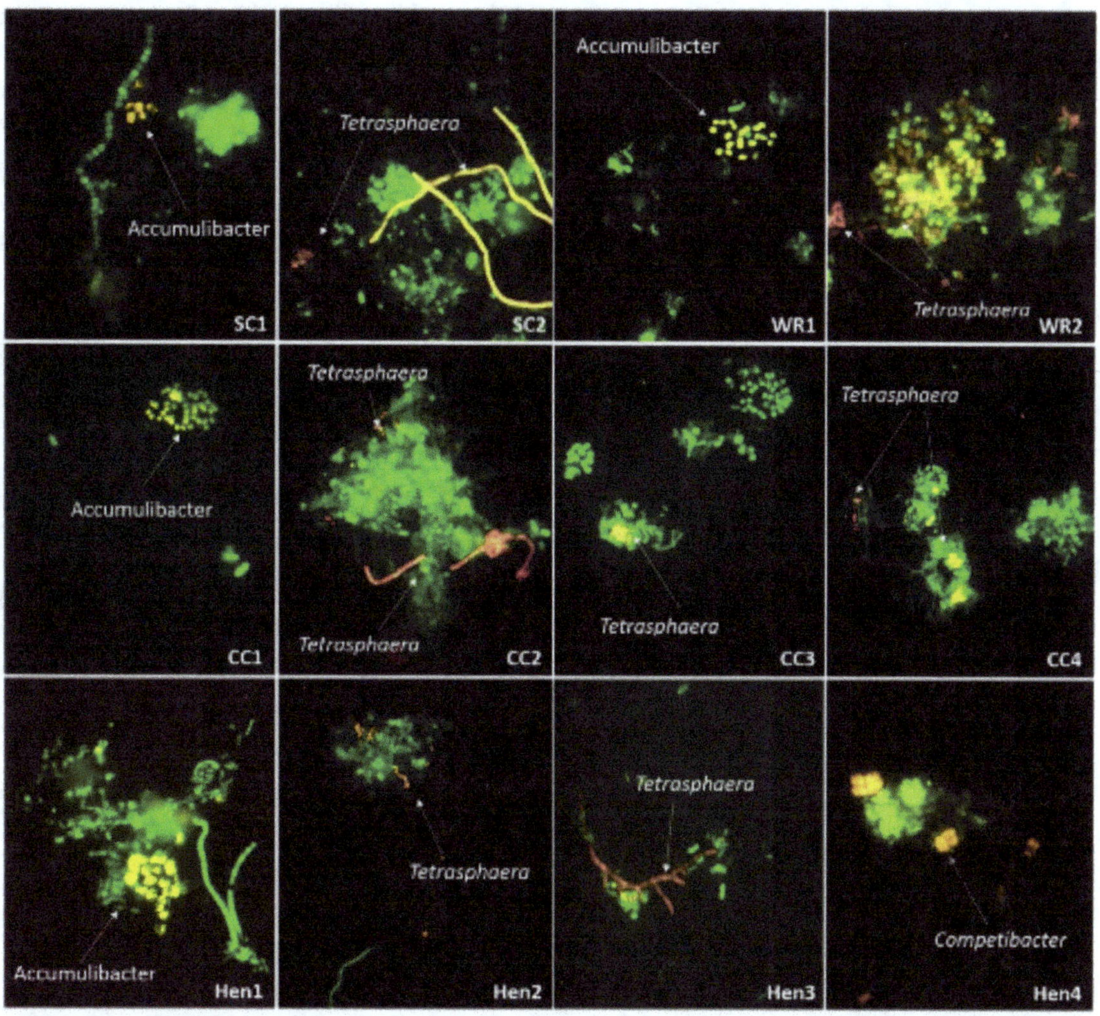


**FIGURE 6.3** FISH Images of PAOs (*Accumilibacter*—orange color, *Tetrasphaera*—orange and red color) and GAOs (*Competibacter*—orange color) Found in EBPR Facilities (green color indicates mix of total viable bacterial cells) (Adapted from WRF, 2014)

sufficient carbon sources to be used under aerobic conditions with simultaneous uptake of orthophosphate. First, attention will be focused on carbon sources used by responsible microorganisms and how these carbon sources influence phosphorus release, formation of PHAs, and the influence on overall EBPR performance. Other external factors that can influence EBPR performance are C/P, secondary release, dissolved oxygen, pH, HRT, SRT, presence of nitrate and nitrite, and temperature.

## 4.1 Short-Chain Volatile Fatty Acids and Non-Short-Chain Volatile Fatty Acids as Carbon Sources

### *4.1.1 Effect of Carbon Source on Phosphorus Release*

Availability of carbon is a primary factor for controlling the EBPR process because the most favorable substrate would cause the optimum release of phosphorus in the anaerobic phase. The EBPR process tends to use substantially more influent COD than the conventional activated sludge process, which is likely because EBPR organisms accumulate storage products that require carbon alone for synthesis (Harper & Jenkins, 2003). The form of COD, or the form of carbon that is available to organisms, is another crucial factor determining the performance of the EBPR process. Carbonaceous oxygen demand must have a sufficient portion of short-chain VFAs, or a readily biodegradable fraction that ferments into VFAs in sewer lines or in fermentation units, to achieve EBPR (Randall et al., 1992). Many laboratory-scale EBPR reactors have been successfully operated with acetate as the significant carbon source. If the retention time of wastewater in collection systems is long enough for fermentation, a significant portion of organic matter in the wastewater will be fermented to soluble VFAs (mainly acetic acid) before being transported to the facility (Mino et al., 1998). Thus, acetate is an important substrate for EBPR. Propionate has also been gaining significant interest in recent studies as a complementary substrate in EBPR; certain long-chain VFAs have also been used. Table 6.1 shows the types of VFAs typically observed in fermented wastewater.

### *4.1.2 Effect of Carbon Source on Storage Products*

When acetate is the only carbon source available, 3HB represents the significant component in the PHA that is formed (Satoh et al., 1992; Smolders et al., 1994); this is then called *PHB*. Experimental studies with PAO- and GAO-enriched sludge have revealed that PAOs produce mainly PHB when fed with acetate, with little PHV production (generally less than 10% [C-mol basis]) (Mino et al., 1998; Satoh et al., 1992; Smolders et al., 1994). As shown in Figure 6.4 (Satoh et al., 1994), the composition of PHAs

TABLE 6.1 Volatile Fatty Acids Typically Found in Fermented Wastewater

| Volatile fatty acid | Chemical formula | Phosphorus uptake/VFA COD consumed |
|---|---|---|
| Acetic acid | $CH_3$-COOH | 0.37 |
| Propionic acid | $CH_3$-$CH_3$-COOH | 0.10 |
| Butyric acid | $CH_3$-$CH_3$-$CH_3$-COOH | 0.12 |
| Isobutyric acid | $CH_3$-$CH_3$-COOH-$CH_3$ | 0.14 |
| Valeric acid | $CH_3$-$CH_3$-$CH_3$-$CH_3$-COOH | 0.15 |
| Isovaleric acid | $CH_3$-$CH_3$-COOH-$CH_3$-$CH_3$ | 0.24 |

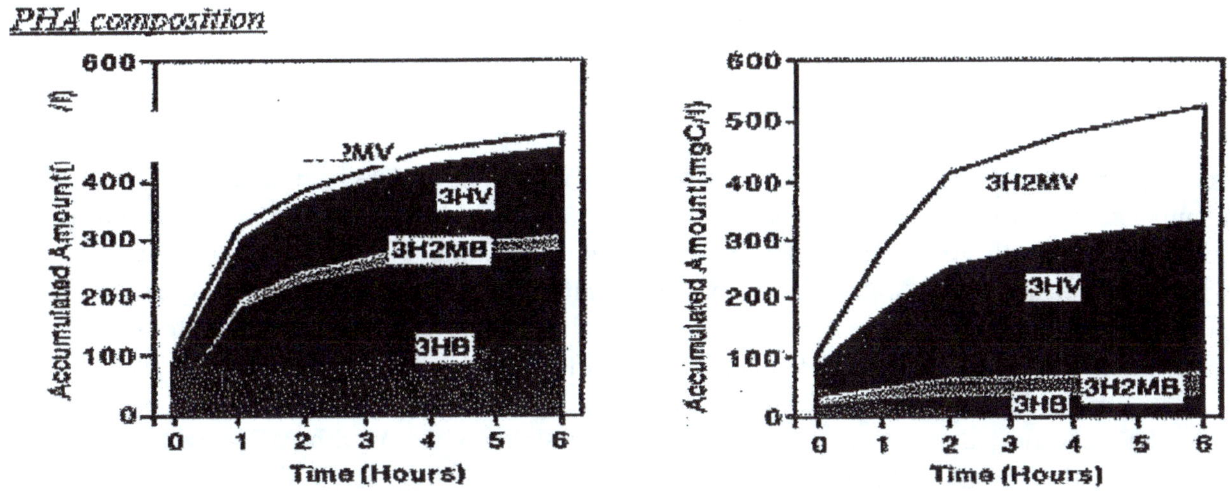


FIGURE 6.4 Polyhydroxyalkanoate Composition for Acetate Update (left) and Propionate Uptake (right)

for both acetate and propionate uptake and the fractions include hydroxy butyrate, hydroxy methyl butyrate, hydroxyvalerate, and hydroxy methyl valerate.

### 4.1.3 *Effect of Carbon Source on Enhanced Biological Phosphorus Removal Performance*

Whereas carbon availability in the form of readily biodegradable COD or VFAs is critical to successful EBPR, carbon type, particularly relative to process stability, is also important (Gebremariam et al., 2011). Choice of substrate or carbon source is critical to EBPR process economics as influenced by the cost of the carbon source and selective use of the carbon source of PAOs against GAOs (Puig et al., 2008). Metabolism of GAOs was first reported by Cech and Hartman (1990), and the main experimental variable in their study, leading to the discovery of GAOs, was feed composition. Glycogen-accumulating organisms were found to be developed in a reactor fed with acetate and glucose, whereas no GAOs were found to be present in the reactor fed with acetate only. This implied that glucose is a preferred substrate for GAOs. Acetate-fed EBPR systems have demonstrated good EBPR performance with PAO enrichment; however, proliferation of GAOs is possible with acetate as the substrate if other operational conditions (e.g., higher COD-to-phosphorus ratio [COD/P] and lower pH) become favorable for GAOs.

Although acetate is the most prevalent VFA that is used as substrate in EBPR facilities, propionate can also be present in substantial quantities (up to 45% of the total) (Burow et al., 2007). Most EBPR studies use acetate as the substrate; however, there has been an increased interest in the effect of propionate on EBPR performance. A number of laboratory studies have used a strategy of alternative feed between acetate and propionate to obtain highly enriched PAO cultures, as those evidences suggest that *Competibacter* GAOs are less competitive with PAOs for propionate uptake compared to acetate uptake alone (Saunders et al., 2007; Zhang et al., 2007; Zhou et al., 2008). Thus, propionate is commonly accepted as an efficient substrate to select against GAOs, especially *Competibacter*. Thomas (2008) reported increased stability of a full-scale facility in Australia when propionate was supplied in an adequate amount relative to acetate. However, this strategy was contradicted by later studies in which some *Alphaproteobacteria* GAOs were observed to be more efficient in propionate uptake compared to acetate uptake (Kong et al., 2006; Oehmen, Yuan, et al., 2005).

## 4.2 Effect of Influent Carbon-to-Phosphorus Ratio

### *4.2.1 Effect of Influent Carbon-to-Phosphorus Ratio on Enhanced Biological Phosphorus Removal Populations*

For the removal of each phosphorus unit, there is a stoichiometric requirement of COD. Above this required carbon loading, at a higher than stoichiometrically required COD-to-phosphorus ratio, an advantage is provided to the non-phosphorus-removing population (GAOs). Previous studies on sequencing batch reactors (SBRs) showed that higher ratios of influent COD to phosphorus create a phosphorus-limiting condition, thus favoring GAO proliferation. Oehmen et al. (2007) stated that PAOs tend to dominate at a COD/P of 10 to 20 mg COD/mg P, whereas GAOs tend to dominate at a COD/P greater than 50 mg COD/mg P. In the intermediate levels of COD/P, both populations can coexist. Gu et al. (2008) postulated that, to a certain extent, this coexistence of GAOs with PAOs might be beneficial to the EBPR process in that GAOs could potentially act as a buffer against higher carbon loading and prevent the system from deteriorating. However, the presence of GAOs also presents the system with such vulnerability that, because of some operational fluctuations, if the overall conditions become favorable to GAOs, the system might collapse. Thus, maintenance of optimum COD/P and good control over operational conditions are required to use the competition for substrate between PAOs and GAOs to provide a positive outcome. Any operational condition that results in a reduction of COD before the biological treatment system must be closely monitored to allow an appropriate COD/P in the EBPR system's feed. Seasonal variations in biochemical oxygen demand (BOD)-to-COD ratios and wastewater VFA content must be closely investigated as part of the wastewater characterization preceding EBPR design and as part of the process control strategy of an operating facility.

### *4.2.2 Effect of Influent Carbon-to-Phosphorus Ratio on Enhanced Biological Phosphorus Removal Performance*

Municipal wastewater fermented in the collection system is generally a good source of VFAs for EBPR operation. Advanced primary treatment practiced at some WRRFs must be looked at carefully because, at facilities where EBPR is going to be implemented, adequate quantities of organic material must be supplied to support PAO functions. In some instances in which sufficient carbon substrate is not available, carbon augmentation of biological reactor influent is practiced. Facility recycles also have significant bearing in EBPR systems operation because they can also contain some VFAs that can be sufficient to ameliorate the phosphorus functions. The influent COD/P or BOD/P (influent COD/P or influent C/P) is crucial for proper design and operation of phosphorus removal systems. Whether a system is limited by COD (or BOD) or phosphorus determines the extent to which PAOs can function, and the amount of excess phosphorus can be taken up from the solution. Previous research has shown that the influent COD/P correlated well with the EBPR biomass total phosphorus content and phosphorus removal functions (Kisoglu et al., 2000; Liu et al., 1997; Schuler & Jenkins, 2003). Because EBPR systems rely on enrichment of PAOs to be able to select the phosphorus accumulators, the system must be fed with at least a minimum amount of phosphorus that can be taken up in correlation with the amount of substrate COD available in the feed. Figures 6.5 and 6.6 show the effect of the influent COD/P on mixed liquor PAO enrichment and sludge phosphorus content.

## 4.3 Secondary Release of Phosphorus

As described in previous sections, release of phosphorus under anaerobic conditions with simultaneous uptake of acetate, and its storage as PHB, is called *primary release* (Barnard, 1984). However,

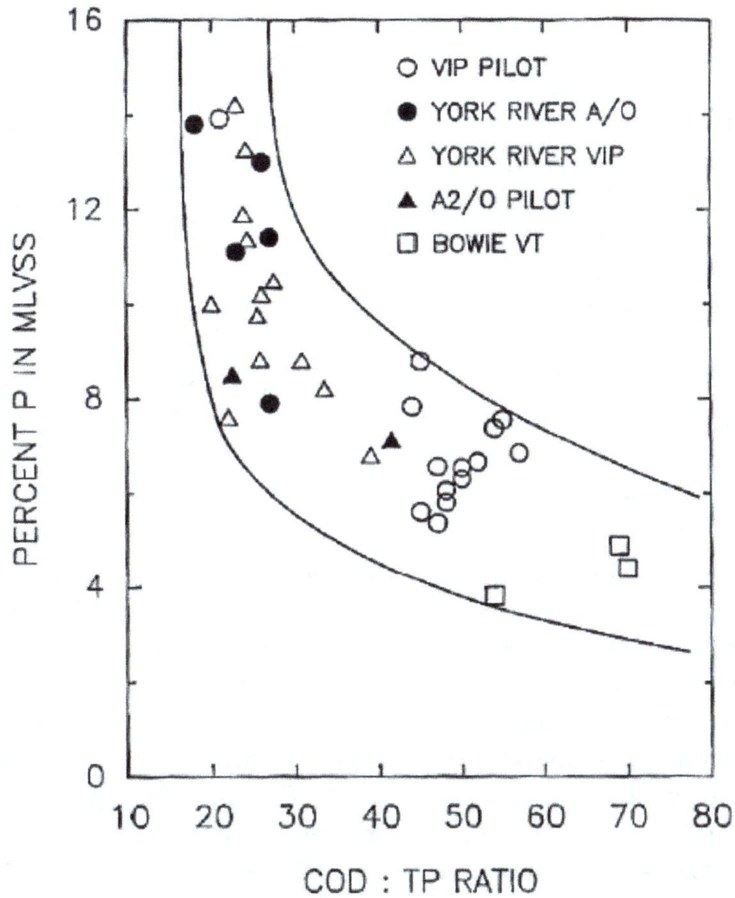


FIGURE 6.5 Effect of COD-to-Total-Phosphorus Ratio on Mixed Liquor PAO Enrichment and Phosphorus Storage at Different EBPR Facilities

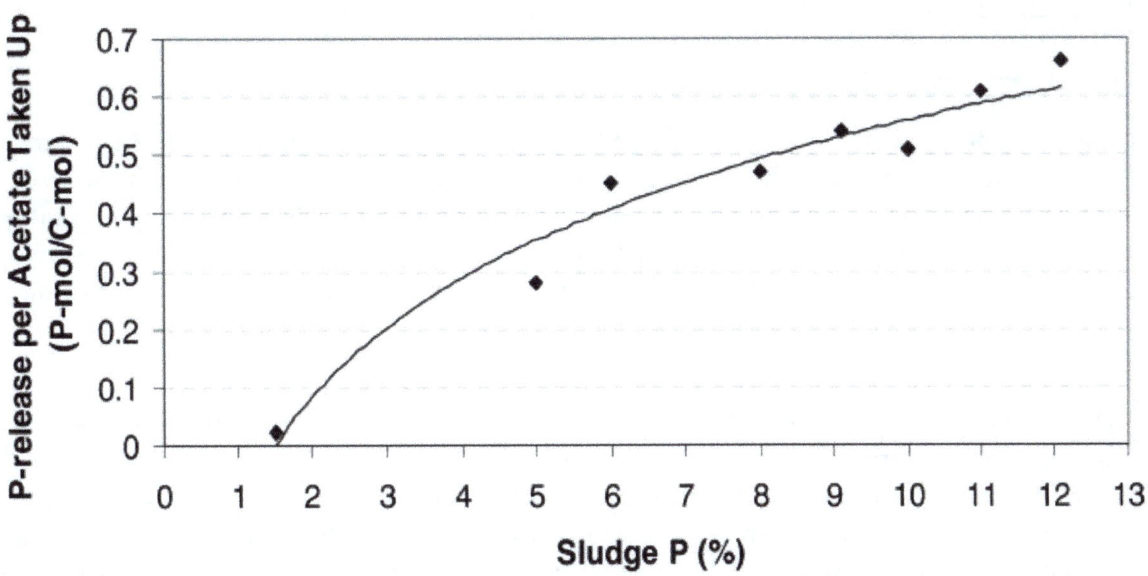


FIGURE 6.6 Using Pilot-Scale EBPR Facilities Fed With VFAs at Various Feed COD-to-Phosphorus Ratios (Data Collected by Liu et al. [1997])

phosphorus is also released from stored polyphosphate under anaerobic conditions even in the absence, or inadequate presence, of VFAs for the microorganisms to derive sufficient energy for maintaining their metabolic activities. This is called *secondary release* of phosphorus (Barnard & Scruggs, 2003). Secondary release takes place at a much slower rate than primary release. Phosphorus resulting from the secondary release mechanism is not removed under a subsequent aerobic or anoxic environment because of the unavailability of stored PHB. Therefore, secondary release should be avoided as much as possible because it adversely affects the performance of an EBPR process.

A common location for secondary release of phosphorus in an EBPR facility is in the secondary clarifier(s). If the sludge blanket in a secondary clarifier is deep, or the sludge is not removed from the clarifier bottom on a regular basis, the resulting anaerobic environment leads to secondary release.

Another possible location is in the upfront anaerobic compartment of an EBPR activated sludge facility. If the anaerobic compartment volume is too large and HRT is much longer than the time needed for fermentation of the incoming COD, then secondary release can take place in the anaerobic compartment. On the other hand, if the compartment volume is too small, then the incoming COD may not have adequate HRT for fermentation to acetic acid, which is a precursor to the formation of PHB as the storage product. In this instance, the stored PHB may not be adequate to supply the energy necessary for subsequent aerobic uptake of orthophosphate over and above that released by the primary mechanism. An inadequate amount of PHB limits the capacity of the PAO cells to uptake all the released orthophosphate in the subsequent aerobic or anoxic compartment.

Organic matter contained in wastewater can get ample opportunity for fermentation when it is conveyed to the facility through a large collection system that provides a long retention time. However, the effectiveness of fermentation depends on a number of factors, including temperature, wastewater characteristics, and characteristics of the sewer system (refer to Section 5.2 for a detailed discussion).

Conversely, fresh wastewater may not be completely fermented if the combined retention time in the collection system and the anaerobic compartment is inadequate for complete fermentation of incoming COD. This was experienced in several facilities in South Africa during the early stages of development of the EBPR process (Uys, R. [1984]. Municipal chemist, Witbank RSA. Personal communication with J. L. Barnard). Such situations may be resolved by providing a prefermenter to process the settled primary solids and adding the resulting VFAs to the anaerobic compartment. Some facilities with inadequate influent VFAs also resort to direct addition of acetic acid in the anaerobic compartment.

## 4.4 Influence of External pH

The pH affects EBPR activity. At a pH lower than 5.4, EBPR activity is completely inhibited, whereas EBPR mechanisms can operate in the pH range of 8.5 to 9, according to *Nutrient Removal* (Water Environment Federation, 2010). Recent research by Filipe et al. (2001a, 2001b) documented an optimum pH range of 7.0 to 7.5 for PAOs, with decreasing activity at lower pH values. In the following subsections, the influence of pH on anaerobic and aerobic reactions and on the overall EBPR process is presented.

### *4.4.1 Effect of pH on Anaerobic Reactions*

Smolders et al. (1995) observed a direct linear relationship (fPO4, rel = 0.18 pH – 0.81, in gP/gCOD units) between pH measured in the bulk liquid and phosphorus release to the acetate-uptake ratio (fPO4, rel) in EBPR systems. In their study, the phosphorus release acetate-uptake ratio was found to increase from 0.25 to 0.75 mol-P/mol-C when pH increased from 5.5 to 8.2. However, whereas an increase in phosphorus release and the phosphorus release rate with increasing pH was observed, the acetate-uptake rate showed no pH dependency. In addition, higher energy demand is required for acetate uptake at higher pH levels.

In reality, a pH of less than 6 should never be allowed to occur in a biological nutrient removal (BNR) facility. Liu, Mino, Matsuo, and Nakamura (1996) observed a more complex pH dependency of the acetate-uptake rate. Between pH 5 and 6.5, the acetate-uptake rate increased linearly from 0 to approximately 50 mg C/g volatile suspended solids (VSS)/h. Between pH 6.5 and 8, no pH dependency was observed, which was in accordance with Smolders et al. (1994). Above pH 8, the acetate-uptake rate started to decrease. Release rates significantly lower than those reported by Smolders et al. (1994) were observed for orthophosphate. Between pH 5.0 and 6.5, the orthophosphate release rate ranged from 20 to approximately 50 mg P/g VSS/h; higher than pH 6.5, this value continuously increased. From pH 8.5 and higher, a decrease in orthophosphate release rate was observed .

More recent studies have shown that the rate of acetate uptake is independent of pH between pH 6.5 and 8, which is the case with PAOs (in accordance with both Smolders et al. [1994] and Liu, Mino, Matsuo, and Nakamura [1996]) but strongly dependent on pH for GAOs. The rate of acetate uptake by GAOs is significantly decreased when the pH of the medium is increased (Filipe et al., 2001a, 2001b).

### 4.4.2 Effect of pH on Aerobic Reactions

Aerobically, a series of batch tests has shown that phosphorus uptake, PHA use, and biomass growth were all inhibited by a low pH (6.5). The batch tests suggested that a higher aerobic pH (7 to 7.5) would be more beneficial for PAOs (Filipe et al., 2001a).

### 4.4.3 Effect of pH on Enhanced Biological Phosphorus Removal Performance

Lower than pH 6.9, the EBPR process has been shown to decline in efficiency. This may be attributable to competition with GAOs. Filipe et al. (2001b) found that GAOs grow faster than PAOs at a pH of less than 7.25. This is attributable to the rate at which PAOs and GAOs can take up VFAs under anaerobic conditions; that is, GAOs take up acetate faster at a pH lower than 7.25, whereas PAOs are faster at a pH higher than this value (Filipe et al., 2001b). At higher pH, the energy needed for acetate uptake is higher; therefore, PAOs, having an extra energy source in polyphosphorus compared to GAOs, can use it to meet this higher energy demand and successfully outcompete GAOs for VFAs (Filipe et al., 2001b).

Several studies have, indeed, shown higher phosphorus removal when the anaerobic and/or aerobic pH level was increased (from pH 7 to 7.5 to 8.5) (Schuler & Jenkins, 2002; Serafim et al., 2002). The reason for the improved performance was hypothesized to be from a shift in microbial competition from GAOs to PAOs. This hypothesis has been supported by assessing population changes in the microbial community. In a study by Zhang et al. (2005), a complete loss of phosphate-removing organisms was observed when the pH was decreased from 7.0 to 6.5; at the same time, a shift in the microbial population was also observed. With the decrease in pH, the phosphorus removal efficiency went from 99.9% of phosphorus to 17% of phosphorus in just 14 days after the pH was lowered to pH 6.5. Clearly, the results show that pH strongly influences PAO-GAO competition and suggest pH as a performance parameter that can be manipulated to favor PAOs to optimize phosphorus removal.

In some studies, a decrease in the VFA uptake, phosphorus release, and phosphorus-uptake rates has been observed at pH values above 8.0 (Liu, Mino, Matsuo, & Nakamura, 1996; Oehmen, Vives, et al., 2005; Schuler & Jenkins, 2002). Therefore, an upper limit for pH should also be considered. It should also be noted that the fraction of phosphorus removed via biologically induced chemical precipitation increases with increasing pH (Maurer et al., 1999).

## 4.5 Influence of Oxidation-Reduction Potential

Use of oxidation-reduction potential (ORP) to determine the aerobic or anaerobic state of wastewater has found widespread application. Measurement of ORP in a wastewater environment is not precise because of the limitation of the electrolytic cells to respond completely to most of the species undergoing biochemical oxidation and reduction reactions. Because of this uncertainty in obtaining precise values, ranges of ORP values have been established that are widely used in wastewater treatment practice as indicators of many important microbiological processes, including EBPR. Typically, ORP values of –50 to –200 mV are accepted as the regime for anaerobic polyphosphate breakdown, whereas +50 to +150 mV is considered as the range for aerobic phosphorus uptake.

## 4.6 Dissolved Oxygen

Few studies have tried to quantify the effect of dissolved oxygen on EBPR processes. Brdjanovic, van Loosdrecht, Hooijmans, Mino, et al. (1998) demonstrated that when dissolved oxygen is too high, phosphorus uptake stops because of a gradual depletion of PHB. If organic substrate is introduced to the system, phosphorus release takes place; however, the released phosphorus cannot be taken up fully again because the PHB content limits the uptake rate. The authors suggest that this phenomenon might explain the observed deterioration of EBPR activity after heavy rainfall or on weekends. Griffiths et al. (2002) also hypothesized that the dissolved oxygen concentration affects competition between PAOs and GAOs, therefore affecting EBPR performance. In this study, the dissolved oxygen concentration was adjusted in numerous full-scale WRRFs, and changes in process performance were identified. The abundance of PAOs and tetrad-forming organisms (TFOs) in the sludge were assessed using staining techniques. Poor phosphorus removal performance and a high number of TFOs were observed more frequently at high dissolved oxygen concentrations of 4.5 to 5.0 mg/L, whereas dissolved oxygen concentrations of approximately 2.5 to 3.0 mg/L seemed to correlate with a greater abundance of PAOs. When a stream containing high dissolved oxygen (i.e., 5 mg/L) is returned from the aerobic zone to the anaerobic zone, it may adversely affect EBPR performance regardless of PAO and GAO competition.

Further study is necessary to clarify the effect of dissolved oxygen concentration on PAO-GAO competition. This is increasingly relevant because there is a general trend today to operate full-scale BNR facilities at lower dissolved oxygen concentrations (0.5 to 1.5 mg/L) to both reduce aeration costs and improve nitrogen removal through simultaneous nitrification and denitrification.

## 4.7 Temperature

Effects of temperature on the efficiency and kinetics of EBPR systems have been investigated for the past few decades, with studies yielding contradictory results. Early researchers such as Barnard et al. (1985) and Ekama et al. (1984) reported that EBPR efficiency was unchanged over the range of 5 to 24 °C. Other studies by Beatons et al. (1999) and Choi et al. (1998) showed that cold temperatures adversely affect EBPR performance.

Contradictory to previous findings, Helmer and Kunst (1997) and Erdal (2002) reported that, despite slowing reaction rates, EBPR performance can be significantly greater at 5 °C compared to 20 °C. This shows that better system performance can be achieved as a result of reduced competition for substrate in the anaerobic zones and increased population of PAOs. The phosphorus content of EBPR biomass achieved by Erdal (2002) was up to 50% VSS in the end of the aerobic zone. Figure 6.7 illustrates the importance of acclimation and the resulting improved cold temperature operation of an EBPR system.

Good EBPR performance can be achieved as long as total SRT values of 16 and 12 days are provided for 5 and 10 °C, respectively. System performance was not affected for SRTs between 16 and 24 days and 12 and 17 days for 5 and 10 °C, respectively. Low temperatures can also lower phosphorus uptake, although this is not an issue in well-operated and properly acclimatized facilities.

High SRT operations increase endogenous glycogen use, thereby consuming the available reducing power used for PHA formation in anaerobic stages. Glycogen metabolism was found to be the most

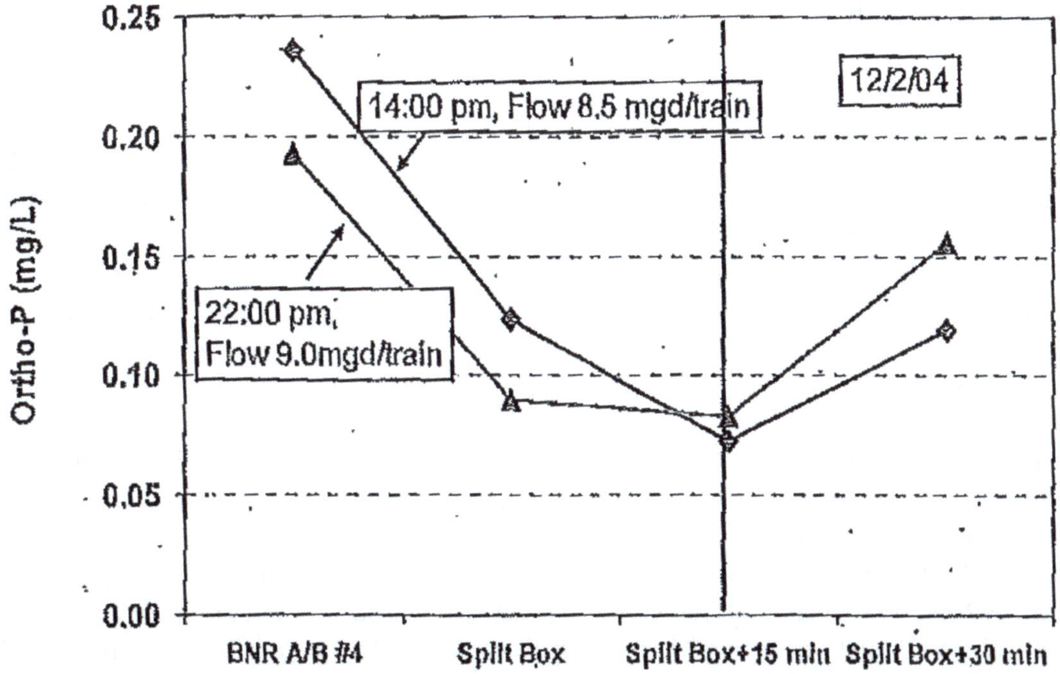


FIGURE 6.7 Residual Orthophosphate Measured at a BNR Basin, Downstream Split Box, and After It Was Further Aerated for 30 Minutes via Laboratory Aeration Testing (Gu et al., 2006)

rate-limiting step in EBPR biochemistry at temperatures below 15 °C. The pilot EBPR systems removed phosphorus until the complete shutdown of glycogen use and replenishment was observed. Despite the presence of available energy sources (polyphosphate and PHA), shutdown of the glycogen metabolism was the significant reason for washout to occur. The shutdown of glycogen use through the anaerobic stage in washout SRTs prevented acetate use and PHA formation. Whereas PAOs wash out of the system, ordinary heterotrophs can continue to grow, using the acetate passing through the anaerobic stages unconsumed and into aerobic stages.

High temperatures adversely affect phosphorus removal because of the increasing effect of GAOs. Modeling studies have shown that GAOs can predominate at higher temperatures because of their increased ability to uptake acetate at those temperatures compared to PAOs (Whang & Park, 2002). At temperatures greater than 28 °C, phosphorus removal will generally be impaired, apparently by the predominance of GAOs.

Li et al. (2010) investigated temperature effects on intracellular absorption and extracellular phosphorus removal by extracellular polymeric substances (EPS) in EBPR processes with the help of a laboratory-scale SBR. Studies conducted at temperatures of 5, 15, and 25 °C revealed that lower temperatures were favorable to removal by EPS, primarily because of precipitation as magnesium phosphate, the maximum of which occurred at 5 °C. Maximum intracellular absorption was observed at 15 °C.

Based on the aforementioned studies, a general conclusion can be made that higher temperatures are detrimental to EBPR performance, primarily because of a shift in population from PAOs to GAOs. The EBPR process is favored by, or is indifferent to, low temperatures down to 5 °C.

## 4.8 Solids Retention Time

Solids retention time is one parameter that can influence EBPR performance and the competition between PAOs and GAOs; however, only a few studies have explored the relationship between SRT and EBPR performance.

It has been demonstrated that an increase in SRT can lead to the decrease of biomass yield and excess sludge discharged, which reduces the phosphorus removed by discharging excess sludge (U.S. Environmental Protection Agency [U.S. EPA], 1987). However, Randall et al. (1992) found that phosphorus content in biomass increases but that phosphorus removal efficiency does not change as SRT increases. Tremblay et al. (1999) also addressed the fact that PAOs take predominant roles in EBPR systems at a long SRT. Additionally, Barnard (1983) reported that SRT plays a smaller role in phosphorus removal in practice than is expected in EBPR. Mamais and Jenkins (1992) showed that there is a washout SRT for all temperatures higher than 10 to 30 °C.

Erdal et al. (2003, 2006) investigated mechanisms leading to washout or cessation of EBPR activity before other heterotrophic functions halt. After examining underlying biochemical methods, they showed that the main effect of system SRT in EBPR systems is on PHA and glycogen polymerization reactions.

A more recent study by Li et al. (2008) reported performance degradation and worse sludge settleability when the SRT was increased from 8 to 16 days. In summary, there are still a lot of contradictions about the effect of SRT on EBPR performance.

There are also few studies that have tried to determine the effect of SRT on the competition between PAOs and GAOs. Seviour et al. (2003) reported that GAOs could successfully compete with PAOs at a long SRT, which resulted in the decrease of phosphorus removal in the EBPR system. Rodrigo et al. (1999) concluded that shorter SRTs are beneficial for PAOs after observing that the EBPR biomass activity decreased as the SRT was extended, suggesting that GAOs may tend to dominate at longer SRTs.

In an acetate-fed laboratory-scale reactor operated at 30 °C and pH 7.5, Whang and Park (2006) observed a switch in the dominant microbial population from an enriched-GAO to an enriched-PAO culture when lowering the applied SRT from 10 to 3 days. Through a model-based analysis, Whang et al. (2007) inferred that, under the operating conditions applied by Whang and Park (2006), GAOs had a lower net biomass growth rate than PAOs and, therefore, were outcompeted after the SRT was shortened. However, those studies do not provide further details about the effect of SRT on the population dynamics and microbial and biochemical mechanisms involved. Moreover, considering the temperature of 30 °C applied by Whang and Park (2006), their observations may not represent the scenario at ambient temperatures similar to those at full-scale facilities. Although studies regarding the minimum anaerobic and aerobic SRTs of PAOs are available in literature (Brdjanovic, Van Loosdrecht, Hooijmans, Alaerts, et al., 1998; Matsuo, 1994), no data concerning the effect of SRT on GAO cultures have been reported. However, Onnis-Hayden et al. (2011) recently reported the low abundance and absence of the most commonly found GAOs (*Competibacter*-type) in a full-scale system operating at an SRT of less than 4 days.

## 4.9 Hydraulic Retention Time

Both anaerobic and aerobic retention times affect phosphorus uptake and storage by PAOs. Sufficient time should be allowed for formation of the VFA and its storage as PHA, with simultaneous release of orthophosphate. If the anaerobic retention time is too short, the phosphorus uptake in the aerobic zone will be lower than achievable because the stored PHA in the anaerobic zone is insufficient. The effects of HRT in both anaerobic and aerobic zones are discussed in the following subsections.

### *4.9.1 Anaerobic Hydraulic Retention Time*

Excessive retention in an anaerobic environment gives rise to secondary release of phosphorus. Optimum anaerobic HRT is typically between 0.25 and 1 hour (Barnard & Fothergill, 1998). The hypothesis of secondary release and the effect of anaerobic HRT on this phenomenon were supported by the findings of Stephens and Stensel (1998) based on results of bench-scale SBR tests. However, results obtained from a recent study by Coats et al. (2011) contradicted the earlier findings. The authors conducted tests on bench-scale SBRs consisting of feed, anaerobic, aerobic, settle, and decant steps in each operating cycle. The wastewater was obtained from the influent of a Moscow, Idaho, WRRF. Settled primary sludge was fermented externally to feed the SBR reactors as the source of VFA. The feed consisted of 90% wastewater and 10% fermented liquor. Three reactors were operated with an overall HRT of 12 hours, with an anaerobic HRT of 1, 2, and 3 hours. Coats et al. (2011) observed that the rate of phosphorus release decreased with increasing anaerobic HRT, with the reactor having 3-hour anaerobic HRT exhibiting approximately half of the rate of the reactor with 1-hour HRT. During the subsequent aerobic step, however, complete removal of phosphorus was achieved by each of the three SBRs. Interestingly, the phosphorus uptake by the reactors with 2- and 3-hour anaerobic HRT were faster (less than 30 minutes) compared to the one with 1-hour anaerobic HRT (more than 30 minutes). Each reactor completely consumed the VFAs and demonstrated no secondary release of phosphorus.

Based on this study, the authors concluded that long anaerobic HRT does not necessarily adversely affect EBPR performance. Additionally, longer anaerobic HRT enriches the concentration of PAOs in the microbial population. These results contradict observations made by earlier researchers and, therefore, warrant further study.

### 4.9.2 Aerobic Hydraulic Retention Time

Enhanced biological phosphorus removal performance is a function of aerobic HRT. Although long aeration time facilitates uptake of orthophosphate by PAOs, excessive aeration time can be detrimental because it induces secondary release (Stephens & Stensel, 1998). In a full-scale study conducted at the Las Vegas Water Pollution Control Facility (Las Vegas, Nevada), Gu et al. (2006) investigated the effect of HRT on soluble phosphorus concentration in the treated effluent from the BNR trains. The authors observed significantly lower (approximately half, or less) phosphorus concentration at the secondary clarifier effluent compared to the aeration basin effluent. By extending the aerobic HRT by 15 minutes downstream of the secondary clarifier splitter box, an even lower orthophosphate concentration (less than 0.10 mg/L) could be achieved in the final effluent. However, further extension of the aerobic HRT by 30 minutes resulted in a rise of orthophosphate concentration because of the onset of secondary release (Figure 6.7). The extended aeration condition for the secondary clarifier was simulated by aerating the clarifier splitter box samples by 15 and 30 minutes, respectively.

In a bench-scale experimental SBR study, Stephens and Stensel (1998) reported that, after extending the aeration step from 3 hours to 7 hours in a given cycle, phosphorus release and phosphorus uptake were not adversely affected during the same cycle. However, the rate of phosphorus release in the anaerobic step and the rate of uptake in the aerobic step in the subsequent cycle were much slower.

Extending the aerobic HRT beyond a critical point induces secondary release as a result of endogenous decay and cell lysis. Both anaerobic and aerobic HRT affect EBPR performance. Neethling et al. (2005) observed that a ratio of 3 to 4 for aerobic to anaerobic HRT led to optimal EBPR performance.

## 4.10 Nitrate

### 4.10.1 Decrease of Enhanced Biological Phosphorus Removal Performance

Poor phosphorus removal under nitrate-rich conditions in the anaerobic zone has been attributed to the disruption of anaerobic conditions by nitrate (Barnard, 1976), consumption of fatty acids by denitrifying non-polyphosphate heterotrophs (Barker & Dold, 1996), and the inhibition of PAOs by nitrite as a result of incomplete denitrification (Jiang et al., 2006; Zhou et al., 2007). Previous studies indicated that there are two types of PAOs based on their ability to use nitrate and/or nitrite as electron acceptors; these are referred to as *DPAOs* as opposed to aerobic PAOs (non-DPAOs) (Kern-Jespersen & Henze, 1993; Kuba et al., 1993). It was suggested that incorporation of an anoxic zone in EBPR can take advantage of DPAOs that can use the intracellular PHA's storage for simultaneous removal of nitrogen and phosphorus, thereby leading to reduced aeration requirements and sludge production of the system (Kuba et al., 1993). However, little is currently known about the factors that govern their abundance and functions.

### 4.10.2 Proliferation of Denitrifying Polyphosphate-Accumulating Organisms

There is controversy regarding whether DPAOs are different from conventional aerobic PAOs or if DPAOs and PAOs are the same organisms that have the capability to induce different enzymes under different electron acceptor conditions (Gebremariam et al., 2011). A number of researchers have hypothesized the former based on observations of lower phosphorus uptake under anoxic versus aerobic conditions (Freitas et al., 2005; Kern-Jespersen & Henze, 1993; Kuba et al., 1993; Shoji et al., 2003; Smolders et al., 1995). A number of earlier studies, through a comparison of EBPR activities (e.g., phosphate-uptake rates) and the abundance of *Accumulibacter*-like PAOs (quantified via FISH methods), indicated that only a fraction of *Accumulibacter* PAOs are able to use nitrate (DPAO phenotype) and that there are likely two different subgroups of PAOs possessing different denitrification capabilities (Kong et al., 2004; Zeng et al., 2003). He et al. (2007) investigated the phylogeny of *Accumulibacter* PAOs with finer-resolution FISH probes and identified different subclusters within the *Accumulibacter* group (differentiated by 16s rRNA nucleotides). A recent study from the same group (Flowers et al., 2009) demonstrated that different subgroups within the *Accumulibacter* PAOs (Clade IA and IIA) have different nitrate reduction capabilities, which was later confirmed by Oehmen et al. (2010). Additional evidence supporting the hypothesis that there are two separate subgroups of *Accumulibacter* that possess different abilities to use nitrate and nitrite as the electron acceptor is the metagenomic study by Martin et al. (2006), which revealed that the particular *Accumulibacter*-like PAOs (*Accumulibacter* Clade IIA) enriched in their laboratory-scale acetate-fed EBPR reactor lacked the gene that encodes for the nitrate reductase enzyme.

Factors that affect the enrichment of DPAOs versus non-DPAOs are not fully understood. Carvalho et al. (2007) and Oehmen et al. (2010) showed that carbon source type may affect the enrichment of DPAOs versus non-DPAOs because they observed different denitrifying performance and identified different *Accumulibacter* clades within two SBR-EBPR systems fed with acetate and propionate as carbon sources, respectively. Another factor that was suggested to possibly affect DPAO activities is the level of carbon source in the anoxic condition, which affects the kinetic competition between heterotrophic denitrifiers and denitrifying PAOs under anoxic conditions (Ahn et al., 2002; Chuang et al., 1996).

## 4.11 Septage and Sidestreams

Sidestreams consist of return streams from thickeners, sludge tanks (digested and undigested), and dewatering devices. Among these, streams returning after digestion of sludge are the most concentrated in nutrients. Approximately 60% of phosphorus contained in the biomass is released as orthophosphate upon anaerobic digestion (Phillips et al., 2006). In EBPR facilities, where the waste activated sludge (WAS) contains high concentrations of stored phosphorus, these streams can have high phosphorus concentrations. The characteristics of these streams are highly variable (City College of New York, 1994; U.S. EPA, 1994). When recycled and blended with influent for treatment as a combined stream, the resulting loads can upset EBPR performance because of altered carbon-to-phosphorus ratios in the influent to the bioreactor system, as described in Section 4.2.

Because of this high variability and the strength of the streams, it is important to control the introduction of these streams in the facility. This can be addressed through an equalization tank, which helps dampen the large fluctuations in concentrations and minimizes any adverse effect on the treatment process, or by treating the combined recycle stream separately to remove phosphorus by a separate sidestream treatment before returning it to the facility. In many instances, sidestreams from several facilities in a region are collected and transported to a larger facility capable of sidestream treatment.

The reduction of phosphorus in a separate sidestream facility in not generally recommended for centrate streams from anaerobically digested processes (Phillips et al., 2006). This is primarily attributable to low concentrations of readily biodegradable chemical oxygen demand (rbCOD) in anaerobically digested sludge because most VFAs are converted to methane in the digester. The VFA concentration of digested sludge centrate may be inadequate to support phosphorus removal by a separate sidestream EBPR process. Chemical precipitation with the help of a ferric chloride ($FeCl_3$) or alum [$Al_2(SO_4)_3$, $xH_2O$] solution is the preferred option if a separate sidestream treatment is planned for return streams containing large fractions of centrate. Another option is removal of phosphorus as struvite ($Mg/NH_4{\cdot}PO_4$, $6H_2O$) using processes such as Ostara. Chapter 10 provides a detailed discussion of chemical precipitation.

### 4.12 Recalcitrant Phosphorus

Dissolved acid hydrolyzable fractions and organic phosphorus fractions of influent phosphorus can adversely affect the effluent soluble phosphorus concentration. Pilot studies at the Westerly Wastewater Treatment Facility in Marlborough, Massachusetts (Lancaster & Madden, 2008), determined that only 35% to 39% of these phosphorus species are removed by EBPR. The remaining part is biologically recalcitrant, which also did not respond to any commercially available physicochemical tertiary treatment process. Based on these findings, the authors recommended that the presence of these species of phosphorus in the influent should be considered while setting practical limits of phosphorus in the treated effluent.

## 5.0 FERMENTATION

Fermentation of complex organics to short-chain VFA end products provides a key input to EPBR. Effective EPBR requires VFA concentrations as acetic acid that are 5 to 7 times the concentration of the influent phosphorus (as phosphorus). The following subsections summarize critical considerations that must be taken into account to ensure that these fermentation products are available to the EPBR process.

### 5.1 Fermentation Process Fundamentals

As stated previously, short-chain VFAs are absorbed by PAOs and stored as PHBs within the cell of these microorganisms. Volatile fatty acids are generated from complex particulate and soluble organic molecules during the fermentation process, which consists of the following three steps: hydrolysis, acidogenesis, and acetogenesis. A schematic of the various processes involved in anaerobic metabolism is shown in Figure 6.8.

#### *5.1.1 Hydrolysis*

Hydrolysis is an enzyme-mediated reaction in which a chemical bond of the complex organic molecule is cleaved and ionized water ($H^+$ and $OH^-$) is inserted to the two ends of the severed bond. Through

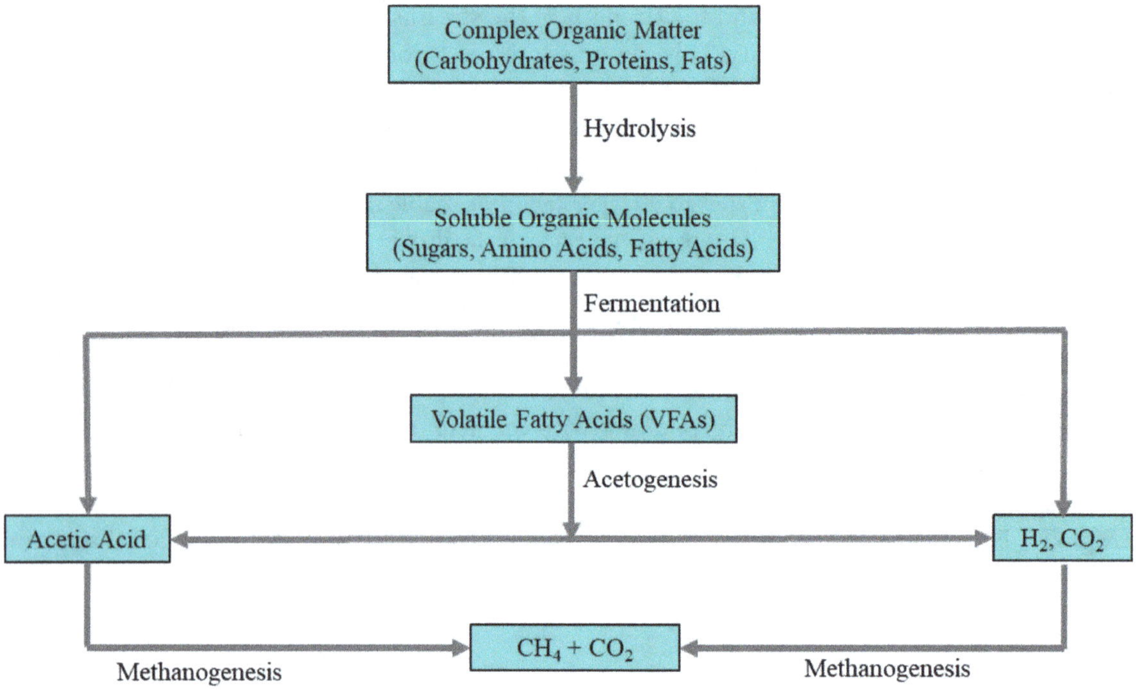


FIGURE 6.8 Chemical Reactions That Occur in Anaerobic Digestion

successive hydrolysis reactions, carbohydrates, lipids, and other carbon-based molecules are ultimately converted to glucose or other similar, shorter-chain organic molecules. Proteins and other nitrogen-based complex organic molecules are hydrolyzed to amino acids.

Hydrolysis is mediated by extracellular enzymes and is the first step of heterotrophic bacteria in preparation for substrate transfer across the cell membrane to support chemoheterotrophic metabolism. These hydrolytic enzymes, or hydrolase, are exerted by microorganisms found external to the surface of heterotrophic bacteria and are critical to cell metabolism. Different hydrolase are specific to the substrate bonds on which they are effective. Burgess and Pletschke (2008) published a review of the importance of hydrolase in wastewater sludge treatment.

Hydrolysis occurs in aerobic, anoxic, and anaerobic environments. The rate appears to be dependent on the oxidizing environment; however, there is some evidence to suggest that the presence of electron acceptors is of minimal importance in complex mixtures such as activated sludge (Goel et al., 1998).

### 5.1.2 Acidogenesis

The second stage of short-chain VFA production is acidogenesis, in which complex soluble organics are converted to VFAs, hydrogen, alcohols, ketones, and carbon dioxide. Again, this process is catalyzed by extracellular enzymes that facilitate the biochemical reactions necessary for the conversion. Various degradation pathways are potentially available, ultimately leading to the predominant VFA produced being acetic acid. Not only do these reactions make the carbon substrate more amenable to cross-membrane transport, but also energy is produced through the cleavage of bonds that activates the enzymes.

### 5.1.3 Acetogenesis

A companion process to acidogenesis is acetogenesis, wherein longer-chain VFAs and alcohols are converted to acetic acid and hydrogen by acetogenic bacteria. Generally, this conversion is achieved by obligate anaerobic bacteria that generate elemental hydrogen and acetic acid from the degradation of the more complex molecules. Because these bacteria are inhibited by high concentrations of elemental hydrogen, they only exist in a synergistic relationship with bacteria that consume hydrogen ions, such as non-VFA consuming, sulfur-reducing bacteria (*Desulfovibrio* sp.) or methanogens that convert hydrogen ($H_2$) and carbon dioxide to methane. Although a few species of acetogenic bacteria have been isolated (Lorowitz et al., 1989; McInerney et al., 1981), there are difficulties in growing these organisms in pure culture because of inhibition caused by hydrogen accumulations.

The key to effective fermentation is truncating the anaerobic conversion process so that methanogenesis, which is the conversion of acetic acid and hydrogen to methane, does not occur. Generally, SRT is controlled in acid-fermentation processes so that methanogenic bacteria are washed out of the system. However, it is also key that hydrogen is removed so that it does not inhibit acetogenic bacteria. Additionally, it is critical that other strains of bacteria that do not consume acetic acid but do scavenge the hydrogen for energy, such as *Desulfovibrio* sp. noted previously, coexist in a fermentation reactor.

Sulfur-reducing bacteria preferentially use sulfate as their electron acceptor, converting it to sulfide. Not all SRB use hydrogen as the electron donor in their metabolism. A number of these strains (*Desulfobacter, Desulfobulbus, Desulfococcus,* etc.) use VFAs as the electron donor, with the end products being sulfide and carbon dioxide (Bruser et al., 2000). These bacteria will consume short-chain VFAs generated by acidogenesis and acetogenesis. Middleton and Lawrence (1977) studied the kinetics of acetic-acid-consuming SRB and determined that the minimum SRT for these bacteria would be temperature dependent, varying from 2.9 days at 20 °C to 4.7 days at 10 °C. Given that these minimum SRTs are similar to those used in fermentation, consumption of VFAs by these bacteria is an issue that needs further study.

Sulfur-reducing bacteria outcompete methanogenic bacteria for acetate in most instances. They also outcompete other syntropic bacteria for propionate (Chen et al., 2008). However, SRB only convert propionate to acetate; hence, SRB are important microorganisms that participate in acetogenesis reactions.

## 5.2 Naturally Occurring Volatile Fatty Acid Production in Wastewater Collection Systems

Fermentation occurs in collection systems and primary treatment before an EPBR secondary treatment system. Furthermore, there are some industrial discharges that will inherently include high naturally occurring VFA concentrations that will persist in the wastewater until entry into the WRRF.

Fermentation of wastewater organics occurs in a wastewater collection system (Vollertsen et al., 2011). The degree of complex COD removal depends on the characteristics of the wastewater, the characteristics of the collection system that affect wastewater oxygen concentration (extent, slope, etc.), the presence of force mains, wastewater temperature, and the presence of other inhibiting substances. Collection system fermentation is dynamic because these characteristics fluctuate because of external influences. For instance, conversion of complex organics to short-chain VFAs will be substantially reduced during wet weather events because of the reduced retention time in the system, higher rates of reaeration caused by increased turbulence, and lower wastewater temperatures that often accompany wet weather flows.

In communities in which the weather is relatively warm, fermentation in the upstream wastewater collection system often generates sufficient VFAs to sustain EPBR, especially where the retention time in the collection system is relatively long. For communities in more temperate climates, VFAs may have to be generated within the treatment process to augment those arriving from the collection system.

There are few effective operating parameters that can be manipulated in the collection system to enhance VFA production. Some facilities have used upstream trunk sewers as equalization basins to reduce flow variations delivered to the facility. Retention times in these ad hoc basins are often sufficient to cause fermentation, especially as solids deposited along the bottom during relatively quiescent times have SRTs greater than liquid retention times. However, VFA production from this type of system is inconsistent because, during high-flow periods, solids are flushed from the system and HRTs are short.

One interaction of primary importance to VFA production in collection systems is odor control. Chemicals added to inhibit sulfide production or to combine with sulfides so they do not evolve into a gaseous state seriously affect fermentation in sewers. Oxidizing chemicals such as chlorine, oxygen, potassium permanganate, nitrates, or hydrogen peroxide are often added to wastewater flows to suppress sulfate reduction. The oxidizing conditions created by these chemical additions inhibit fermentation, especially those reactions that are obligate anaerobic.

The other method used to control odor emissions is the addition of compounds that will precipitate sulfides, that is, the addition of ferrous, or ferric, salts, predominantly. However, the addition of ferrous salts to wastewater also results in some precipitation of VFAs in ferro-organic compounds, which can inhibit acetogenesis (Ahmed et al., 2001) and, hence, the production of acetic acid.

TABLE 6.2 Summary of EBPR System Fundamentals, Common Operational Problems, Causes, and Possible Solutions

| Process problem | Possible cause | Possible solutions or steps |
|---|---|---|
| Oxygen intrusion in anaerobic zone | Any process structures that include mixing vortexes, cascade type aeration, and type of pumps (air lift pump, screw pump, etc.) | • Reduce vortex, turbulence, or cascade to avoid unintentional oxygen entrainment.<br>• Select type of pumps related to EBPR process that cannot introduce unwanted oxygen. |
| Nitrate and nitrite intrusion in anaerobic zone | Incomplete denitrification | Depending on the type of EBPR process configurations, return activated sludge or internal recirculation requires nitrate removal (via denitrification) before reaching anaerobic zone. |
| VFA limitation in anaerobic zone | Influent wastewater VFA limitation, inhibition in anaerobic zone due to dissolved oxygen, nitrate intrusion promoting denitrification and depleting VFA | Generate supplemental VFA by fermentation process using RAS, PS, or mixed liquor suspended solids. |
| Secondary P release in anaerobic zone | VFA depletion due to oversized anaerobic zone or longer HRT | Prevent secondary P release by minimizing the HRT. |
| Secondary P release in pre-anoxic zone | Nitrate depletion due to oversized anoxic zone or longer HRT | Maintain minimum necessary HRT; maintain nitrate concentration between 0.5 and 1.0 mg/L to avoid anaerobic conditions. |
| Secondary P release in post-anoxic zone | Lack of nitrates inducing anaerobic conditions and releasing P | Maintain minimum necessary HRT. |
| Decreased P-uptake in aerobic zone | Longer SRT in aerobic zone decreasing the BioP efficiency | Minimize SRT to avoid secondary release of P and maintain nitrification process. |
| Secondary P release in aerobic zone | Longer SRT inducing P release due to cell lysis and requiring higher COD/P for the bacteria to take up the available P | Maintain minimum necessary SRT to enhance EBPR performance while maintaining required nitrification. |

In many WRRFs, EBPR processes suffer because of the inadequacy of VFAs in the influent. To maintain EBPR without interruption, those facilities resort to the addition of acetic acid from external sources or self-generation of VFAs by fermenting parts of primary sludge in dedicated tanks that are used as fermenters. Primary sludge is a ready source of carbon substrate that is not accompanied by substantial amounts of nitrogen and phosphorus. Short-chain VFA production by fermenting this material has been used by a large number of facilities to facilitate EBPR. Whereas primary sludge is commonly used as the source of VFAs, secondary solids also have been used by a few facilities. Typically, facilities that are not equipped with primary clarifiers resort to the fermentation of return activated sludge (RAS) or WAS as an alternative source. A detailed discussion of various types of fermenters and their operating parameters is presented in Chapter 8.

## 6.0 OPERATIONAL SUMMARY

Table 6.2 summarizes a list of EBPR process common operational problems, causes, and potential steps that can be taken to avoid biological upset of the EBPR process.

## 7.0 REFERENCES


Ahmed, Z., Ivanov, V., Hyun, S.-H., Cho, K. M., & Kim, I. S. (2001). Effect of divalent iron on methanogenic fermentation of fat-containing wastewater. *Environmental Engineering Research, 6*(3), 139–146.

Ahn, J., Daidou, T., Tsuneda, S., & Hirata, A. (2002). Transformation of phosphorus and relevant intracellular compounds by a phosphorus-accumulating enrichment culture in the presence of both the electron acceptor and electron donor. *Biotechnology & Bioengineering, 79*, 83–93.

Ahn, J., Schroeder, S., Beer, M., McIlroy, S., Bayly, R. C., May, J. W., Vasiliadis, G., & Seviour, R. J. (2007). Ecology of the microbial community removing phosphate from wastewater under continuously aerobic conditions in a sequencing batch reactor. *Applied and Environmental Microbiology, 73*, 2257–2270.

Barker, P. S., & Dold, P. C. (1996). Denitrification behavior in biological excess phosphorus removal activated sludge systems. *Water Research, 30*(4), 769–780.

Barnard, J. L. (1976). A review of biological phosphorus removal in the activated sludge process. *Water SA, 2*(3), 136–144.

Barnard, J. L. (1983). Design consideration regarding phosphate removal in activated sludge plants. *Water Science & Technology, 15*, 319–328.

Barnard, J. L. (1984). Activated primary tanks for phosphate removal. *Water SA, 10*(3), 121–126.

Barnard, J. L., & Fothergill, S. (1998, October 3–7). Secondary phosphorus release in biological phosphorus removal systems. *Proceedings of the 71st Annual Water Environment Technical Exposition and Conference*, Orlando, FL, United States. Water Environment Federation.

Barnard, J. L., & Scruggs, C. L. (2003). Biological phosphorus removal. *Water & Environmental Technologies, 15*(2), 27–33.

Barnard, J. L., Stevens, G. M., & Leslie, P. J. (1985). Design strategies for nutrient removal plant. *Water Science & Technology, 17*(11/12), 233–242.

Beatons, D., Vanrolleghem, P. A., van Loosdtrecht, M. C. M., & Hosten, L. H. (1999). Temperature effects in bio-P removal. *Water Science & Technology, 39*(1), 215–225.

Blackall, L. L., Crocetti, G., Saunders, A. M., & Bond, P. L. (2002). A review and update of the microbiology of enhanced biological phosphorus removal in wastewater treatment plants. *Antonie Van Leeuwenhoek, 81*(1/4), 681–691.

Bond, P. L., Keller, J., & Blackall, L. L. (1999). Anaerobic phosphate release from activated sludge with enhanced biological phosphorus removal. A possible mechanism of intracellular pH control. *Biotechnology & Bioengineering, 63*(5), 507–515.

Brdjanovic, D., van Loosdrecht, M. C. M., Hooijmans, C. M., Alaerts, G. J., & Heijnen, J. J. (1998). Minimal aerobic sludge retention time in biological phosphorus removal systems. *Biotechnology & Bioengineering, 60*, 326–332.

Brdjanovic, D., van Loosdrecht, M. C. M., Hooijmans, C. M., Mino, T., Alaerts, G. J., & Heijnen, J. J. (1998). Effect of polyphosphate limitation on the anaerobic metabolism of phosphorus-accumulating microorganisms. *Applied Microbiology and Biotechnology, 50*, 273–276.

Bruser, T., Lens, P. N. L., & Truper, H. G. (2000). Chapter 3—the biological sulphur cycle. In *Environmental Technologies to Treat Sulfur Pollution*. IWA Publishing.

Burgess, J. E., & Pletschke, B. I. (2008). Hydrolytic enzymes in sewage sludge treatment: A mini-review. *Water SA, 34*(3), 343–350.

Burow, L. C., Kong, Y., Nielsen, J. L., Blackall, L. L., & Nielsen, P. H. (2007). Abundance and ecophysiology of *Defluviicoccus* spp., glycogen-accumulating organisms in full-scale wastewater treatment processes. *Microbiology, 153*, 178–185.

Carvalho, G., Lemos, P. C., Oehmen, A., & Reis, M. A. M. (2007). Denitrifying phosphorus removal: Linking the process performance with the microbial community structure. *Water Research, 41*, 4383–4396.

Cech, J. S., & Hartman, P. (1990). Glucose induced break down of enhanced biological phosphate removal. *Environmental Technology, 11*(7), 651–656.

Cech, J. S., & Hartman, P. (1993). Competition between polyphosphate and polysaccharide accumulating bacteria in enhanced biological phosphate removal systems. *Water Research, 27*(7), 1219–1225.

Chen, Y., Cheng, J. J., & Creamer, K. S. (2008). Inhibition of anaerobic digestion process: A review. *Bioresource Technology, 99*(10), 4044–4064.

Choi, E., Rhu, D., Yun, Z., & Lee, E. (1998). Temperature effects on biological nutrient removal system with municipal wastewater. *Water Science & Technology, 37*(9), 219–226.

Chuang, S. H., Ouyang, C. F., & Wang, Y. B. (1996). Kinetic competition between phosphorus release and denitrification on sludge under anoxic conditions. *Water Research, 30*(12), 2961–2968.

City College of New York. (1994). Chapter 2—characterization and treatment of New York City sludge dewatering centrate. In *Recycles from Centrate in New York City Wastewater Treatment Plants*. City College of New York.

Coats, E. R., Watkins, D. L., Brinkman, C. K., & Loge, F. R. (2011). Effect of anaerobic HRT on biological phosphorus removal and the enrichment of phosphorus accumulating organisms. *Water Environment Research, 83*(5), 461–469.

Crocetti, G. R., Hugenholtz, P., Bond, P. L., Schuler, A., Keller, J., Jenkins, D., & Blackall, L. L. (2000). Identification of polyphosphate-accumulating organisms and design of 16S rRNA-directed probes for their detection and quantitation. *Applied and Environmental Microbiology, 66*, 1175–1182.

Ekama, G., Marais, G., & Siebritz, I. (1984). Biological excess phosphorus removal. In *Theory, Design, and Operation of Nutrient Removal Activated Sludge Processes*. Water Research Commission.

Erdal, U. G., Erdal, Z. K., & Randall, C. W. (2003). The competition between PAOs (phosphorus accumulating organisms) and GAOs (glycogen accumulating organisms) in EBPR (enhanced biological phosphorus removal) systems at different temperatures and the effects on system performance. *Water Science & Technology, 47*, 1–8.

Erdal, U. G., Erdal, Z. K., & Randall, C. W. (2006). The mechanism of enhanced biological phosphorus removal washout and temperature relationships. *Water Environment Research, 78*(7), 710–750.

Erdal, Z. K. (2002). *The biochemistry of EBPR: Role of glycogen in biological phosphorus removal and the impact of the operating conditions on the involvement of glycogen.* Department of Civil and Environmental Engineering, Virginia Polytechnic Institute and State University.

Filipe, C. D. M., & Daigger, G. T. (1999). Evaluation of the capacity of phosphorus-accumulating organisms to use nitrate and oxygen as final electron acceptors: A theoretical study on population dynamics. *Water Environment Research, 71*(6), 1140-1150.

Filipe, C. D. M., Daigger, G. T., & Grady, C. P. L. (2001a). A metabolic model for acetate uptake under anaerobic conditions by glycogen accumulating organisms: Stoichiometry, kinetics, and the effect of pH. *Biotechnology & Bioengineering, 76*(1), 17–31.

Filipe, C. D. M., Daigger, G. T., & Grady, C. P. L. (2001b). pH as a key factor in the competition between glycogen-accumulating organisms and phosphorus-accumulating organisms. *Water Environment Research, 73*(2), 223–232.

Flowers, J. J., He, S., Yllmaz, S., Noguera, D. R., & McMahon, K. D. (2009). Denitrification capabilities of two biological phosphorus removal sludges dominated by different "*Candidatus Accumulibacter*" clades. *Environmental Microbiology Reports, 1*(6), 583–588.

Forbes, C. M., O'Leary, N. D., Dobson, A. D., & Marchesi, J. R. (2009). The contribution of "omic"-based approaches to the study of enhanced biological phosphorus removal microbiology. *FEMS Microbiology Ecology, 69*(1), 1–15.

Freitas, F., Temudo, M., & Reis, M. A. M. (2005). Microbial population response to changes of the operating conditions in a dynamic nutrient-removal sequencing batch reactor. *Bioprocess and Biosystems Engineering, 28*(3), 199–209.

Gebremariam, S. Y., Beutel, M. W., Christian, D., & Hess, T. F. (2011). Research advances and challenges in the microbiology of enhanced biological phosphorus removal—a critical review. *Water Environment Research, 83*(3), 195–219.

Goel, R., Mino, T., Satoh, H., & Matsuo, T. (1998). Enzyme activities under aerobic and anaerobic conditions in activated sludge sequencing batch reactor. *Water Research, 32*, 7.

Griffiths, P. C., Stratton, H. M., & Seviour, R. J. (2002). Environmental factors contributing to the "G bacteria" population in full-scale EBPR plants. *Water Science & Technology, 46*(4–5), 185–192.

Gu, A. Z., Hughes, T., Fisher, D., Swartzlander, D., Dacko, B., Ellis, W. G., He, S., McMahon, K. D., Neethling, J. B., Wei, H. P., & Chapman, M. (2006, October 22–25). The devil is in the details: Full scale optimization of the EBPR process at the city of Las Vegas WPCF. *Proceedings of the 79th Annual Water Environment Federation Technical Exhibition and Conference* [CD-ROM] (pp. 5110–5130), Dallas, TX, United States. Water Environment Federation.

Gu, A. Z., Saunders, A., Neethling, J. B., Stensel, H. D., & Blackall, L. L. (2008). Functionally relevant microorganisms to enhanced biological phosphorus removal performance at full-scale wastewater treatment plants in the United States. *Water Environment Research, 80*, 688–698.

Guisasola, A., Qurie, M., Vargas, M. D., Casas, C., & Baeza, J. A. (2009). Failure of an enriched nitrite-DPAO population to use nitrate as an electron acceptor. *Process Biochemistry, 44*(7), 689–695.

Harper, W. F., & Jenkins, D. (2003). The effect of an initial anaerobic zone on the nutrient requirements of activated sludge. *Water Environment Research, 75*(3), 216–224.

He, S., Gall, D. L., & McMahon, K. D. (2007). "Candidatus Accumulibacter" population structure in enhanced biological phosphorus removal sludges as revealed by polyphosphate kinase genes. *Applied and Environmental Microbiology, 73*, 5865–5874.

He, S., Gu, A. Z., & McMahon, K. D. (2008). Progress toward understanding the distribution of accumulibacter among full-scale enhanced biological phosphorus removal systems. *Microbial Ecology, 55*, 229–236.

Helmer, C., & Kunst, S. (1997). Low temperature effects on phosphorus release and uptake by microorganisms in EBPR plants. *Water Science & Technology, 37*(4–5), 531–539.

Hesselmann, R., Werlen, C., Hahn, D., van Der Meer, J., & Zehnder, A. (1999). Enrichment, phylogenetic analysis and detection of a bacterium that performs enhanced biological phosphate removal in activated sludge. *Systematic and Applied Microbiology, 22*, 454–465.

Jiang, Y., Wang, B., Wang, L., Chen, J., & He, S. (2006). Dynamic response of denitrifying poly-P accumulating organisms batch culture to increased nitrite concentration as electron acceptor. *Journal of Environmental Science and Health, 41*(11), 2557–2570.

Kern-Jespersen, J. P., & Henze, M. (1993). Biological phosphorus uptake under anoxic and aerobic conditions. *Water Research, 27*, 617–624.

Kisoglu, Z., Erdal, U. G., & Randall, C. W. (2000, October 14–18). The effect of COD/TP ratio on intracellular storage materials, system performance and kinetic parameters in a BNR system. *Proceedings of the 73rd Annual Water Environment Federation Technical Exposition and Conference* [CD-ROM], Anaheim, CA, United States. Water Environment Federation.

Kong, Y. H., Nielsen, J. L., & Nielsen, P. H. (2004). Microautoradiographic study of rhodocyclus-related polyphosphate-accumulating bacteria in full-scale enhanced biological phosphorus removal plants. *Applied and Environmental Microbiology, 70*(9), 5383–5390.

Kong, Y. H., Nielsen, J. L., & Nielsen, P. H. (2005). Identity and ecophysiology of uncultured actinobacterial polyphosphate-accumulating organisms in full-scale enhanced biological phosphorus removal plants. *Applied and Environmental Microbiology, 71*(7), 4076–4085.

Kong, Y. H., Xia, Y., Nielsen, J. L., & Nielsen, P. H. (2006). Ecophysiology of a group of uncultured gammaproteobacterial glycogen-accumulating organisms in full-scale enhanced biological phosphorus removal wastewater treatment plants. *Environmental Microbiology, 8*(3), 479–489.

Kuba, T., Smolders, G., van Loosdrecht, M. C. M., & Heijnen, J. J. (1993). Biological phosphorus removal from waste-water by anaerobic–anoxic sequencing batch reactor. *Water Science & Technology, 27*(5–6), 241–252.

Kuba, T., van Loosdrecht, M. C. M., & Heijnen, J. J. (1996). Phosphorus and nitrogen removal with minimum COD requirement by integration of denitrification in a two-sludge system. *Water Research, 30*(7), 1702–1710.

Kuba, T., Wachtmeister, A., van Loosdrecht, M. C. M., & Heijnen, J. J. (1994). Effect of nitrate on phosphorus release in biological phosphorus removal systems. *Water Science & Technology, 30*(6), 263–269.

Lancaster, C. B., & Madden, J. E. (2008, October 18–22). Not so fast: The impact of recalcitrant phosphorus on the ability to meet low phosphorus limits. *Proceedings of the 81st Annual Water Environment Federation Technical Exhibition and Conference* [CD-ROM] (pp. 3531–3543), Chicago, IL, United States. Water Environment Federation.

Levin, G. V., & Shapiro, J. (1965). Metabolic uptake of phosphorus by waste water organisms. *Journal (Water Pollution Control Federation), 37*(6), 800–821.

Li, N., Ren, N.-Q., Wang, X.-H., & Kang, H. (2010). Effect of temperature on intracellular phosphorus absorption and extra-cellular phosphorus removal in EBPR process. *Bioresource Technology, 101*(15), 6265–6268.

Li, N., Wang, X., Ren, N., Zhang, K., Kang, H., & Youa, S. (2008). Effects of solid retention time (SRT) on sludge characteristics in enhanced biological phosphorus removal (EBPR) reactor. *Chemical and Biochemical Engineering Quarterly, 22*(4), 453–458.

Liu, W.-T., Mino, T., Matsuo, T., & Nakamura, K. (1996). Biological phosphorus removal processes—Effect of pH on anaerobic substrate metabolism. *Water Science & Technology, 34*(1–2), 25–32.

Liu, W.-T., Mino, T., Nakamura, K., & Matsuo, T. (1996). Glycogen accumulating population and its anaerobic substrate uptake in anaerobic-aerobic activated sludge without biological phosphorus removal. *Water Research, 30*(1), 75–82.

Liu, W.-T., Nakamura, K., Matsuo, T., & Mino, T. (1997). Internal energy-based competition between polyphosphate- and glycogen-accumulating bacteria in biological phosphorus removal reactors—Effect of P/C feeding ratio. *Water Research, 31*(6), 1430–1438.

Lorowitz, W. H., Zhao, H., & Bryant, M. P. (1989). *Syntrophomonas wolfei* subspecies *saponavida* subsp. *nov.,* a long chain fatty-acid-degrading subspecies *Wolfei* subsp *nov.*, and amended descriptions of the genus and species. *International Journal of Systematic and Evolutionary Microbiology, 39*, 122–126.

Mamais, D., & Jenkins, D. (1992). The effects of MCRT and temperature on enhanced biological phosphorus removal. *Water Science & Technology, 26*(5/6), 955–965.

Martin, H. G., Ivanova, N., Kunin, V., Warnecke, F., Barry, K. W., McHardy, A. C., Yeates, C., He, S. M., Salamov, A. A., Szeto, E., Dalin, E., Putnam, N. H., Shapiro, H. J., Pangilinan, J. L., Rigoutsos, I., Kyrpides, N. C., Blackall, L. L., McMahon, K. D., & Hugenholtz, P. (2006). Metagenomic analysis of two enhanced biological phosphorus removal (EBPR) sludge communities. *Nature Biotechnology, 24*(10), 1263–1269.

Matsuo, T., Mino, T., & Sato, H. (1992). Metabolism of organic substances in anaerobic phase of biological phosphate uptake process. *Water Science & Technology, 25*(6), 83–92.

Matsuo, Y. (1994). Effect of the anaerobic solids retention time on enhanced biological phosphorus removal. *Water Science & Technology, 30*(6), 193–202.

Maurer, M., Abramovich, D., Siegrist, H., & Gujer, W. (1999). Kinetics of biologically induced phosphorus precipitation in waste-water treatment. *Water Research, 33*(2), 484–493

McInerney, M. J., Bryant, M. P., Hespell, R. B., & Costerton, J. W. (1981). *Syntrophomonas wolfei gen. nov.* sp. *nov.,* An anaerobic syntrophic, fatty acid-oxidizing bacterium. *Applied and Environmental Microbiology, 41*(4), 1029–1039.

Middleton, A. C., & Lawrence, A. W. (1977). Kinetics of microbial sulfate reduction. *Journal Water Pollution Control Federation, 49*(7), 1659–1670.

Mino, T., van Loosdrecht, M. C. M., & Heijnen, J. J. (1998). Microbiology and biochemistry of the enhanced biological phosphate removal process. *Water Research, 32*(11), 3193–3207.

Neethling, J. B., Blake, B., Benisch, M., Gu, A. Z., Stephens, H., Stensel, D., & Moore, R. (2005). Factors influencing the reliability of enhanced biological phosphorus removal (Report 01-CTS-3). Water Environment Research Foundation.

Nielsen, P. H., Kragelund, C., Seviour, R. J., & Nielsen, J. L. (2009). Identity and ecophysiology of filamentous bacteria in activated sludge. *FEMS Microbiology Reviews, 33*(6), 969–998.

Nielsen, J. L., Nguyen, H., Meyer, R. L., & Nielsen, P. H. (2012). Identification of glucose-fermenting bacteria in a full-scale enhanced biological phosphorus removal plant by stable isotope probing. *Microbiology, 158*(7), 1818–182.

Oehmen, A., Carvalho, G., Freitas, F., & Reis, M. A. M. (2010). Assessing the abundance and activity of denitrifying polyphosphate accumulating organisms through molecular and chemical techniques. *Water Science & Technology, 61*(8), 2061–2068.

Oehmen, A., Lemos, P. C., Carvalho, G., Yuan, Z. G., Keller, J., Blackall, L. L., & Reis, M. A. M. (2007). Advances in enhanced biological phosphorus removal: From micro to nacro scale. *Water Research, 41*(11), 2271–2300.

Oehmen, A., Vives, M. T., Lu, H. B., Yuan, Z. G., & Keller, J. (2005). The effect of pH on the competition between polyphosphate-accumulating organisms and glycogen-accumulating organisms. *Water Research, 39*(15), 3727–3737.

Oehmen, A., Yuan, Z. G., Blackall, L. L., & Keller, J. (2005). Comparison of acetate and propionate uptake by polyphosphate accumulating organisms and glycogen accumulating organisms. *Biotechnology & Bioengineering, 91*(2), 162–168.

Onnis-Hayden, A., Majed, N., Schramm, A., & Gu, A. Z. (2011). Process optimization by decoupled control of key microbial populations: Distribution of activity and abundance of polyphosphate-accumulating organisms and nitrifying populations in a full-scale IFAS-EBPR plant. *Water Research, 45*(13), 3845–3854.

Phillips, H. M., Kobylinski, E., Barnard, J., & Wallis-Lage, C. (2006, October 21–25). Nitrogen and phosphorus rich streams: Managing the nutrient merry-go-round. *Proceedings of the 79th Annual Water Environment Federation Technical Exhibition and Conference* [CD-ROM] (pp. 5282–5304), Dallas, TX, United States. Water Environment Federation.

Puig, S., Coma, M., Monclús, H., van Loosdrecht, M. C. M., Colprim, J., & Balaguer, M. D. (2008). Selection between alcohols and volatile fatty acids as external carbon sources for EBPR. *Water Research, 42*(3), 557–566.

Randall, C. W., Barnard, J. L., & Stensel, H. D., Eds. (1992). *Design and retrofit of wastewater treatment plants for biological nutrient removal.* Vol. 5, Water Quality Management Library. Technomic Publishing.

Rodrigo, M. A., Seco, A., Ferrer, J., & Penya-Roja, J. M. (1999). The effect of the sludge age on the deterioration of the enhanced biological phosphorus removal process. *Environmental Technology, 20*(10), 1055–1063.

Rodrigo, M. A., Seco, A., Penya-Roja, J. M., & Ferrer, J. (1996). Influence of sludge age on enhanced phosphorus removal in biological systems. *Water Science & Technology, 34*(1–2), 41–48.

Satoh, H., Mino, T., & Matsuo, T. (1992). Uptake of organic substrates and accumulation of polyhydroxyalkanoates linked with glycolysis of intracellular carbohydrates under anaerobic conditions in the biological excess phosphate removal processes. *Water Science & Technology, 26*(5–6), 933–942.

Satoh, H., Mino, T., & Matsuo, T. (1994). Deterioration of enhanced biological phosphorus removal by the domination of microorganisms without polyphosphate accumulation. *Water Science & Technology, 30*(6), 203–211.

Saunders, A. M., Mabbett, A. N., McEwan, A. G., & Blackall, L. L. (2007). Proton motive force generation from stored polymers for the uptake of acetate under anaerobic conditions. *FEMS Microbiology Letters, 274*(2), 245–251.

Schuler, A. J., & Jenkins, D. (2002). Effects of pH on enhanced biological phosphorus removal metabolisms. *Water Science & Technology, 46*(4–5), 171–178.

Schuler, A. J., & Jenkins, D. (2003). Enhanced biological phosphorus removal from wastewater by biomass with different phosphorus contents, part I: Experimental results and comparison with metabolic models. *Water Environment Research, 75*(6), 485–498.

Serafim, L. S., Lemos, P. C., Levantesi, C., Tandoi, V., Santos, H., & Reis, M. A. M. (2002). Methods for detection and visualization of intracellular polymers stored by polyphosphate-accumulating microorganisms. *Journal of Microbiological Methods, 51*(1), 1–18.

Seviour, R. J., & McIlroy, S. (2008). The microbiology of phosphorus removal in activated sludge processes—the current state of play. *Journal of Microbiology, 46*(2), 115–124.

Seviour, R. J., Mino, T., & Onuki, M. (2003). The microbiology of biological phosphorus removal in activated sludge systems. *FEMS Microbiology Reviews, 27*(1), 99–127.

Shoji, T., Satoh, H., & Mino, T. (2003). Quantitative estimation of the role of denitrifying phosphate accumulating organisms in nutrient removal. *Water Science & Technology, 47*(11), 23–29.

Smolders, G. J. F., van der Meij, J., van Loosdrecht, M. C. M., & Heijnen, J. J. (1994). Model of the anaerobic metabolism of the biological phosphorus removal process—Stoichiometry and pH influence. *Biotechnology & Bioengineering, 43*(6), 461–470.

Smolders, G. J. F., van der Meij, J., van Loosdrecht, M. C. M., & Heijnen, J. J. (1995). A structured metabolic model for the anaerobic and aerobic stoichiometry and kinetics of the biological phosphorus removal process. *Biotechnology & Bioengineering, 47*(3), 277–287.

Stephens, H. L., & Stensel, D. (1998). Effect of operating conditions on biological phosphorus removal. *Water Environment Research, 70*(3), 362–369.

Thomas, M. P. (2008). The secret to achieving reliable biological phosphorous removal. *Water Science and Technology, 58*(6), 1231-1236.

Tremblay, A., Tyagi, R. D., & Urampalli, S. R. Y. (1999). Practice periodical of hazardous, toxic, and radioactive waste. *Management, 10*, 183.

U.S. Environmental Protection Agency. (1987). *Design manual: Phosphorus removal* (EPA-625/1-87-001).

U.S. Environmental Protection Agency. (1994). *Decentralized systems technology fact sheet septage treatment disposal.* Municipal Technology Branch, U.S. Environmental Protection Agency.

Vollertsen, R. E., Hvilved-Jacobsen, J., & Nielsen, T. (2011). Anaerobic transformations of organic matter in collection systems. *Water Environment Research, 83*(6), 532–540.

Water Environment Federation. (2010). *Nutrient removal* (WEF Manual of Practice No. 34).

Water Research Foundation. (2014). *Optimization and design of a side-stream EBPR process as a sustainable approach for achieving stable and efficient phosphorus removal* (WRF Project No. U1R13/4869).

Whang, L. M., Filipe, C. D. M., & Park, J. K. (2007). Model-based evaluation of competition between polyphosphate- and glycogen-accumulating organisms. *Water Research, 41*(6), 1312–1324.

Whang, L. M., & Park, J. K. (2002). Competition between polyphosphate- and glycogen-accumulating organisms in biological phosphorus removal systems—Effect of temperature. *Water Science & Technology, 46*(1–2), 191–194.

Whang, L. M., & Park, J. K. (2006). Competition between polyphosphate- and glycogen-accumulating organisms in enhanced-biological-phosphorus-removal systems: Effect of temperature and sludge age. *Water Environment Research, 78*(1), 4–11.

Wong, M., Mino, T., Seviour, R., Onuki, M., & Liu, W. (2005). In situ identification and characterization of the microbial community structure of full-scale enhanced biological phosphorous removal plants in Japan. *Water Research, 39*(13), 2901–2914.

Xia, Y., Kong, Y. H., & Nielsen, P. H. (2007). In situ detection of protein-hydrolysing microorganisms in activated sludge. *FEMS Microbiology Ecology, 60*(1), 156–165.

Zeng, R. J., Saunders, A. M., Yuan, Z. G., Blackall, L. L., & Keller, J. (2003). Identification and comparison of aerobic and denitrifying polyphosphate-accumulating organisms. *Biotechnology & Bioengineering, 83*(2), 140–148.

Zhang, C., Chen, Y., & Liu, Y. (2007). The long-term effect of initial pH control on the enrichment culture of phosphorus- and glycogen-accumulating organisms with a mixture of propionic and acetic acids as carbon sources. *Chemosphere, 69*(11), 1713–1721.

Zhang, T., Liu, Y., & Fang, H. H. P. (2005). Effect of pH change on the performance and microbial community of enhanced biological phosphate removal process. *Biotechnology & Bioengineering, 92*(2), 173–182.

Zhou, Y., Pijuan, M., & Yuan, Z. (2007). Free nitrous acid inhibition on anoxic phosphorus uptake and denitrification by poly-phosphate accumulating organisms. *Biotechnology & Bioengineering, 98*(4), 903–912.

Zhou, Y., Pijuan, M., Zeng, R. J., Lu, H., & Yuan, Z. (2008). Could polyphosphate-accumulating organisms (PAOs) be glycogen accumulating organisms (GAOs)? *Water Research, 42*(10–11), 2361–2368.

## 8.0 SUGGESTED READINGS

Oehmen, A., Yuan, Z., Blackall, L. L., & Keller, J. (2004). Short-term effects of carbon source on the competition of polyphosphate accumulating organisms and glycogen accumulating organisms. *Water Science & Technology, 50*(10), 139–144.

Rassie, W. H., & Pretorius, W. A. (2001). A review of characterisation requirements for in-line prefermenters paper 1: Wastewater characterisation. *Water SA, 27*(3), 405–412.

Snoeyink, V. L., & Jenkins, D. (1980). *Water chemistry.* Wiley & Sons.

Vollertsen, J., Petersen, G., & Borregaard, V. R. (2005, September 19–21). Hydrolysis and fermentation of activated sludge to enhance biological phosphorus removal. *Proceedings of the IWA Specialized Conference on Nutrient Management in Wastewater Treatment Processes and Recycle Streams*, Krakow, Poland. International Water Association.

# Chapter 7

# Combined Nitrifying and Denitrifying Systems

## 1.0 INTRODUCTION

Combined nitrification and denitrification (NDN) treatment systems achieve 5-day biochemical oxygen demand (BOD) removal and NDN in a single-sludge configuration using linked reactors or zones in series and a common process for solids separation (clarifiers or membranes). Combining these processes in one system provides the potential for initial economic capital and ongoing operating costs. It also presents a significant operational challenge related to proper balancing that is sometimes required between the different process environments.

Combined NDN systems are classified in various ways, including fixed versus suspended growth, continuous-flow versus batch treatment, staging of process sequences, and method of aeration. A common thread that is consistent throughout all these processes is the alternating environment that is used sequentially to nitrify and denitrify.

## 2.0 BASIC CONSIDERATIONS

Many process options exist to accomplish NDN in a combined system. The most commonly applied configurations of these are described in Section 3.0. The right process configuration for any particular facility is influenced by several factors, the most significant of which are described in the following subsections.

### 2.1 Effluent Requirements

Fundamental to the selection of the appropriate process configuration is the level of treatment that must be provided and the process effluent quality required. Effluent total nitrogen requirements have tended to be implemented in the following tiers, based on a general and common understanding of the relative difficulty and costs of compliance: (1) a relatively easily attainable concentration in the range of 8 to 10 mg/L, (2) a more stringent concentration in the range of 3 to 8 mg/L, and (3) a very stringent concentration of less than 3 mg/L. As described in Section 3.0, different process configurations are capable of meeting different effluent total nitrogen results; achieving lower effluent total nitrogen requires the use of more complex and costly process configurations.

### 2.2 Economics

Economic effects should be considered during both design and operation when selecting which process configuration to use. Achieving sufficient nitrification (typically complete nitrification in total nitrogen removal systems) requires having enough of the bioreactor volume under aeration to achieve this throughout the diurnal cycle of wastewater flow and loading and with a sufficient safety factor to protect the sensitive nitrifying organism population from toxic loads, low temperature, and other factors that can cause loss of the population. The amount of bioreactor volume under aeration typically the biggest factor in operating costs. Because of the inability to fully characterize the variability of most wastewater influents and the erratic sensitivity of nitrifiers, there is a tendency to have an excessive volume under aeration. In addition to higher costs, over-aeration of mixed liquor can interfere with efficient denitrification. Therefore, it is important to accurately define the operational parameters affecting nitrification for each facility as much as possible before deciding on aerated volumes. Future operational parameters are frequently difficult to accurately define and predict; as such, many facilities are equipped with zones capable of being operated in either aerated or anoxic mode. These zones are referred to as *swing zones* and can provide operators with a great deal of flexibility to fine-tune the anoxic and oxic volume ratio in response to loading variation, thereby potentially reducing operating costs. Swing zones can increase capital costs because the swing zone must be equipped with both mixers and aeration diffusion equipment. Denitrification volumes are easier to define because of the stability of denitrifying organisms; however, the efficiency of the denitrification process is dependent on a number of complex and interrelated factors that must be continuously considered and evaluated by operators.

## 2.3 Operational Considerations

Operational control of the NDN process is important to achieve permit compliance and prevent operational issues with other aspects of the facility, such as chlorine dosage control and unwanted denitrification in the final clarifiers. The most important operational considerations include the following:

- Establishing a reliable and repeatable sampling program with the most important factors being chemical oxygen demand (COD)—or BOD—and total Kjeldahl nitrogen (TKN) concentrations entering the bioreactor;
- Providing sufficient aeration for complete nitrification while avoiding over-aeration of the process;
- Finding the optimum mixed liquor and return activated sludge (RAS) recycle rates and avoiding excessive dissolved oxygen in the mixed liquor recycle;
- Operating the process at a high enough aerobic solids retention time (aSRT) to ensure complete nitrification while avoiding a buildup of nuisance filamentous or foaming organisms;
- Selecting a swing-zone operating mode (aerobic or anoxic) to provide sufficient nitrification or denitrification capacity based on loading;
- Maintaining sufficient alkalinity in the bioreactor and monitoring pH; and
- Providing supplemental carbon to the process if required.

## 2.4 Sustainability

Sustainability, that is, using approaches that balance current environmental, economic, and social considerations without compromising the ability of future generations to meet their needs, is an increasingly vital criterion that affects the design and operational approaches of achieving biological nutrient removal (BNR). Original design of the facility should carefully consider and incorporate sustainable approaches to conservation of materials, minimizing site effects caused by construction and controlling initial costs. Operation of BNR facilities should continuously consider ways to improve sustainability by balancing the sometimes-competing goals of optimizing treatment, conserving energy, reducing chemical use, and minimizing sludge production. Refer to Chapter 11 for a detailed discussion of optimization of BNR systems.

## 2.5 Safety

Safety is always of paramount importance in the operation of any water resource recovery facility (WRRF). One of the biggest operational safety concerns for NDN facilities is the use of methanol or other hazardous chemicals as a supplemental carbon source. See Section 2.5 for additional safety information for methanol use.

# 3.0 PROCESS CONFIGURATIONS

## 3.1 Modified Ludzack–Ettinger Process

### *3.1.1 Process Description*

In this process configuration, nitrate in the mixed liquor recycle and RAS flow streams are mixed with influent wastewater and reduced to nitrogen gas in a pre-denitrification reactor (anoxic zone) upstream of the main aerobic reactor. An earlier process configuration, the Ludzack–Ettinger process, was initially developed to take advantage of the carbonaceous substrate available in influent wastewater by placing the anoxic zone upstream of the nitrification reactor, as shown in Figure 7.1a (Ludzack & Ettinger, 1962). However, the total nitrogen removal efficiency of the Ludzack–Ettinger process is limited by the quantity of nitrate recycled back to the anoxic zone in RAS flow. In response to that limitation, Barnard developed an improvement to the Ludzack–Ettinger process, identified as the modified Ludzack–Ettinger (MLE) process, which adds the recirculation of mixed liquor recycle (MLR) from the end of the aerobic zone to the beginning of the anoxic zone, as shown in Figure 7.1b (Barnard, 1974).

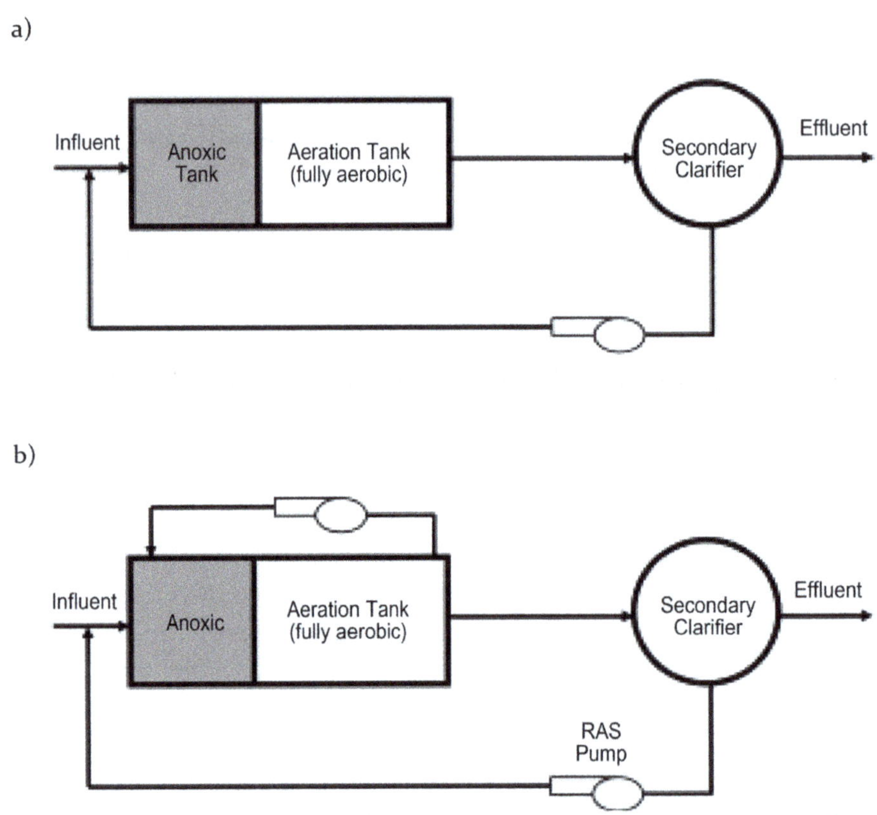


FIGURE 7.1 (a) Ludzack–Ettinger and (b) MLE Process Schematics

Figure 7.2 depicts the amount of denitrification (nitrate removal) that can be achieved in a pre-anoxic zone of an MLE process based on MLR flow as a percentage of influent flow. It is important to note that Figure 7.2 indicates the maximum denitrification that can be achieved by way of MLR recycle, if other factors such as substrate availability or kinetic limitations do not limit performance.

### 3.1.2 Expected Process Performance

Effluent total nitrogen from an MLE process is typically in the range of 8 to 10 mg/L. This is a function of the relationship presented in Figure 7.2, typical process influent TKN in the 40- to 50-mg/L range, and the common determination that a MLR flow of approximately 4 times influent flow is typically the feasible upper limit.

### 3.1.3 Operational Features

Table 7.1 lists key operating parameters of the MLE process. The presence of dissolved oxygen in the anoxic zone is detrimental to the denitrification rate. Therefore, dissolved oxygen inputs from sources such as MLR and back-mixing from the aerobic reactor need to be minimized. The optimum MLR rate will be dependent on the dissolved oxygen included in the MLR flow and the oxygen demand (COD or BOD) of the influent wastewater.

Weaker-strength wastewater and/or higher MLR dissolved oxygen concentrations may limit the optimum MLR rate to less than 200%; higher-strength wastewater and low MLR dissolved oxygen concentrations may allow beneficial use of MLR rates of 400% or higher. Often, MLR rates greater than 400% of influent flow are not beneficial to improving total nitrogen removal because of insufficient substrate availability, dissolved oxygen recirculation, limits on anoxic zone volume, and substrate dilution that reduces denitrification kinetics.

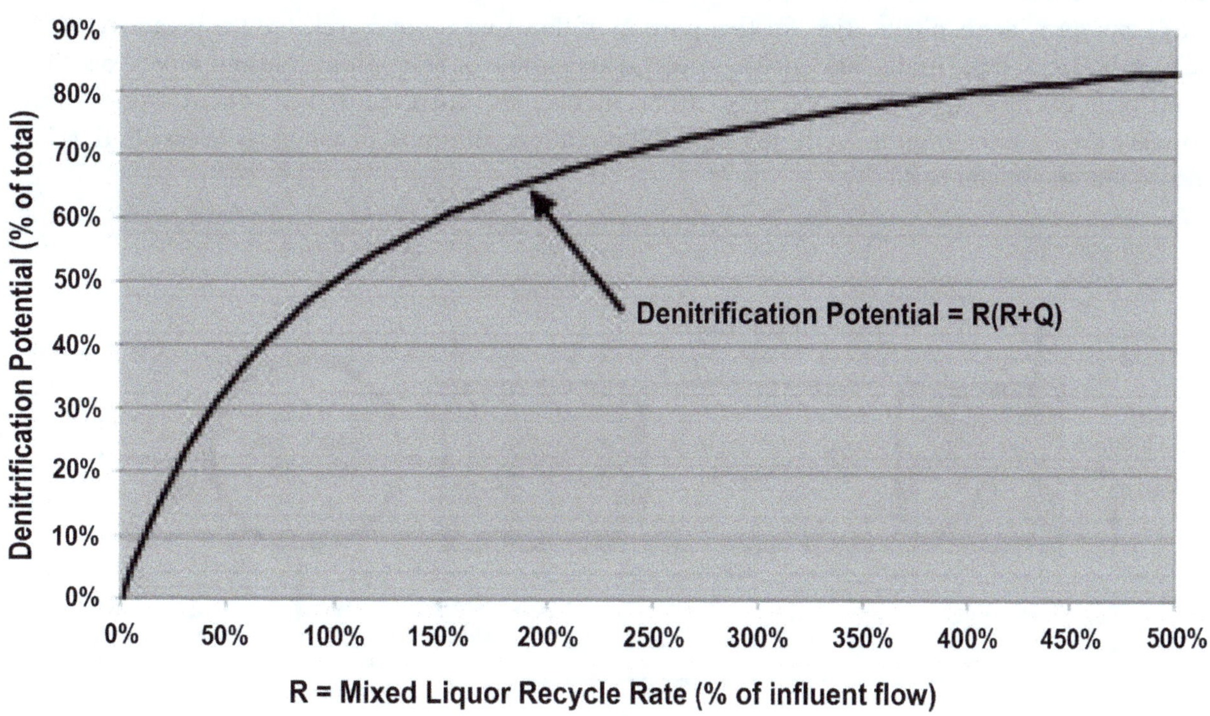


FIGURE 7.2 Denitrification Potential of the MLE Process

TABLE 7.1 Key Operating Parameters of the MLE Process

| Reactor | Parameter | Rationale |
|---|---|---|
| Anoxic | Influent carbon (BOD or COD) load | Inadequate carbon load limits denitrification. |
| | Dissolved oxygen | Dissolved oxygen in zone will reduce denitrification rate. |
| | Nitrate | High nitrate recycled back to aerobic zone may cause filamentous bulking. |
| | Mixed liquor recycle | Controls nitrate load to anoxic and back to aerobic zone. |
| | | High dissolved oxygen in aerobic zone (and therefore in MLR) may inhibit upstream denitrification. |
| Aerobic | Dissolved oxygen | Low dissolved oxygen may inhibit nitrification. |
| | Alkalinity, pH | Nitrification consumes alkalinity. |

## 3.2 Four-Stage Bardenpho Process

The four-stage Bardenpho process, shown schematically in Figure 7.3, incorporates the principles used for the MLE process but also includes a second anoxic zone to achieve a higher level of total nitrogen removal, and a second aerobic zone to increase dissolved oxygen and strip nitrogen has bubbles upstream of secondary clarification.

### *3.2.1 Process Description*

The first two stages of the Bardenpho process function similarly to an MLE process. The pre-anoxic zone in Bardenpho facilities is often sized large enough to consistently accommodate at least a 400% MLR rate without nitrate "bleed-through." This preanoxic zone removes most nitrate generated in the process (per the relationship shown in Figure 7.2), and the secondary anoxic zone, located outside the MLE "loop," provides denitrification for that portion of flow that is not recycled to the preanoxic zone. The fourth reactor zone in the Bardenpho process is an aerobic or reaeration zone and serves to strip any nitrogen gas formed in the second anoxic zone, increase the dissolved oxygen concentration before secondary clarification, improve activated sludge flocculation, minimize phosphorus release in the clarifier, and reduce effluent turbidity.

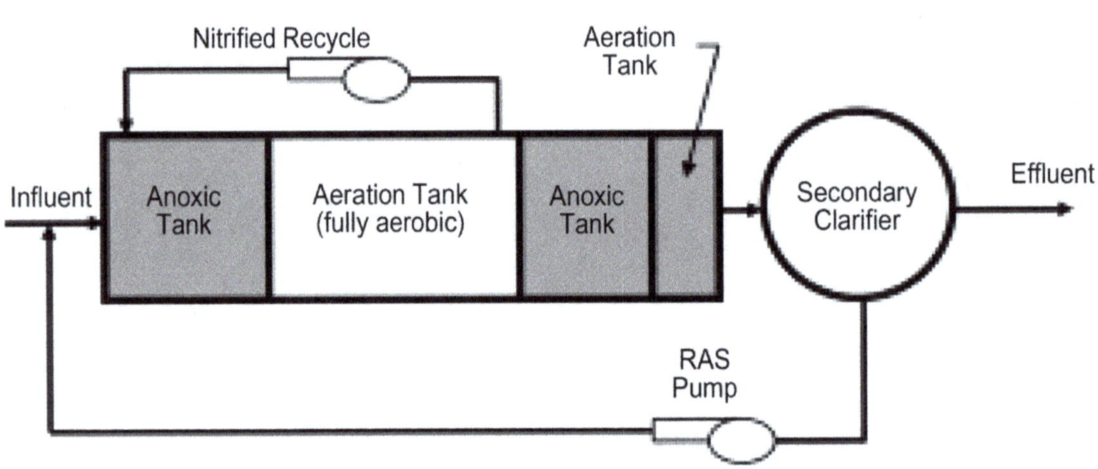


FIGURE 7.3 Four-Stage Bardenpho Process Schematic

### 3.2.2 Expected Process Performance

The Bardenpho process can generally achieve greater than 80% (and often better than 90%) total nitrogen removal (U.S. Environmental Protection Agency [U.S. EPA], 1993, 2010a). With the addition of supplemental carbon in the second anoxic zone, effluent total nitrogen from a Bardenpho process can be low, limited only by refractory dissolved organic nitrogen (rDON) and the organic nitrogen associated with effluent total suspended solids (TSS). The ability to achieve effluent total nitrogen in this range depends on the ratio of oxidizable nitrogen to carbon in the influent to the biological process and use of supplemental carbon.

### 3.2.3 Operational Features

The Bardenpho process can be operated with many reactor configurations, including plug flow, complete-mix, and oxidation-ditch reactors. Table 7.2 lists the key monitoring requirements at different stages of the four-stage Bardenpho process (U.S. EPA, 1993).

## 3.3 Sequencing Batch Reactors

Sequencing batch reactors (SBRs) represent a variable-volume, suspended growth treatment technology that uses time sequences to perform various treatment operations that continuous treatment processes accomplish in different tanks or zones. The first activated sludge processes were actually SBRs, but the lack of automation steered a shift toward continuous treatment systems that required less operator intervention. Modern automatic monitoring and control systems have resulted in SBRs being a commonly used nitrogen-removal process.

### 3.3.1 Process Description

Sequencing batch reactors proceed through a series of phases in a timed cycle and typically complete four to six cycles per day for domestic wastewater treatment. Typically, 50% to 75% of the liquid volume of the SBR containing settled biomass is retained at the end of every cycle, although a higher percentage

TABLE 7.2 Key Operating Parameters of the Bardenpho Process

| Reactor | Parameter | Rationale |
|---|---|---|
| First anoxic | Influent carbon (BOD or COD) load | Inadequate carbon load limits denitrification. |
| | Dissolved oxygen | Will reduce denitrification rate. |
| | Nitrate | High nitrate recycled back to aerobic zone may cause filamentous bulking. |
| | Mixed liquor recycle | Controls nitrate load to anoxic and back to aerobic zone. |
| | | High dissolved oxygen in aerobic zone (and therefore in MLR) may inhibit upstream denitrification. |
| Aerobic | Dissolved oxygen | Low dissolved oxygen may inhibit nitrification. |
| | Alkalinity, pH | Nitrification consumes alkalinity. |
| Second anoxic | Available carbon | Inadequate available carbon substrate limits denitrification capacity. |
| | Nitrate | High nitrification in aerobic zone may overwhelm endogenous denitrification capacity, resulting in $NO_3$ in effluent. |
| | Dissolved oxygen | High dissolved oxygen will inhibit endogenous denitrification. |

may be necessary depending on effluent requirements. The minimum liquid level determines the volume of sludge inventory that can be retained in the SBR reactor and the maximum volume of influent that can be accommodated for any individual cycle. Sequencing batch reactors cycle through the following four basic phases (refer to Figure 7.4):

1. Fill: Wastewater is added to the SBR, raising the liquid level from the minimum level to a depth that corresponds to the amount of influent received during the fill phase period. In SBRs operated for nitrogen removal, the initial filling phase is generally under anoxic conditions (anoxic fill). Depending on the overall cycle structure and process need, a portion of the fill phase can also be aerobic.
2. React: Biological processes are performed, including under aerobic and nonaerated mixing (anoxic) regimes, depending on process goals. In this way, SBRs can be operated to mimic the process conditions of a MLE, four-stage Bardenpho, or other configuration.
3. Settle: Aeration and mixing are stopped, and the biomass is allowed to settle.
4. Decant: Clarified effluent is removed from the basin, aeration and mixing are off, and biomass is wasted as necessary during the decant phase.

For nitrogen removal, the fill and react phases are broken down into static fill, mixed (anoxic) fill, and mixed react (aerobic). Nitrification occurs in the aerobic fill and react phases, denitrification happens in the anoxic fill and react phases, and carbon oxidation happens in all phases. Denitrification results from selecting anoxic fill, and anoxic react periods that are long enough to meet process goals.

### 3.3.2 Expected Process Performance

With sufficient volume to provide for the necessary reaction time, SBRs can achieve greater than 90% total nitrogen removal. A survey was conducted of 10 SBR systems in the northeast United States performing nitrogen removal (Young et al., 2008) that showed that facility effluent total nitrogen levels ranged from 2.5 to 9.5 mg/L. These total nitrogen levels were all achieved without the use of supplemental carbon, although it should be noted that the facilities were operating well below design loads.

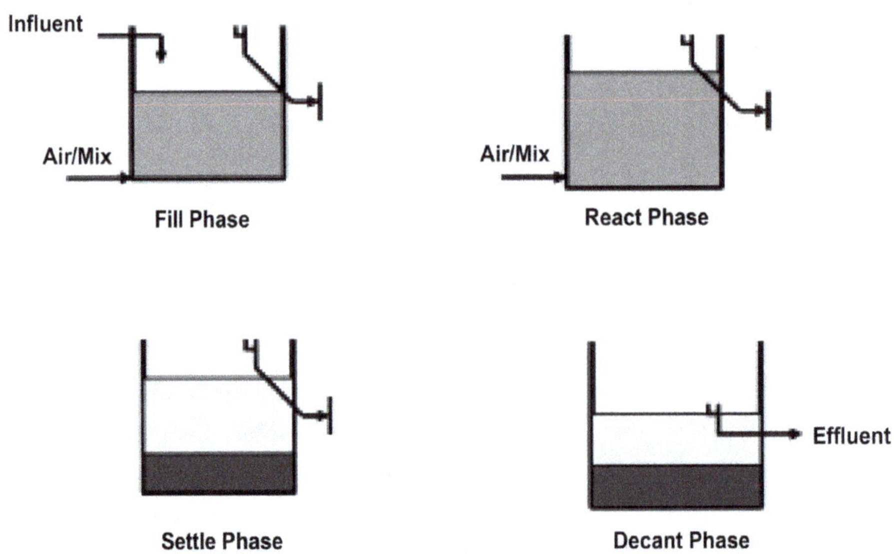


FIGURE 7.4 Sequencing Batch Reactor Process Phases

### 3.3.3 Operational Features

Sequencing batch reactors can provide very effective nitrogen removal. They are operated using a programmable logic controller (PLC) that has user-input cycle times that control the time of the fill, react, settle, and decant cycles. One important aspect of the SBR is the ability to change these cycle times to dial-in on performance. An average effluent total nitrogen concentration of less than 5 mg/L can be reliably and routinely expected depending on how the cycle's phases are controlled and the influent wastewater characteristics. Effluent total nitrogen concentration of less than 3 mg/L are frequently achieved at some facilities. The greatest challenges related to SBRs are related to the significant hydraulic grade-line drop through the system, the difficulty removing floating material from the SBR tanks, and intermittent, high-rate decant that generally warrants equalization before downstream processes such as filtration and disinfection.

## 3.4 Simultaneous Nitrification and Denitrification

Biological removal of nitrogen relies primarily on two mechanisms: aerobic nitrification and anoxic denitrification. Generally, the two treatment regimes are carried out in physically separated aerobic and anoxic treatment zones (on in the case of SBRs, different phases). However, when the dissolved oxygen concentration is sufficiently low, it has been demonstrated that heterotrophs can denitrify and autotrophs can nitrify in the same reactor. This is termed the *simultaneous nitrification and denitrification* (SND) process.

### 3.4.1 Process Description

Studies by Trivedi and Heinen (2000) have demonstrated that SND can occur in the same reactor, arising from the following mechanisms in suspended growth systems (Barnard et al., 2004; Kaempfer et al., 2000, Satoh et al., 2003; Stensel, 2001):

- Presence of anoxic and oxic micro-environments within sludge flocs—aeration promotes the dissolution of oxygen into water. The dissolved oxygen subsequently diffuses into the flocs and, along the diffusion path, is consumed by organisms. The diffusion and consumption cause a gradient of dissolved oxygen concentration in the flocs and, at suitably low dissolved oxygen, create flocs that nitrify in the oxic outer layer and denitrify in the anoxic inner core. It is important to note that similar processes have been demonstrated to occur across the dissolved oxygen gradient in denitrifying biofilm processes and in the aerobic granular sludge process. Operational approaches to target specific SND performance have not been firmly established. A discussion of SND in biofilms is not within the scope of these standards.
- Presence of macroscopic anoxic and oxic zones within bioreactors—such zones are commonly formed as a result of nonhomogeneous mixing and aeration, particularly in bioreactors with surface aerators in which dissolved oxygen is high near the aerators and low or zero away from the aerators.
- Presence of novel microorganisms—Robertson et al. (1988) reported that *Thiosphaera pantotropha* could simultaneously nitrify and denitrify under aerobic conditions. In addition, several bacteria were found to perform aerobic denitrification (Davies et al., 1989).

Oxidation ditches, as described in Section 3.6, are often configured to provide SND. Different configurations of oxidation ditches may provide low dissolved oxygen zones or channels as part of the reactor or sections that operate in a swing mode to vary with daily load conditions.

There has recently been a strong interest and increasing practice in achieving SND by operating a low–dissolved oxygen process for achieving SND (and indeed, some enhanced biological phosphorus removal as well) while reducing energy usage. Reducing operating dissolved oxygen concentration from the historical practice of 2 to 0.5 mg/L can reduce aeration costs by up to 30% (Miller et al., 2022), and it has been shown that the expected reduction in nitrification rate with low dissolved oxygen is at least partially recovered by the acclimation to low dissolved oxygen conditions and an observed decrease in the half-saturation constants for the nitrifying bacteria (ammonia-oxidizing organisms and nitrite oxidizing organisms) (Kisling et al., 2022).

### 3.4.2 Expected Process Performance

Because of the lack of dependence on mixed liquor recycle for achieving denitrification, use of SND can result in higher total nitrogen removal than MLE systems, although accomplishing SND typically requires longer solids retention time (SRT) and larger bioreactor volumes than staged processes. Tight dissolved oxygen control must be maintained because SND can be lost if the dissolved oxygen concentration is allowed to climb too high or drop too low.

### 3.4.3 Operational Features

This process offers an advantage by potentially eliminating the MLR stream and reducing energy use because of increased oxygen-transfer efficiency. However, Jenkins et al. (2003) found the process to use influent carbon less efficiently than the MLE or Bardenpho processes and determined that low dissolved oxygen conditions resulted in filamentous bulking. However, Schuyler et al. (2009) found no or few low dissolved oxygen filaments when the dissolved oxygen was consistently maintained below 0.4 mg/L.

## 3.5 Cyclic Aeration

Development of cyclic aeration technology began in Australia as early as 1965 (Goronszy, 1979). Many activated sludge reactor basins can be easily converted from continuous aeration to cyclical aeration by cycling mechanical aerators on and off or by alternating diffused aeration zones by electrically or pneumatically actuated valves.

### 3.5.1 Process Description

The following are process features of cyclic aeration systems:

- Short cycle times (4 hours and less), introduction of biological selector zones using transverse partial baffle walls, fill sequence regulation for filamentous sludge bulking control, and a means of regulating the kinetics of oxidation-reduction potential depletion for enhanced biological phosphorus removal (Goronszy, 1979);
- Process control using in-basin respiration rates, which allowed permanent control of the metabolic activity of the biomass and, consequently, changed the principles of process operation from a time-based control to a demand-oriented process control;
- A clear water withdrawal system for high-rate decanting of up to 2.5 m of solids-free effluent within a short time from basins having surface areas of up to 8000 $m^2$ (i.e., multiple basins and multiple decanters); and
- Establishment of adjustable normal- and high-flow operating protocols.

### 3.5.2 Expected Process Performance

A properly operating cyclic aeration process typically achieves 85% to 90% total nitrogen removal.

### 3.5.3 Operational Features

A key factor that will likely determine if cyclic aeration can function successfully is if the existing system has the ability to maintain a sufficient aerobic SRT to be able to consistently nitrify with intermittent aeration. If the aerobic SRT (total SRT multiplied by the percentage of time aerating) is more than is necessary to maintain nitrification, the cyclical aeration process can reduce effluent total nitrogen.

## 3.6 Oxidation Ditches

### 3.6.1 Process Description

Oxidation-ditch processes use looped channels that provide a continuous circulation of mixed liquor. They are often referred to as "closed-loop" or "racetrack" reactors. The high recycle ratio is implemented by configuring the entire channel flow to recirculate in a closed loop, with the influent and effluent flow representing only a small fraction of the main channel flow. Because of the high recycle rate, oxidation ditches can be efficient NDN systems with properly configured anoxic and aerobic environments. Anoxic and aerobic environments are achieved by establishing a dissolved oxygen profile along the flow path rather than with discrete compartments. Different oxidation-ditch systems use either an aeration disc for oxygen transfer and mixing, turbine aerators, or a combination of vertical-drum mixers and diffused air to maintain the dissolved oxygen profile around the entire oxidation ditch. The resulting intermediate low dissolved oxygen environments create conditions in which simultaneous NDN can occur, as described in Section 3.4. Figure 7.5 presents a typical oxidation-ditch configuration with an external secondary clarifier.

### 3.6.2 Expected Process Performance

Oxidation ditches typically provide up to 90% total nitrogen removal. Most oxidation ditches operate at 6 to 10 mg/L effluent total nitrogen (Sen et al., 1990).

### 3.6.3 Operational Features

The oxidation ditch bioreactor is characterized by an ovular or "racetrack" shape. At certain locations along the racetrack, oxygen is transferred into the mixed liquor by surface aerators that also provide the motive force to move the mixed liquor around the track. The mixed liquor may go around the track 50+ times before discharging to the secondary clarifiers, and the varying dissolved oxygen profile around the

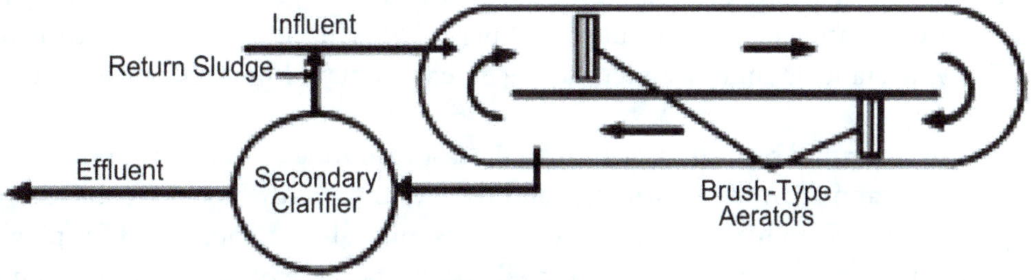


FIGURE 7.5 Oxidation-Ditch Process Schematic

track (highest at the aerator locations, lowest just upstream of the aerator locations) provides the alternating aerobic-anoxic conditions needed for nitrogen removal. In some systems, dedicated anoxic zones are added to the racetrack arrangement to achieve denitrification goals.

## 3.7 Step-Feed Configuration

Use of a step-feed activated sludge process configuration with anoxic zones (step-feed BNR) has been a popular method to achieve nitrogen removal without internal recirculation and minimal capital upgrades in existing WRRFs. This process entails feeding a portion of the influent to multiple points downstream of the front end of the reactor; each of the feed points has an anoxic zone for denitrification.

### *3.7.1 Process Description*

The step-feed configuration with anoxic zones provides conditions for denitrification of nitrate produced in the aerobic zones using the organic carbon in the step-fed wastewater without the need for internal nitrate recirculation and external carbon addition (Larrea et al., 2001). Typically, there are three or four "passes," although some facilities have as many as six passes. Small, mixed, anoxic zones are provided at each feed point, making raw wastewater carbon available to the nitrate created in the previous pass. This configuration is well suited for activated sludge facilities consisting of multiple passes in the aeration tanks. In general, by splitting the flow to several influent feed locations and directing RAS to the beginning of the first pass, a higher system SRT can be held than in a plug flow system with the same basin volume. The increase in SRT can be obtained without increasing the mixed liquor suspended solids (MLSS) concentration; as such, solids loading to the clarifiers is not increased.

### *3.7.2 Expected Process Performance*

Implementation of a step-feed BNR system for the removal of total nitrogen has proven successful at several WRRFs in achieving effluent total nitrogen concentrations in the 6 to 8 mg/L range (deBarbadillo et al., 2002; Fillos et al., 1996; Goodwin et al., 2005).

### *3.7.3 Operational Features*

Control of influent flow split is imperative for a successfully operating step-feed BNR system and can affect step-feed BNR system optimization in several ways, depending on which aspect of the BNR process is limiting. When the BNR system is fully nitrifying, denitrification may be limiting. Total nitrogen removal can be increased by adjusting the flow-splitting regime to minimize the quantity of unused carbon in the anoxic zones of certain passes. The RAS flow also has an effect on the performance of step-feed systems. Typically, RAS is discharged to the beginning of the first pass. If settling characteristics allow, increasing or reducing the RAS flow can affect total nitrogen removal. At lower RAS flows, the solids are more concentrated, which results in higher MLSS concentrations in the early passes and an increased SRT. A postanoxic zone is sometimes included to obtain lower effluent total nitrogen concentrations.

Minimizing carryover of dissolved oxygen from the last oxic zone of one pass to the anoxic zone of the next pass improves anoxic zone performance and reduces the readily biodegradable carbon consumed during reduction of residual dissolved oxygen. This must also be considered for plug flow BNR systems; however, it is even more critical for step-feed systems because of the need to minimize dissolved oxygen entering the anoxic zones of each pass.

Some facilities have experienced frothing events while operating in step-feed BNR mode at a high MLSS concentration (Chandran et al., 2003). Foam-control considerations for step-feed systems are similar to those associated with plug flow systems except that foaming can be more problematic with high MLSS concentrations in the early passes. Additional passes and zones in a step-feed system complicate control of foaming because there may be more locations for the foam to become trapped in the aeration tanks. The New York City Department of Environmental Protection found surface wasting, surface chlorination, and RAS chlorination effective in controlling froth (Chandran et al., 2003).

## 3.8 Integrated Fixed-Film Activated Sludge

### *3.8.1 Process Description*

As described in Chapter 4, integrated fixed-film activated sludge (IFAS) processes incorporate some type of media within a suspended growth activated sludge process to provide a fixed surface area for biofilm growth to supplement the suspended-phase solids inventory. The principal behind the IFAS process is to expand treatment capacity or to upgrade the level of treatment by supplementing the suspended biomass by growing additional biomass (biofilm) on media contained within the aeration tank. The additional biomass allows a higher effective rate of treatment within the existing process tanks, thus making other tank volume available to incorporate anoxic zones for denitrification or anaerobic zones for biological phosphorus removal within the same reactor volume. Advantages of using this type of process to expand or upgrade WRRFs are as follows:

- Additional biomass for treatment without increasing solids loading on final clarifiers;
- Higher-rate treatment, thus allowing greater treatment in a smaller space;
- SND;
- Similar operation to conventional activated sludge (CAS);
- Improved resistance to inhibitory compounds and washout;
- Increased robustness in colder temperatures because of an increased total SRT; and
- "Self-regulating" biofilm responds to process requirements and sudden load changes.

The IFAS systems also have disadvantages compared to suspended growth systems, including high capital costs for media and other system components, issues with media handling, and—most importantly—the potential vulnerability to media loss because of hydraulic limitations. Leaf et al. (2011) discuss issues with media loss and recommend methods for reducing risks.

Although media types vary, they can be grouped into two categories: moving bed and fixed media. Moving bed media include primarily plastic wheels and also sponge cuboids that are designed to remain in suspension and mixed using conventional aeration. Examples of plastic wheel are shown in Figure 7.6.

Fixed media include rope or hybrid buoyant media installed in cages or frames or plastic structured sheet media supported by spaced piers, which are placed within an activated sludge basin. Examples of fixed-type media are shown in Figure 7.7.

All types of IFAS media produce the same result of increasing the capacity of the treatment process within the existing reactor volume, with each type having specific advantages and disadvantages. These are summarized in Table 4.2 in Chapter 4.

FIGURE 7.6 Plastic Wheel (Moving Bed) Media (Courtesy of Headworks International Inc.)

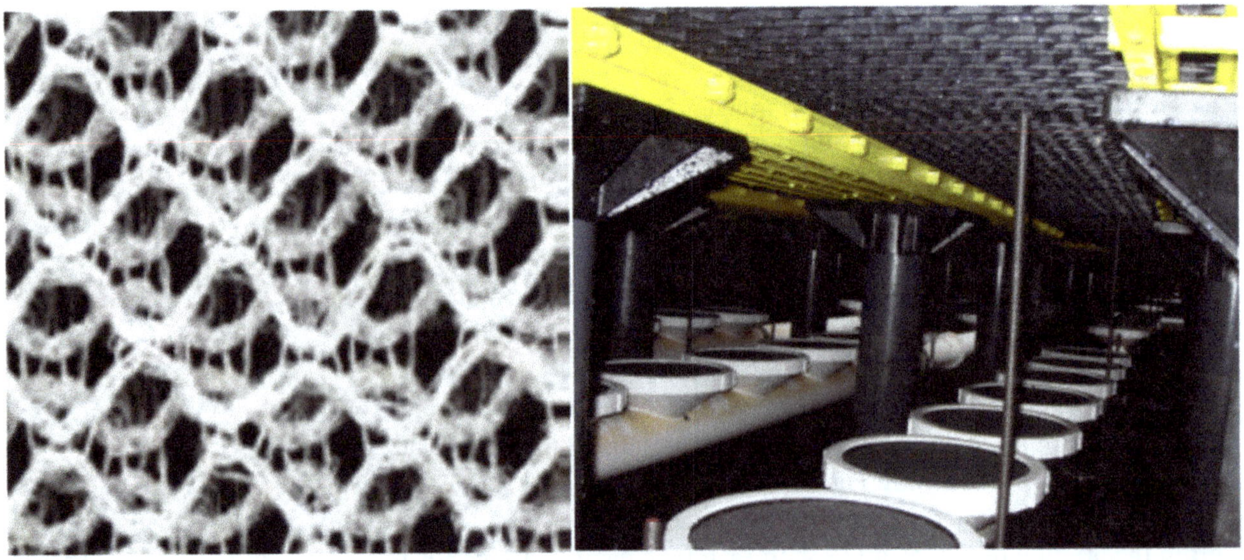

FIGURE 7.7 Hybrid Buoyant Media (Fixed) (Courtesy of Entex Technologies, Inc.) and Plastic Structured Sheet With Pier Support (Fixed) (Courtesy of Brentwood Industries, Inc.)

### 3.8.2 Expected Process Performance

Integrated fixed-film activated sludge installations effectively increase the treatment capacity of an activated sludge process within the existing tank volume without the need for additional secondary clarifiers. The actual increase in nitrification is directly related to the fill fraction or surface area of IFAS media that are added to a basin and the additional biomass that can be grown on the media. At the Broomfield, Colorado, WRRF, which was the site of the first installation of IFAS in North America, the addition of IFAS media doubled the nitrification capacity of the facility from 15 ML/d (4 mgd) up to 30 ML/d (8 mgd) within the same reactor volume (Rutt et al., 2006). Although optimal fill fractions of floating media are site dependent, they are in the range of 30% to 60% fill by volume. The Broomfield, Colorado, WRRF was originally expanded up to 30 ML/d (8 mgd) using a fill fraction of 48% during phase 1. However, the fill fraction was reduced to 30% for phase 2 (up to 45 ML/d [12 mgd]) based on full-scale operation data that were collected during phase 1 (Phillips et al., 2008).

Simultaneous nitrification and denitrification have been observed in many IFAS facilities because low dissolved oxygen in the inner layers of the biofilm results in anoxic conditions. The degree of reliable denitrification that occurs in these inner layers is difficult to anticipate. As a result, this SND is often considered a "bonus" of IFAS processes rather than an operational requirement.

### 3.8.3 Operational Features

#### 3.8.3.1 Reactor Configuration

The IFAS process schematic may be configured similarly to many of the process configurations described earlier in this chapter (e.g., MLE and Bardenpho), with media installed in the aerobic zone(s) of the process. Often, the most efficient configuration is to have more than one aerobic reactor in series, with no media installed in the first aerobic zone. The media provide the most benefit in achieving nitrification, and using the first zone as a carbonaceous BOD removal zone, followed by a media-filled zone for nitrification, increases the nitrification rate of the biofilm.

#### 3.8.3.2 Media Retention, Influent Screens, and Foam Management

Integrated fixed-film activated sludge systems require process modifications that are unique to the media type being considered. Floating media require the installation of wedge-wire media retention screens through the cell dividing walls (see Figure 4.4, Chapter 4) to prevent floating media from entering the secondary clarifiers and other treatment processes. The cylindrical screens installed through the wall are designed to encourage contact with floating plastic media to agitate and clean any accumulated biogrowth or debris off of the screens and floating media. The top of the wall screens retains the media while allowing the passing of foam and floatables on to the secondary clarifiers for removal. Although fixed-type media do not require retention screens, they have media-specific fine influent screen requirements to prevent debris from accumulating on the fixed-media structures within the basins.

Influent screening and foam management are important components to consider in IFAS systems. In most applications of free-floating media, media retention screens are a wedge-wire configuration with a smaller effective passing orifice than influent bar screens that can allow long, fibrous debris to pass through the bars. Floatables and debris that pass through the headworks screens can accumulate on media retention screens and create an operational challenge. Therefore, installing finer headworks screens or a debris removal method within aeration basins should be considered. Similarly, the top of wall screens can impede the passing of foam on to the secondary clarifiers. Foam-causing organisms that are retained can proliferate and cause increased foaming compared to conventional systems. Therefore, foam removal and management needs to be evaluated when considering an IFAS system.

#### 3.8.3.3 Aeration Systems and Dissolved Oxygen Concentrations

The biological treatment feature of an IFAS system is a biofilm grown on fixed media. To achieve equal performance to a suspended growth system and effectively activate most of the biomass within the layers of the biofilm, residual dissolved oxygen concentrations of 3 to 4 mg/L need to be maintained (Stricker et al., 2009). Aeration system requirements vary between IFAS media types. Although less efficient than fine-bubble diffusers, coarse-bubble diffusers may be required if a method for removing the media to perform diffuser maintenance is not available. Coarse-bubble aeration also increases agitation of the media and reduces debris accumulation and excess biogrowth on the exterior of media carriers. Operation costs can, therefore, be much greater for IFAS systems than suspended growth systems that operate at lower bulk dissolved oxygen concentrations and with more efficient aeration systems. Operating at higher dissolved oxygen concentrations can also interfere with efficient denitrification if oxygen is recycled into an anoxic zone.

#### 3.8.3.4 Media Management

Maintenance requirements for aeration systems or other equipment can require a basin to be taken out of service, thereby requiring the removal of IFAS media. Fixed-type media are often installed in modules that can be removed with cranes or other lifting equipment. Some floating media such as sponge-type cuboids can be pumped to an adjacent basin or holding tank. Other types of floating media such as plastic carriers are more difficult to remove from a basin. Less-maintenance-intensive aeration systems such as coarse-bubble diffusers should be considered when a media management method is not readily available. Proper media management systems should be evaluated and coordinated with each manufacturer being considered in an IFAS application. A summary of the benefits and drawbacks of each IFAS media type is provided in Table 3.2 of Chapter 3.

## 3.9 Membrane Bioreactor

### 3.9.1 Process Description

Membrane bioreactors (MBRs) are a form of activated sludge treatment that use membrane filters instead of clarifiers to separate the mixed liquor in the bioreactor from treated effluent. The most common form of MBRs uses a membrane tank, which has suction-type membranes submerged directly in the mixed liquor in a tank separated from the bioreactor. The treated water or "permeate" is pulled through the outside of the membranes and transported inside the membranes ("outside-in" using low-pressure pumps). The first version of the MBR and several current versions use pressure-type ("inside-out") membranes located outside the bioreactor, with current versions using larger-diameter membranes to avoid plugging issues associated with the early pressure-membrane MBR. Membranes used are either in the microfiltration (0.1- to 0.4-μm openings) or ultrafiltration (0.04- to 0.1-μm openings) categories. There are two basic types of suction membranes that are used: hollow fiber and flat panel.

With final clarifier solids loading not being a concern, MBRs are able to operate at high MLSS concentrations. Typical MBR design is for 8000 mg/L MLSS in the bioreactor and a maximum of 10 000 mg/L in the membrane tank. Because of the emphasis on energy use, there is a trend to design MBRs for lower MLSS concentrations because of the lower oxygen-transfer efficiency at higher MLSS (if space requirements allow for the construction of larger bioreactors).

Coarse-bubble aeration and periodic chemical cleaning with chlorine and acid are used to maintain the design flow (flux) rate through the membranes. The energy required for aeration of the membrane tank is one of the biggest disadvantages of an MBR, resulting in continuing advancements designed to reduce this energy requirement by using more efficient aeration systems.

Current thinking in MBR design is to size the bioreactor for an aerobic SRT that it is sufficient to produce a fully nitrified effluent. Because the aerobic SRT required for full nitrification varies with wastewater temperature, the design aerobic SRT for an MBR will vary similar to any activated sludge process designed for full nitrification.

Membrane integrity is monitored using online turbidity meters. If a membrane becomes ruptured, the effluent turbidity will suddenly increase. Most membrane systems tend to be "self-healing" to a degree; however, membrane repair or replacement may be necessary if a permanent increase in turbidity occurs. Membrane life is typically 5 to 10 years or even longer.

Nitrogen removal configurations for MBR systems are generally the same as those for conventional systems, as described previously, although they typically have somewhat different recycle processes and rates. The one significant difference is related to the RAS rate in an MBR. When designed for typical MLSS values of 8000 mg/L in the bioreactor, a RAS rate of 4 times the influent flow is required to keep the solids concentration below the required 10 000 mg/L in the membrane tank. Because this is a typical mixed liquor recirculation rate for a BNR configuration for total nitrogen removal, the RAS stream of the MBR can also provide the necessary mixed liquor recirculation for nitrogen removal in a BNR process.

The RAS flow from the membrane tank can contain high concentrations of dissolved oxygen (typically from 2 to 6 mg/L) because of membrane scouring. This dissolved oxygen will reduce the soluble BOD necessary for the denitrification process. For this reason, one (of many) design and operating approaches involves returning the RAS stream of an MBR to the aeration tank, depleting the dissolved oxygen in that reactor, and then using separate MLR pumps for the return of the lower dissolved oxygen mixed liquor to the anoxic zones, similar to conventional WRRFs. This also lowers aeration requirements in the bioreactor by using the dissolved oxygen in RAS for process oxygen. What is lost with this approach are the more concentrated solids in the membrane tank and RAS, which results in a higher solids concentration in the anoxic zones, thereby increasing nitrogen removal efficiency.

Because of these factors, design and operational troubleshooting for nitrogen-removal MBR facilities should be based on process modeling to properly evaluate all of these factors. Figure 7.8a shows the simplest configuration of a nitrogen-removal MBR that is typically appropriate for meeting limits of 8 mg/L total nitrogen. Figure 7.8b shows the nitrogen-removal MBR with internal recycle that is capable of achieving slightly lower total nitrogen limits, depending on the dissolved oxygen in the membrane tank. If dissolved oxygen in the membrane tank is 6 mg/L or higher, this configuration will provide slightly better total nitrogen removal; however, at 4 mg/L or less of dissolved oxygen in the membrane tank (as is achievable in some of the more modern MBR designs), the single RAS configuration will provide equivalent total nitrogen removal; therefore, the cost of operating a separate mixed liquor recycle pump would not be justified.

### 3.9.2 Expected Process Performance

Because of the high degree of filtration, MBRs can achieve effluent total nitrogen concentrations equivalent to that achievable with denitrification filters without the use of external carbon sources. Second-stage anoxic zones following the aerated portion of the bioreactor are required to achieve these low-effluent total nitrogen concentrations.

### 3.9.3 Operational Features

The most significant operational issue with MBRs is the potential for membrane biofouling. The higher the MLSS concentration and the higher the flux rate, the greater the potential for fouling. Additionally, as the potential for fouling increases, the frequency of cleanings that are required increases. The need to

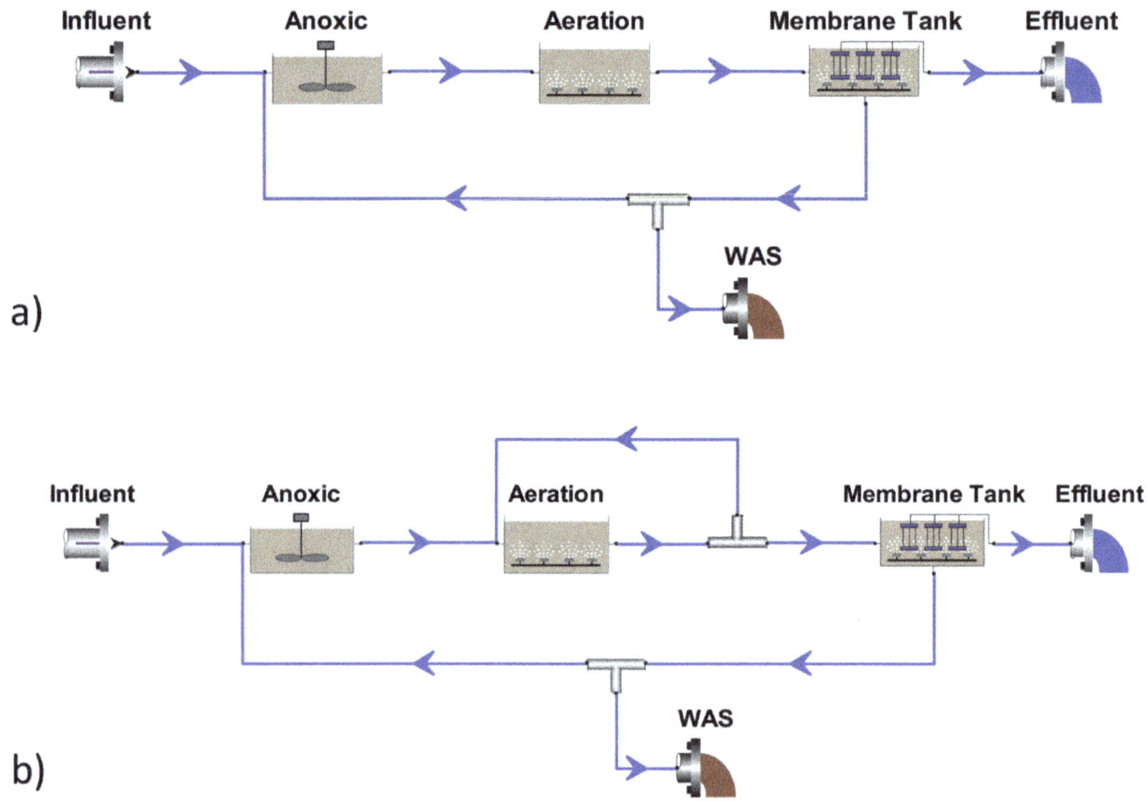


FIGURE 7.8 (a) Basic Total Nitrogen Removal MBR Configuration and (b) Separate Mixed Liquor Recycle Total Nitrogen Removal MBR Configuration (shown in BioWin™ as developed by EnviroSim Associates Ltd.)

clean a membrane is determined by the pressure differential across the membrane. The initial buildup of solids on the membrane surface improves performance; however, as the solids continue to accumulate, the pressure loss across the membrane increases. Increasing the suction pressure to overcome pressure loss may maintain the desired flowrate for a limited period but cannot be sustained long-term.

There are several operational changes that can be implemented to minimize biofouling. Fouling may be reduced by increasing the turbulence induced by the coarse-bubble aeration system beneath the membranes, which reduces the thickness of the biofilm. There is evidence that the primary cause of biofouling are extracellular polymeric substances (EPS) that form on the surface of the bacteria that comprise the biofilm. Extracellular polymeric substances tend to clog the smaller pores in the cake layer surrounding the membrane. Reducing the amount of EPS reduces filtration resistance. The amount of EPS is influenced by SRT, with the degree of EPS formation increasing at low and high SRTs. Improvements can also be realized by adding a flocculant, which creates larger particles in the cake layer, thereby providing larger pores in the biofilm and reducing filtration resistance.

There are several cleaning cycles required to maintain MBR operation. Typically, there is a frequent but short duration backwash using permeate or filtered wastewater to flush solids from the membrane surface. Periodically, a more extensive backwash is performed with chemical cleaning agents. A more extensive cleaning operation in which the membrane is removed from service and washed in a chemical bath is typically only required once every 6 to 12 months depending on the waste and operating conditions. Most backwash and cleaning operations can be automated.

Because of the trapping nature of the membrane filtration process, foaming is one of the most serious operational issues with MBR facilities. Excessive foam can result in a large amount of old sludge in the system and the release of cellular nitrogen and phosphorus into the effluent. The most reliable method of preventing excessive foaming is incorporating a surface wasting method to the design and operation of the MBR. Skimming pipes or weirs are used in areas where the mixed liquor passes through a narrow channel, such as the channels between the bioreactor and the membrane tank. Either all or a part of the waste activated sludge (WAS) is then taken from the surface of the tank, preventing the buildup of foam. Because of the high solids concentration in an MBR, downstream solids handling processes will be loaded with similar concentrations as that of WAS from a conventional facility.

### 3.10 Process Comparison Summary

Table 7.3 presents a summary of the process configurations described in this section.

## 4.0 RELATED SYSTEMS AND EQUIPMENT

### 4.1 Primary Clarification

Operation of primary clarifiers is important when operating nitrogen removal facilities because the overall capacity of the nitrification process and the efficiency of the denitrification process are directly related to the amount of COD in the influent to the bioreactor. Primary clarification will typically remove 30% to 50% of raw wastewater COD without the addition of coagulant chemicals and more than 80% COD with the addition of coagulant chemicals. This reduced COD loading will result in less waste activated sludge generation and thereby increase the capacity of the aerobic process (nitrification). However, typically less than 10% of total nitrogen is removed during primary clarification; as such, the COD-to-total-nitrogen ratio is significantly lowered by primary clarification, which reduces the efficiency of the denitrification process. These effects must be considered in the management of primary clarifier operation.

### 4.2 Secondary Clarifiers

Based on empirical measurements by McCarty (1970), typical effluent TSS from biological treatment units contain approximately 12% nitrogen and 2.2% phosphorus. This translates into 1.2 mg/L of total nitrogen for every 10 mg/L of effluent TSS. With the exception of some of the biological filter processes and the MBR, the biological nitrogen removal processes presented above depend on the efficient capture of biological solids by secondary clarifiers. It is critical that secondary clarifiers are operated to achieve a consistently high level of treatment to both protect the water environment and avoid the high cost of downstream tertiary treatment. Although many of the basic principles of suspended and biofilm growth secondary clarification are the same, the units have some key differences related to upstream biological processes.

#### *4.2.1 Clarifiers in Suspended Growth Systems*

Suspended growth (activated sludge) reactors and clarifiers must be designed and operated as a coordinated system to provide a consistently high level of performance. It is crucial that the operations staff responsible for process control decisions enhance their understanding of clarifier capacity and operating procedures. Incorporation of nitrogen removal to an activated sludge facility will typically result in improved settling characteristics because of the selection of floc-forming organisms and the suppression of filamentous organisms in the anoxic zones. With the exception of non-nitrifying facilities, BNR facilities will result in lower concentrations of nitrate in the clarifier and decreased problems associated with

TABLE 7.3 Process Comparison Summary

| Process | Features, Benefits, and Drawbacks |
|---|---|
| Modified Ludzack–Ettinger | Two-stage process: anoxic-aerobic; internal mixed liquor recycle recycles nitrate from the aerobic zone to the anoxic zone at a rate of 3 to 5 times the influent flow; achieves moderate total nitrogen removal (effluent total nitrogen: 6 to 8 mg/L); low tank volume requirement compared to other multiple-stage nitrogen removal processes. |
| Four-stage Bardenpho | Four-stage process: anoxic 1, aerobic 1, anoxic 2, and reaeration; internal mixed liquor recycle recycles nitrate from aerobic 1 to anoxic 1 at a rate of 3 to 5 times the influent flow; achieves excellent total nitrogen removal (effluent total nitrogen: 2 to 4 mg/L); larger footprint required compared to fewer stage flow through nitrogen removal processes. |
| Sequencing batch reactor | Single reactor tanks that cycle between aerobic and anoxic conditions to achieve nitrogen removal; wastewater influent cycles between reactor tanks to balance influent organic and nitrogen loading; each reactor cycles between four to six phases depending on the application and nutrient limits required; secondary clarifiers not required; secondary settling occurs within the SBR reactor; relatively large reactor volumes; typical application: WRRFs with a capacity range of 3.8 to 38 ML/d (0.1 to 10 mgd). |
| Simultaneous nitrification-denitrification | Single reactor tank that operates at a low dissolved oxygen concentration, sufficient to maintain aerobic nitrification but low enough to denitrify; significant energy savings due to reduced aeration costs. |
| Cyclic aeration | Process used to achieve nitrification and denitrification within the same reactor volume. Alternating aerobic conditions and anoxic conditions per a timed schedule by the use of controlled air distribution to balance both processes and achieve total nitrogen removal. |
| Oxidation ditches | Bioreactors configured in a carousel, where flow continually flows around the ditch with influent and effluent weirs controlling flow based on water-level elevations; continual flow of nitrate from the effluent to the influent end of the ditch eliminates the need for internal recycle that is common with other flow-through nitrogen removal processes; achieves moderate total nitrogen removal (effluent total nitrogen: 3 to 7 mg/L). |
| Step feed | Includes the addition of a portion of influent flow at multiple points in a flow-through treatment process; addition of carbon from influent wastewater eliminates the need for internal recycle; lower volume requirements compared to CAS. |
| Integrated fixed-film activated sludge | Supplemental process used to increase the treatment capacity of an existing BNR process; addition of fixed or floating media to the aerobic zones of bioreactors to encourage biofilm growth and increase the solids inventory; most applicable for WRRF locations that are space limited, subject to a wide range of influent loadings, colder climates, or subject to toxic shock because of inhibitory substances in the influent; higher operations costs compared to CAS. |
| Membrane bioreactor | Solids separation process applicable to any flow-through BNR process; eliminates the need for secondary clarification; typical MLSS: 8000 to 12 000 mg/L; smaller bioreactor volumes because of increased MLSS; higher operations costs compared to CAS; nitrogen removal similar to CAS. |

denitrification in the sludge blanket. Generally, reducing nitrate to less than 10 mg/L will help minimize denitrificaion issues in the sludge blanket.

The Water Environment Research Foundation Clarifier Research Technical Committee's activated sludge secondary clarifier evaluation protocol defines a step-by-step procedure for evaluating secondary clarifier capacity and troubleshooting poor clarifier performance (WERF, 2001). Poor activated sludge secondary clarifier performance has been associated with one or more of the following conditions (WERF, 2001):

- Sludge blanket denitrification;
- Sludge particle flocculation problems;
- High sludge blankets caused by solids overload; and
- Poor clarifier hydraulics.

With the possible exception of clarifier hydraulics, operators have corrective action measures that can be used to address these conditions. The actual measures undertaken depend on site-specific conditions that exist at the time of the problem. (Refer to Appendix A for a troubleshooting and process optimization guide.)

### *4.2.2 Clarifiers in Biofilm Systems*

Secondary clarifiers for biofilm systems have not experienced the same level of interest or research as those for activated sludge systems. Two of the conditions identified as challenges to suspended growth system clarifier performance, flocculation and hydraulic problems, have been identified as the primary cause of problems with biofilm system clarifier performance as well (Water Environment Federation [WEF], 2010). For moving bed biofilm reactor (MBBR) systems, dissolved air flotation clarifiers are commonly used as an alternative to conventional clarifiers because of the light nature of the MBBR floc. (Refer to Chapter 4 for a detailed discussion of this topic.)

## 4.3 Aeration Systems

In the context of nitrogen removal, the fundamental purpose of aeration systems is to provide the process oxygen necessary for nitrification as the first step of the nitrogen removal process. The sensitivity of nitrifying organisms to environmental conditions makes this the "weak link" in the nitrogen removal process, and maintaining sufficient dissolved oxygen in the bioreactor is critical to ensuring a healthy population of nitrifiers. Additionally, the nitrification rate is directly related to the dissolved oxygen concentration, with the nitrification rate dropping quickly as the dissolved oxygen concentration drops below 1 mg/L. With the discovery of nitrifying organisms such as *Nitrospira* that can efficiently nitrify at lower dissolved oxygen concentrations, the traditional practice of maintaining a dissolved oxygen concentration at 2 mg/L or greater may not be necessary and in some cases may be detrimental to overall nutrient removal effluent quality. This is a critical concern because the implementation of nitrogen removal in WRRFs has caused a fundamental shift in operational control strategies of aeration systems. Whereas historically the emphasis has been on maintaining an excess of dissolved oxygen in the bioreactor, this philosophy can be detrimental to the efficiency of denitrification processes. This is because of the reliance of denitrification processes on internal recycle streams within the bioreactor.

With up to 4 times the influent flow typically being recirculated to anoxic zones, the concentration of dissolved oxygen at the point in the bioreactor where the recirculation stream is being taken must be kept to a minimum to prevent "oxygen poisoning" of the anoxic zone. This can be done by either incorporating a dissolved oxygen control zone (i.e., a zone designed to reduce dissolved oxygen in the recycle stream) or providing a mixer at the end of the bioreactor to allow the operator to not have to depend on the aeration system for mixing at a point at which oxygen demand is low.

## 4.4 Anoxic Zone Mixing

Mixing of anoxic zones and the portions of aerobic zones where aeration is decreased below the minimum value required for mixing is a critical feature of nitrogen removal processes. It is required to keep solids in suspension and/or "degassing" mixed liquor before secondary clarification. The following are general guidelines to consider with anoxic zone mixing:

- Mixers should be configured such that no oxygen is entrained into the mixed liquor.
- Position of mixers should minimize localized backflow into the preceding zones.
- Baffling and reactor configuration must be considered in the mixing evaluation (WEF, 2023).

Mixing can be accomplished by a variety of means and equipment including submersible horizontal propeller-type mixers, pumped-jet mixing, large-bubble compressed air, vertical turbine mixers, and hyperboloid mixers. Variable-intensity systems (such as a variable-frequency drive [VFD]-equipped mixer) provide an opportunity to "dial in" mixing intensity while saving energy. Anoxic zone mixing requirements vary for each application and are specific to the reactor configuration and process requirements. The features of each mixing type are discussed in the following subsections.

### *4.4.1 Submersible Horizontal Propeller*

Submersible horizontal propeller mixers are installed on the sidewalls of an anoxic zone reactor and rotate to produce turbulence and keep the MLSS in suspension (Figure 7.9). Propeller mixing has been

FIGURE 7.9 Submersible Propeller Mixer

successfully used in anoxic zones for many years and at less capital cost than other types of commonly used mixing technologies. However, submersible horizontal propellers often have higher maintenance requirements because of submerged motor and debris accumulation on the propellers. Operation costs for propeller mixing are also higher than other mixing technologies because of the higher power requirements, depending on the reactor configuration.

### 4.4.2 Jet Mixing

Jet mixing consists of a mixing pump that can be submerged in the basin or located on top of the basin that discharges to a header containing a series of jet nozzles to produce turbulence and keep solids in suspension. The advantage of jet mixing is that the pumps can be installed outside of the basin, requiring no routine maintenance within the tank. Operators are often already familiar with maintenance requirements because the pumps used in jet-mixing systems are common to other processes in a WRRF.

### 4.4.3 Large-Bubble Compressed Air

Large-bubble mixing systems consist of a compressor, receiving tank, solenoid valve panel(s), and large-bubble diffuser grid(s). Solenoid valves operate in a defined sequence pattern to release pulses of air into a distribution header with large-bubble diffusers. The diffusers produce large bubbles ranging from marble to softball size that keep the MLSS in suspension (Figure 7.10). Although bubbles are introduced into nonoxygenated zones, anoxic environments are maintained that are fully compatible with BNR requirements. Full-scale oxygen-transfer studies have been conducted between large-bubble systems and conventional submersible propeller-type mixers to confirm that the use of large bubbles does not impede the anoxic reaction because of oxygen entrainment (U.S. EPA, 2010b). Large-bubble mixing systems have lower power consumption and maintenance requirements than other forms of mixing. Capital costs of large-bubble mixing systems are often higher than other conventional mixing alternatives, although the more separate zones that require mixing, the more cost effective large-bubble systems become. Actual power requirements and operation costs are site specific and should be carefully evaluated for each application.

FIGURE 7.10 Large-Bubble Mixing (Courtesy of EnviroMix, Inc.)

### 4.4.4 Vertical Turbine

Vertical turbine mixers are mounted on platforms above the anoxic zone reactor and consist of a drive motor and gearbox, vertical shaft, and impeller (or turbine). Flow is directed toward the floor of the tank, where it is redirected up the sides of the reactor to mix the tank contents. Compared to submersible-style mixers, vertical turbine mixers offer the advantage of having the motor and electrical components located above the tank rather than being submerged.

### 4.4.5 Hyperboloid

Hyperboloid mixers are vertical mixers with a drive motor installed on top of the tank connected to a vertical shaft and hyperboloid-shaped mixer body that is installed relatively close to the tank bottom. Hyperboloid mixers offer the advantages of reduced power consumption compared to other vertical and horizontal propeller- or turbine-type mixers and a reduction in operation and maintenance requirements. The hyperboloid shape and configuration also reduce the potential for debris accumulation.

## 4.5 Internal Recycle Pumping

Internal recycle pumping is critical to denitrification and BNR systems. It is used to return nitrate to the upstream anoxic zone to use influent carbon in the raw wastewater to fuel denitrification and alkalinity recovery, as discussed in Chapter 5. The denitrification process also produces a net oxygen demand reduction because carbon (COD or BOD) is being degraded under anoxic conditions rather than aerobically with aeration energy. Internal recycle rates are dependent on the process goals and configuration. Two types of pumps are commonly used for internal recycle: axial flow propeller pumps and submersible nonclog pumps. Internal recycle pumping systems equipped with VFDs are able to optimize recycle rate based on performance of the pre-anoxic zone.

### 4.5.1 Axial Flow Propeller Pumps

Axial flow propeller-type pumps are known as "high flow–low head" pumps and represent the most common type of internal recycle pump. These pumps typically operate in a head range of 0.9 to 1.8 m (3 to 6 ft). Axial flow pumps are less costly and have lower power consumption and operation costs compared to submersible nonclog pumps because of the low head requirement. However, axial flow pumps have been known to have ragging issues in facilities where only coarse (6-mm [.25-in.]) or greater) influent screens are installed or where ragging is known to be an operational issue.

### 4.5.2 Submersible Nonclog Pumps

Submersible nonclog-type pumps have been used in wastewater applications for many years in influent pumping, RAS, and internal recycle because of their ability to pass solids that are found in raw wastewater and sometimes in mixed liquor. Submersible nonclog pumps typically operate in a head range of 5 to 15 m (15 to 35 ft). Therefore, their power consumption and operating costs are higher than axial flow propeller pumps. Submersible nonclog pumps have the advantage of passing solids and typically have less maintenance requirements than axial flow propeller pumps in an application in which ragging issues are known to exist. The pumping properties of submersible nonclog pumps also offer greater flexibility in flow variation when a VFD is added compared to an axial flow propeller pump. With VFD, operators can adjust the flow as needed based on process requirements and environmental conditions.

## 4.6 Staging

Staging is used in nitrogen removal systems through the implementation of baffles for a variety of purposes, such as the following:

- Separation of aerated and unaerated zones;
- Separation of diffuser zones in a tapered aeration aerobic zone; and
- Nitrified mixed liquor recycle pump baffle to reduce the amount of oxygen recycled to the anoxic zone (WEF, 2023).

Baffles can be constructed of a variety of materials including cast-in-place and precast concrete, high-density polyethylene, stainless and painted carbon steel, wood, fiber-reinforced plastic, and hanging curtains. Baffles used to separate reactor and diffuser zones should be placed such that they result in minimal head loss across the baffle. Operation and maintenance of diffusers and other reactor components should be considered when installing baffle walls. Baffle layout should not result in capture and trapping of surface scum. Drain ports are needed in the bottom of the baffle to allow for complete draining of the tank for maintenance purposes so that no water is retained between baffled zones. Openings installed in baffle walls should be sized for operator access between zones without having to exit each baffled zone through the top of the reactor on a ladder or stairwell.

## 4.7 Supplemental Carbon Systems

The process requirement for the provision of supplemental carbon to improve denitrification is described in detail in Chapter 5. Once it is determined that supplemental carbon is required to meet process goals, the next step is to determine which of many available sources of supplemental carbon are best suited for a particular application (WERF, 2010).

Components of supplemental carbon storage and feed systems are similar to most other chemicals used in wastewater treatment processes: one or more bulk storage tanks, one or more chemical feed pumps and associated piping, and a means of controlling chemical dose. Two of the most commonly used sources of supplemental carbon (methanol and ethanol) are hazardous and highly flammable; therefore, storage and feed systems must be carefully designed, constructed, maintained, and operated to comply with all necessary safety standards and procedures.

Methanol has historically been the most economical alternative carbon source and is typically used to increase denitrification rates in second-stage anoxic zones and tertiary denitrification processes. Handling and feeding methanol must take into account the dangers associated with its use. Methanol represents an explosion hazard and is toxic when ingested. Methanol should be stored in mild or stainless-steel tanks; plastic, galvanized, and aluminum tanks are not suitable. Grounding is important to protect against static electricity. Carbide-tipped clamps (to ensure good contact through paint) and dip-tube filling are generally used to guard against ignition from static electricity. Proper venting of tanks is important because of the volatility of methanol, and tanks should be painted white to reduce methanol vapors.

Piping for these chemicals must comply with Section 30 of the National Fire Protection Association Standards, and all indoor spaces through which the piping passes must be suitably equipped with fire-detection and prevention systems. Prolonged or repeated breathing of methanol vapors should be avoided at all times. Proper ventilation is required to ensure safe working conditions. The type of ventilation will depend on such factors as dead air spaces, temperature, convection currents, and wind direction and must be considered when determining equipment location, type, and capacity. If mechanical ventilation is used, spark-proof fans should be implemented. Because of the explosive nature of methanol, smoking and welding around methanol facilities must be avoided.

Other, less hazardous options include glycerin-based products and acetic acid (refer to Chapter 5).

## 4.8 Froth Control and Scum Removal

Froth, foam, and scum can be common nuisances at WRRFs. Scum and foam accumulation occurs because of a variety of causes including grease, soaps, scum, detergents, and other surfactants in the influent, operating at a solids retention time that is conducive to proliferation of foam-causing bacteria, and trapping and recycling of foam-causing bacteria in the process. The following techniques are commonly used for foam- and scum-control measures in the operation of BNR systems:

- Careful placement of over and under baffles, inverted chimneys to direct the flow to the bottom of the following basin, and peak-flow overtopping weirs, as appropriate, in the partitions between the various zones of a bioreactor. There should be no back-mixing from the aerobic to the anoxic zone, and foam should spill freely from the anoxic to aerobic zones. Partitions within any zones should be just below the surface to avoid contact areas for scum to grow.
- Implementation of continuous selective wasting for the preferential removal of foam and scum organisms from the aeration basin as part of the WAS stream. Recycling of this foam to the facility should be minimized because it will reseed the foaming organisms.
- Elimination of dead ends, dead corners, and other quiescent zones in channels or bioreactors where there is a potential for foam and scum to accumulate.
- Installation of an effective scum removal system on primary and secondary clarifiers.
- Avoidance of opportunities for recycling foam and scum organisms to the mainstream treatment train from sidestream solids processing facilities.
- Use of chlorine sprays, as necessary, at localized points of foam and scum collection and/or accumulation to kill these organisms and prevent them from causing problems in either mainstream or sidestream treatment processes. Early microscopic detection is important to minimize the extent of the problem. Alternatively, there are some polymers that could be sprayed on the foam.
- Effective destruction of foam and scum removed from primary and secondary processes in the solids handling process through digestion, incineration, composting, or other solids handling methods.

## 4.9 Alkalinity Addition

Alkalinity is a measure of a wastewater's acid-neutralizing capacity. Alkalinity and pH are closely related and are important in BNR facilities. The nitrification process consumes alkalinity, which, in turn, causes the wastewater pH to drop. As the pH drops, the rate of nitrification can decrease and stop at a pH of approximately 6. Alkalinity supplementation may be needed to support nitrification in some WRRFs or to support nitrification at certain times of the year. Alkalinity consumption for nitrification must be accounted for to determine if alkalinity supplementation is necessary.

When alkalinity supplementation is required, there are several chemicals that can be used. The preferred chemical is influenced by local conditions, local chemical prices, and operator preferences. Chemicals that can be used as an alkalinity supplement include the following:

- Sodium hydroxide (caustic soda)—$NaOH$;
- Calcium hydroxide (lime)—$Ca(OH)_2$;
- Calcium oxide (quick lime)—$CaO$;

- Magnesium hydroxide—$Mg(OH)_2$;
- Sodium carbonate (soda ash)—$Na_2CO_3$; and
- Sodium bicarbonate—$NaHCO_3$.

### 4.9.1 Sodium Hydroxide

Sodium hydroxide (commonly called *caustic* or *caustic soda*) is used for alkalinity supplementation because of its ease of handling. Sodium hydroxide is often not the lowest-cost chemical for alkalinity addition; however, it is easy to use, and annual maintenance costs for the caustic storage and feed system tend to be low. Many utility operators believe that the ease of handling of sodium hydroxide far outweighs the higher cost of the chemical. Sodium hydroxide can be purchased at a 50% by weight solution strength, or in some locations, it is available at 20% or 25% by weight solution strengths. It is classified as a strong base and, if overdosed, can raise the pH much higher than expected or desired. An automatic system to control chemical dose based on pH can be used to prevent overdosing and optimize chemical usage. Once the liquid temperature drops below 12.8 °C, sodium hydroxide will begin to crystallize out of solution. Heat-tracing systems may be used on the storage tank and piping system components to prevent crystallization. It is difficult to redissolve crystallized sodium hydroxide. If sodium hydroxide is delivered to the site at 50% strength and then diluted on-site with facility water or potable water, scaling will occur at the point of mixing. The local pH will rise well above pH 10, and calcium carbonate will form immediately.

### 4.9.2 Calcium Hydroxide

Calcium hydroxide, or hydrated lime, has been slaked by the manufacturer and is sold as a dry material. Hydrated lime must be slurried before use. The slurry tank will be susceptible to scaling if facility water or potable water is used because the hardness in these waters will precipitate. Calcium hydroxide is typically prepared in a 3% to 5% by weight slurry. A small fraction of the calcium hydroxide will dissolve and raise the pH to approximately 12, which is high enough to be hazardous to operators. Calcium hydroxide should be added at a point of turbulence to ensure sufficient mixing. Dosing point should be selected to prevent solids deposition.

### 4.9.3 Quicklime

Before being used, calcium oxide, or quicklime, must be slaked, a process in which the calcium oxide is mixed with water and allowed to react. Slaker operating temperatures are typically between 120 and 180 °C and the elevated temperatures increase the amount of scaling that can form. Quicklime is typically cheaper than hydrated lime. However, slaking is labor-intensive process, and operation costs must be considered in a life-cycle analysis.

### 4.9.4 Magnesium Hydroxide

Magnesium hydroxide is increasingly being used in BNR processes because it will not raise the pH above approximately 10.5, which correlates to its precipitation point range of pH 10.2 to 10.5. However, the effects of magnesium addition on other facility processes such as struvite formation in anaerobic digestion operations must be carefully considered. The costs are regionally driven; therefore, before selecting magnesium hydroxide for long-term use, delivery costs should be investigated. Magnesium hydroxide is generally sold as a slurry, although a dry product can also be purchased. Scaling will begin above pH 8. Dosing point should be selected to prevent solids deposition.

### 4.9.5 Sodium Carbonate

Sodium carbonate or soda ash is generally available as a dry product; therefore, it must be dissolved on-site. The maximum pH level of a sodium carbonate solution is approximately 12.

### 4.9.6 Sodium Bicarbonate

Sodium bicarbonate use is not widespread because other chemicals are easier to handle and cost less. Sodium bicarbonate is generally available as a dry product and, therefore, must be dissolved on-site. The maximum pH level of a sodium bicarbonate solution is 8.3.

## 5.0 OPERATIONS SUMMARY

The following are key operating take-aways for combined nitrifying and denitrifying systems:

- Establish a reliable and repeatable sampling program that provides sufficient data on which to base process control decisions.
- Nitrogen removal in conventional systems requires stable nitrification. It is vital to operate the process at a high enough aSRT to maintain sufficient populations of nitrifying organisms. When operating aSRT, wastewater temperature must be considered.
- If the design provides flexibility, selection of the best process configuration (e.g., MLE) should first consider effluent goals, and then optimization of aeration settings, chemical addition, and other process setpoints.
- Track carbon (BOD or COD), oxygen, and nitrogen throughout the process to determine opportunities to improve and optimize process efficiency. These data will help guide decisions regarding dissolved oxygen setpoint(s), internal recycle pumping rates, chemical addition requirements, and biomass inventory.

## 6.0 REFERENCES

Barnard, J. L. (1974). Cut N and P without chemicals. *Water and Wastes Engineering, 11*, 41–44.

Barnard, J., Steichen, M., & deBarbadillo, C. (2004, October 2–6). Interaction between aerator type and simultaneous nitrification and denitrification. *Proceedings of the 77th Annual Water Environment Federation Technical Exhibition and Conference* [CD-ROM], New Orleans, LA, United States. Water Environment Federation.

Chandran, K., Pape, R., Ezenekwe, I., Stinson, B., & Anderson, J. (2003, October 11–15). Froth control and prevention strategies for step-feed BNR at NYC WPCP. *Proceedings of the 76th Annual Water Environment Federation Technical Exhibition and Conference* [CD-ROM], Los Angeles, CA, United States. Water Environment Federation.

Davies, K. J. P., Lloyd, D., & Boddy, L. (1989). The effect of oxygen on denitrification in *Paracoccus denitrificans* and *Pseudomonas aeruginosa. The Journal of General Microbiology, 135*, 2445–2451.

deBarbadillo, C., Carrio, L., Mahoney, K., Anderson, J., Passarelli, N., Streett, F., & Abraham, K. (2002). Practical considerations for design of a step-feed biological nutrient removal system. *Florida Water Resource Journal, 54*(1), 18–35.

Fillos, J., Diyamandoglu, V., Carrio, L., & Robinson, L. (1996). Full-scale evaluation of biological nitrification/denitrification in a step-fed activated sludge process. *Water Environment Research*, *68*(2), 131–142.

Goodwin, S., Johnson, B., Daigger, G., & Crawford, G. (2005, May 28–June 1). A comparison between the theory and reality of full-scale step-feed nutrient removal systems. *Proceedings of the European Federation of Clean Air and Environmental Protection Associations (Warsaw, Poland) General Assembly Meeting and Conference,* Krakow, Poland.

Goronszy, M. C. (1979). Intermittent operation of the extended aeration process for small systems. *Journal Water Pollution Control Federation, 51*, 274.

Jenkins, D., Richard, M. G., & Daigger, G. T. (2003). *Manual on the causes and control of activated sludge bulking and foaming*. Lewis Publishers.

Kaempfer, H., Daigger, G., & Adams, C. (2000, October 14–18). Characterization of the floc microenvironment in dispersed growth systems. *Proceedings of the 73rd Annual Water Environment Federation Technical Exposition and Conference* [CD-ROM], Anaheim, CA, United States. Water Environment Federation.

Kisling, T., Khatiwada, S,, McCullough, K., Klaus, S., Wilson, C., Rauch-Williams, T., & Bott, C. (2022, October 8–12). Mechanistic evaluation of the adaptation of nitrifiers and PAOs to low DO operation. *WEFTEC 2022 Conference Proceedings,* New Orleans, LA, United States. Water Environment Federation.

Larrea, L., Larrea, A., Ayesa, E., Rodrigo, J. C., Lopez-Carrasco, M. D., & Cortacans, J. A. (2001). Development and verification of design and operation criteria for the step feed process with nitrogen removal. *Water Science & Technology, 43*(1), 261-268.

Leaf, W. R., Boltz, J. P., McQuarrie, J. P., Menniti, A., Daigger, G., & Adams, C. (2011, October 15–19). Overcoming hydraulic limitations of the integrated fixed-film activated sludge (IFAS) process. *Proceedings of the 84th Annual Water Environment Federation Technical Exhibition and Conference* [CD-ROM], Los Angeles, CA, United States. Water Environment Federation.

Ludzack, F. T., & Ettinger, M. B. (1962). Controlling operation to minimize activated sludge effluent nitrogen. *Journal Water Pollution Control Federation, 34*, 9.

McCarty, P. L. (1970, June 25–27). Phosphorus and nitrogen removal in biological systems. *Proceedings of the Wastewater Reclamation and Reuse Workshop* (p. 226), Lake Tahoe, CA, United States. University of California.

Miller, M. W., Srinivasan, V., Al-Omari, A., Esping, D., Regmi, P., Bauhs, K., Sturm, B., Downing, L, & Jimenez, J. (2022, October 8–12). More than just energy savings: Understanding the benefits of low DO operation. *WEFTEC 2022 Conference Proceedings,* New Orleans, LA, United States. Water Environment Federation.

Phillips, H. M., Maxwell, M., Johnson, T., Barnard, J., Rutt, K., Seda, J., Corning, B., Grebenc, J. M., Love, N., & Ellis, S. (2008, October 18–22). Optimizing IFAS and MBBR designs using full-scale data. *Proceedings of the 81st Annual Water Environment Federation Technical Exhibition and Conference* [CD-ROM], Chicago, IL, United States. Water Environment Federation.

Robertson, L. A., van Niel, E. W. J., Torremans, R. A. M., & Gijs Kuenen, J. (1988). Simultaneous nitrification and denitrification in aerobic chemostat cultures of *Thiosphaera pantotropha. Applied and Environmental Microbiology, 54*(11), 2812–2818.

Rutt, K., Seda, J., & Johnson, C. H. (2006, October 21–25). Two-year case study of integrated fixed film activated sludge (IFAS) at Broomfield, CO, WWTP. *Proceedings of the 79th Annual Water Environment*

*Federation Technical Exhibition and Conference* [CD-ROM], Dallas, TX, United States. Water Environment Federation.

Satoh, H., Nakamura, Y., Ono, H., & Okabe, S. (2003). Effect of oxygen concentration on nitrification and denitrification in single activated sludge flocs. *Biotechnology & Bioengineering, 83*(5), 604–607.

Schuyler, R. G., Tamburini, S. H., & Staggs, R. (2009). How low is too low? *Water & Environmental Technologies, 21*(6), 32–39.

Sen, D., Randall, C., & Grizzard, T. (1990). *Biological nitrogen and phosphorus removal in oxidation ditch and high nitrate recycle systems* (Pub. CBP/TRS 47/90). U.S. Environmental Protection Agency, Chesapeake Bay Program.

Stensel, H. D. (2001, October 13–17). Biological nutrient removal: Merging engineering innovation and science. *Proceedings of the 74th Annual Water Environment Federation Technical Exhibition and Conference* [CD-ROM], Atlanta, GA, United States. Water Environment Federation.

Stricker, A., Barrie, A., Maas, C. L. A., Fernandes, W., & Lishman, L. (2009). Comparison of performance and operation of side-by-side integrated fixed-film and conventional activated sludge processes at demonstration scale. *Water Environment Research, 81*(3), 219–232.

Trivedi, H., & Heinen, N. (2000, October 14–18). Simultaneous nitrification/denitrification by monitoring NADH fluorescence in activated sludge. *Proceedings of the 73rd Annual Water Environment Federation Technical Exposition and Conference* [CD-ROM], Anaheim, CA, United States. Water Environment Federation.

U.S. Environmental Protection Agency. (1993). *Nitrogen control manual* (EPA-625/R-93-010).

U.S. Environmental Protection Agency. (2010a). *Nutrient control design manual* (EPA-600/R-10-100).

U.S. Environmental Protection Agency. (2010b). *Evaluation of energy conservation measures for wastewater treatment facilities* (EPA-832/R-10-005).

Water Environment Federation. (2010). *Biofilm reactors* (WEF Manual of Practice No. 35).

Water Environment Federation. (2023). *Standards of design of water resource recovery facilities* (WEF 8-23).

Water Environment Research Foundation. (2001). *WERF/clarifier research technical committee (CRTC) protocols for evaluating secondary clarifier performance.*

Water Environment Research Foundation. (2010). *Protocol to evaluate alternative external carbon sources for denitrification at full-scale wastewater treatment plants.*

Young, T., Crosswell, S., & Wendle, J. (2008, October 18–22). Comparison of nitrogen removal performance in SBR systems. *Proceedings of the 81st Annual Water Environment Federation Technical Exhibition and Conference* [CD-ROM], Chicago, IL, United States. Water Environment Federation.

## 7.0 SUGGESTED READINGS

U.S. Environmental Protection Agency. (2015). *Case studies on implementing low-cost modifications to improve nutrient reduction at wastewater treatment plants* (EPA-841/R-15-004).

Water Environment Federation. (2010). *Technical submission to EPA: Removal of nutrients with currently available technologies.*

Water Research Foundation. (2023). *Guidelines for optimizing nutrient removal plant performance* (Project No. 4973).

# Chapter 8

# Enhanced Biological Phosphorus Removal Systems

# 1.0 INTRODUCTION

This chapter will build on the fundamentals of enhanced biological phosphorus removal (EBPR) reviewed in Chapter 6 and will primarily cover the components of a functional EBPR system. Many of the topics covered in this chapter are also discussed elsewhere in these standards; therefore, the relevant chapters are cross referenced.

# 2.0 BASIC CONSIDERATIONS

Reference is made to Section 2.0 of Chapter 7, which discusses fundamental considerations related to the design and operation of nutrient removal facilities. These basic considerations are also applicable to this chapter, and include effluent requirements, economics, operational considerations, sustainability, and safety.

# 3.0 SYSTEM CONFIGURATION

Several process configurations are available to achieve EBPR with or without concurrent nitrogen removal. Commonly used process schemes are summarized in Table 8.1 and are briefly described in the following subsections (Water Environment Federation [WEF], 2010). These can be categorized as plug flow or completely mixed systems.

## 3.1 Plug Flow Systems

In theory, plug flow is characterized by fluid particles passing through the reactor with little or no longitudinal mixing and exiting the reactor with their relative positions remaining the same as when they entered. Consequently, all particles have the same retention time. Because of hydraulic inefficiencies, true plug flow is difficult to achieve in full-scale applications. Tanks are designed with high length-to-width ratios or compartmentalized to approximate plug flow conditions.

Figure 8.1 presents process flow diagrams of different EBPR processes.

### *3.1.1 Anaerobic and Oxic*

The anaerobic and oxic (A/O) configuration represents the basic EBPR scheme and incorporates the two environmental conditions needed for phosphorus release and uptake reactions. It was originally developed as the Phoredox system in South Africa in 1974 (Barnard, 1974). In North America, it is now known as the A/O process. The anaerobic zone receives secondary influent and return activated sludge (RAS). This provides polyphosphate-accumulating organisms (PAOs) in RAS the first opportunity to use the influent carbon source. This critical step selects PAOs and provides them with a competitive advantage to perform EBPR in the aerobic zone. Chapter 6 contains a detailed description of the EBPR. The main indicator of the presence of a well-established PAO population is the typical release and uptake pattern of the EBPR process described in Chapter 6.

For the A/O process to be effective, it is important to protect the integrity of the anaerobic zone by eliminating dissolved oxygen in the influent and nitrate in RAS. Secondary influent typically has low or no dissolved oxygen. However, upstream unit operations such as preaeration, use of screw pumps, open discharge from centrifugal pumps, and free-fall over weirs can incorporate dissolved oxygen and should

TABLE 8.1 Key Features of Available EBPR Alternatives (WEF, 2010)

| Process | Zones | Internal recycles | Key features |
|---|---|---|---|
| Anaerobic–oxic or phoredox | Anaerobic–oxic | None | Total phosphorus removal, mainstream anaerobic and oxic zones |
| PhoStrip | Oxic–anaerobic | None | Total phosphorus removal, mainstream oxic and sidestream anaerobic zones, biological and chemical removal |
| Anaerobic–anoxic–oxic ($A^2/O$) | Anaerobic–anoxic–oxic | Nitrate (oxic to anoxic) | Total phosphorus and total nitrogen removal |
| University of Cape Town | Anaerobic–anoxic–oxic | Nitrate (oxic to anoxic)<br>Anaerobic (anoxic to anaerobic) | Total phosphorus and total nitrogen removal, return activated sludge (RAS) to anoxic, anaerobic zone protected, larger anaerobic because of lower mixed liquor volatile suspended solids |
| Modified University of Cape Town | Anaerobic–anoxic1–anoxic2–oxic | Nitrate (oxic to anoxic1)<br>Anaerobic (anoxic2 to anaerobic) | Total phosphorus and total nitrogen removal, RAS to anoxic1 greater protection of anaerobic zone, larger anaerobic because of lower mixed liquor suspended solids (MLSS) |
| Virginia Initiative Process | Anaerobic–anoxic1–anoxic2–oxic | Nitrate (oxic to anoxic1)<br>Anaerobic (anoxic2 to anaerobic) | Similar to modified University of Cape Town except nitrate recycle mixed with RAS and anaerobic and anoxic zones are staged |
| Five-stage Bardenpho | Anaerobic–anoxic1–oxic–anoxic2–reaeration | Nitrate (oxic to anoxic1) | Total phosphorus and total nitrogen removal, second anoxic (anoxic2) |
| Oxidation ditch | Anaerobic–oxic | None | Total phosphorus and total nitrogen removal, anaerobic zone precedes oxidation ditch, simultaneous nitrification and denitrification |
| Johannesburg | Preanoxic–anaerobic–oxic | Nitrate (oxic to anoxic) | Total phosphorus and total nitrogen removal, preanoxic zone for RAS denitrification and protect anaerobic zone |
| Westbank | Preanoxic–anaerobic–anoxic–oxic | Nitrate (oxic to anoxic) | Total phosphorus and total nitrogen removal, preanoxic zone for RAS denitrification and protect anaerobic zone |
| Schreiber | Anaerobic–oxic (temporal zone distribution) | None | Total phosphorus removal, air is cycled to achieve anaerobic–oxic conditions, incidental total nitrogen removal |
| Sequencing batch reactor | Anaerobic–oxic (temporal zone distribution) | None | Total phosphorus removal, air is cycled to achieve anaerobic–oxic conditions, incidental total nitrogen, no clarifiers |
| Phased isolation ditch (BioDenipho) | Anaerobic–oxic (temporal zone distribution) | None | Total phosphorus removal, air is cycled to achieve anaerobic–oxic conditions, incidental total nitrogen removal |

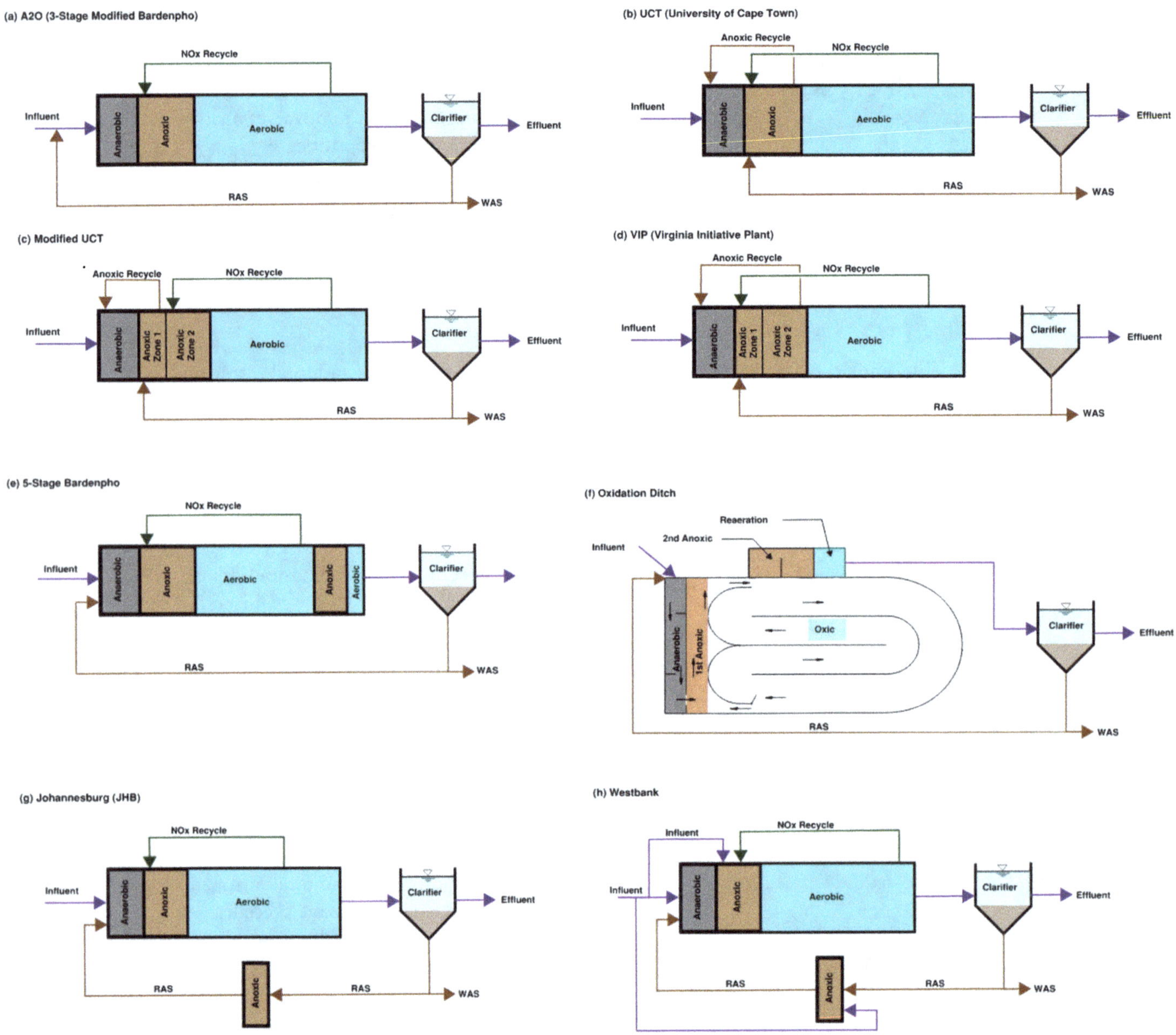


FIGURE 8.1 Process Flow Diagrams of Various EBPR Processes

be avoided. In nitrifying systems, nitrate in RAS will turn part of the anaerobic zone into an anoxic zone for denitrification in the presence of influent carbon. If this was not considered during sizing of the anaerobic zone, EBPR performance could be affected because of reduced anaerobic volume.

### *3.1.2 Anaerobic–Anoxic–Oxic Variations*

Achieving EBPR requires cycling the microbial consortium (sludge) between anaerobic and aerobic conditions for the appropriate length of time. An anoxic zone is added to accomplish simultaneous nitrogen and phosphorus removal. Several process configurations are available and are outlined in Chapter 9.

#### 3.1.3 Oxidation-Ditch Configurations

Traditional oxidation ditches are constructed in a racetrack setup in which the secondary influent and mixed liquor are circulated around a center barrier. Vendor-specific variations of the traditional oxidation-ditch concept have been developed. As reported by Daigger and Littleton (2000), full-scale operating experience has shown that simultaneous nitrogen and phosphorus removal can be accomplished in these systems via two mechanisms: (1) biological reactor (bioreactor) mixing patterns that allow the anoxic and anaerobic zones necessary for EBPR to develop, referred to as the *bioreactor macroenvironment,* and (2) development of anoxic and anaerobic zones within the floc, referred to as the *floc microenvironment.* Tight dissolved oxygen control, good mixing, and adequate solids inventory are important to achieving EBPR in oxidation ditches. In some configurations, an anaerobic zone is added upstream of the oxidation ditch to select PAOs.

### 3.2 Complete-Mix Systems

In a complete-mix system, as the fluid particle enters the reactor, it is immediately dispersed and uniformly distributed throughout the reactor. Complete-mix conditions are typically established in a round or square reactor. The following subsections present examples of some complete-mix systems.

#### 3.2.1 Sequencing Batch Reactor

In a sequencing batch reactor (SBR), the anaerobic and aerobic environments required for EBPR are created in the same tank but not at the same time, which is the case in a plug flow reactor. An anaerobic reaction period can be developed during the fill and initial react periods by removing sufficient nitrate following the aeration period. It is also possible to minimize nitrate concentration by incorporating several alternating aerobic and anoxic periods during the react period.

## 4.0 FACTORS AFFECTING ENHANCED BIOLOGICAL PHOSPHORUS REMOVAL PERFORMANCE

Enhanced biological phosphorus removal is a complex process encompassing different environmental conditions and complementing and competing biochemical reactions. This section discusses the many factors that could singly or in combination affect EBPR efficiency.

### 4.1 Influent Characteristics

The key factor that determines the amount of phosphorus stored in activated sludge is the amount of readily biodegradable organic matter in the anaerobic zone. There must be a large excess beyond that needed to deplete electron acceptors (dissolved oxygen and nitrogen oxide) in the anaerobic zone because the bacteria will preferentially metabolize the organic matter and reduce the amount of stored polyhydroxyalkanoates (PHAs), thereby affecting PAO selection and EBPR performance.

The type of organic compound in the anaerobic zone is also important; it must be soluble and readily biodegradable. Because biodegradability is a function of the length of the carbon chain, short-chain volatile fatty acids (SCVFAs) are ideal sources. In particular, acetic (two carbons) through valeric acid (five carbons) are considered to be the prototype of readily available organic compounds for EBPR. Although all of these volatile fatty acids (VFAs) are usable for EBPR, their relative effectiveness varies. Of all the VFAs commonly generated during municipal wastewater fermentation, acetic acid is the most efficient based on mass of phosphorus uptake per carbonaceous oxygen demand (COD) consumed. It is important to note that EBPR cannot be accomplished with one-carbon compounds such as formic acid because their polymerization is thermodynamically unfavorable for the bacteria.

The minimum ratios typically used to indicate the EBPR capability of a system are summarized in Table 8.2 (WEF, 2010). These ratios refer to bioreactor influent and should account for recycle loads and removals in primary clarifiers. Typically, recycle flows represent additional phosphorus load, whereas primary clarifiers remove 5-day carbonaceous biochemical oxygen demand ($CBOD_5$). As a result, primary effluent (bioreactor influent) will contain a lower substrate-to-phosphorus ratio relative to raw influent. The $CBOD_5$-to-total-phosphorus ratio can be used as a first approximation of the adequacy of carbon substrate for EBPR. It is now thought that readily biodegradable COD (rbCOD) is the most accurate measure of substrate availability because it represents influent VFAs and organic compounds that could potentially be fermented to VFAs in the anaerobic zone of the bioreactor. The rbCOD is an estimate of truly soluble COD.

The composition of influent phosphorus affects EBPR efficiency. In particular, corrosion inhibitors used by water resource recovery facilities (WRRFs) can be a significant source of polyphosphate, which is not readily reactive and is difficult to remove. For example, when the Xenia, Ohio, WRRF switched its corrosion control chemical from a 25 polyphosphate and 75 orthophosphate formulation to one containing 75 polyphosphates and 25 orthophosphates, the city's two wastewater facilities discharged elevated levels of total phosphorus (Jeyanayagam, 2007). This was corrected by switching the corrosion inhibitor back to the original formulation. It should be noted that the effluent orthophosphate values were consistently low, indicating that the higher effluent total phosphorus was mainly because of increased influent polyphosphate.

Municipal wastewater fermented in the collection system is typically a good source of VFAs for EBPR operation. Seasonal variation of influent VFAs can be tracked by observing the $CBOD_5$-to-COD ratio over a 12-month period. Sewer fermentation is inhibited by dissolved oxygen entrainment, wet weather high flows, the type of odor control strategy used, and cold temperatures, including those from the effect of snowmelt. Although adequate VFAs may be available in the facility influent to meet summer phosphorus limits, sewer system fermentation may not be a reliable and consistent source of carbon substrate throughout the year. In such instances, the anaerobic zone may need to be sized to support additional fermentation. Facility recycles, such as supernatant from primary sludge gravity thickening, also may contain sufficient VFAs for EBPR. Other sources of carbon augmentation include on-site VFA

TABLE 8.2 Minimum Substrate-to-Phosphorus Requirements for EBPR (WEF, 2010)

| Substrate measure | Substrate-to-phosphorus ratio | Remarks |
|---|---|---|
| Five-day carbonaceous BOD | 25:1 | Provides a rough and initial estimate. Based on typically available facility data |
| Five-day soluble BOD | 15:1 | Better indicator than $cBOD_5$ |
| Carbonaceous oxygen demand | 45:1 | More accurate than CBOD; not measured by all facilities |
| Volatile fatty acids | 7:1 to 10:1 | More accurate than COD; involves specialized laboratory analysis |
| Readily biodegradable carbonaceous oxygen demand | 15:1 | Most accurate; measures VFA formation potential; accounts for VFA formation in the anaerobic zone; specialized laboratory analysis |

generation via sludge prefermentation, industrial wastewater containing VFAs, or acetic and propionic acids. Often, it may be simpler and more cost effective to provide a standby chemical feed system.

## 4.2 Integrity of the Anaerobic Zone

The anaerobic zone of an EBPR bioreactor is expected to perform two functions. The primary function is PAO selection, which is a relatively rapid reaction if adequate rapidly biodegradable substrate is available. In some instances, the anaerobic zone is also required to perform a secondary function, that is, VFA generation through fermentation. This is a slower reaction requiring a longer anaerobic hydraulic retention time (HRT).

An anaerobic condition is one in which free (dissolved) and combined oxygen are not available for bacterial metabolism. The oxidation-reduction potential (ORP) can also be used to confirm the presence of anaerobic conditions. Typical ORP values in anaerobic reactors are in the range of −300 mV or less. Field testing, however, is required to establish site-specific ORP for anaerobic conditions. Ideally, anaerobic conditions must be maintained throughout the anaerobic volume for reliable EBPR. Sources of dissolved oxygen include preaeration, influent screw pumps, free-fall over weirs, excessive turbulence, aggressive anaerobic zone mixing, backflow from the aerobic zone, and internal mixed liquor recycle. Typical sources of nitrate include RAS and internal mixed liquor recycle (Jeyanayagam, 2007).

Introduction of nitrate or dissolved oxygen to the anaerobic zone causes a reduction in the actual anaerobic volume. Consequently, the effective anaerobic solids retention time (SRT) and HRT will be reduced. This will decrease the anaerobic contact time between PAOs and the substrate (VFAs), which could potentially compromise phosphorus removal (Jeyanayagam, 2007). In addition, the presence of nitrate and dissolved oxygen will provide competing organisms with access to the substrate. For example, 1.0 mg of nitrate-nitrogen will take up readily biodegradable organics needed for the removal of 0.7 mg of phosphorus by supporting denitrification. Likewise, the presence of 1.0 mg of dissolved oxygen will use up the substrate needed for the removal of 0.3 mg of phosphorus by facilitating normal heterotrophic activity (biochemical oxygen demand oxidation).

## 4.3 Solids Capture

Effluent total phosphorus consists of two components: soluble phosphorus and particulate phosphorus. A well-designed and operated EBPR system with adequate VFAs can reduce the effluent soluble phosphorus to approximately 0.1 mg/L. Particulate phosphorus, which represents solids-associated phosphorus, depends on effluent total solids and its phosphorus content. For example, if the effluent total suspended solids (TSS) is 10 mg/L (75% volatile suspended solids [VSS]) and the phosphorus content of the mixed liquor is 0.06 mg/mg VSS (6%), then the effluent particulate phosphorus concentration would be 0.45 mg/L. Hence, controlling effluent solids is important to achieve low effluent total phosphorus.

## 4.4 Recycle Loads

Return streams from sludge operations (e.g., dewatering) can significantly affect EBPR performance. The quantity and quality of these streams vary based on the technology used in solids processing operations. For example, anaerobic digestion is likely to release more phosphorus than aerobic digestion. Sludge dewatering using belt filter presses typically generates 2 times more recycle flow (filtrate) compared to centrifuge dewatering because of the amount of wash water used in the dewatering operation. The total recycle flow can amount to 1% to 3% of the facility influent flow and can reduce process HRT.

The recycle nitrogen and phosphorus mass loads can represent 15% to 40% of influent total nitrogen and 5% to 40% of influent total phosphorus loads, respectively. The effect of recycle streams is discussed further in Section 7.0.

## 4.5 Mixing

The goal of mixing in an anaerobic zone is to keep the MLSS suspended while minimizing surface turbulence that could transfer oxygen from the atmosphere. Therefore, the minimum power input necessary to keep the solids suspended should be used, and the stirring mechanism should be designed to avoid vortexing. Alternatives include submersible, vertical, and pulsed air mixers. Power input should be sufficient to maintain a velocity of 0.3 m/s (1.0 ft/sec) throughout the zone. Because horizontal velocities can be difficult to quantify in a reactor with a single-point mixing device, a more practical criterion is complete turnover of the cell contents every 20 minutes based on the primary pumping rate of the mixer.

## 4.6 Baffles

Baffles play an important role in enhancing the EBPR process. The following types of baffles are used in EBPR bioreactors: interzone and intrazone.

### *4.6.1 Interzone Baffles*

In EBPR bioreactor configurations, interzone baffles are provided for separating the anaerobic and anoxic or oxic zones to protect the integrity of the anaerobic zone. Interzone baffle design should consider the following:

- Because of density differences, the water surface in the aerobic zone will be considerably higher than the water surface in the nonaerated (anaerobic) zone. This could initiate a backflow of a high dissolved oxygen stream into the anaerobic zone, which will reduce the available substrate for EBPR and could potentially stimulate filamentous growths and cause *Nocardiaform* foaming. In addition, foam will be trapped in the zone upstream of the aerated zone because of the increase in water surface.
- Head loss, when available, is the most effective way to segregate zones by providing a drop in water surface across the top baffle. Baffle openings should be sized to ensure a forward flow of 0.15 m/s (0.5 ft/sec) at minimum flow. Fully submerged baffles with the top approximately 25 mm (1 in.) below the water surface would allow the free flow of surface scum and foam without an opportunity to accumulate. Alternatively, when baffles extend well above the water surface, slots should be cut at the top for the scum and foam to pass.
- Bottom baffle openings must be provided to facilitate tank draining. The opening should be small enough to avoid significant forward flow.
- To facilitate tank cleaning, removable walk-through baffle segments should be considered to allow operators to move from zone to zone without having to climb back out.
- Provisions should be made for confined space activities by including a lifeline and retrieval device.

### *4.6.2 Intrazone Baffles*

Intrazone baffles typically are provided within a zone to enhance reaction kinetics within a zone. In an EBPR system, intrazone baffles are provided to improve plug flow conditions. An anaerobic stage divided into discrete stages would result in more rapid substrate uptake kinetics (because of a higher initial food-to-microorganism ratio [F/M]) than a completely mixed zone (Stensel, 1991). Consequently, the total anaerobic volume can be decreased by staging. Similar benefits have been reported by staging the aerobic zone (Jeyanayagam, 2007; Narayanan et al., 2006).

### 4.7 Supplemental Chemical Addition

Enhanced biological phosphorus removal entails the complex interaction of design and operational factors that, singly or in combination, can affect process efficiency. The flexibility to add metal salts when necessary will ensure process reliability and regulatory compliance. However, indiscriminate use of metal salts could potentially interfere with the recovery of the EBPR process and could facilitate the eventual conversion from backup to exclusive chemical addition. This is because the soluble phosphorus that is chemically precipitated is unavailable for uptake and storage (as energy-rich volutin granules) by PAOs in the anoxic and aerobic zone. Consequently, PAOs will have less internal energy available to them in the anaerobic zone for the uptake of SCVFAs and will lose their competitive advantage over other heterotrophs. This is the reason overdosing with chemical precipitants results in a washout of the EBPR function.

The recommended strategy for chemical addition is to dose only as much metal salt as is needed to precipitate the phosphorus that the PAOs cannot take up. This could be accomplished by measuring the soluble phosphorus (mostly orthophosphate) toward the end of the bioreactor and adding metal salt in direct proportion to it (e.g., a 2-to-1 molar ratio, to be confirmed by jar testing as outlined in Chapter 10). The orthophosphate at the end of the bioreactor represents phosphorus that the organisms are unable to take up at that point in time. The chemical could be added to a well-mixed section of the mixed liquor channel, just upstream of the secondary clarifiers. However, it should be noted that the unreacted metal salt that is returned with RAS may continue to remove phosphorus after chemical addition is terminated. For this reason, a preferred first option might be to chemically treat the sidestream to remove recycled phosphorus before adding supplemental chemical to the EBPR process.

## 5.0 WASTING STRATEGIES

The parameter of fundamental importance to the design and control of the activated sludge system is SRT. Operationally, the design SRT governs the mass of sludge wasted daily from the system. In nutrient removal systems, SRT, rather than the F/M or organic loading, should be used for process control. By definition, total system SRT in days is

$$\text{Total SRT} = \frac{\text{Mass of sludge in the bioreactor}}{\text{Mass of sludge wasted per day}} \tag{8.1}$$

In EBPR systems, anaerobic and aerobic SRTs can be calculated by replacing the numerator in the aforementioned equation with the mass of solids in the anaerobic and aerobic zones, respectively. It should be noted that the SRT calculation presented in Equation 8.1 considers the solids inventory in the bioreactor only and could underestimate the actual SRT in facilities where significant solids inventory is held outside of the bioreactor in final clarifiers and excessively large mixed liquor and RAS channels. The mass of sludge wasted per day is equal to the daily sludge production. *Wastewater Engineering: Design of Water Resource Recovery Facilities* (WEF, 2023) contains a discussion of the effect of SRT and wastewater characteristics on daily sludge production.

Available sludge wasting strategies, shown in Figure 8.2, include wasting from the final clarifier underflow (conventional) and hydraulic wasting directly from the bioreactor. The recommended strategy is to waste continuously, when possible.

## 5.1 Final Clarifier Underflow Wasting

The most common method of sludge wasting is from the return sludge line, sometimes called *conventional wasting*. For this wasting strategy, using the aforementioned mathematical expression, SRT may be computed as follows:

$$\text{SRT} = \frac{V \times X}{(Qw \times Xr + Qe \times Xe)} \tag{8.2}$$

Where

$V$ = bioreactor volume, $m^3$;
$X$ = MLSS concentration, mg/L;
$Qw$ = waste sludge flowrate from the return sludge line, $m^3/d$;
$Xr$ = RAS solids concentration, mg/L;
$Qe$ = effluent flowrate from secondary clarifier, $m^3/d$; and
$Xe$ = effluent TSS concentration, mg/L.

If effluent solids are neglected, the wasting rate approximates to

$$Qw \approx \frac{V \times X}{\text{SRT} \times Xr} \tag{8.3}$$

## 5.2 Hydraulic Wasting

The hydraulic control of sludge wasting was first proposed and implemented by Garrett in 1958. In essence, this approach entails wasting MLSS ($X$) directly from the bioreactor. If effluent solids are ignored, the SRT is calculated as follows:

$$\text{SRT} \approx \frac{VX}{QwX} = \frac{V}{Qw} \tag{8.4}$$

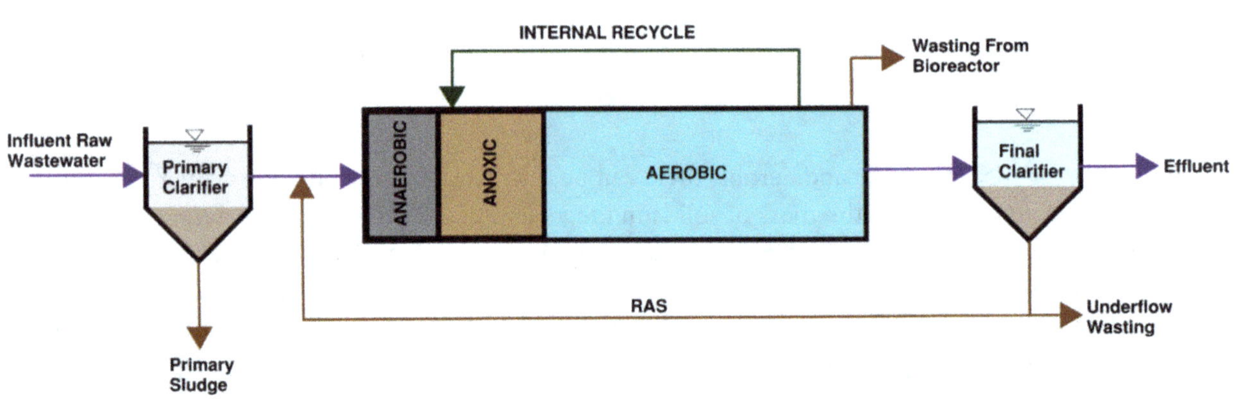


FIGURE 8.2 Alternate Wasting Strategies

Additionally, the waste flowrate from the bioreactor ($Qw$) is independent of MLSS and RAS characteristics, as follows:

$$Qw \approx \frac{V}{\text{SRT}} \tag{8.5}$$

Hydraulic wasting offers the following advantages for the EBPR process:

- Ease of determining the amount of daily wasting by simply dividing the volume of the bioreactor by the desired SRT, as shown in the aforementioned equation.
- Because the solids recycle rate does not enter the calculation of SRT, the RAS flowrate can be chosen to give proper operation of the final clarifier, thereby ensuring that almost all biomass is returned to the bioreactor (Grady et al., 2011).
- Wasting aerobic sludge has a lower tendency to release phosphorus during subsequent handling than underflow from a secondary clarifier. In addition, the wasted MLSS stream can be thickened continuously in a dissolved air flotation unit, which is the preferred method of thickening phosphorus-rich EBPR solids. Use of other thickening methods often requires sludge storage to allow intermittent operation. This practice could potentially lead to secondary phosphorus release during storage.
- Because of lower solids concentration than clarifiers under flow, continuous wasting is viable and more aligned with continuous biological synthesis and growth in the bioreactor.

The main drawback of hydraulic wasting is the need to size subsequent solids handling equipment for the higher hydraulic load.

Hydraulic wasting also can be used for wasting clarifier underflow if the RAS flow and influent flow are metered. The Nine Springs facility in Madison, Wisconsin, has successfully used hydraulic wasting of underflow with biological phosphorus removal for approximately 15 years. A mass balance on the aeration/clarifier system using influent flow ($Qi$), RAS flow ($Qr$), MLSS concentration ($X$), and RAS concentration ($Xr$) yields

$$(Qi + Qr)X = QrXr \tag{8.6}$$

or

$$Xr = (Qi + Qr)X/Qi \tag{8.7}$$

Substituting for $Xr$ in Equation 8.3, the expression for wasting ($Qw$) becomes

$$Qw = VQi/[\text{SRT}(Qi + Qr)] \tag{8.8}$$

With known influent and recycle flowrates, the waste sludge rate can be set hydraulically for wasting underflow. If the return rate is constant with a varying influent flow, the wasting rate must be varied to maintain a constant solids loading on a downstream thickener. If the return rate is varied proportionally to influent flow, the waste sludge flowrate can be constant throughout the day. This method will work successfully if metering is accurate. And, as long as thickening is not attempted in the final clarifier, secondary release of phosphorus in the clarifier is not likely to be a problem.

## 6.0 APPLICATION OF MASS-BALANCE PRINCIPLES IN ENHANCED BIOLOGICAL PHOSPHORUS REMOVAL OPERATIONS

Mass balances provide a convenient way to evaluate the performance of a unit operation. Using mass-balance concepts, it is also possible to perform a reality check of facility data. Because of the dynamic nature of facility operating conditions, average data should initially be used for mass-balance calculations. This section presents how mass balances can be used to make operating decisions.

The best way to monitor the oxygen and nitrate returned to the head of EBPR systems is the "mass-balance method." The mass of phosphorus, oxygen, and nitrate can be calculated using the flow and the respective concentration as shown in Equation 8.9. Once the mass going in and coming out are calculated for each reactor, the change in the reactor can be calculated from the difference between the two, as follows:

$$\text{Mass difference} = \text{Mass in} - \text{Mass out} \qquad (8.9)$$

$$= [C_{in} \times Q_{in}] - [C_{out} \times Q_{out}]$$

Where

$M$ = mass of the specific constituent (lb/d or kg/d),
$C$ = concentration of the specific constituent (mg/L), and
$Q$ = flowrate (mgd or ML/d).

Figure 8.3 illustrates the results of mass-balance calculations performed for a University of Cape Town–type EBPR system that consists of two anaerobic (An), two anoxic (Ax), and three aerobic (Ae) reactors. The figure also illustrates the effect of the mixed liquor recycle from the end of the aeration zone (Ae3) to the first anoxic zone (Ax1). The recycle contains nitrate and oxygen, which contributed to the electron acceptor budget in the anoxic zone.

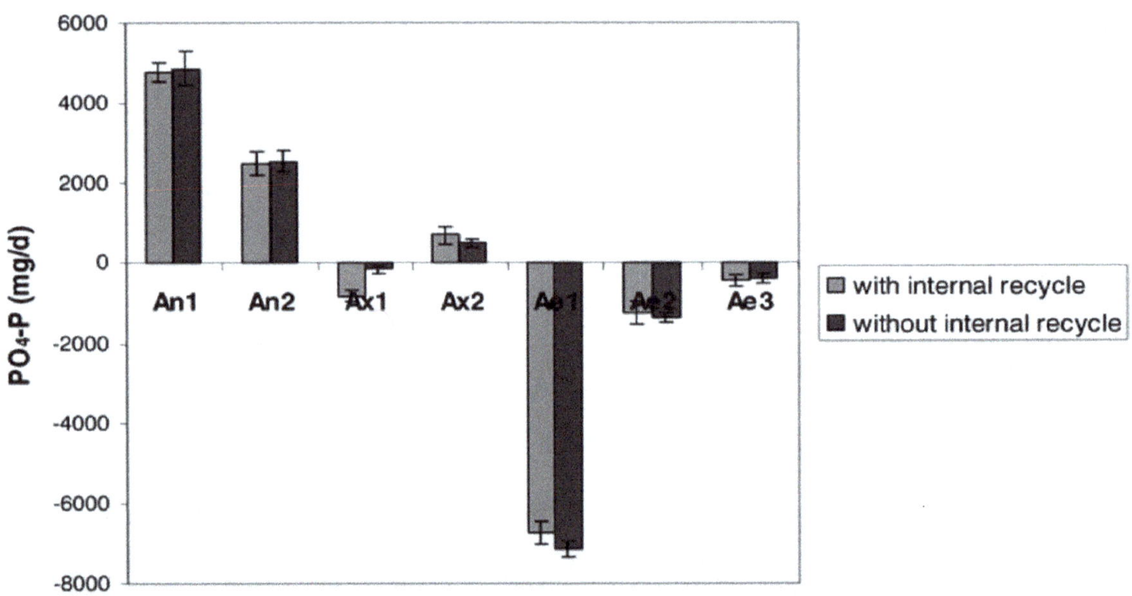


FIGURE 8.3 Phosphorus Mass Balance Performed for Phosphorus in a Pilot-Scale, Virginia Initiative Process–Type EBPR System (Erdal et al., 2002). Positive values indicate phosphorus release, whereas negative values indicate phosphorus uptake (WEF, 2010).

Once the recycle was taken out of service, anoxic phosphate uptake decreased significantly, leading to an increase of aerobic uptake. By using mass-balance calculations, one can avoid confusion that can be caused by using concentration values as a monitoring tool. Because the recycle flows also have a dilution effect on facility influent at different points of the treatment system, mass loading must be used instead of concentrations.

Nitrate in RAS can be reduced by providing a wide spot in the RAS line to allow the nitrate to dissipate. A small amount of primary effluent may be added to enhance the denitrification rate. Another strategy is to operate at the lowest possible RAS rate. With RAS rates of 50% or less, little nitrate remains when RAS arrives at the anaerobic zone. Whereas it is possible to reduce the RAS nitrate load by promoting denitrification in the final clarifiers, this is not a recommended practice, particularly if the clarifier effluent nitrate-nitrogen is more than approximately 5 mg/L. At these higher nitrate levels, denitrification in the clarifier can release enough nitrogen gas to affect sludge settleability and the ability of the facility to meet TSS permit limits.

## 7.0 PRIMARY SLUDGE FERMENTATION FOR INTERNAL CARBON SOURCE PRODUCTION

The availability of adequate carbon is crucial to reliable EBPR. Broadly categorized, carbon sources for EBPR include *internal sources*, which refer to the generation of short-chain fatty acids through sludge fermentation of the influent rapidly biodegradable organic matter. Acid fermentation is the process by which complex particulate and soluble substrates present in the sludge and wastewater are aerobically broken down to form SCVFAs. Because they are soluble and readily biodegradable, SCVFAs are a highly effective carbon source for EBPR and denitrification. The fundamental mechanism of fermentation is presented in Chapter 6.

In this section, the various primary sludge fermenter configurations are outlined. It should be noted that VFA production also occurs naturally in wastewater collection systems under the right conditions.

Sludge fermenters are configured to accomplish three basic functions: acid fermentation, solids separation, and elutriation. In a laboratory-scale investigation, Lilley et al. (1990) found VFA production from primary sludge fermentation to involve first-order reaction kinetics. The maximum potential VFA yield at 20 °C was reported to be 0.17 mg SCVFA (as COD)/mg of primary sludge (as COD) applied. Oldham and Abraham (1994) reported on the SCVFA yield from the complete-mix fermenter in Penticton, British Columbia; the single-stage static fermenter in Kelowna, British Columbia; and the two-stage complete-mix fermenter and thickener in Kalispell, Montana. These fermenters added the equivalent of 17, 21, and 58 mg/L of SCVFA, respectively, to the wastewater entering EBPR processes. Practical experience suggests that higher SCVFA yields are observed in fermenters receiving primary sludge from wastewaters with relatively minimal natural fermentation in the collection system.

### 7.1 Primary Sludge Fermentation Configurations

The available flow sheets are shown in Figure 8.4 and briefly described in the following subsections. Table 8.3 summarizes the differentiating features of the various fermenter configurations.

#### *7.1.1 Activated Primary Sedimental Tank*

The primary clarifier is operated as a fermenter by continuously recycling the primary sludge directly to the inlet or through an elutriation tank. Fermentation occurs in the accumulated sludge blanket. Generally, activated primary tanks require purposely designed primary clarifiers that allow sufficient

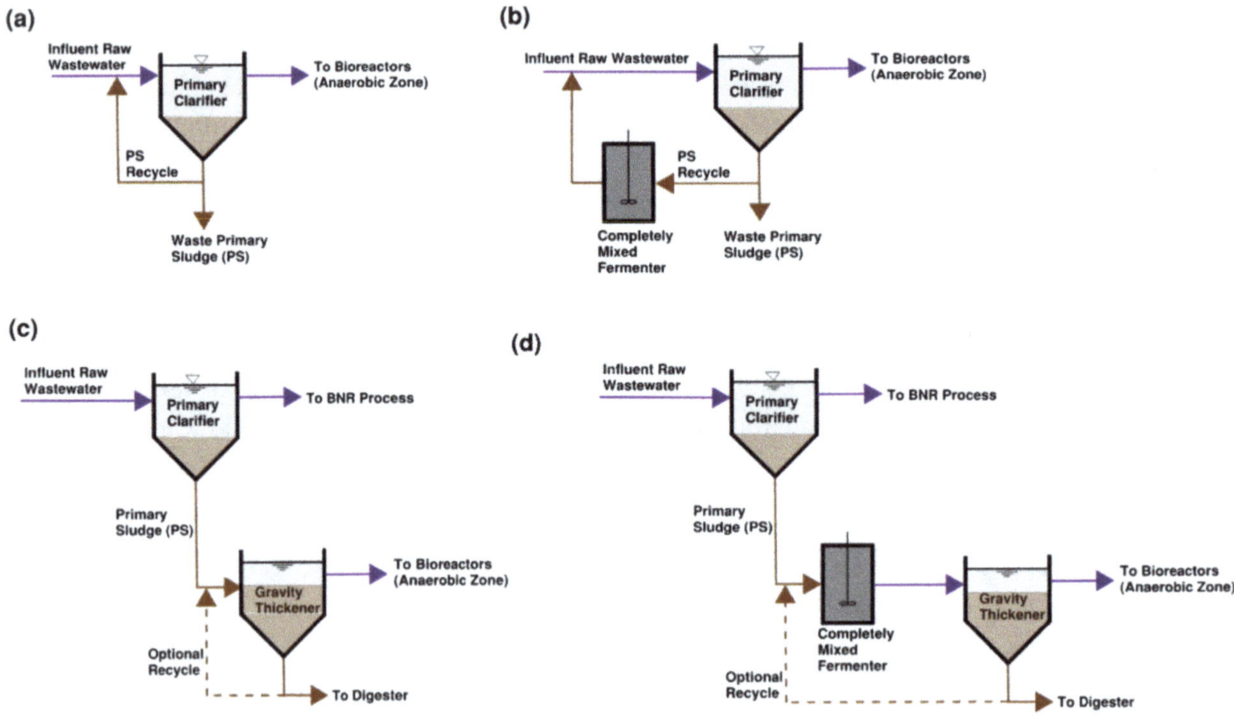


FIGURE 8.4 Various Primary Sludge Fermentation Configurations

sludge blanket depth to be maintained. The SCVFAs generated in the sludge mass are released into the bulk liquid by elutriation with influent wastewater and conveyed to the EBPR process together with the primary effluent. The sludge inventory and blanket height are controlled by wasting a fraction of the primary sludge to the solids handling system. Barnard (1984) first proposed this fermentation concept. A modified operating strategy, called *alternating activated primaries*, involves two primary clarifiers. Primary sludge is retained and fermented in one and then pumped to another for SCVFA elutriation. Raw wastewater is settled in both tanks at the same time. This design was used at the Baviaanspoort Wastewater Treatment Works in South Africa, and the operation allowed for the process to be reversed (Randall et al., 1992).

### 7.1.2 Complete-Mix Fermenter/PST

Complete-mix fermenters enable the fermentation process to be removed in large part from the primary treatment process, although the solids loadings to primary clarifiers are significantly reduced. Complete mixing of the fermenter also provides better contact between the fermenting solids and the bulk liquid.

TABLE 8.3 Differentiating Features of the Various Fermenter Configurations

| Configuration | Fermentation | Solids separation | VFA stream |
|---|---|---|---|
| Active PST | Primary clarifier | Primary clarifier | Primary effluent |
| Completely mixed fermenter/PST | Separate completely mixed tank | Primary clarifier | Primary effluent |
| Gravity thickener fermenter | Gravity thickener | Gravity thickener | Gravity thickener supernatant |
| Two-stage fermenter/thickener | Separate completely mixed tank | Gravity thickener | Gravity thickener supernatant |

Primary sludge is continuously pumped to a mechanically mixed tank where acid fermentation occurs. The complete-mix fermenter tank overflow is returned to the primary clarifier inlet. The SCVFAs generated in the fermenter are elutriated though mixing with influent wastewater and conveyed to the EBPR process together with the primary effluent. The fermenter SRT is controlled by wasting a fraction of the fermented sludge to the solids handling system. The fermenter HRT is controlled either by adjusting the complete-mix tank volume or by adjusting the primary sludge pumping rate. The SCVFAs elutriated in the primary clarifier are conveyed to the EBPR process together with the primary effluent. This fermenter configuration was first proposed by Rabinowitz et al. (1987).

### 7.1.3 Gravity Thickener Fermenter

Static fermenters simplify the fermentation process, removing it completely to a separate unit process. In addition, supernatant from this process can be directed to the EPBR bioreactor independent of the primary effluent so that it may be applied where it is most appropriate in the process. This fermenter is a lowly loaded gravity thickener with an increased side-water depth to allow a fermenting sludge inventory to accumulate in the bottom of the thickener. Primary sludge is continuously pumped into a center well and allowed to settle and thicken in the fermenter and thickener. The SCVFA-rich supernatant is conveyed directly to the EBPR bioreactor, where it can be optimally used for EPBR and/or denitrification. A source of nitrate-rich elutriation water is typically added to the incoming primary sludge or directly to the thickener to minimize sulfide and methane formation in the sludge blanket and to help convey SCVFAs to the EBPR bioreactor. The fermenter SRT and sludge inventory are controlled by wasting a portion of the sludge inventory, as thickened sludge, to the solids handling system to maintain a target sludge blanket height. This fermentation concept was first applied at the Kelowna's WRRF in British Columbia in the early 1980s (Oldham & Stevens, 1984).

### 7.1.4 Two-Stage Fermenter/Thickener

This configuration combines complete-mix and static fermenters into a single unit process. Primary sludge is pumped to a mechanically mixed tank, where acid fermentation occurs. The complete-mix tank overflows to a gravity thickener for liquid–solids separation. The SCVFA-rich supernatant is directed to the EPBR bioreactor independent of the primary effluent so that it may be applied where it is most appropriate in the process. Furthermore, the complete-mix basin allows intimate contact between the fermenting solids and the bulk liquid, optimizing the transfer of SCVFAs to the supernatant. Primary sludge is continuously pumped from the mainstream primary clarifier to a mechanically mixed tank where acid fermentation occurs. The complete-mix tank overflows to the gravity thickener for liquid–solids separation. Thickened sludge is continuously recycled from the bottom of the thickener to the complete-mix tank inlet. The VFA-rich thickener supernatant is conveyed directly to the EBPR bioreactor, where it can be directed to an anaerobic or anoxic zone to optimally be used for EPBR and/or denitrification. A source of elutriation water is typically added to the gravity thickener inlet or into the thickener blanket to minimize sulfide and methane formation in the sludge blanket and to help convey SCVFAs to the EBPR bioreactor. The fermenter SRT and sludge inventory are controlled by wasting a portion of the thickened sludge to the sludge handling system and maintaining a target sludge blanket depth. This fermentation concept was first applied at the Kalispell Wastewater Treatment Plant in Montana in the early 1990s (Oldham & Abraham, 1994).

The design of these systems typically accounts for the solids inventory in both the complete-mix tank and the thickener. An option exists to exchange the thickener for a mechanical thickening process such as a centrifuge, rotary drum thickener, or screw press. If this option is implemented, the entire sludge inventory must be contained within the complete-mix tank.

Table 8.4 summarizes the crucial design and operational parameters for primary sludge fermenter.

**TABLE 8.4** Summary of Design Criteria for Primary Sludge Fermentation (Adapted from Metcalf & Eddy, Inc./AECOM, 2013)

| Mode of primary sludge fermentation process | SRT in fermenter (days) | Elutriation recycle ratios: Raw influent | Elutriation recycle ratios: Thickener influent | Contact tank (days) | Thickener feed fraction of raw influent flow | Thickener loading (Kg/m$^2$·d) |
|---|---|---|---|---|---|---|
| Activated PST | 2–4 | 0.05–0.10 | n/a | n/a | n/a | n/a |
| Completely mixed fermenter/PST | 4–6 | 0.05–0.10 | n/a | 0.25–0.50 | n/a | n/a |
| Gravity thickener fermenter | 4–6 | n/a | 0.10–0.20 | n/a | 0.04–0.08 | 20–40 |
| Two-stage fermenter/thickener | 4–6 | n/a | 0.30–0.50 | 2.0–4.0 | 0.02–0.04 | 100–150 |

*Note.* PST—primary settling tank; n/a—not applicable

## 7.2 Factors Affecting Primary Sludge Fermentation

The following factors affect the operation of all primary sludge fermenters regardless of configuration. Consequently, these factors should also be considered during design.

### 7.2.1 *Solids Retention Time*

Solids retention time is a key operating parameter for primary sludge fermentation systems because it defines the groups of fermentative organisms that will grow in the system. If the SRT is too short, many of the bacteria that produce the enzymes necessary for hydrolysis and acidogenesis, as well as acetogenesis bacteria, will be washed out of the system and a stable population of acid-formers will not develop. Conversely, if the SRT is too long, either sulfur-reducing bacteria that convert VFAs to $CO_2$ (rather than to acetate) or a population of methane formers will develop that will consume the VFAs generated and produce methane gas. The growth rate of fermentative bacteria is highly temperature dependent. For this reason, primary sludge fermenters are generally operated at an SRT of 3 to 5 days during the summer months when operating temperatures are above 16 °C, and at an SRT of 4 to 8 days during the winter months when operating temperatures are below 15 °C (Water Environment Research Foundation [WERF], 2011). Given the relatively unique wastewater characteristics and operating conditions found at most facilities, establishment of an optimal SRT is generally based on experience; however, it is typically within the aforementioned limits for various operating temperatures.

Determining the sludge inventory is critical to calculating the SRT of a fermenter system. In systems that use a completely mixed reactor, inventory can be determined from the total volume of the fermenter and suspended solids concentration. However, in a fermenter that uses a gravity thickener, stratification of solids and liquids requires estimation of the sludge blanket depth in the thickener. This measurement can be done directly with sludge blanket probes or sludge blanket instruments. However, use of the latter is difficult because of the low light transmittance of the thickener supernatant. Although other methods using mechanism torque have been used as a surrogate measure of sludge inventory, these measurements can vary seasonally as sludge densification characteristics change in the thickener.

### 7.2.2 *Hydraulic Retention Time*

Fermenter HRT is a function of tank volume and the primary sludge pumping rate. In general, HRT is not a critical parameter used in the design of primary sludge fermenters, although it is often used to fine-tune fermenter operation. Complete-mix fermenters are generally sized with an HRT between 6 and

12 hours while treating between 4% and 8% of incoming facility flow. The primary sludge pumping rate to two-stage, complete-mix, and gravity thickener fermenters is generally lower (i.e., approximately 2% to 4% of the incoming facility flow) so that the HRT of the complete-mix tank is typically between 12 and 24 hours. Other fermenters are designed to maintain a specific sludge inventory and SRT in the system. In these instances, the fermenter HRT can also be controlled by varying the primary sludge pumping rate and the elutriation water flowrate when added directly to a gravity thickener.

### 7.2.3 pH

The bulk liquid pH is extremely important in anaerobic digestion because methanogenesis proceeds at a high rate when the pH is maintained in the neutral range, with most methane formers inhibited at pH values below 6.3 and above 7.8. However, acid formers are significantly less sensitive to high or low pH values. Hence, acid fermentation typically prevails over methane formation at pH values below 6.5. Primary sludge fermenters tend to operate at ambient pH values between 4.8 and 6.0 because of acid formation reactions that are part of fermentation; raising the fermenter pH value does not improve VFA production. The lower pH is beneficial because it inhibits growth of methane-forming bacteria. Sulfanogenic bacteria are also relatively immune to low pH values and can reduce sulfate to sulfides at pH values below 4.

### 7.2.4 Elutriation Water

In addition to generating VFAs through conversion of the particulate primary solids to soluble organics, these VFAs must be separated from the fermenting solids and conveyed to the EBPR process. Partitioning of the VFAs to the portion of the flow directed to EBPR rather than to sludge disposal is enhanced by introducing elutriation water. In addition, it is useful to provide some level of mixing to enhance diffusion of VFAs from the sludge mass to the liquid fraction.

Use of a water source with a significant level of nitrate has been practiced to control the sulfur-reducing bacteria population. Heterotrophic bacteria, including sulfate-reducing bacteria (SRB), prefer nitrate to sulfates as their terminal electron acceptor. Although SRB that use hydrogen as their electron source are critical to acetogenesis, the proliferation of those bacteria will begin consuming VFAs at a rate in excess of their generation. The addition of some nitrate-rich stream (secondary effluent is most common) can be used to reduce, but not eliminate, SRB dominance, especially at long SRTs. This addition has the secondary benefit of inhibiting the activity of methanogenic bacteria, although intermittent aeration has also been used with some effect.

The key parameter used to control the operation of primary sludge fermenters is the system SRT or sludge inventory. Volatile fatty acid production tends to be low at short SRTs because of the difficulty of establishing a stable population of acid formers that complete the hydrolysis, acidogenesis, and acetogenesis reactions. However, at long-fermenter SRTs, sulfanogenic and methanogenic bacteria will begin to dominate and the generated VFAs will be consumed by methanogenic activity that produces methane gas or sulfanogenic activity that generates hydrogen sulfide.

## 8.0 SIDESTREAM EBPR PROCESS FOR INTERNAL CARBON SOURCE PRODUCTION

Fermentation can be applied to several flow streams at a WRRF including raw sewage, primary sludge, RAS, and mixed liquor. The target of fermentation is the generation of VFAs through the hydrolysis of particulate organic material. Increased VFA concentration improves the efficiency of EBPR by ensuring sufficient readily biodegradable carbon is available to satisfy carbon needs of PAOs. Return activated sludge and mixed liquor fermentation differ from primary sludge and raw influent fermentation in that the fermentation process also provides an environment that promotes the growth of PAOs over glycogen accumulating organisms (GAOs). In a traditional treatment system, both the PAOs and GAOs grow under similar anaerobic and aerobic conditions competing for influent VFAs. Process conditions that allow GAOs to outcompete PAOs for available VFAs lead to phosphorus removal instability.

Effective EBPR can be achieved in processes in which mixed liquor or RAS is fermented to form rbCOD or SCVFAs. Fermentation in anaerobic zones has always been an integral component of EBPR systems. However, configurations have been advanced that depend on fermentation of the secondary sludge, provided that only a portion of the secondary sludge is fermented and that the rbCOD or VFAs formed are made available to PAOs in the mainstream EBPR process. Secondary sludge fermenters typically consist of completely or intermittently mixed reactors that are incorporated to the EBPR process train. These reactors can either be in the main process stream, a sidestream of the process (e.g., on the RAS line), or on a recycle loop that is withdrawn from and returned to the anaerobic zone. Secondary sludge fermentation has been successfully implemented in several EBPR facilities where manipulation of the aeration and mixing systems has created mixed liquor or RAS fermentation zones within the EBPR process train (Barnard et al., 2010).

Clarke and Neethling (2009) switched off the mixer in one out of two anaerobic zones of a facility with insufficient VFAs in the incoming wastewater and successfully removed phosphorus from approximately 8 mg/L in the influent to less than 0.5 mg/L through in-basin fermentation. A similar strategy at a Henderson, Nevada, facility, where nitrate added to the sewerage system destroyed the natural VFAs, resulted in effluent orthophosphate concentrations below 0.1 mg/L (Barnard et al., 2010).

Four different types of fermentation process are commonly practiced in WRRFs. Two configurations use RAS fermentation, and the other two configurations use MLSS fermentation. The key difference between various fermentation configurations is HRT, SRT, and MLSS concentration. Figure 8.5 presents a process flow diagram of the four sidestream EBPR (S2EBPR) configurations.

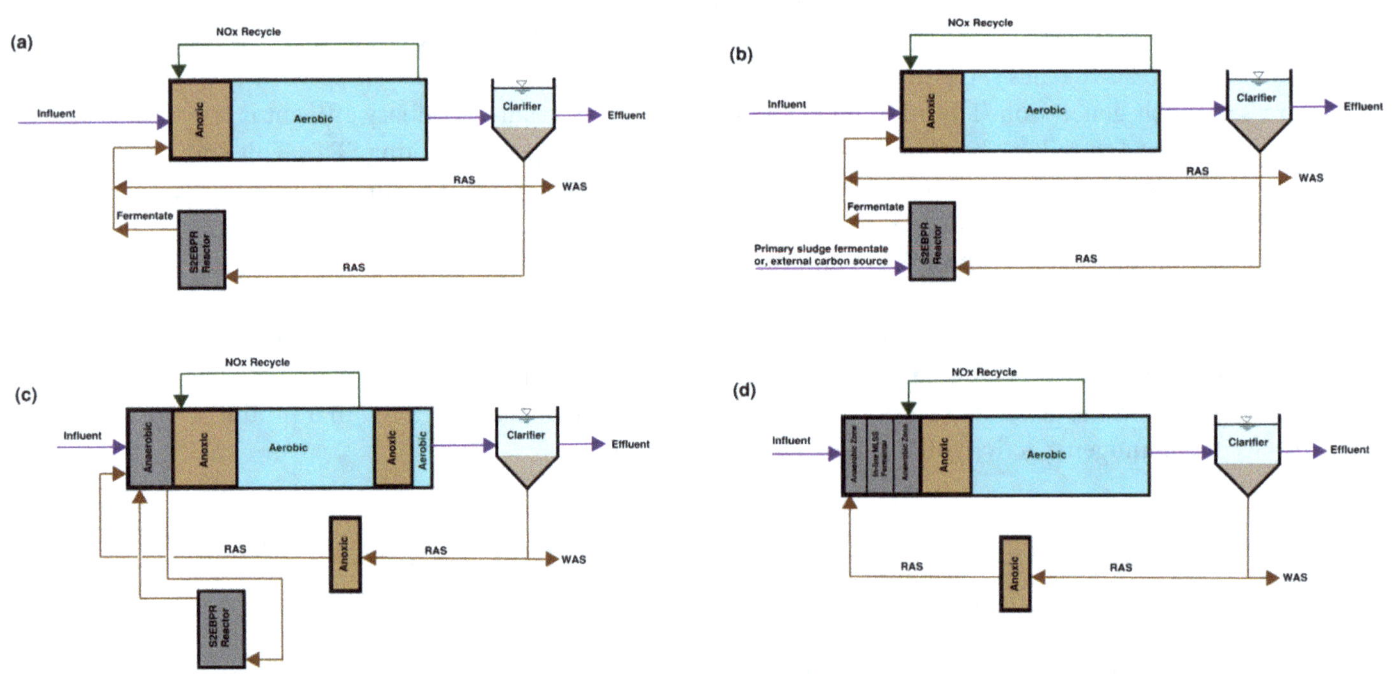


FIGURE 8.5 Process Flow Diagram of Different S2EBPR Configurations: (a) RAS Fermentation, (b) RAS Fermentation With Supplemental Carbon Addition, (c) MLSS Fermentation, and (d) Unmixed Inline Fermentation

Performing RAS or MLSS fermentation can promote PAO self-fermentation, which can help the PAOs to outcompete GAOs. Polyphosphate-accumulating organisms also grow faster than GAOs, so a lower SRT will help to flush out GAOs while PAOs are retained. Providing fermentation tanks with deeply anaerobic conditions will allow the PAOs to ferment the substrate and provide advantages to compete with the GAOs to achieve efficient EBPR.

## 8.1 Sidestream RAS Fermentation

Fermentation of a portion of RAS has been used for VFA generation in EBPR processes (WERF, 2011). Although this fermentation option can be used at any EBPR process, it particularly applies to facilities that operate secondary treatment processes without primary clarifiers. There are several variations of the configuration in operation in North Carolina. In a patented sidestream EBPR process using RAS fermentation, a portion of RAS from the secondary clarifier is first directed to a sidestream anaerobic zone. A portion of the RAS from the anaerobic zone is then directed to a sidestream fermentation zone (Lamb, 1994). Effluent from the sidestream fermentation zone is sent back to the anaerobic zone as a VFA source. The sidestream RAS fermentation zone is similar in operation to any anaerobic or anoxic zone in the mainstream EBPR process. As an alternative to the sidestream biological phosphorus removal configuration, the fermentation zone effluent could be sent directly to the anaerobic zone of the mainstream EBPR process. The common practice for RAS fermentation is to send a portion of RAS (5% to 30%) to a sidestream fermentation reactor with HRT and SRT range of 16 to 36 hours and 1 to 2 days, respectively (Water Research Foundation [WRF], 2019).

Although secondary sludge fermentation is a relatively new concept and the biochemical mechanisms involved are not yet well defined, it is regarded as a viable alternative to primary sludge fermentation, primarily because it has few of the operating problems associated with primary sludge fermenters (i.e., odors, corrosion, blockages, abrasion, etc.) (Barnard et al., 2010).

A significant disadvantage of fermenting secondary sludge is the high concentration of ammonia and phosphorus released into the fermentate. Ammonia release is a result of the anaerobic metabolism of protein material in the sludge, whereas phosphorus is released by PAOs under anaerobic conditions in the absence of SCVFAs. However, practice has shown that the overall benefits of additional SCVFA provided through secondary sludge fermentation can outweigh the negative effect of these additional nutrient loads. The key to successful application of secondary sludge fermentation is to ferment a relatively small portion of the mixed liquor or RAS, thereby fermenting sufficient secondary sludge to generate the VFAs needed to drive EBPR, but not so much that the released masses of nitrogen and phosphorus overwhelm the process (Andreasen et al., 1997; Barnard et al., 2010; Vale et al., 2008; Vollertsen et al., 2006).

Care must be taken to ensure that the VFA-rich fermentate does not come into contact with nitrate-rich mixed liquor because the VFAs generated will be preferentially used for denitrification, thus rendering them unavailable to drive the EBPR process.

## 8.2 Sidestream RAS Fermentation With Supplemental Carbon Addition

Sidestream RAS fermentation with supplemental carbon addition is a modification of RAS fermentation process. In this process, RAS is mixed with supplemental carbon source such as supernatant (overflow) of primary sludge fermenter. If primary sludge fermenter is not available, then molasses or expired food grade beverages can be used as other supplemental carbon source. External carbon source enhances the fermentation process. Before RAS is being introduced to fermentation reactor, RAS can be passed through a small anoxic reactor with HRT of 5 to 10 minutes to avoid any nitrate and dissolved oxygen entering the fermentation reactor. The typical HRT of the sidestream RAS fermentation reactor with supplemental carbon source is 1 to 4 hours and requires continuous mixing.

The advantage of this process configuration is that the addition of supplemental carbon source provides shorter HRT in the fermentation reactor. To implement this sidestream fermentation process, internal carbon source can be added to the fermenter using a primary sludge fermentation reactor. In addition, diversion of primary effluent to main-stream anoxic zone is required, so that VFA produced in the sidestream RAS fermentation reactor can be diverted to the main-stream anoxic zone. A primary sludge fermenter also provides additional carbon source that can be used for denitrification as well as for the sidestream fermenter. The disadvantage of this process configuration is that if there is no primary clarifier, then this sidestream fermentation process cannot be implemented unless capital cost is considered for external carbon source infrastructure systems.

## 8.3 Sidestream MLSS Fermentation

Sidestream MLSS fermentation is gaining attractions in S2EBPR systems. A fraction of MLSS (5% to 15%) from the anaerobic zone is diverted to the sidestream fermentation zone (reactor) with infrequent mixing (once or twice per day for 1 to 5 minutes). A key benefit of intermittently mixed fermenters is that these reactors allow the mixed liquor to settle, thus increasing the SRT of the fermenting sludge while physically separating it from the nitrate in the supernatant. Flow enters the fermentation zone through a floor-level distribution grid and flows upward through a sludge blanket, elutriating VFA that has formed there. Flow (supernatant) from the fermentation zone is returned to the mainstream anaerobic or anoxic zone. In this configuration, approximately, thickening in the fermentation tank is accomplished by decoupling the SRT and HRT with a target HRT of 5 to 15 hours (typical 12 hours) and an SRT of 2 days (typical). The differential in SRT and HRT is accomplished by infrequent mixing and alternating tank feed to allow settling of MLSS. The benefit of MLSS fermentation is that it provides more fermentation activity using adsorbed particulate COD attached with biomass cell surface. Thus, more VFA production is achievable. The anaerobic conditions and relatively long HRT and high solids concentration lead to considerably high hydrolysis and fermentation activity. This results in the production of sufficient VFA to promote the EBPR and support nitrogen removal. Another advantage of this configurations is to decouple the sidestream reactor HRT and SRT using intermittent mixing for optimum performance. One disadvantage of this configurations is to generate odors from fermentation reactors, thus increasing operational complexity.

## 8.4 Unmixed In-line Fermentation

The unmixed in-line fermentation (UMIF) process is similar to MLSS fermentation configuration with the exception that this process uses a portion of existing anaerobic (main-stream) tank space. The mixing in this in-line fermentation zone is significantly decreased (minimal) from the other zone of the anaerobic reactor. This allows the sludge blanket to form at the bottom of the basin. The minimal mixing decouples the SRT and HRT in the anaerobic zone and increases the fermentation activity in settling zone. Because this process is integrated into the main-stream anaerobic zone, estimation of SRT is difficult, and there is no definitive range to use.

The main advantages of the UMIF process are that it can be implemented in any existing EBPR facilities and does not require any sidestream reactor. The drawback of this process is that it provides less operational flexibility due to integration in the main-stream anaerobic reactor. In addition, maintaining proper hydraulics can be an issue and may cause different levels of short-circuiting in the reactor where UMIF process is located and may change the process performance.

Table 8.5 summarizes the critical design and operational parameters for various S2EBPR processes.

TABLE 8.5 Summary of Design and Operational Criteria for Various S2EBPR Facilities (Modified from WRF, 2019)

| Parameters | Feed type | SRT (days) | HRT (hours) | Sludge diversion (%) | MLSS concentration (mg/L) | Mixing type |
|---|---|---|---|---|---|---|
| Sidestream RAS fermentation | RAS | 1–2 | 16–36 | 5–30 | >6000 | Intermittent |
| Sidestream RAS fermentation with supplemental carbon addition | RAS with supplemental carbon (PS fermenter supernatant or other source such as molasses or expired food-grade beverages) | n/a | 1–4 | 30–100 | >6000 | Continuous |
| Sidestream MLSS fermentation | MLSS | 1–2 | 5–15 | 5–15 | 10 000–13 000 | Intermittent |
| Unmixed in-line fermentation | MLSS | n/a | n/a | n/a | >6000 | Minimal intermittent mixing |

## 9.0 ENHANCING EBPR PROCESS USING EXTERNAL CARBON SOURCES

Acetic acid ($CH_3COOH$) is the most common external carbon source used for EBPR. Acetic acid is a colorless liquid with a strong vinegar odor and is commonly delivered as glacial (approximately 100% solution), 84% or 56% solution. Although not as volatile as methanol, glacial acetic acid has a relatively low flash point (40 °C [104 °F]) and has the added complication of having a 17 °C (62 °F) freezing point. Consequently, continuous heating is required under most climate conditions. However, as acetic acid is diluted with water, the freezing point decreases, and all dilutions below 85% have a freezing point at or below that of water. Similar to methanol, meeting code requirements for a flammable liquid is a vital component of design (unless very dilute solutions are used, nearing the properties of water), and measures must be taken to avoid freezing. Because of flammability concerns, systems designed for use of high concentrations of acetic acid generally have similar requirements to those of methanol systems. Storage tanks, piping, and appurtenances should be constructed of metal materials. Because of the corrosivity of acetic acid, Type 316 stainless steel is typically used. If glacial acetic acid is used in warm climates, it may be necessary to consider an inert gas blanket or floating cover because of the relatively low flash point.

An external carbon feed of pure acetic acid can promote the growth of GAOs, which compete directly with PAOs, thus increasing the chances of EBPR upset. In contrast, a mixture of acetic and propionic acids has been found to be more beneficial to EBPR (Randall & Chapin, 1997).

In addition to acetic and propionic acids, there are other options for carbon source supplementation. Some WRRF operators have looked to industrial waste products as sources of supplemental carbon. In addition, there are a number of possibilities including sugar wastes, molasses, and waste acetic acid solution from pharmaceutical manufacturers. It is important to ensure that these sources are free of contaminants and debris. Finally, the design of chemical feed facilities must comply with all applicable

National Fire Protection Association, U.S. Department of Transportation, Occupational Safety and Health Administration, and local code requirements.

# 10.0 PROCESS MONITORING AND CONTROL

## 10.1 Dissolved Oxygen

Dissolved oxygen is a measurement of oxygen dissolved in a liquid stream. In the EBPR process, adequate dissolved oxygen should be available for rapid phosphorus uptake in the aerobic zone. If dissolved oxygen is present in the anaerobic zone, however, it could potentially inhibit VFA production and PAO selection. Dissolved oxygen in the aerobic zones should be continuously monitored to ensure sufficient oxygen is provided during all flow and loading conditions. The preferred aeration basin layout for maximum removal efficiency is plug flow. In practice, the aeration basin is often divided into several baffled zones or compartmentalized to approach plug flow conditions. The greater the number of zones, the closer the flow regime is to plug flow. However, cost and practical considerations often limit the number of aerated zones.

When dissolved oxygen is not limiting, the phosphorus uptake rate is typically high in the initial oxic zone of the process, as discussed in Chapter 10. Facility operating data have shown that orthophosphate can be reduced to a level as low as 0.1 mg/L when sufficient oxygen is supplied in the first oxic zone of the EBPR process and all other conditions for biological phosphorus removal are met. Whereas it is possible to achieve EBPR at dissolved oxygen concentrations less than 2 mg/L, the initial oxic zone (first 25% to 33%) should be designed to operate at higher than 2.0 mg/L dissolved oxygen to support rapid phosphorus uptake, meet nitrogenous demand, and guard against the growth of low dissolved oxygen filaments such as *Microthrix parvicella*. If the facility has a completely mixed aeration tank, then phosphorus removal efficiency may be increased by partitioning the tank to cost-effectively maintain high dissolved oxygen in the first oxic cell. Because oxygen demand toward the end of a plug flow aeration basin is primarily associated with endogenous respiration, tapering air delivery and reducing the target dissolved oxygen in the subsequent oxic zones will meet the process requirements and reduce the mass of dissolved oxygen in the mixed liquor recycle (MLR) streams. Too much dissolved oxygen introduced by MLR to the anoxic zone can inhibit denitrification. However, it is important to note that the recommended minimum airflow rates for mixing may make it difficult to maintain low dissolved oxygen concentrations at the end of the aeration basin. Dissolved oxygen in the MLR recycle can also be minimized by providing a small deoxygenation zone (nominal HRT of 20 to 30 minutes) before the MLR pumps. The need to mechanically mix the deoxygenation zone should be assessed if problems related to solids settling are experienced.

On the contrary, sufficient dissolved oxygen should be maintained in the bioreactor effluent to avoid problems in final clarifiers such as phosphorus release in the sludge blanket and denitrification (in nitrifying facilities). If both objectives (reduced recycle dissolved oxygen load and adequate bioreactor effluent dissolved oxygen) must be met, the bioreactor effluent could be aerated in the final clarifier distribution box or in mixed liquor channels to the clarifiers.

For systems approaching complete-mix conditions (such as an oxidation ditch), it is possible to control the airflow with one dissolved oxygen probe that is strategically located. Plug flow reactors encounter significant oxygen demand gradients. As treatment occurs, oxygen requirements progressively decrease along the length of the bioreactor. As a result, a tapered diffuser arrangement with multiple oxic zones with independent aeration grids may be considered. For these configurations, more sophisticated control systems may be necessary. One approach is to provide a common header system equipped with multiple control valves to regulate the airflow to each grid with dedicated dissolved oxygen probes.

Many facilities use online instrumentation for continuous monitoring of dissolved oxygen in aeration zones. Often, this information is used to control aeration and to reduce energy costs. The dissolved oxygen concentrations in the bioreactor influent, anaerobic zone, internal recycle, and RAS streams typically are not monitored but should be checked if EBPR is reduced. When instrumentation and control equipment are limited, it is possible to control the aeration system with one strategically placed dissolved oxygen probe in combination with a manually controlled valve to each of the aeration grids to avoid underaeration or overaeration.

## 10.2 Oxidation-Reduction Potential

Oxidation-reduction potential is a measurement of the oxidation or reduction potential of a liquid. The ORP measures all ionic species in the fluid and cannot be used to make fine adjustments based on subtle changes. Chapter 13 includes a detailed description of process monitoring and control using ORP.

## 10.3 Online Nutrient Analyzers

Increasingly, sophisticated analyzers are available to measure nutrients directly in the process. Analyzers can be used for monitoring ammonia, nitrate, nitrite, and phosphate through the treatment process. Many of these analyzers are costly to purchase and maintain but provide invaluable information on dynamic variations in nutrient concentrations through the process and can be used for troubleshooting.

## 10.4 Solids Retention Time

Solids retention time is an important operational parameter in a biological system because it provides a measure of the time a solid particle (organism) spends in the biological reactor. Controlling the SRT manages the microbial makeup of the mixed liquor. Ideally, the goal is to operate the system close to an SRT that ensures reliable treatment and reduces operating cost.

In an EBPR system, because the emphasis is on the anaerobic storage of the readily available substrate and subsequent consumption of stored carbon, anaerobic and aerobic SRT gain importance to allow just enough time for the PAOs to proliferate and for the EBPR reactions to take place. In addition to SRT, the ratio of feed COD to phosphorus (COD/P) also shapes the microbial composition of the EBPR sludge and effluent phosphorus levels that can be attained. Figure 8.6 illustrates the combined effect of the observed yield, as dictated by system SRT, and the feed COD/P on the phosphorus content of the biomass. As the observed yield increases for a constant COD/P value, biomass phosphorus content decreases. In other words, at lower SRT values, less phosphorus is stored in the biomass. Conversely, as the SRT is increased, the sludge will be more robustly enriched by PAOs, as reflected in higher biomass phosphorus values. If, at the same SRT and observed yield value, the feed COD/P is lowered (i.e., increased phosphate or decreased COD in the feed), the biomass phosphorus content will be observed to increase. As feed phosphate increases, more of it can be stored in the biomass, given that the feed COD is sufficient, resulting in an improvement in biomass PAO enrichment. Again, because of the strong effect of feed characteristics and COD/P on the biomass enrichment and microbial composition, results from two different facilities cannot be reliably compared in terms of EBPR performance.

Through years of field and laboratory experience, EBPR systems have been shown to operate at SRTs greater than 3 days. At SRTs between 3 and 4 days, effluent quality declines, and chemical polishing may be needed. At SRTs greater than 4 days and at temperatures greater than 15 °C, nitrification will start to occur and process configurations that include anoxic zones for denitrification of nitrate in system recycles must be used. As the SRT is increased to a level at which endogenous reactions become significant (i.e., increased biomass decay), secondary release of phosphorus may lead to decreased performance at

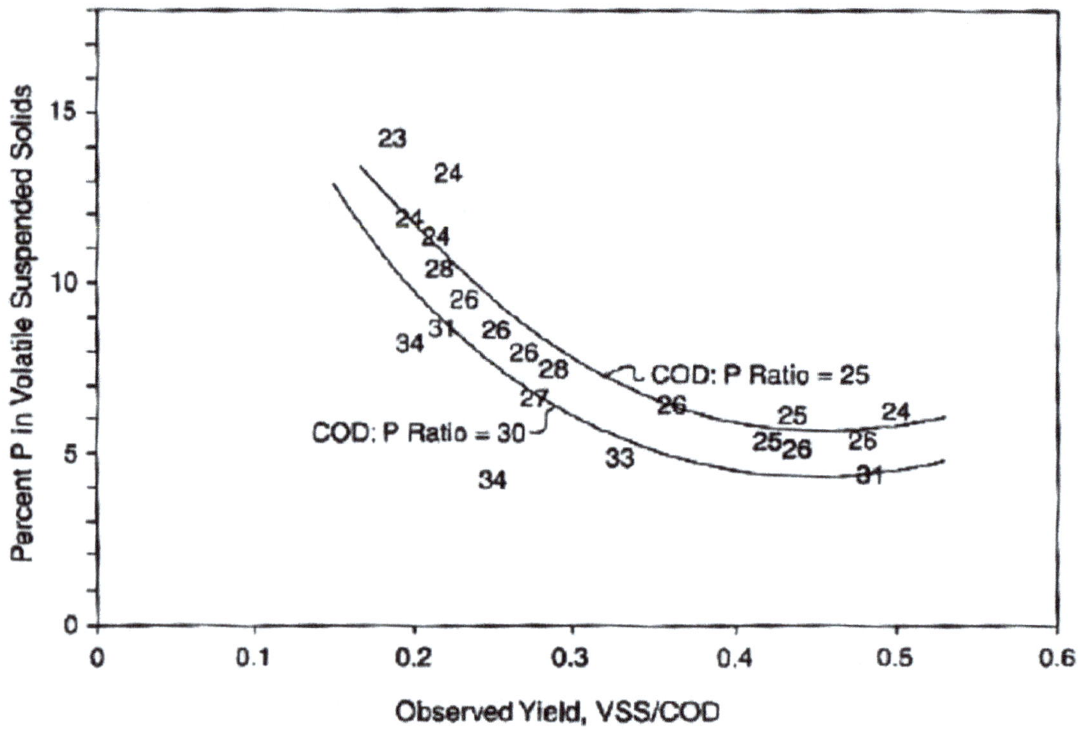


FIGURE 8.6 Effect of EBPR Biomass Observed Yield and SRT on Mixed Liquor Volatile Suspended Solids Phosphorus Content (WEF, 2010)

given feed VFA and COD values, and phosphorus removal at these higher SRT values can be increased if the feed COD can be increased. Thus, seasonal variations in feed COD and phosphorus can be handled by varying the operational conditions; operators, in turn, can determine the best operational strategy for their influent quality and WRRF setup.

To explain SRT and EBPR performance, a number of researchers conducted experiments under various operational conditions. McClintock et al. (1992) showed that, at a temperature of 10 °C and an SRT of 5 days, EBPR function of a given activated sludge system would "wash out" before other heterotrophic functions. The washout SRT is a design parameter that defines a critical SRT point, below which no growth of biomass occurs (Grady et al., 2011). Mamais and Jenkins (1992) showed that there is a washout SRT for all temperatures over the range of 10 to 30 °C. Their study clearly indicates that if the SRT-temperature combination is below a critical value, EBPR ceases before other heterotrophic functions. Lu et al. (2007) demonstrated the effect of SRT on the number of bacterial species present in different biological systems by a comprehensive, simplified microbial model, which could be used in the future for SRT control on biodiversity and treatment performance.

Commercially available programs can automatically control SRT by using sophisticated control algorithms that take into account real-time variations in solids inventory in the aeration tanks and in solids to be wasted. This type of system has been used in various WRRFs such as the San Jose and Santa Clara Water Pollution Control Plant, in the City of Oxnard, California, and in Oakland, California.

In EBPR systems designed for both nitrogen and phosphorus removal, individual zones should be designed to have variable sizing with provisions for mixing and aeration zones that can be expanded or reduced to achieve desired effluent phosphorus concentrations, depending on influent characteristics and operational conditions. At full-scale facilities, VFA uptake is a relatively rapid reaction, requiring an anaerobic zone SRT as low as 0.3 to 0.5 days.

Table 8.6 summarizes critical design and operational parameters for different EBPR configurations.

TABLE 8.6 Typical Design and Operational Parameters for Various EBPR Configurations (Adapted from Metcalf & Eddy, Inc./AECOM, 2013)

| Process type | SRT (days) | MLSS (mg/L) | HRT (hours) Anaerobic zone | HRT (hours) Anoxic zone | HRT (hours) Aerobic zone | RAS % of influent | Internal recycle % of influent |
|---|---|---|---|---|---|---|---|
| A/O | 2–5 | 3000–4000 | 0.5–1.5 | n/a | 1–3 | 25–100 | n/a |
| A2/O | 5–25 | 3000–4000 | 0.5–1.5 | 1–3 | 4–8 | 25–100 | 100–400 |
| 5-Stage modified Bardenpho | 10–20 | 3000–4000 | 0.5–1.5 | 1–3 (pre-anoxic), 2–4 (post-anoxic) | 4–12 (1st stage), 0.5–1 (2nd stage) | 50–100 | 200–400 |
| University of Cape Town | 10–25 | 3000–4000 | 1–2 | 2–4 | 4–12 | 80–100 | 200–400 (NRCY), 100–300 (RAS) |
| Virginia Initiative Plant | 5–10 | 2000–4000 | 1–2 | 1–2 | 4–6 | 80–100 | 100–200 (anoxic), 100–300 (aerobic) |
| SBR | 20–40 | 3000–4000 | 1.5–3 | 1–3 | 2–4 | n/a | n/a |

*Note.* n/a—not applicable; NRCY—NOx recycle

## 11.0 MINIMIZING FACILITY-WIDE EFFECTS

The biological cycle of anaerobic phosphorus release and aerobic phosphorus uptake in an EBPR facility can significantly affect both upstream and downstream unit operations. These effects can be significant and must be considered during design and operation. Many of the effects are caused by the re-release of phosphorus in anaerobic conditions, such as in thickening or dewatering operations.

Other significant effects may come from filamentous bulking or foaming. Filamentous bulking is rare in biological phosphorus removal facilities because the incorporation of an anaerobic zone for phosphorus release acts like a selector, which will favor the predominance of nonbulking organisms. The anaerobic selector does not necessarily inhibit growth of all types of filamentous organisms. Organisms such as *Gordona amaerae* and *Microthrix parvicella* are capable of adapting to alternating anaerobic and aerobic cycles. Other low-F/M filaments such as Types 0041 and 0675 may also be able to proliferate in the anaerobic zone. *Microthrix parvicella*, in particular, are capable of storing polyphosphates and to compete successfully for SCVFAs in anaerobic selectors if wastewater characteristics and environmental conditions are favorable.

### 11.1 Primary Clarification

In secondary treatment facilities, use of primary clarifiers reduces aeration requirements in the downstream activated sludge tanks. With EBPR processes, excessive organics removal in primary treatment can potentially result in lowering the primary effluent COD/P and can reduce EBPR efficiency. This would only happen if the anaerobic zone were sufficiently large to affect VFA generation from settleable solids, which typically may not be the case. Another possibility is that primary polymer addition would remove colloidal solids that would otherwise be fermented to SCVFAs in the anaerobic zone. In such instances, primary sludge fermentation could be included to increase the VFA concentration entering the anaerobic zone.

One strategy for thickening waste activated sludge (WAS) is to co-thicken it in primary clarifiers along with primary sludge. In general, this strategy should not be used for EBPR because the combination of primary sludge and WAS will result in significant and rapid secondary phosphorus release. Work by Skalsky and Daigger (1995) showed that combining WAS and primary sludge resulted in more rapid fermentation of the primary sludge than would occur without the addition of WAS. As investigated by Chaparro and Noguera (2003), it may be possible to use this characteristic of co-thickening to affect phosphorus release and create a phosphorus-rich recycle stream for a struvite harvesting process.

Another effect to consider in the primary clarifier is the fate of iron addition to a recycle stream for recycle phosphorus control. If ferric or ferrous chloride is added to precipitate phosphorus from a thickening or dewatering recycle stream, the precipitate formed will settle in the primary clarifiers. The effectiveness of this strategy will be reduced because a pH of approximately 6.5 in the anaerobic environment of the primary clarifier, or in a downstream gravity thickener, will cause a portion of the ferric phosphate precipitated at a higher pH to be reduced to ionized form. This may result in some additional carryover of iron and phosphorus to the secondary process, although it may be insignificant.

## 11.2 Final Clarification

Separation of biological solids is of primary importance in EBPR. Phosphorus uptake by bacteria will result in biological solids with 2% to 5% phosphorus. The effluent phosphorus concentration is, therefore, highly dependent on the capture of the solids. For example, if the effluent solids concentration is 20 mg/L and the biological solids are 4% phosphorus, the effluent phosphorus concentration will be 0.8 mg/L just from the solids, plus additional soluble phosphorus remaining after treatment. This level of effluent TSS would make it difficult to reliably comply with a 1-mg/L total phosphorus standard in the effluent. Fortunately, the anaerobic selector in an EBPR facility will typically result in good settling sludge that also flocculates well. Effluent TSS concentrations less than 10 mg/L are common in EBPR facilities. With appropriately designed clarifiers, an EBPR facility can produce 1.0 to 0.50 mg/L of total phosphorus in the effluent, and with filtration, total phosphorus concentrations less than 0.30 mg/L are possible without supplemental chemical addition, provided conditions for efficient EBPR are satisfied. The potential for high flows, development of a deep blanket, and washout of solids must be considered. If the clarifiers do not have sufficient capacity for high-flow conditions, high-effluent phosphorus values may result.

Because mixed liquor solids will generally settle and compact well in an EBPR facility, it is often possible to reduce the return rate and provide a higher degree of thickening in the secondary clarifier. This allows for a lower waste sludge flowrate. Caution must be exercised because a deep sludge blanket may result in secondary release of phosphorus and higher soluble phosphorus levels in the effluent. Improvements to the sludge removal mechanism can be beneficial if the sludge blanket builds up despite good settling and compaction.

## 11.3 Secondary Release

Bacteria generated in an EBPR process contain two forms of phosphorus. One form is polyphosphate temporarily stored during aerobic respiration, which then allows energy generation by phosphorus release with VFA uptake. The second form is the normal metabolic phosphorus product from microbial growth and reproduction. *Secondary release* is the term for release of temporary phosphorus storage products in anaerobic conditions without the concurrent uptake of VFAs. This release will not be taken up again in the aerobic zone and will result in elevated effluent phosphorus levels. Table 8.7 shows some of the potential causes of secondary phosphorus release.

TABLE 8.7 Location and Cause of Secondary Phosphorus Release

| Environment/operation | Reason for secondary phosphorus release |
|---|---|
| Anaerobic zone | VFA depletion because of oversized anaerobic zone |
| Anoxic zone | Nitrate depletion because of oversized anoxic zone |
| Second anoxic zone | Lack of nitrates resulting in anaerobic conditions |
| Aerobic zone | Long solids retention time leading to cell lysis |
| Final clarifier | Septic conditions caused by deep sludge blanket and low return rate |
| Return sludge piping | Septic conditions from long detention time in pipe |
| Sludge storage | Septic conditions from poorly or unaerated sludge storage |
| Aerobic digestion | Cell lysis because of long aeration time |
| Anaerobic digestion | Cell lysis because of long anaerobic detention time |

Secondary release might be caused in the secondary system by long aerobic, anoxic, or anaerobic detention times. Long SRTs in the aerobic system may also result in secondary release. Long SRT operation also requires higher COD/P for the bacteria to take up the available phosphorus. Because of these factors, it is generally best to operate an EBPR system as close to minimum SRTs as possible and to maintain minimum necessary anoxic and anaerobic detention times to meet process goals. If nitrification is required, this may limit the minimum SRT based on minimum winter wastewater temperatures. If effluent phosphorus concentrations rise and jeopardize effluent limitations, it may be necessary to collect samples profiling the secondary process to determine whether the problem may be caused by secondary release.

## 11.4 Recycle Streams

Recycle streams result from sludge thickening and dewatering operations. Thickening of primary sludge may actually benefit the EBPR process because VFAs will be generated in the thickening process. This is similar to inclusion of a dedicated primary sludge fermentation process to enhance EBPR.

Thickening of WAS may have little effect on the EBPR process unless the sludge is held too long, resulting in secondary release. In addition, relatively high levels of recycle phosphorus to the secondary process may result.

Gravity co-thickening of WAS and primary sludge is typically not recommended for an EBPR facility because this will result in the recycle of high concentrations of soluble phosphorus released from WAS. As described for primary clarification, gravity co-thickening of primary sludge and WAS will result in enhanced production of VFAs in the primary sludge and significant release of polyphosphate from WAS.

High concentrations of soluble phosphorus will result from the release of stored polyphosphates and cellular phosphorus in both anaerobic and aerobic digestion. Thickening or dewatering of these streams will result in a stream that is high in soluble phosphorus (200 to 300 mg/L and higher) and high in ammonia-nitrogen (700 to 1000 mg/L and higher). Recycled phosphorus load can represent 5% to 40% of influent total phosphorus load and may alter the COD/P such that successful EBPR to low levels may no longer be possible. Strategies to minimize the effects of digested sludge recycle streams include operating thickening or dewatering continuously to minimize slug loading because of recycled phosphorus and the addition of iron salts to the recycle stream to reduce the soluble phosphorus recycle.

## 11.5 Bulking and Foam Control

Filamentous bulking and foaming are potential operating problems associated with EBPR facilities. These issues negatively affect effluent quality and create serious housekeeping and odor problems. An in-depth examination of the topic may be found in literature by Eikelboom (2000), Jenkins et al. (2003), and Wenner (1994). Detailed information on microscopic evaluation and identification of activated sludge characteristics, levels, and types of filamentous organisms is presented in literature by Jenkins et al. (2003). The following subsections provide a brief discussion of the causes and potential corrective strategies of bulking and foaming in EBPR systems.

### *11.5.1 Filamentous Bulking*

Excessive filaments often lead to poor settling sludge and high effluent solids. A variety of operating conditions, singly or in combination, can cause growth of filamentous organisms in EBPR systems. These include low dissolved oxygen, low or high F/M, sulfides, and low pH. Filament identification is the first step in resolving the problem. Typically, operational controls focus on removing conditions responsible for bulking or killing filamentous organisms to control their number. Some common strategies include

- using selectors to provide growth advantage to floc formers;
- chlorinating RAS;
- adding nutrients;
- correcting the dissolved oxygen concentration in the bioreactor; and
- correcting the pH.

Generally, the anaerobic zone in an EBPR system will select the floc-forming PAOs and minimize settling problems. However, if for any reason VFAs are not totally consumed in the anaerobic selector and enter the aeration tank, septicity-type filaments can grow.

### *11.5.2 Filamentous Foaming*

Although *Gordona* (formerly known as *Nocardia*) *amarae* and other nocardiaforms are the most commonly found organisms responsible for filamentous foaming, others (such as *Microthrix parvicella* and Type 1863) also can cause foaming. The surface appearance of tanks with foaming problems may be similar regardless of the organism causing the foaming problem. The presence of some foam in the activated sludge bioreactor is normal. In a well-operated process, 10% to 25% of the bioreactor surface may be covered with a 50- to 80-mm (2- to 3-in.) layer of light tan foam. Under certain operating conditions, foam can become excessive and affect operations.

Three types of problem-causing foams are stiff, white foam; brown and dark-tan foam, often incorporating scum; and dark-brown or black foam. The most common foaming problem in EBPR systems is brown and dark-tan foam associated with nocardiaforms or *Microthrix parvicella.* If this greasy, thick, scummy foam builds up and is conveyed to secondary clarifiers, it will tend to build up behind influent baffles and create additional cleaning requirements. In colder climates and extreme cold conditions, it can build up on scum-removal ramps, hang up the scum-removal skimmers, and cause sludge removal mechanisms to overload. This can result in loss of effluent solids or secondary phosphorus release. It can also plug scum-removal piping. Unmanaged, scum can even overflow aeration tank walls and flow onto walkways and roadways. Waste activated sludge containing foaming organisms can also result in serious foaming problems in anaerobic or aerobic digesters.

*Nocardia* foaming is typically associated with warmer temperatures, lower pH, grease, oil, fats, and long SRT. *Microthrix parvicella* foaming is more commonly associated with mixed liquor temperatures less than 16 °C and long SRTs. *Microthrix* spp. also prefer grease, oils, and fats; these microorganisms are better able to compete for these substrates at colder temperatures.

Because foaming is a surface phenomenon, it typically has a longer SRT than the underlying MLSS. Facilities prone to foaming often receive oil and grease waste from restaurants with poorly performing or missing grease traps, have poor or no primary scum removal or recycle scum, and have bioreactors and final clarifiers that are not properly designed to remove scum and foam. The most effective strategy to deal with foaming is to eliminate conditions that encourage growth and to prevent accumulation within the process. This is not always easy, however, because exact cause-and-effect relationships have not been fully established. The following is a listing of foam- and scum-control methods that can be implemented during design of EBPR systems:

- Design interzone and intrazone baffles to promote free-flow surface foam and scum.
- Eliminate dead ends, sharp corners, and quiescent zones in channels or bioreactors where there is a potential for foam and scum to accumulate.
- Design secondary sedimentation tank inlet wells and flocculation wells to allow for passage of floating material. Provide effective foam and scum collection mechanisms (such as full-radius collectors).
- Incorporate provisions for selectively wasting foam and scum from the surface of the aeration basin. Collected foam should not be recycled to avoid reseeding the foam-causing organisms. Parker et al. (2003) reported several successful applications of selective wasting in controlling activated sludge foaming. The overall objective of this strategy is to reduce the SRT of the foam-causing organisms while minimizing the effect on process SRT.
- Consider including selectors; however, selectors have been found to have marginal effectiveness in preventing the predominance of foam-causing organisms.

If foaming occurs, corrective action should be taken quickly. Microscopic analysis of mixed liquor will allow for identification of the predominant causative organisms. This will provide clues as to which operational strategies are likely to be successful. Some examples include the following:

- Operate at the lowest SRT required to reliably meet process goals.
- Avoid opportunities for recycling foam and scum to the mainstream treatment train to prevent recurrence of the problem.
- Apply chlorine (0.5% to 1% solution) spray at localized points of foam and scum collection or accumulation to kill nocardiaforms and to prevent them from causing problems in either mainstream or sidestream treatment processes. The chlorine dose should be carefully controlled to avoid EBPR inhibition, which can take several days to recover.
- Consider RAS chlorination. However, this has limited effectiveness in controlling foam-causing organisms.
- Add polymers to destroy the hydrophobic properties of the foam and to allow it to mix with the sludge so that it can be removed with waste sludge.
- Addition of polyaluminum chloride has also been shown to be effective in controlling *Microthrix parvicella* foaming (Melcer et al., 2009).

## 12.0 SIDESTREAM PHOSPHORUS RECOVERY

In WRRFs, a significant amount of nitrogen and phosphorus load can be generated internally through reject water of sidestream treatment processes. The anaerobic digestion and centrate/filtrate return from sludge dewatering processes are critical sidestream sources that produce highly concentrated nutrients (nitrogen and phosphorus) stream. Most of these sidestreams flow are conventionally returned back to the headworks of WRRFs, where they are mixed with influent wastewater and directed to a mainstream liquid treatment process. This sidestream flow contributes only 1% to 3% of the forward flow at a WRRF and accounts for almost 10% to 20% and 20% to 30% of the facility's influent nitrogen and phosphorus load, respectively.

Implementation of EBPR in the mainstream liquid treatment process further increases the sidestream phosphorus loading. This loading can account for up to 50% influent total phosphorus loading to the facility's mainstream treatment process. In addition, the EBPR process also reduces dewaterability (3% to 5%) because phosphate stabilizes the water/sludge matrix resulting in reduction in cake solids. Thus, the polymer dosage can also be increased (up to 30%) for the phosphorus-containing-sludge dewatering processes.

Struvite precipitation ($MgNH_4PO_4$) is one of the most common problems in using EBPR facilities. The deposition of struvite causes significant problems in the biosolids handling treatment processes, including anaerobic digestion and post-digestion processes. The struvite formation is caused by a high concentration of ammonium and phosphate released in anaerobic digestion process. The precipitation of struvite can create significant operational and maintenance problems in WRRFs, including scaling, fouling, and clogging issues in piping, valves, dewatering pumps, within digesters, heat exchangers, and digester mixing equipment. Any uncontrolled struvite formation can increase the operational and maintenance cost of the facility. Scaling and clogging issues in plumbing and piping can decrease the hydraulic capacity of the facility due to a decrease in the diameter of the piping. Struvite can precipitate in the pH range of 7.5 to 9.0.

Benefits of the sidestream phosphorous recovery process include

- decreasing sidestream phosphorus loading;
- reducing scaling or fouling of equipment from phosphorus precipitation;
- decreasing the negative effect of EBPR sludge on dewaterability;
- helping WRRFs meet effluent nutrient limits;
- generating a sustainable and beneficial byproduct; and
- the opportunity for revenue generation through marketable and saleable products such as fertilizer for land application.

Table 8.8 summarizes different phosphorus recovery technologies that are established and proven for full-scale applications.

In current practice, phosphorus sequestration is gaining more attraction than recovery because of recovery products' quality variation, the final product polishing process cost, and recovery products' marketability. In the phosphorus sequestration process, small struvite crystals are formed so that the solids from the phosphorus sequestration reactor can be easily transferred to a dewatering process. During the dewatering process, struvite crystals are discharged through biosolids (cake solids) rather than heading back to the headworks of the facility, resulting in elimination of detrimental struvite deposition in the system. In a phosphorus recovery system, struvite is formed in larger crystals, which can be easily recovered. The recovery process can easily be added for any phosphorus sequestration reactor.

TABLE 8.8 Summary of Sidestream Phosphorus Recovery Technologies

| Phosphorous recovery technology | Form of phosphorus precipitation | Applicability | Chemical and other components addition to phosphorus recovery reactor | Manufacturer |
|---|---|---|---|---|
| Ostara Pearl process | Struvite | Pre- and post-digestion liquors | Magnesium salts, NaOH | Ostara/Evoqua Water Technologies |
| MagPrex | Struvite | Post-digestion sludge | Magnesium chloride, and air stripping | Centrisys/CNP |
| Crystalactor | Struvite, Vivianite | Dewatering liquors, digested sludge | Addition of sand as seed material, calcium, or magnesium salt | Royal Haskoning DHV, Netherlands |
| NuReSys | Struvite | Centrate, centrate + filtrate, digested sludge, digested sludge + centrate | Magnesium chloride, and air stripping | Nutrient Recovery Systems/Schwing Bioset |
| CalPrex | Vivianite | Pre-digestion liquor or centrate | Calcium hydroxide, pH adjustment | Centrisys/CNP |

## 13.0 TROUBLESHOOTING ENHANCED BIOLOGICAL PHOSPHORUS REMOVAL

Enhanced biological phosphorus removal is one of the most widely studied biological wastewater treatment processes because of its role in controlling nutrients discharged to the aquatic environment in a manner that is cost effective and free from chemical usage. Although many full-scale facilities consistently demonstrate excellent phosphorus removal performance under stringent discharge limits, they are not free from occasional process upsets, potentially resulting in permit violations. From this standpoint, it is essential to analyze the causes of such upsets and initiate urgent troubleshooting measures to minimize the risk of such permit violations and to restore the facility to normal running conditions. In this section, three case studies are presented as examples of problem analyses and solutions at selected full-scale WRRFs. Chapter 16 is devoted to case studies only and presents the details of eight operating facilities. The purpose of the case studies presented in this chapter is to demonstrate how the underlying principles of EBPR, as described in the preceding sections, have been used by facilities to achieve their respective nutrient control goals.

### 13.1 Kelowna Wastewater Treatment Plant, British Columbia, Canada

The bioreactor system at the Kelowna Wastewater Pollution Control Centre (42 mL/d) in British Columbia, Canada, was designed and operated as a five-stage Bardenpho process to achieve high levels of nitrogen and phosphorus removals required to comply with the permit limits of 6 mg/L total nitrogen and 0.25 mg/L total phosphorus. Oldham and Stevens (1984) conducted a study to investigate the reasons for the failure of the facility to achieve the desired total phosphorus concentration in the final effluent. At the time of this study, total flow was handled by two parallel, multicell, plug flow reactor trains with an HRT of 24 hours each. The authors observed that the total phosphorus concentration reached about 0.15 mg/L at about 80% of the total length of the aerobic stage, after which it climbed above 1 mg/L through the rest of the treatment train and in the final effluent.

After changing the mode of operation to only one train in line, with an HRT in the bioreactor system of 11 hours, Oldham and Stevens (1984) achieved effluent total phosphorus of ~0.1 mg/L. It was revealed that, in the two-train operation COD oxidation, nitrification and EBPR were complete in the front 80% and denitrification was occurring in the tail end of the aerobic stage, leaving no nitrate for denitrification in the anoxic stages. Secondary release of phosphorus was taking place in the absence of any dissolved oxygen or nitrate in the second anoxic zone, resulting in a higher-than-expected total phosphorus concentration in the effluent. After reducing the HRT in the one-train operating mode, the denitrification reaction was moved to the second anoxic zone. This, combined with reaeration in the final stage, helped to stop the secondary release, which demonstrates that excessive operating HRT may be detrimental to effective EBPR performance.

## 13.2 Black River Falls Wastewater Treatment Plant, Wisconsin

This case study was conducted in a small (~0.76 mL/d) community WRRF in Black River Falls, located in the east-central part of Wisconsin (Basu et al., 2008). The biological treatment process includes both nitrogen and phosphorus removal in an oxidation ditch, with brush-type aerators. The RAS returns to the outer ditch and mixes with the influent. The facility has occasionally experienced EBPR process failure despite maintaining dissolved oxygen concentrations of 3 to 4 mg/L in the aerobic zone and the presence of adequate VFAs in the influent. Microbiological analysis of the mixed liquor indicated healthy populations of PAOs and an absence of GAOs. However, the biomass indicated proliferation of filamentous microorganisms, primarily Type 0092. The root cause of this was traced back to the operation under excessively long SRT (~30 days), which caused high effluent phosphorus because of secondary release and poor settleability, resulting in carryover of excessive solids with the effluent. Gradual reduction of SRT to the level still adequate for nitrification enabled elimination of the bulking problem and restoration of a normal EBPR process.

## 13.3 Eagles Point Wastewater Treatment Plant, Cottage Grove, Minnesota

The 30-mL/d Eagles Point Wastewater Treatment Plant in Cottage Grove, Minnesota, was started up in September 2002. The treatment train includes screening and grit removal, primary settling, secondary treatment, and secondary clarification. The biological treatment process has been designed for both phosphorus and nitrogen removal, with the flexibility of operating in plug flow or step-feed modes. Primary solids are fermented in a gravity thickener, as needed, to supplement for VFAs to support EBPR. Lindeke and Barnard (2005) reported on the experience gained from operating the facility for the first 2.5 years since its startup.

During the period of this study, the average influent BOD, TSS, ammonia-nitrogen, and total phosphorus were 248, 220, 27.2, and 6.8 mg/L, respectively. Corresponding effluent characteristics were 4, 4, 2.4, and 0.68, respectively. The operating MLSS and SRT of the facility were maintained at 2000 to 2500 mg/L and 6.8 days (for two aeration tanks together), respectively.

The operational experience revealed three potential challenges to EBPR performance of this facility. These were seasonal unavailability of sufficient VFAs, carryover of excessive solids with the clarifier effluent, and operation of the biological reactor, as discussed in the following paragraphs.

Volatile fatty acid concentrations (expressed in terms of COD) across the primary sedimentation tank at the facility increased by approximately 50% annually because of fermentation in the sludge blanket. The VFA species were primarily acetic acid, with little propionic acid and no heavier acids. However, the fermenter overflow demonstrated 5 times as much VFA concentration in terms of COD as that in the primary effluent, with propionic acid about half that of acetic acid and a minor presence of isomers of butyric and valeric acids. During this study, facility staff monitored influent VFAs generated in the collection system as a function of wastewater temperature. The data presented in Figure 8.7

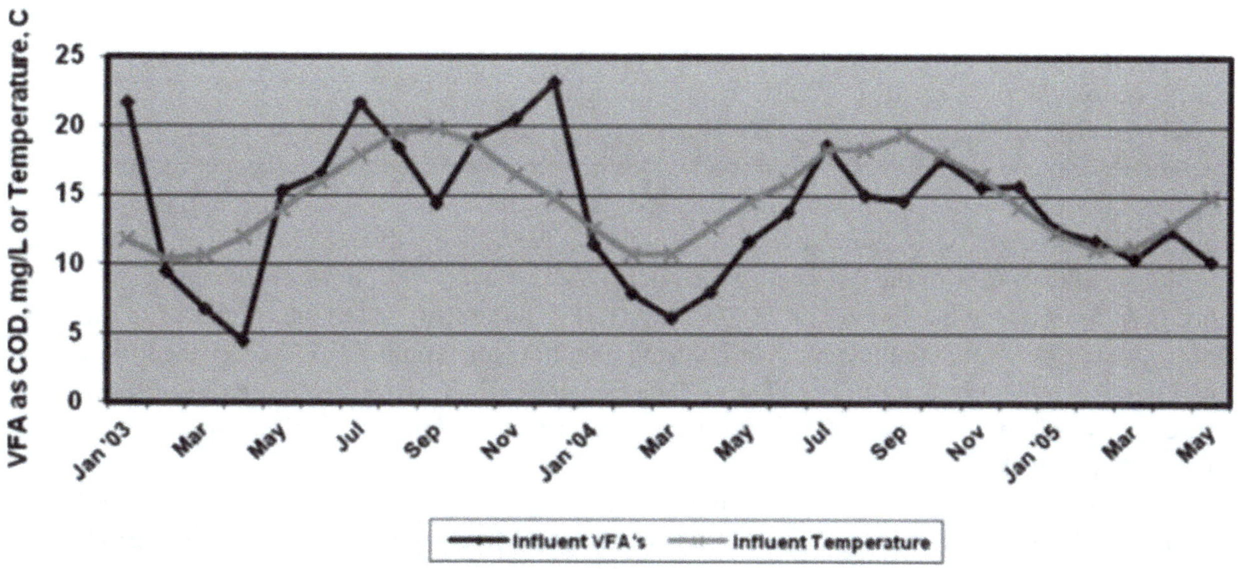


FIGURE 8.7 Seasonal Profile of Temperature Versus VFAs at Eagles Point Wastewater Treatment Plant

demonstrate the strong influence of temperature on influent VFAs, which can also potentially affect the EBPR performance of the facility because of its dependence on the presence of adequate amounts of VFAs. Accordingly, the operating staff supplemented any deficiency by adding fermenter overflow to the influent during winter months.

The facility effluent TSS spiked several times during the study period because of process upsets, as shown in Figure 8.8. Twice, it was caused by the proliferation of microorganisms that caused poor

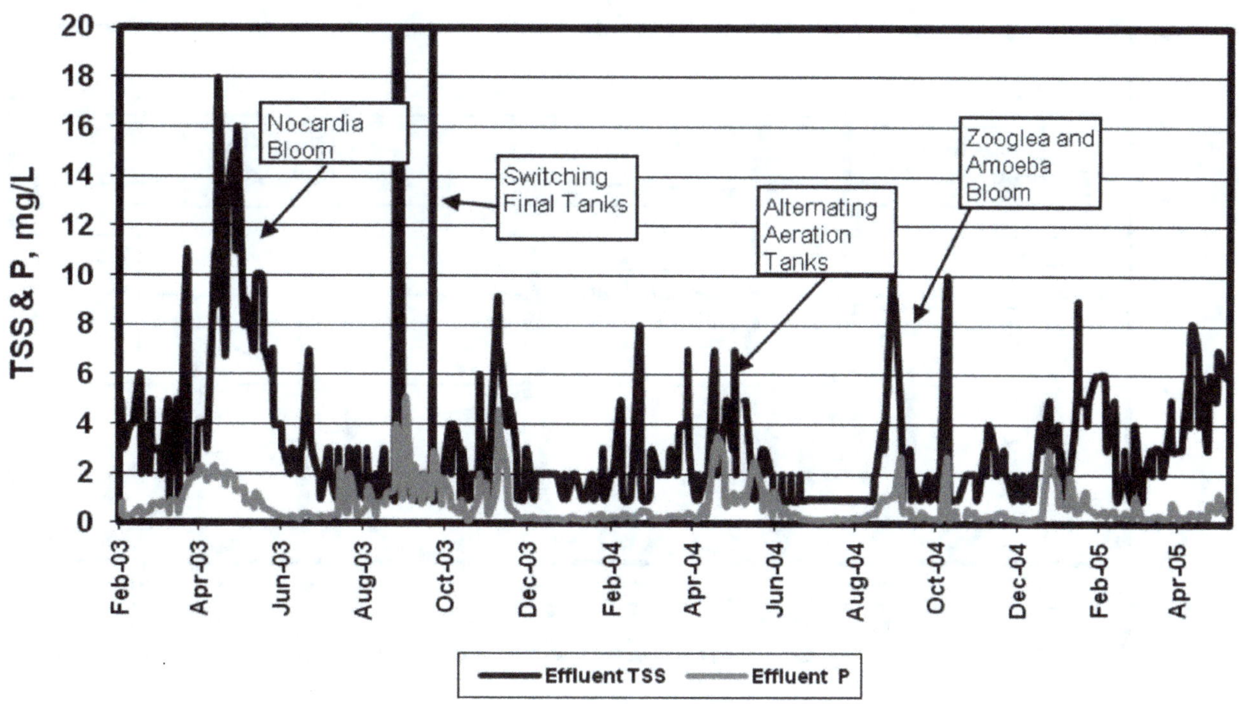


FIGURE 8.8 Effect of Effluent TSS on Effluent Total Phosphorus at Eagles Point Wastewater Treatment Plant

settling in secondary clarifiers as a result of difficulty of formation of biological flocculation. Two other times, the primary reason was operational changes. The effect of effluent TSS on the effluent total phosphorus concentration is clearly demonstrated in this figure.

The facility demonstrated superior EBPR performance during the plug flow mode of operation of the bioreactor compared to the step-feed mode. This is primarily attributable to the fact that during the step-feed mode, primary effluent containing rbCOD is introduced to the reactor through multiple inlet points along its length. This results in a significant part of VFAs being directed to different points along the biological reactor, with the front end, which needs VFAs most for EBPR, receiving only a small fraction. This increases the likelihood of VFA deficit for EBPR in the step-feed mode compared to the plug flow mode of operation. On the other hand, during plug flow operation, all VFAs enter the front end of the reactor with the primary effluent. Figure 8.9 demonstrates the poor performance of EBPR processes between August and December 2003 operating in step-feed mode. After this period, the reactor was changed to a plug flow mode of operation.

Under the circumstances of EBPR failure because of VFA deficiency, facility staff dosed rbCOD in the form of molasses, which supplemented VFAs by fermentation and restored EBPR. The periods of molasses dosing and the effect on effluent total phosphorus are clearly shown in Figure 8.9.

The case studies presented herein demonstrate that there is no single solution to resolve all the problems leading to EBPR failure. Indeed, each situation requires careful analysis on a case-by-case basis for effective troubleshooting.

## 14.0 OPERATIONS SUMMARY

There are several operational considerations for the EBPR process for removing phosphorus, including but not limited to those indicated in Table 8.9.

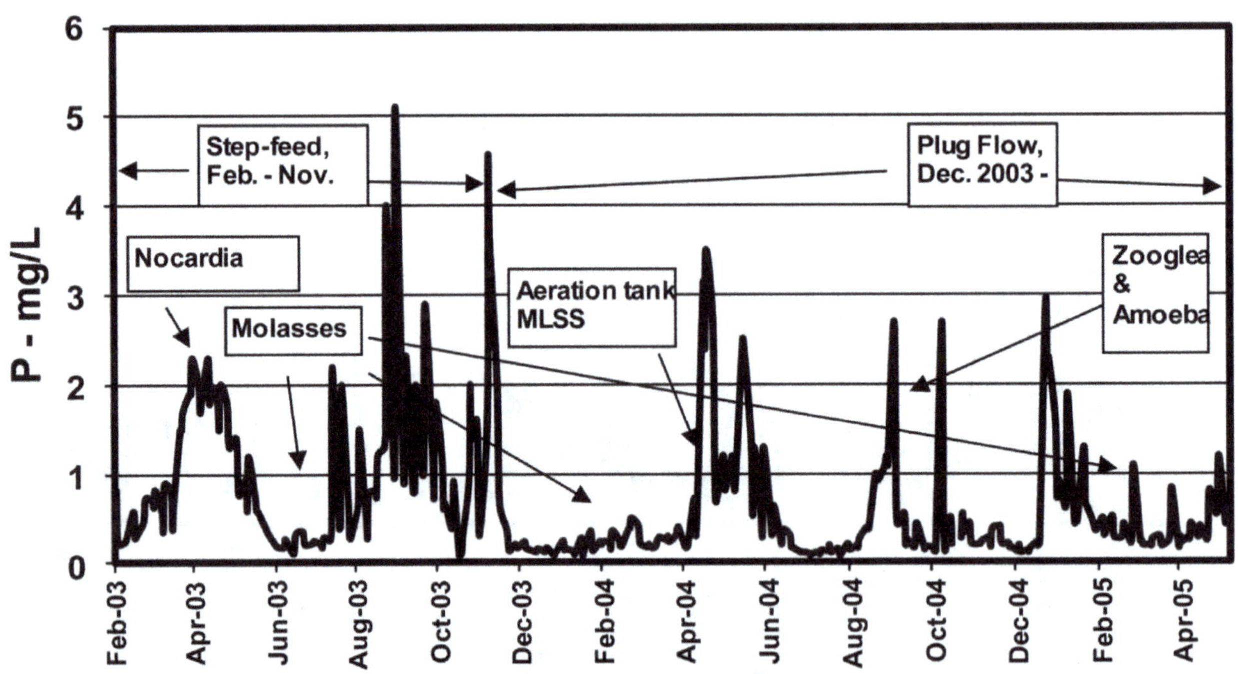


FIGURE 8.9 Effects of Step-Feed Versus Plug Flow Operation of the Bioreactor on Effluent Total Phosphorus at Eagles Point Wastewater Treatment Plant

TABLE 8.9 Typical Operational Considerations for EBPR Facilities

| Operations | Possible cause | Check, verify, or monitor | Possible solutions or steps |
|---|---|---|---|
| **Secondary influent (influent of EBPR) TP** | | | |
| High TP | If recycle stream TP load is higher than approximately 25% of influent TP loading, then EBPR process efficiency goes down.<br>High TP in recycle streams due to septic conditions from long detention time in pipe<br>High TP in the recycle line due to longer aeration time in aerobic digester causing cell lysis<br>High TP in the recycle line due to longer anaerobic time in the anaerobic digester causing cell lysis<br>Unaerated sludge storage causing secondary P release<br>Septic conditions in the final clarifier due to a deep sludge blanket can increase TP in the recycle line<br>Insufficient metal salt addition | Ortho-P, TP, in secondary influent<br>Check sludge blanket in final clarifier.<br>Check sludge storage condition.<br>Verify aeration time in aerobic digester.<br>Verify anaerobic time in anaerobic digester. | Decrease recycle flow.<br>Increase metal salt addition to enhance P removal.<br>Avoid high sludge blanket in final clarifier.<br>Provide aeration for sludge storage.<br>Optimize aeration time in aerobic digester.<br>Optimize anaerobic time in anaerobic digester. |
| Low TP | Excess metal salt addition for P removal | Ortho-P, TP in secondary influent and bioreactor MLSS | Reduce metal salt addition for P removal. |
| **Effluent** | | | |
| High TP | Intrusion of dissolved oxygen in anaerobic zones<br>Biological process upset<br>Insufficient metal salt addition | Profile of P removal (TP, Ortho-P)<br>BOD5 in influent<br>Soluble BOD in anaerobic zone<br>P removal chemical dosage<br>Check if the upstream process includes any mixing vortexes, cascade type aeration, or the type of pumps (air lift, screw, etc.) related to the EBPR process. | Adjust dissolved oxygen setpoint as necessary.<br>Increase supplemental carbon (VFA production).<br>Increase MLSS flowrate.<br>Avoid any turbulence, cascade, or the type of pumps that can introduce oxygen and entrain in anaerobic zone.<br>Avoid a high sludge blanket in secondary clarification (this avoids sludge settling problems due to nitrogen gas production via denitrification).<br>Avoid floating sludge in the sand filtration/ultrafiltration process. |
| Low TP | Secondary influent does not have sufficient TP to meet nutrient requirements for microorganism growth.<br>Biological process upset | Profile of P removal (TP, Ortho-P)<br>Measure the specific oxygen uptake rate, sludge volume index, and microbiological investigation on mixed liquor. | Decrease (or stop) metal salt addition to avoid overdosing and maintain phosphorus nutrient requirements for biomass growth. |

## 15.0 REFERENCES


Andreasen, K., Petersen, G., Thomsen, H., & Strube, R. (1997). Reduction of nutrient emission by sludge hydrolysis. *Water Science & Technology, 35*(10), 79-85.

Barnard, J. L. (1974). Cut P and N without chemicals. *Water, Waste and Wastewater Engineering, 11*, 33–36.

Barnard, J. L. (1984). Activated primary tanks for phosphate removal. *Water SA, 10*(3), 121–126.

Barnard, J., Houweling, D., Analla, H., & Stachen, M. (2010, October 2–6). Fermentation of mixed liquor for phosphorus removal. *Proceedings of the 83rd Annual Water Environment Federation Technical Exhibition and Conference* [CD-ROM], New Orleans, LA, United States. Water Environment Federation.

Basu, S., Wiesner, T., & Page, J. (2008, October 18–22). Failure and subsequent restoration of ebpr capability of a small community wastewater treatment plant. *Proceedings of the 81st Annual Water Environment Federation Technical Exhibition and Conference* [CD-ROM], Chicago, IL, United States. Water Environment Federation.

Chaparro, S. K., & Noguera, D. R. (2003). Controlling biosolids phosphorus content in enhanced biological phosphorus removal reactors. *Water Environment Research*, *73*(3), 254–262.

Clarke, K., & Neethling, J. B. (2009, October 10–14). *Operation of 5-stage Bardenpho process with clarifier/filters to achieve low effluent phosphorus.* 82nd Annual Water Environment Federation Technical Exhibition and Conference [Workshop W216], Orlando, FL, United States. Water Environment Federation.

Daigger, G. T., & Littleton, H. X. (2000). Charaterization of simultaneous nutrient removal in staged, closed-loop bioreactors. *Water Environment Research, 72*(3), 330–339.

Eikelboom, D. K. (2000). *Process control of activated sludge plants by microscopic investigation.* IWA Publishing.

Erdal, U. G., Erdal, Z. K., & Randall, C. W. (2002, September 28–October 2). The effect of temperature on EBPR system performance and bacteria community. *Proceedings of the 75th Annual Water Environment Federation Technical Exposition and Conference* [CD-ROM], Chicago, IL, United States. Water Environment Federation.

Grady, C. P. L., Daigger, G. T., Love, N. G., & Filipe, C. D. M. (2011). *Biological wastewater treatment* (3rd ed.). CRC Press.

Jenkins, D., Richard, M. G., & Daigger, G. T. (2003). *Manual on the causes and control of activated sludge bulking, foaming and other solids separation problems* (3rd ed.). IWA Publishing.

Jeyanayagam, S. S. (2007). So, you want to remove phosphorus? Part 1: Enhanced biological phosphorus removal. *Buckeye Bulletin, 4*(80), 22–29.

Lamb, J. (1994). *Wastewater treatment with enhanced biological phosphorus removal and related purification processes* (U.S. Patent US 5288405A). U.S. Patent and Trademark Office.

Lilley, I. D., Wentzel, M. C., Lowenthal, R. E., & Marais, G. v. R. (1990). *Acid fermentation of primary sludge at 20 °C* (Research Report W64). Department of Civil Engineering, University of Cape Town.

Lindeke, D., & Barnard, J. (2005, October 29–November 2). The role and production of VFA's in a highly flexible BNR plant. *Proceedings of the 78th Annual Water Environment Federation Technical Exhibition and Conference* [CD-ROM], Washington, DC, United States. Water Environment Federation.

Lu, T., Saikaly, P. E., & Oerther, D. B. (2007). Modeling competition of aerobic heterotrophs for complementary nutrient in a biofilm reactor: Effect of hydraulic retention time on coexistence. *Water Science & Technology, 55*(8–9), 227–235.

Mamais, D., & Jenkins, D. (1992). The effects of MCRT and temperature on enhanced biological phosphorus removal. *Water Science & Technology, 26*(5–6), 955.

McClintock, S., Randall, C. W., & Pattarkine, V. (1992). The effects of temperature and mean cell residence time on enhanced biological phosphorus removal. *Environmental Engineering, the Proceedings of the 1991 Specialty Conference on Environmental Engineering*, ASCE, 319–324.

Melcer, H., Robinson, W., Jue, P., Doty, J., Yanasak, J., & Land, G. (2009, June 28–July 11). The application of PAX compounds for *Microthrix parvicella* foam control in BNR systems. *Proceedings of the 2009 Water Environment Federation Specialty Conference—Nutrient Removal*, Washington, DC, United States. Water Environment Federation.

Metcalf & Eddy, Inc./AECOM. (2013). *Wastewater engineering—treatment and resource recovery.* McGraw Hill Education.

Narayanan, B., Johnson, B., Baur, R., & Mengelkoch, M. (2006, October 21–25). Uptake in biological phosphorus removal. *Proceedings of the 79th Annual Water Environment Federation Technical Exhibition and Conference* [CD-ROM], Dallas, TX, United States. Water Environment Federation.

Oldham, W. K., & Abraham, K. (1994, October 15–19). Overview of full-scale fermenter performance. *Proceedings of the 67th Annual Water Environment Federation Technical Exposition and Conference* [CD-ROM], Chicago, IL, United States. Water Environment Federation.

Oldham, W. K., & Stevens, G. M. (1984). Initial operating experience of a nutrient removal process (modified Bardenpho) at Kelowna, British Columbia. *Canadian Journal of Civil Engineering, 11*, 474–479.

Parker, D., Geary, S., Jones, G., McIntyre, L., Oppenheim, S., Pedregon, V., Pope, R., Richards, T., Voigt, C., Volpe, G., Willis, J., & Witzgall, R. (2003). Making classifying selectors work for foam elimination in the activated-sludge process. *Water Environment Research, 75*(1).

Rabinowitz, B., Koch, F. A., Vassos, T. D., & Oldham, W. K. (1987, October). A novel design of a primary sludge fermenter for use with the enhanced biological phosphorus removal process. *Proceedings of the IAWQ Specialized Conference on Biological Phosphate Removal from Wastewaters.* Rome, Italy. International Association on Water Quality.

Randall, C. W., Barnard, J. L., & Stensel, H. D. (Eds.). (1992). *Design and retrofit of wastewater treatment plants for biological nutrient removal* (Vol. 5). Water Quality Management Library, Technomic Publishing.

Randall, C. W., & Chapin, R. W. (1997). Acetic acid inhibition of biological phosphorus removal. *Water Environment Research, 60*(5), 955–960.

Skalsky, D., & Daigger, G. (1995). Wastewater solids fermentation for volatile acid production and enhanced biological phosphorus removal. *Water Environment Research, 67*, 230.

Stensel, D. (1991). Principles of biological phosphorus removal. In R. Sedlak (Ed.), *Phosphorus and nitrogen removal from municipal wastewater—principles and practices.* Lewis Publishers.

Vale, P., Barnard, J., Thomas, D., & Dold, P. (2008, October 18–22). RAS fermentation to enhance biological phosphorus removal. *Proceedings of the 81st Annual Water Environment Federation Technical Exhibition and Conference* [CD-ROM], Chicago, IL, United States. Water Environment Federation.

Vollertsen, J., Petersen, G., & Borregaard, V. R. (2006). Hydrolysis and fermentation of activated sludge to enhance biological phosphorus removal. *Water Science & Technology 53*(12), 55-64.

Water Environment Federation. (2010). *Nutrient removal* (WEF Manual of Practice No. 34).

Water Environment Federation. (2023). *Wastewater engineering: Design of water resource recovery facilities* (MOP 8 series).

Water Environment Research Foundation. (2011). *Fermenters for biological phosphorus removal carbon augmentation* (WERF Nutrient Challenge Project).

Water Research Foundation. (2019). *Optimization and design of a sidestream EBPR process as a sustainable approach for achieving stable and efficient phosphorus removal* (WRF Project No. U1R13/4869).

Wenner, J. (1994). *Activated sludge bulking and foaming control.* Technomic Publishing.

## 16.0 SUGGESTED READINGS

Reynolds, T. D. (1982). *Unit operations and processes in environmental engineering.* Brooks/Cole Engineering Division, Wardsworth.

Thomas, M., Wright, P., Blackall, L., Urbain, V., & Keller, J. (2003). Optimisation of Noosa BNR plant to improve performance and reduce operating costs. *Water Science & Technology, 47*(12), 141–148.

Watts, J. B., Parrott, I., Shaw, A., & Mason, L. K. (2000, October 14–18). On-line monitoring instrument for the measurement of nitrification and denitrification rates. *Proceedings of the 73rd Annual Water Environment Federation Technical Exposition and Conference* [CD-ROM], Anaheim, CA, United States. Water Environment Federation.

# Chapter 9

# Combined Nitrogen and Enhanced Biological Phosphorus Removal Processes

# 1.0 INTRODUCTION

Chapter 7 presented processes for nitrification and denitrification (NDN) to accomplish ammonia and nitrate removal. In Chapter 8, the mechanism of enhanced biological phosphorus removal (EBPR) was explained, and the possible interference of nitrate in the mechanism for EBPR was discussed. This chapter focuses on water resource recovery facilities (WRRFs) that combine both NDN and EBPR, and it builds on the content presented in Chapters 7 and 8. Whereas phosphorus removal in a combined nitrogen and phosphorus removal facility can be achieved either chemically and biologically, the biological alternative has a number of advantages, such as considerably lower operating costs, less sludge production, and little or no added chemicals in the sludge. However, enhanced biological phosphorus removal requires more skillful operation.

Total nitrogen removal in a combined nitrogen and phosphorus removal biological nutrient removal (BNR) facility is most commonly achieved in a two-zone system through nitrification under aerobic conditions and denitrification under anoxic conditions. Because of the interference of nitrate on the anaerobic conditions required for the EBPR process, three stages or zones (anaerobic, anoxic, and aerobic) are typically required to provide the different environmental conditions necessary for more reliable EBPR. Additionally, interferences between nitrogen and phosphorus removal processes can require the use of additional zones for more effective phosphorus removal performance (e.g., for the removal of nitrate in return activated sludge [RAS]) (Keller et al., 2001). Very stringent effluent nitrogen and phosphorus limits will require additional zones to reduce effluent total nitrogen and total phosphorus below 3.0 and 1.0 mg/L, respectively. The second anoxic zone (post-anoxic) in a five-stage Bardenpho process is an example of this.

In this chapter, combined nitrogen and phosphorus removal systems are discussed, along with operational conditions, process control strategies, and troubleshooting.

# 2.0 BASIC CONSIDERATIONS

## 2.1 Effluent Requirements

The effluent requirements to be met will dictate the level of complexity and cost required to meet them. Effluent requirements considered to be only moderately stringent are in the range of 10 to 15 mg/L for nitrogen (either as nitrate-nitrogen, total inorganic nitrogen, or total nitrogen) and 1.0 mg/L for total phosphorus. Effluent requirements considered to be highly stringent for combined nitrogen and phosphorus facilities range from 3 to 5 mg/L for nitrogen species and less than 0.1 to 0.2 mg/L for total phosphorus.

## 2.2 Economics

Facilities required to meet the fairly stringent combined nitrogen and phosphorus requirements merit a process design that includes sufficient process volumes to achieve the competing objectives of denitrification and EBPR, but the facilities also need to be flexible to adapt to changing seasonal conditions and loadings over time. This flexibility can include the use of aerated/unaerated swing zones and design of internal recycle pumping and blowers for a wide range of possible operating conditions. These design features typically add capital cost to a project and additional training for personnel.

Facilities needing to meet the most stringent limits typically also need to dose either supplemental carbon, metal salt coagulants, or both unless the influent contains significant fractions of readily degradable carbon. Dosing external chemicals adds significant operating costs not only for the chemical purchases but also for the increased sludge production and handling costs. Guidance for chemical phosphorus removal is provided in Section 5.4.

## 2.3 Operational Considerations

Operation of a combined biological nitrogen and phosphorus facility requires adept operations personnel who understand the differing mechanisms for nitrogen removal versus EBPR. Some key factors for optimizing combined NDN and EBPR processes are as follows:

- Complete nitrification should be achieved in aeration zones with as short a solids retention time (SRT) as possible. Longer SRTs will increase microbial decay, thereby increasing nitrogen and phosphorus release.
- Denitrification should be optimized either in single-stage anoxic zones or multiple-stage anoxic zones (pre-anoxic and post-anoxic) to achieve lower effluent nitrate levels returned in the RAS to the anaerobic zone. Nitrate is often a significant fraction of effluent total nitrogen. Dissolved oxygen must be excluded from anoxic zones. If dissolved oxygen is present, bacteria will preferentially use oxygen as the terminal electron acceptor, thereby reducing the efficiency of denitrification.
- Sufficient nitrate recycle should be provided. Because internal recycle flow is taken from the end of the aeration zone, it will also contain dissolved oxygen in addition to nitrate. Residual dissolved oxygen should be maintained at a lower level (such as around 0.3 to 0.5 mg/L) at the internal recycle flow withdrawal point compared to the rest of the aeration zones.
- For facilities using EBPR instead of chemical phosphorus removal, dissolved oxygen and nitrate return to anaerobic zones should be minimized to increase EBPR performance and resiliency. Otherwise, ordinary heterotrophic bacteria will outcompete polyphosphate-accumulating organisms (PAOs) for volatile fatty acids (VFAs) by using the available dissolved oxygen and nitrate.
- Secondary phosphorus release in an EBPR system should be prevented in anoxic zones by operating at an optimized SRT. Monitoring nitrate levels or oxidation-reduction potential (ORP) in anoxic zones can help operators to limit the occurrence of secondary phosphorus release.
- Highly effective suspended solids removal should be provided for particulate nitrogen and phosphorus removal when low levels of effluent nutrient levels are required.

Combining nitrogen removal and EBPR is particularly challenging due to the competition for influent substrate (referred to as "influent carbon" throughout this chapter). For this reason, optimizing aeration control to avoid excessive dissolved oxygen concentrations in undesirable locations is a critical operational consideration.

## 2.4 Sustainability

Denitrification and EBPR nearly always increase energy consumption at facilities because of the additional energy required for mixing and internal recycle pumping. Facilities requiring a high level of nitrogen and/or phosphorus removal also increase consumption of either supplemental carbon, metal salt coagulants, or both, as discussed further in this chapter. Chemical dosing also increases the quantity of sludge produced and handled. Optimization of mixing energy, internal recycle pumping, aeration energy, and chemical dosing is recommended not only to reduce operating expenses but also to increase the sustainability of operating a BNR facility. Refer to Chapter 11 for a detailed discussion of optimization of BNR systems.

## 2.5 Safety

One of the main operational safety concerns for combined nitrogen removal and phosphorus removal facilities is the use of methanol or other hazardous chemicals. More benign alternatives for supplemental carbon such as glycerin are becoming increasingly common because of the health and safety benefits compared to methanol. Metal salt coagulants also vary in their corrosivity and risk to staff health and safety.

# 3.0 COMMONLY USED PROCESS CONFIGURATIONS

This section is intended to present an overview of the most commonly used process configurations for combined nitrogen and phosphorus removal processes to help operations professionals understand why a configuration might be designed and constructed to meet the facility's effluent quality requirements. This section is not intended to cover all potential process configurations, and newer, emerging technologies are not discussed in detail.

## 3.1 Staged Zone Process Configurations

### *3.1.1 The Anaerobic-Anoxic-Oxic Process*

The three-stage anaerobic-anoxic-oxic (A2/O) process consists of a modified Ludzack–Ettinger (MLE) process for nitrogen removal, with an anaerobic zone in front for phosphorus removal (Figure 9.1). The process is called *A2/O* because it consists of anaerobic-anoxic-oxic (aerobic) stages in series. The RAS is recycled to the anaerobic zone; the anaerobic zone is followed by an anoxic zone for nitrogen removal through denitrification. Internal recycle is typically recycled from the aerobic zone to the anoxic zone at a rate of 100% to 400% of the tank influent flow rate; higher recycle rates can yield some further nitrate reductions but at an added operational cost. Nitrogen removal is limited by the internal recycle flow, and

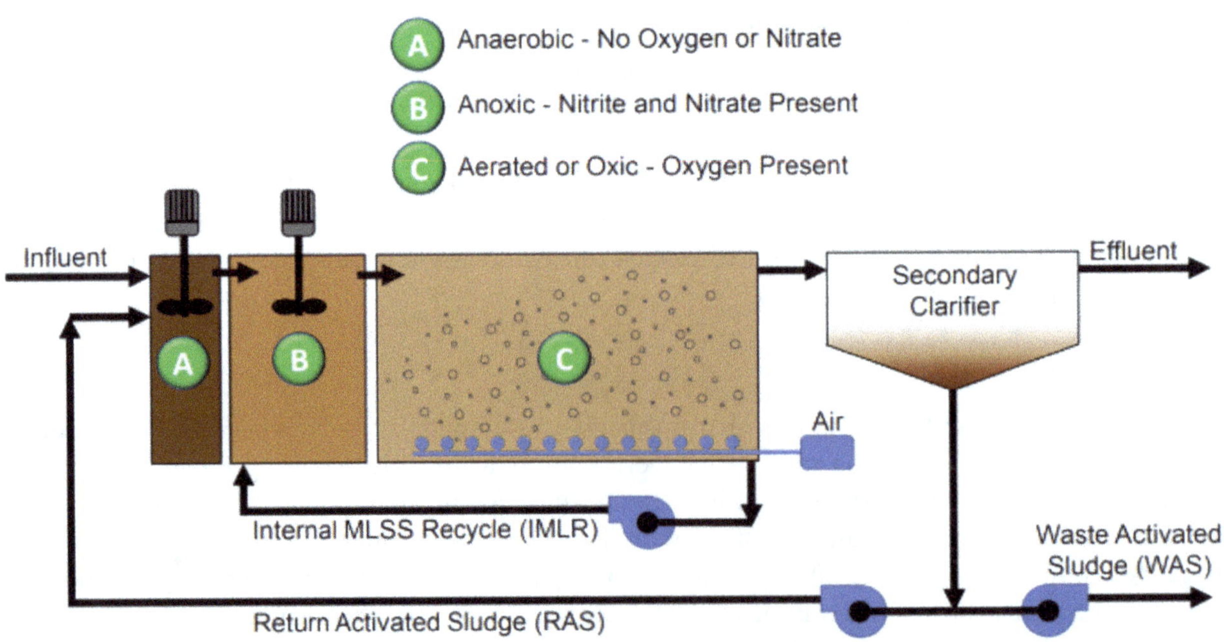


FIGURE 9.1 Three-Stage A2/O Process (Reprinted with permission by Indigo Water Group)

nitrate-nitrogen concentrations achievable may vary between 5 and 15 mg/L depending on the influent nitrogen concentration and the recycle rate. Because this level of nitrate may still be present in RAS, the phosphorus removal capability of A2/O is reduced unless there is a surplus of VFA that can denitrify the nitrate with a sufficient amount remaining to activate the phosphorus removal mechanism. This process has largely been replaced with other configurations that allow for denitrification of nitrate in RAS. This is a relatively simple process to operate. In addition, the process has a small footprint. Existing A2/O BNR facilities may be retrofitted with denitrifying tertiary filters to achieve very stringent limits.

### 3.1.2 Five-Stage Bardenpho

Barnard (1976) proposed a number of configurations for combined nitrogen and phosphorus removal. These configurations all consisted of adding an anaerobic zone ahead of either a high-rate non-nitrifying aeration zone (later called *anaerobic-oxic* [A/O]), an MLE process (later called *A2/O*), or a four-stage Bardenpho process, which was then referred to as the *modified* or *five-stage Bardenpho process*. The goal was to ensure that nitrate in RAS could be reduced to avoid consuming VFAs and therefore interfering with EBPR. The five-stage Bardenpho process is illustrated in Figure 9.2. All recycle rates are defined in terms of the average influent flow rate. The four-stage Bardenpho process described in Chapter 7 may reduce effluent nitrate to less than 2 mg/L when ratios of influent carbon to nitrogen are favorable and sufficient SRT is provided. When the ratio of influent carbon to nitrogen is lower, an external carbon source may be needed to reduce effluent nitrate to less than 2 mg/L. There will thus be little nitrate in RAS to interfere with the mechanism for EBPR.

In the five-stage Bardenpho process, influent wastewater combines with RAS from secondary clarifiers in an anaerobic zone. In this zone, phosphorus is released and VFAs are stored by PAOs in the absence of dissolved oxygen and nitrate. The flow then enters the first anoxic zone, where it combines with the internal recycle stream containing nitrate from nitrified mixed liquor from the end of the first aerobic zone. In this anoxic zone, nitrate is reduced to nitrogen gas while the bacteria are using readily biodegradable substrate. The flow then enters the first aerobic zone, where ammonia is oxidized to nitrate and the released orthophosphate and more orthophosphate present in mixed liquor is taken up by PAOs. Mixed liquor is typically recycled at a rate between 100% to 400% of influent flow from the end of the first aerobic zone to the head end of the first anoxic zone.

The remainder of the mixed liquor enters a second anoxic zone to reduce (or polish) the effluent nitrate concentration from the first aerobic zone. The second anoxic zone can be as large as the first anoxic zone, although less nitrate removal occurs because denitrification occurs via endogenous respiration and degradation of slowly biodegradable substrate. The second anoxic zone should be sized to optimize denitrification and prevent secondary phosphorus release, which can occur after all of the nitrate is denitrified. The mixed liquor then enters a reaeration zone, where nitrogen gas is stripped from the bulk liquid, and then passes to the secondary clarifiers. Phosphorus is removed from the process in

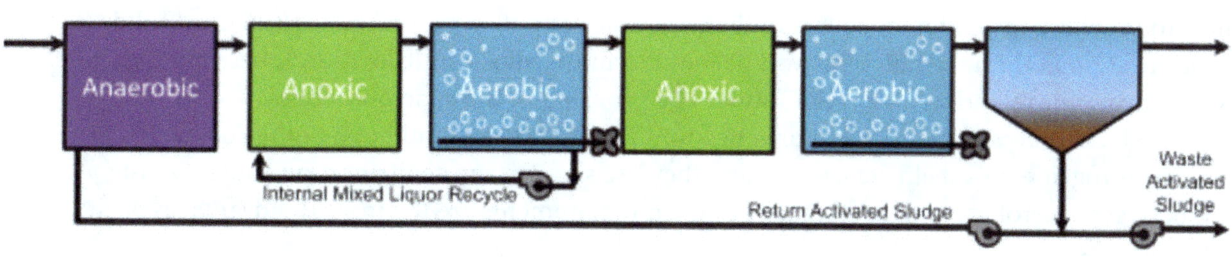


FIGURE 9.2 Modified Five-Stage Bardenpho Process (Reprinted with permission by Indigo Water Group [modified from U.S. EPA, 1993])

accordance with the theory described in Chapter 6, along with the sludge wasted from the secondary clarifier underflow as waste activated sludge (WAS). Even though it helps to take up phosphorus and to prevent the occurrence of anaerobic conditions and the associated release of phosphorus from sludge to the final effluent, the main purpose of the final aerobic zone is to help release nitrogen gas from the preceding denitrification process; hence, it is typically much smaller than the preceding anoxic and aerobic zones. Other purposes of the reaeration zone are to

- raise dissolved oxygen sufficiently in the mixed liquor to prevent secondary phosphorus release in the secondary clarifier as a result of low nitrate concentrations in mixed liquor entering the secondary clarifiers;
- provide nitrification of ammonia released in the second anoxic zone when underloaded with respect to nitrate, resulting in anaerobic conditions; and
- provide oxidation of excess methanol to prevent high biochemical oxygen demand (BOD) in the effluent.

The intent of the original five-stage Bardenpho process was to reduce nitrogen to low levels. At low levels of the effluent total nitrogen concentration, coupled with some unavoidable denitrification in the sludge blanket, there is no danger of recycling high levels of nitrate to the anaerobic zone through RAS. Today, five-stage Bardenpho processes are commonly used in facilities that require low total nitrogen levels coupled with low-effluent total phosphorus limits such as 1.0 mg/L.

Five-stage Bardenpho processes are highly efficient for nitrogen removal, but less efficient for phosphorus removal. Two stages of anoxic zones (i.e., pre-anoxic and post-anoxic) are the main reason for efficient nitrogen removal. On the other hand, the process requires a long SRT, which can diminish the efficiency of phosphorus removal. In some cases, particularly if there is insufficient carbon available, then it is not possible for a five-stage Bardenpho process to lower total nitrogen and total phosphorus to very stringent effluent limits (i.e., less than 3.0 and 1.0 mg/L, respectively). In such cases, metal salt addition for chemical phosphorus removal and external carbon source addition to the second anoxic zone can be implemented.

### 3.1.3 University of Cape Town, Modified University of Cape Town, and Virginia Initiative

The purpose of these configurations is to minimize nitrate input to the anaerobic zone, particularly when the effluent nitrate is high. The University of Cape Town (UCT) process consists of anaerobic, anoxic, and aerobic zones, with RAS directed to the anoxic zone for achieving denitrification (Ekama & Marais, 1984) (Figure 9.3). The influent wastewater flows directly to the anaerobic zone, providing a source of VFAs to PAOs. Nitrified mixed liquor from the aerobic zone is pumped to the anoxic zone at rates of 200% to 400% of influent flow to increase nitrogen removal through denitrification. Denitrified effluent from the end of the anoxic zone is recycled to the anaerobic zone at a rate of 200% of influent flow to provide a mixed liquor source to the anaerobic zone. The mixed liquor suspended solids (MLSS) concentration in the anaerobic zone will be lower than the anoxic and aeration zones, and therefore, a longer hydraulic retention time (HRT) should be provided in the anaerobic zones.

The UCT process can provide more effective EBPR than A2/O, although both processes have similar nitrogen removal efficiencies because they use single-stage denitrification zones. On the other hand, larger anaerobic zones and higher pumping requirements are the main disadvantages of the UCT process.

The Virginia Initiative Process is a high-rate version of the UCT process (Daigger et al., 1988). All zones are staged, consisting of at least two cells in series. The idea behind these processes is that, when

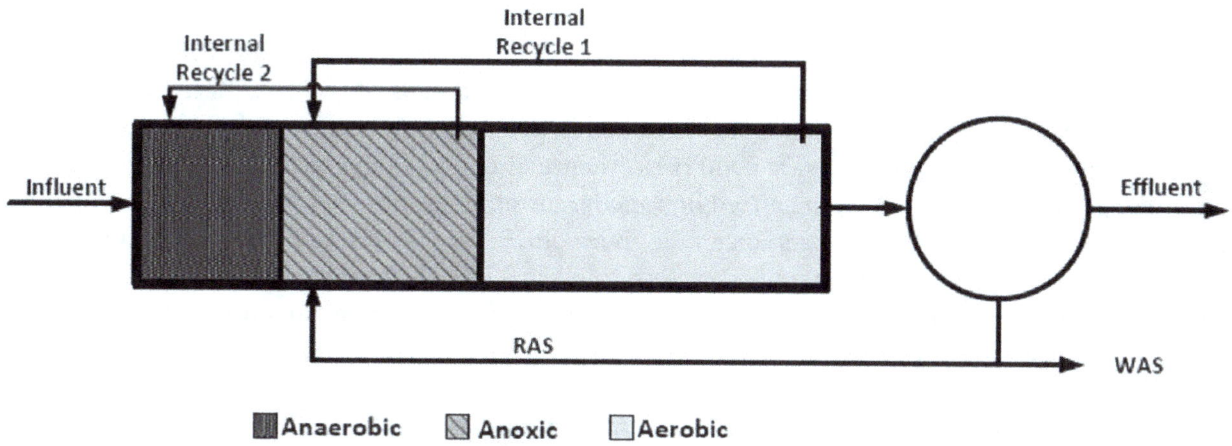


FIGURE 9.3 The UCT and Virginia Initiative Process Configurations

there is insufficient carbon in the influent for both nitrogen and phosphorus removal, preference could be given to phosphorus removal. For example, if the effluent and RAS nitrate is high, the nitrate in the RAS could be denitrified in the anoxic zone before it is recycled back to the anaerobic zone. If the anoxic zone cannot remove all the nitrate, the mixed liquor recycle (MLR) can be reduced to ensure that the nitrate at the end of the anoxic zone will be low, thus ensuring that nitrate will not be recycled to the anaerobic zone. This approach led to the development of the modified UCT (MUCT) process. The MLR flowrate from the aeration to the anoxic zone could be reduced to ensure that all nitrate is removed before passing the mixed liquor to the anaerobic zone. This may then result in higher effluent nitrate. If there is a need for further reduction in nitrate, a tertiary denitrifying system, such as denitrifying sand filters or a moving bed biofilm reactor, could be used with external carbon source addition.

In the MUCT process, the anoxic zone is divided into two cells (Figure 9.4). The first anoxic zone receives the effluent of the anaerobic zone and RAS. The denitrified mixed liquor is then recycled to the anaerobic zone. The first anoxic cell is, therefore, required to reduce only the nitrate in RAS, and can be better controlled. The second anoxic cell receives internal recycle from the aerobic zone, where bulk denitrification occurs. Facility influent mixed with denitrified anoxic recycle from the first anoxic cell is routed into the anaerobic zone, where multi-staged compartments are used to minimize the effects of nitrate recycle to the anaerobic zone. Flow from the anaerobic zone then enters the first anoxic cell

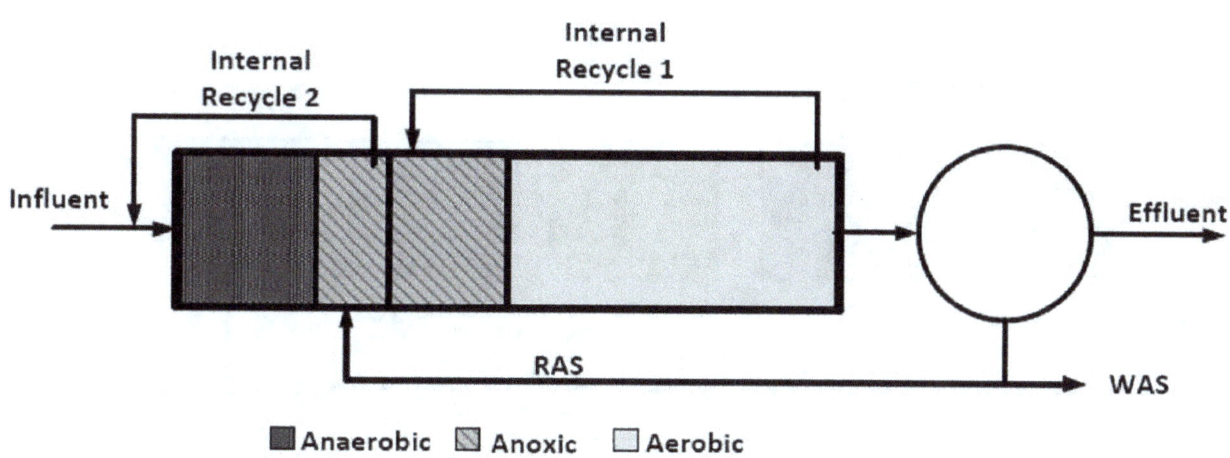


FIGURE 9.4 The Modified UCT Configuration

and then the second anoxic cell. From there, it flows to the aerobic zone, where nitrification occurs. The nitrified aerobic mixed liquor is recycled back to the second anoxic cell for denitrification. It is not vital that MLR to the second anoxic cell be accurately controlled because it will not affect the return of nitrate to the anaerobic zone. This configuration is highly effective, and therefore, a large number of MUCT facilities are in operation, giving good performance of removal of phosphorus to low levels. One disadvantage of the process is that the mixed liquor concentration in the anaerobic tank is lower than that of other processes, requiring larger anaerobic zone volume for the same anaerobic SRT. Another is the operational complexity of maintaining two sets of internal recycle pumping systems and the added complexity of either maintaining separate RAS sludges for each train or providing proper flow distribution of RAS in a one-sludge system.

### 3.1.4 Johannesburg, Modified Johannesburg, and Westbank

The City of Johannesburg, South Africa, solved their EBPR problems caused by excessive nitrate in RAS by providing an endogenous pre-denitrification zone, as illustrated in Figure 9.5 (Pitman, 1992). The MLSS concentration in the pre-anoxic zone (equal to RAS from the secondary clarifier underflow) is typically approximately twice that of the MLSS concentration in subsequent zones. Operating with RAS concentrations in the pre-anoxic zone results in endogenous denitrification taking place at a much higher rate, reducing the nitrate in the RAS before the RAS is introduced to the anaerobic zone. This is known as the *Johannesburg* (JHB) *process*.

Some facilities later modified the JHB process by adding a recycle stream from the end of the anaerobic zone to the preanoxic zone to use readily biodegradable compounds not taken up by PAOs for denitrification (the *modified JHB process*). This recycle stream is typically set at 10% of influent flow, resulting in an inexpensive way of providing more carbon for denitrification before the anaerobic zone. Because of the high concentration of solids in the pre-anoxic zone, the volume of this zone may be as little as 25% of the first anoxic zone of the JHB process. Figure 9.5 shows a flow schematic for the modified JHB process.

The Westbank configuration, which is named after the community in the city of West Kelowna in British Columbia where it was first used, is similar to the Johannesburg process in that it provides a RAS denitrification zone upstream of the mainstream process. The difference is that VFAs generated from the supernatant of a primary sludge fermenter are recycled directly to the anaerobic zone to supplement the carbon available in the reactor influent. Additionally, some Westbank facilities also provide the flexibility to step feed the reactor influent into different zones to manage carbon where it is needed the most to maximize combined nitrate and phosphorus removal.

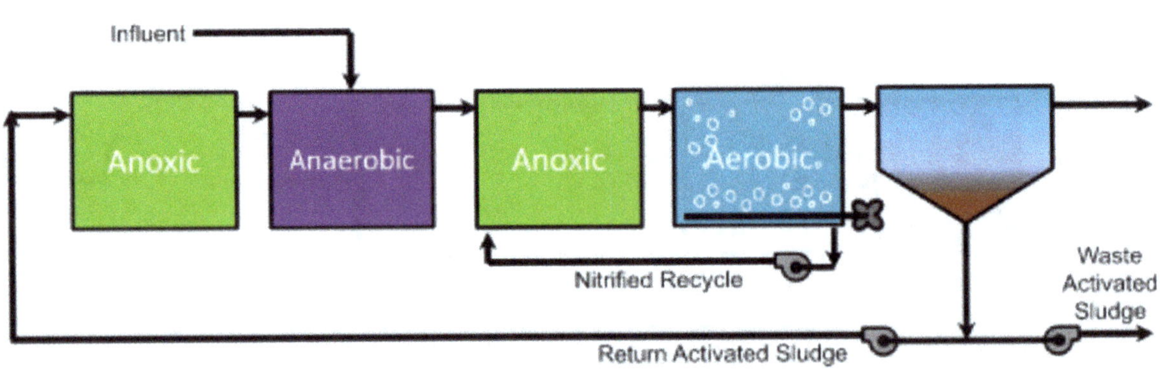


FIGURE 9.5 The JHB Process Configuration (Reprinted with permission by Indigo Water Group [modified from U.S. EPA, 1993])

## 3.2 Process Configurations Using Cyclic Aeration

*Cyclic aeration* refers to alternating cycles of "air on" and "air off" in the bioreactor. Nitrification and denitrification occur in the same tank. During extended air-off cycles, EBPR can occur in cyclic aeration systems when the nitrate is fully consumed and therefore the tank goes into anaerobic conditions. There are several types of processes currently available that use cyclic aeration. These include

- sequencing batch reactors (SBRs);
- phased isolation oxidation ditch; and
- sequential aeration.

*Sequencing batch reactors for phosphorus removal* consist of a large batch reactor with a reseeding anaerobic zone. A typical sequence for phosphorus removal incorporates a number of stages, namely the following:

- An anoxic idle period, during which nitrate in the sludge is reduced;
- Feed to an anaerobic zone, with recycle of sludge to this zone from the anoxic stage;
- Aeration of the main reactor while still feeding to the anaerobic zone;
- Settling; and
- Decanting from the main tank.

Figure 9.6 shows two basins of a multiple-SBR configuration. Each environmental condition occurs at a different time rather than in a different volume compared to conventional, continuous-flow, activated sludge processes. After decanting, the sludge is again kept under anoxic conditions to reduce nitrate. Mixed liquor is pumped from under the decanters to the anaerobic zone, where it is contacted with the feed that contains VFAs and phosphorus is released. The mixed liquor then overflows into the batch reactor during a mixing and no-aeration (anoxic) period, when denitrification takes place. This is followed by the aeration stage, during which nitrification and phosphorus uptake takes place.

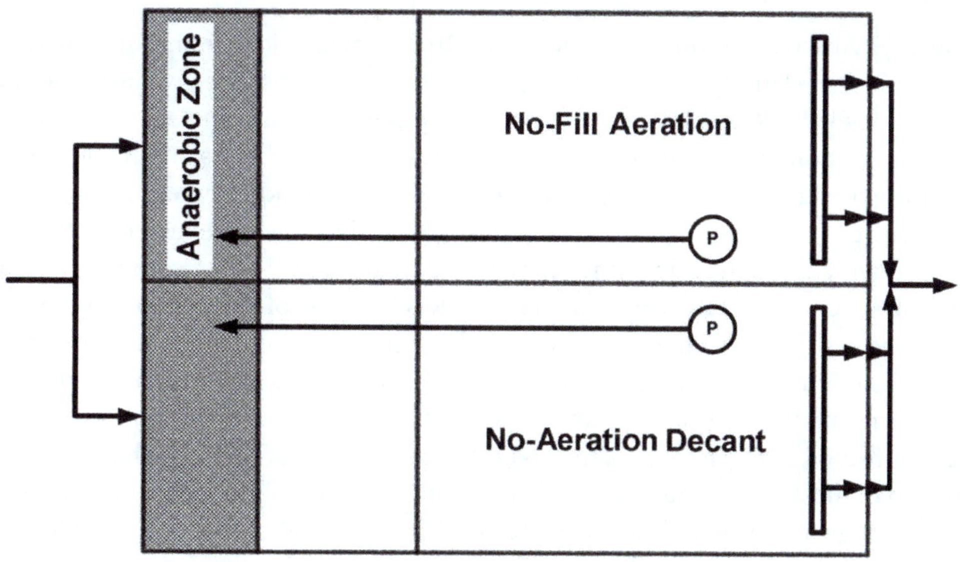


FIGURE 9.6 Biological Nutrient Removal SBR Configuration

Endogenous denitrification takes place during settling and decant. It is essential that the remainder of the nitrate be removed during the settling and decant phases to achieve better phosphate removal in the next cycle (Kazmi & Furumai, 2000).

A *phased isolation ditch system* is a continuous-flow activated sludge process in which the main treatment phases of the process are isolated into separate oxidation ditch trains. Process conditions within the oxidation ditches will alternate or phase between oxic, anoxic, and/or settling. Phased isolation ditches can be designed to operate with or without external clarifiers. The process can be designed to incorporate an anaerobic selector for EBPR. These processes include the proprietary double ditch (D-Ditch) and triple ditch (T-Ditch) processes, which do not include separate clarifiers, or the BioDenitro and BioDenipho processes, which do include separate clarifiers.

The *sequential aeration process* uses a single alternating aeration tank (first-stage aeration) that feeds two parallel alternating aeration tanks (second-stage aeration). The sequential process offered by Aeromod includes a patented in-line clarifier. The clarifier is submerged in the mixed liquor using RAS air lifts, allowing for continuous flow through the process. When EBPR is desired, an anaerobic zone is incorporated before the first-stage aeration tank. One difference between sequential aeration and other alternating aeration processes is that mixing is not provided during the air-off cycles, with no apparent negative effect on treatment.

Alternating aeration processes can have lower operating costs than continuously aerated processes because, when properly designed and operated, only enough air is applied to satisfy the organic and nitrogenous oxygen demands. Air-on and air-off periods will be unique to each facility and are determined by operational experience. A typical cycle is 4 hours, with 2 hours of aeration and 2 hours of no aeration. During periods of no aeration, residual dissolved oxygen will allow the process to continue to nitrify, particularly if fine-bubble aeration is used. Determining aerobic SRT in alternating aeration systems is, therefore, based on system response and the type of aeration diffusers used. Total system SRT tends to be longer in alternating aeration systems to make up for perceived inefficiencies associated with times when conditions are not fully aerobic or anoxic. As more knowledge is gained about simultaneous nitrification and denitrification (SND), particularly organisms capable of nitrification at low dissolved oxygen, it may be possible to reduce the total system SRT in alternating aeration systems.

## 3.3 Process Configurations With Oxidation-Ditch Systems

Oxidation-ditch systems are common, with more than 10 000 installations in the United States and abroad, because of their relatively simple and reliable operation and cost effectiveness (refer to *Wastewater Engineering: Design of Water Resource Recovery Facilities* [Water Environment Federation {WEF}, 2023]). Oxidation-ditch operation is based on the use of mechanical surface aerators, mechanical aerators, brush aerators, and slow-speed mixers to provide flow of mixed liquor around channels in a racetrack configuration. An alternative is to use diffused aeration with slow-speed submersible mixers, which provide more flexibility in controlling aeration and mixing separately. Oxidation ditches are commonly designed as extended aeration systems with HRTs of 24 hours and SRTs in excess of 15 days; however, there is a tendency to reduce the design SRT of these processes to make them more competitive with other processes.

Oxidation ditches that use point-source aerators, such as surface aerators and brush aerators, push the mixed liquor around the racetrack channel, passing it through aerated and unaerated zones, resulting in various degrees of nitrogen removal by SND. A higher degree of SND is achieved in oxidation ditches compared to most continuously aerated conventional activated sludge processes because oxidation ditches create high-to-low dissolved oxygen zones. Theoretically, all that is needed for EBPR is to add an anaerobic zone in front and perhaps a reaeration zone after the ditch channel, as was suggested by Barnard (1976).

For phosphorus removal, an anaerobic zone is added in front of the anoxic zones to result in a three-stage A2/O process, with the oxidation ditch as the aeration zone. Enhanced biological phosphorus removal to reduce effluent phosphorus well below 1 mg/L also requires that the SRT be kept as short as possible. At longer SRTs (in excess of 15 days), there is more endogenous breakdown of sludge; therefore, there is less sludge production and the phosphorus content of the sludge must be higher for the same degree of removal. In addition, some secondary release of phosphorus by PAOs may occur when all the carbon has been removed and the nitrate-nitrogen is less than 2 mg/L.

## 3.4 Process Configurations With Membrane Bioreactors

Biological nutrient removal configurations for membrane bioreactor (MBR) systems are generally the same as those for conventional systems. The one significant difference is related to the RAS rate. When designed for typical MLSS values of 8000 mg/L in the bioreactor, a RAS rate of 400% of the influent flow is required to keep the solids concentration below 10 000 mg/L in the membrane tank. Because this is a typical mixed liquor recirculation rate for a BNR configuration for total nitrogen removal, the RAS stream of the MBR can also provide the necessary mixed liquor recirculation for nitrogen removal in a BNR process.

The RAS of an MBR can also contain high concentrations of dissolved oxygen (typically, 2 to 6 mg/L); as such, this dissolved oxygen will reduce the soluble BOD available for the BNR processes. For this reason, a common practice is to return the RAS stream of an MBR to the aeration tank and to use separate MLR pumps for return of the mixed liquor to the anaerobic and/or anoxic zones, similar to the internal recycles provided in the MUCT, modified JHB, and other BNR configurations described above that use conventional clarifiers. Recycling RAS directly to the aeration zone also lowers the aeration requirement in the bioreactor by using dissolved oxygen in RAS for this purpose. What is lost with this approach is the more concentrated solids in the membrane tank and RAS, resulting in a higher solids concentration in the BNR zones and increasing nutrient removal efficiency.

As an alternative design, the RAS stream can be returned to a deoxygenation tank. Effluent with low dissolved oxygen from the deoxygenation tank enters the anaerobic tank, which also receives the raw influent wastewater. This alternative design allows higher concentrations of mixed liquor in BNR basins with lower pumping requirements to be maintained. A variation to achieve EBPR is incorporation of the anaerobic zone downstream of the anoxic zone, allowing for a single RAS recycle. While the availability of VFAs in the raw wastewater is lost in this configuration, the high MLSS concentration in MBRs enhances the production of VFAs in the anaerobic zone via fermentation. Figure 9.7 shows the simplest configuration of a BNR MBR that is typically appropriate for meeting limits of 8 mg/L of total nitrogen and 1.0 mg/L of total phosphorus (assuming no recycle of nitrogen and phosphorus from sidestreams). Figure 9.7 also shows the more complex BNR MBR configuration capable of achieving slightly lower total phosphorus limits, although achieving equivalent total nitrogen to the simpler configuration requires high recirculation rates.

As a result of high dissolved oxygen in the RAS stream, the MBR has gained a reputation for inefficient EBPR, and as a result, chemical phosphorus removal is more common in MBR facilities. However, the high RAS solids concentration can also enhance EBPR, and as manufacturers work to continually lower dissolved oxygen in the membrane tank, EBPR will become just as appropriate in MBRs as it is in conventional facilities. Even at higher RAS dissolved oxygen, EBPR can be effective, depending on the ratio of influent chemical oxygen demand (COD) to total phosphorus, emphasizing the need for accurate influent characterization and a calibrated process model to properly assess the appropriate method of phosphorus removal in MBRs.

Chemical phosphorus removal can also be effective in MBRs. The total retention of solids in the bioreactor typically results in lower dosages of the coagulant chemical required per gram of phosphorus

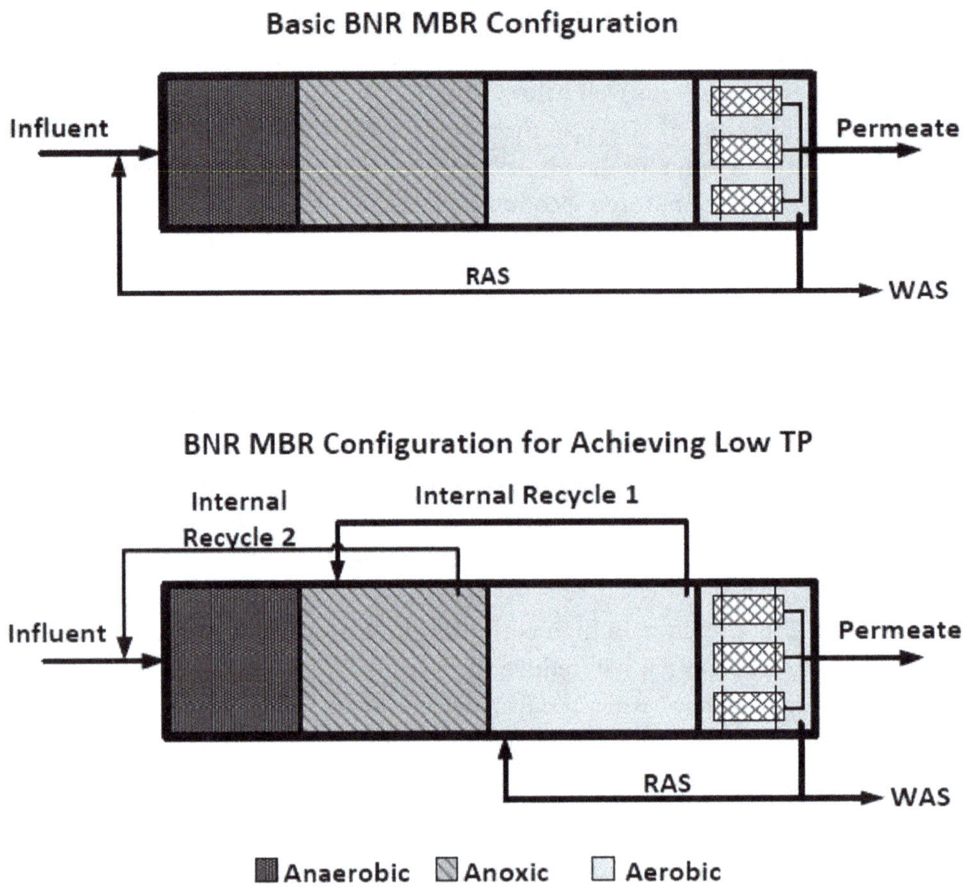


FIGURE 9.7 Biological Nutrient Removal MBR Process Configurations

removed than conventional facilities. Both aluminum and iron coagulants are used. Although iron can result in staining of the membranes, performance is not affected. Ultraviolet disinfection is frequently used following MBRs. It is important to remember that the membranes will pass soluble compounds, and therefore, excessive iron doses should be avoided to prevent interference with the UV system. Membrane bioreactors are also used as a pretreatment step ahead of reverse osmosis membranes; reverse osmosis can be affected by relatively low concentrations of aluminum. It is also important to avoid excessive chemical dosages in MBRs that are achieving some degree of EBPR because the excessive coagulant combined with the high RAS rate of an MBR will result in suppression of PAO growth.

Because MBR effluent is nearly free of solids, very low concentrations of total phosphorus can be achieved using MBRs (i.e., even lower than conventional activated sludge treatment in combination with conventional filters). Because of the high degree of filtration, MBRs can achieve effluent total nitrogen concentrations equivalent to those achievable with denitrification filters without the use of external carbon sources. Second-stage anoxic zones following the aerated portion of the bioreactor are required to achieve very low-effluent total nitrogen concentrations.

## 3.5 Emerging Process Enhancements and Configurations

Although the process configurations in this section are not in widespread use, they are rapidly seeing increased adoption and therefore are presented briefly. Refer to *Wastewater Engineering: Design of Water Resource Recovery Facilities* (WEF, 2023) for more information on these emerging technologies.

### 3.5.1 Simultaneous Nitrification Denitrification and Emerging Nitrate Removal Technologies

Chapter 7 presented the fundamentals of SND. As operating aerobic zones at the low dissolved oxygen concentrations necessary to promote SND is gaining acceptance, some facilities with staged processes with anoxic zones and internal recycle pumps have been experimenting with turning off the recycle pumps and relying only on SND to reduce nitrate levels. Turning off the internal recycle pumps avoids recycle of dissolved oxygen to the pre-anoxic zone and effectively turns an A2/O process into an A/O process with a larger anaerobic zone. Design criteria for the proper SRT and aerobic volume necessary are still being researched and evaluated, but with further development this type of A/O staged process with aerobic SND zones may become more than just an operational experiment and become more commonly employed by design engineers for new facilities.

In addition to simply operating staged continuous flow processes at low dissolved oxygen, there are emerging technologies available that can provide enhanced nitrate removal and/or SND using biofilm processes, thus helping to enhance EBPR performance. One example includes membrane aerated biofilm reactors (MABRs), which can be retrofitted into existing anoxic zones to promote SND on the biofilm that grows on the surface of a submerged membrane that is supplied with air from the membrane interior. Another emerging example is the use of floating media at the final aerobic stage of an existing A2/O, Bardenpho, or similar staged configuration to promote nitrogen removal via annamox bacteria using a partial denitrification-annamox (PDNA) process.

### 3.5.2 Aerobic Granular Sludge and Densified Activated Sludge

Aerobic granular sludge (AGS) is a proprietary wastewater treatment technology that uses a sequencing batch reactor with an anaerobic feeding strategy at the bottom of the tank to promote the formation of dense granular sludge particles, which are defined as particles exceeding 200 μm in diameter and whose sludge settleability after 5 minutes is comparable to the settleability at 30 minutes. The large size of the granules allows for the development of separate aerobic, anoxic, and anaerobic micro-environments inside the floc even though the particles are suspended in a single aerated tank; these micro-environments are illustrated in Figure 9.8. The stratified biomass can achieve nitrification,

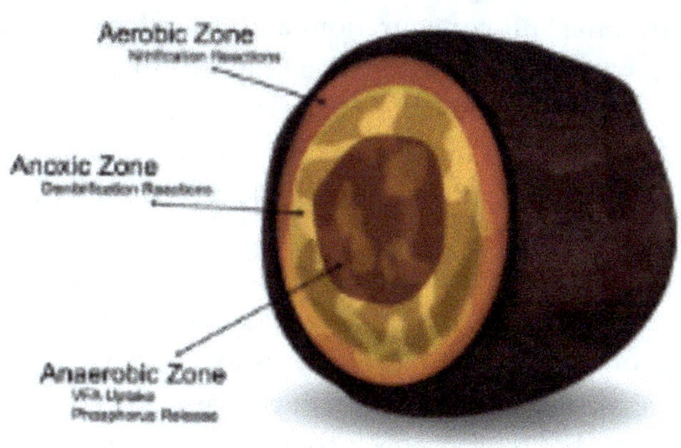


FIGURE 9.8 Illustration of a Typical AGS Granule Showing Stratified Micro-Environments for Simultaneous Nitrogen and Phosphorus Removal
From "A Review of the State of Development of Aerobic Granular Sludge Technology Over the Last 20 Years: Full-Scale Applications and Resource Recovery" by R. Hamza, A. Rabii, F. Ezzahraoui, G. Morgan, & O. Terna Iorhemen, 2022, *Studies in Chemical and Environmental Engineering*, 5, 100173 (https://doi.org/10.1016/j.cscee.2021.100173). 

denitrification and SND, as well as EBPR. Aerobic granular sludge is typically can accomplish effluent total phosphorus below 1 mg/L. The leading AGS technology provider is AquaNereda®.

External physical selectors such as hydrocyclones are an alternative to AGS that can be incorporated into plug flow reactors to promote a densified activated sludge (DAS). Several studies (Avila et al., 2021; Sturm, 2020) have demonstrated that a noteworthy fraction of the sludge can become granular at plug flow facilities. A common thread observed at these plug flow facilities is their use of both high food-to-microorganism (F/M) anaerobic selector zones and an external selector (such as hydrocyclones), which help retain the granules while selectively wasting the lightest portion of the flocculent sludge. This bioreactor design can produce a "densified sludge" that, when incorporated into processes designed for BNR, can retain the microorganisms most important to BNR, thus stabilizing nitrogen and phosphorus removal performance and resiliency.

## 3.6 Advantages and Limitations of Process Configurations

The various aforementioned conventional process configurations can be categorized as staged processes, cyclic aeration processes, oxidation ditches, and MBRs. Each process is capable of achieving low-effluent nitrogen and phosphorus if designed and operated correctly.

### 3.6.1 *Staged Processes*

With the exception of the five-stage Bardenpho configuration, which is specifically designed for lower total nitrogen effluent limits, variations in the other staged processes (i.e., A2/O, JHB, MUCT, Westbank, etc.) have been found to have only minor differences in process performance. This is because the beneficial effect of higher solids concentration in RAS tends to negate the detrimental effects of RAS nitrate on the anaerobic zone. The greatest advantage of using staged processes is that the anaerobic, anoxic, and oxic volumes and recycles can be well defined and independently manipulated, allowing a tailoring of the process to fit the specific influent characteristics and effluent goals. Disadvantages of staged processes are complexity and pumping energies required for internal recycles.

### 3.6.2 *Cyclic Aeration Processes*

Cyclic aeration processes offer the advantage of being able to perform all nutrient removal processes in a simpler tank configuration. In some SBR configurations and phased isolation ditches without clarifiers, no recycles are used and the same biomass is sequentially exposed to anoxic, anaerobic, and oxic conditions. Because the anaerobic cycle must follow the anoxic cycle, EBPR is not as effective in true cyclic aeration processes as in staged processes. This is the reason for the separate-stage anaerobic zone paired with the SBRs shown in Figure 9.6, where nitrification in the settled sludge is used to create a first-stage anaerobic zone.

For phased isolation ditches with clarifiers and sequential aeration, RAS is used to create an anaerobic zone in the same way as staged processes; it is only the denitrification process that differs. The biggest operational advantage of cyclic aeration processes is that, when operated correctly, only enough aeration is provided to achieve nitrification, thereby reducing aeration costs. The denitrification efficiency of cyclic aeration processes has typically been considered inferior to staged processes because of a less concentrated food source for the denitrifiers; however, with increasing evidence of efficiencies associated with SND, cyclic aeration processes may eventually overcome this perceived disadvantage.

### 3.6.3 *Oxidation Ditches*

Ditches are one of the most commonly used processes for BNR for small- to medium-sized facilities, primarily because of operational simplicity and resistance to upsets caused by the long SRT typically used. However, the long SRTs increase the space requirements compared to staged BNR configurations and cyclic aeration facilities. The oxygen-transfer efficiency of some surface aeration equipment is also lower than the diffused aeration equipment typically used in staged and cyclic aeration processes. When phosphorus removal is required, the design SRT for ditches should be decreased to enhance EBPR, making it necessary to supplement surface aerators with diffused aeration or to switch to diffused aeration entirely. The high recirculation rate of ditches can make it difficult to establish distinct anoxic zones, and for this reason, there is a tendency for SND to occur in ditch processes. The exception to this is the concentric-ring ditch design that is designed for isolated anaerobic, anoxic, and oxic zones. Because of the multiple zones and high recirculation rates, properly operated ditches are capable of achieving lower nitrogen and phosphorus concentrations than staged processes.

#### *3.6.4 Membrane Bioreactors*

The basic definition of an MBR is the substitution of membranes for final clarifiers. Membrane bioreactors can be configured with any of the aforementioned BNR configurations, even oxidation ditches, and have two key operational advantages: (1) the ability to use the membrane tank as a mixed liquor concentrator and (2) a high degree of filtration provided by the membranes. The main bioreactors of MBRs have traditionally been designed to operate at a high MLSS concentration of 8000 mg/L, making it necessary to use a RAS rate of 4 times the influent flow. This RAS rate can be detrimental to the anaerobic zone, reducing EBPR efficiency in MBR facilities. If the MBR bioreactor is designed and/or operated at lower MLSS, this allows the RAS rate to be reduced, increasing the efficiency of EBPR. Separate anoxic recycle may be required in this instance to meet effluent total nitrogen limits; however, the need for separate recycles must be weighed against the advantage of the high mixed liquor concentration occurring in the membrane tank. When MBRs are operated at high MLSS, the efficiency of second-stage anoxic zones is increased because of the higher concentrations of endogenous carbon. Combining this with the high degree of filtration provided by the membranes, it is possible for an MBR in a Bardenpho configuration to be capable of achieving the lowest nitrogen and phosphorus concentrations of all the BNR processes. The biggest disadvantage of MBRs is their high operational costs associated with aeration of the membrane tank, although recent advances in the technology are reducing operational costs.

## 4.0 PROCESS VARIABLES

### 4.1 Influent Characteristics

Combined nitrogen and phosphorus removal processes that mainly rely on biochemical reactions (excluding chemical phosphorus removal) are sensitive to certain influent wastewater characteristics because denitrification will compete with phosphorus removal processes for the forms of carbon needed for EBPR.

#### *4.1.1 Influent Flow Variability*

A key aspect of most BNR processes is adequate control of SRT. To a certain extent, this is affected by HRT. When the influent flow variability to these processes is large, then the HRT of each process could be significantly affected to the point that either the influent flow does not have enough contact time with active biomass (which, for example, can cause ammonia breakthrough) or it has too much contact time, which could lead to nitrate depletion and/or secondary phosphorus release. Operationally, the best strategy to manage flow variability is to increase or decrease recycle rates to maintain proper HRTs or

SRTs. Influent flow equalization and/or additional chemical feed systems (carbon or metal salts) can also be used to achieve consistent nitrogen and phosphorus removal when large diurnal influent flow and load variations are experienced.

### 4.1.2 Influent Carbon Requirements

Nitrogen and phosphorus removal processes each have specific substrate (herein referred to the more colloquial term as "influent carbon") requirements to ensure sufficient levels of nutrient removal. Managing influent carbon is the most critical success factor and the most challenging balancing act for operating a combined nitrogen and EBPR facility. There are several carbon-to-nutrient ratios commonly used in the industry for measuring the adequacy of combined nutrient removal processes. Higher influent carbon-to-nitrogen and carbon-to-phosphorus ratios will make the wastewater more amenable to BNR (Grady et al., 2011).

It is important to consider not only the quantity of carbon available but also the type of carbon available, which matters greatly depending on the process objectives and the process configuration. The following sections describe the differences in the types and quantities of carbon needed for biological nitrogen removal versus biological phosphorus removal, and thus how these differences can influence operational choices for a combined nitrogen and phosphorus removal facility.

#### 4.1.2.1 Biological Nitrogen Removal Carbon Requirements

As noted previously, providing adequate carbon for nitrate removal is an important factor for successful EBPR because excess nitrate in an anaerobic zone will limit EBPR performance. However, denitrifying microorganisms can use a more diverse range of carbon sources and types compared to PAO microorganisms.

For facilities employing conventional NDN, it is typically recommended to maintain a minimum ratio of COD to total Kjeldahl nitrogen (TKN) on the order of 8 : 1 to 10 : 1 in the anoxic zone influent. This ratio needed for effective nitrogen removal may also be influenced by pretreatment processes or by instances in which the groundwater contains nitrate that infiltrates the sewers or nitrate is added for odor control or from an industrial source. At lower ratios of COD to TKN, the following recommendations are provided:

- All aeration of the influent should be avoided so that the carbon is not oxidized by influent oxygen.
- Greater reliance on endogenous respiration is needed to optimize use of the available carbon. This could take the form of a second anoxic zone or a preanoxic zone to reduce nitrate through endogenous respiration before entering the anaerobic zone.
- Operating the process at low dissolved oxygen to promote SND has been shown to reduce the COD-to-TKN requirements needed to achieve moderate amounts of nitrate removal (Regmi et al., 2022).

Where the organic carbon content of the influent is not sufficient and good nitrogen and phosphorus removal is required, primary sludge or RAS may be fermented and the supernatant directed to the anaerobic zone. Primary sludge could also be stored in the primary tanks for fermentation. Some VFAs will diffuse into the liquid phase, but the sludge could also be dewatered and the filtrate and centrate returned to the anaerobic zone.

#### *4.1.2.2 Biological Phosphorus Removal Carbon Requirements*

Once adequate nitrate removal upstream of the anaerobic zone is accomplished, operators can focus on providing adequate quantities and types of carbon to the anaerobic zone for EBPR, where VFA are the most readily degradable carbon source for EBPR microorganisms. Other carbon sources need to be fermented before they are suitable for use by PAOs. The ratio of readily biodegradable COD (rbCOD) to phosphorus into biological reactors controls the reliability of phosphorus removal because a significant portion of rbCOD will be converted to VFAs in the anaerobic zone.

Table 8.2 showed the minimum recommended substrate requirements for EBPR. Typically, design engineers use the ratio of VFA to phosphorus as the most commonly used parameter for verifying EBPR efficacy and for sizing the anaerobic zone. When there is a sufficient supply of VFAs in influent flow (or from a supplemental VFA process such as a fermenter), the biology for phosphorus removal is forgiving. Figure 9.9 illustrates the performance of several side-by-side treatment trains at the Douglas L. Smith Middle Basin Wastewater Treatment Facility (WWTF) in Johnson County, Kansas. The split of VFA supplement from the supernatant of a primary sludge fermenter is known to provide less flow to Train 4 relative to the other trains. Figure 9.9 demonstrates the negative effect on soluble phosphate uptake in Train 4, which is attributable to the lower carbon availability from VFA supplement. This operational experience validates the correlation between available VFAs and total phosphorus removal in EBPR systems.

Insufficient levels of VFAs present in the influent of the facility should be evaluated. For example, chemical addition to the collection system for corrosion and odor control can inhibit and hinder the formation of VFAs (Kobylinski et al., 2008). In addition, smaller collection systems in cold climates may also generate lower VFAs.

A recent Water Research Foundation study of EBPR processes that ferment mixed liquor or RAS has also established additional guidelines for operators to use to monitor the carbon availability for EBPR using a carbon mass balance approach to help with troubleshooting EBPR performance issues

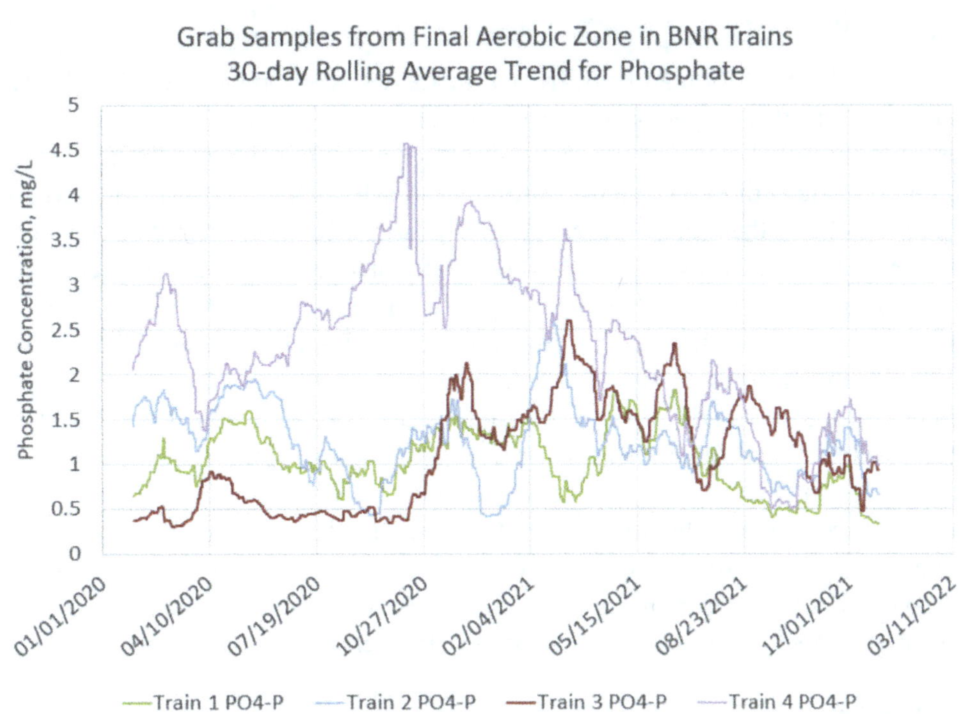


**FIGURE 9.9** Influence of Lower VFA Supplement to Train 4 on Effluent Total Phosphorus in a Modified Johannesburg Process at the Middle Basin WWTF in Overland Park, Kansas

(Downing, 2023). This study also included operational data demonstrating the performance benefits achievable from operating anaerobic zone mixers in an intermittent fashion, thus allowing the sludge to settle and ferment VFAs. This operating mode is referred to as an unmixed inline fermenter and is one of the simplest tools operators can use to generate more VFAs for EBPR at an existing facility, but the operator should proceed with caution to optimize the settling time versus mixing time to avoid excessively anaerobic conditions.

## 4.2 Effluent Characteristics

Effluent quality from combined nutrient removal processes depends on the degree of nutrient treatment designed for the combined system. Combined systems that rely primarily on EBPR may have effluent with average total phosphorus concentrations around 1.0 mg/L and total nitrogen concentrations as low as 3.0 mg N/L when properly designed and operated. Achieving these effluent concentrations can be challenging, particularly in colder climates. Total suspended solids concentration in the effluent is also linked to the total phosphorus concentration in the effluent because roughly 3% to 6% of TSS is typically associated with phosphorus. The effluent total nitrogen concentration is highly dependent on the recalcitrant dissolved organic nitrogen portion that remains in the effluent, which is not removed by commonly used conventional BNR processes discussed in this chapter. Recalcitrant dissolved organic nitrogen is discussed in detail in Chapter 11. Effluent that undergoes chemical phosphorus removal (as standalone treatment for phosphorus or as a polishing step) can achieve total phosphorus concentrations below 0.1 mg/L with a tertiary, highly efficient solids separation step. However, regardless of the amount of chemical dosed, there will be a recalcitrant phosphorus fraction that will not react with metal salts and may constitute a significant portion of the effluent total phosphorus. Recalcitrant dissolved organic nitrogen and recalcitrant total phosphorus vary within each sewershed and should be carefully determined if low nutrient levels are required in the effluent.

## 4.3 Temperature

Temperature effects on nitrification and denitrification are described in Chapters 3 and 5. Conventional wisdom is that EBPR can operate successfully in a temperature range from 5 to 20 °C, with some evidence that glycogen-accumulating organisms (GAOs, which take up VFA but not phosphorus) may successfully compete with PAOs at temperatures approaching 30 °C (WEF, 2010), whereas other researchers have found that these processes are relatively insensitive to temperature changes compared to other biological processes. For example, a facility in Grimstad, Norway, has been reported to remove phosphorus consistently, even though the mixed liquor temperature is approximately 5 °C for approximately 3 months of the year (Ydstebø, 2000). In addition, the Changi WRRF in Singapore has operated an EBPR process at temperatures regularly exceeding 30 °C (Cao et al., 2017). In both cases, having sufficient influent carbon was a critical factor for successful EBPR, as the Norway WRRF has an efficient fermenter and the Changi WRRF uses a step-feed approach to optimize influent carbon feeding. Temperature can have an effect on fermentation rates in the collection system, where colder temperatures may produce fewer VFAs in the influent. On the other hand, some WRRFs find it more difficult to maintain stable EBPR performance in the summer as temperatures increase. This is most likely due to the increased potential for secondary release of phosphorus due to operating at longer SRTs than necessary for the warmer temperatures, which is discussed in the next section.

## 4.4 Solids Retention Times

Solids retention time is the most important operational parameter in any BNR facility. When implementing nutrient removal, it is important to make a distinction between aerobic SRT and total SRT.

Aerobic SRT will determine the ability of the facility to nitrify and is the basis for sizing the aerobic portion of the bioreactor. Total system SRT for BNR facilities is defined as the total mass of solids in the entire bioreactor (aerobic plus anaerobic plus anoxic portions) divided by the sum of the total solids in WAS and the total solids in the effluent. Although total system SRT is occasionally discussed for comparative purposes, it is, in fact, of limited usefulness and can lead to erroneous conclusions. This is because, whereas the required aerobic SRT is a well-defined term based on bioreactor temperature, total system SRT can vary greatly depending on the required sizes of the anaerobic and anoxic zones, as dictated by the ratios of influent COD to total nitrogen to total phosphorus and required effluent total nitrogen and total phosphorus. Therefore, it is important to understand clearly which SRT is being used to control nitrogen and phosphorus removal processes.

A combined NDN and EBPR facility should maintain wasting rates necessary to strike the proper balance between maintaining the minimum aerobic SRT needed for stable nitrification while not allowing the SRT to extend too long to hamper EBPR performance. Longer SRTs are worse for EBPR because they result in lower mass removed in WAS, which lowers the amount of total phosphorus removed through biomass (WEF, 2023).

Additionally, some facilities have observed that providing a more stable SRT is a critical success factor for maintaining reliable EBPR performance. For example, the Upper Blackstone Clean Water facility in Millbury, Massachusetts, observed a significant improvement in EBPR performance after converting from RAS-based wasting to wasting of mixed liquor, which is a much simpler operation and results in more consistent and stable operating SRT. In addition to the inherent benefit of stability for biological populations to thrive, maintaining a stable SRT gave the operators the confidence to experiment with operating at lower SRTs which has been shown to improve EBPR. This change in sludge wasting practices combined with converting from dosing caustic in the RAS to dosing magnesium hydroxide slurry to provide more stable pH yielded tangible benefits in EBPR resilience and stability (Neville et al, 2015).

## 4.5 Aeration and Dissolved Oxygen

As discussed in Chapter 3, a dissolved oxygen concentration of more than 2 mg/L promotes the fastest nitrifier growth. In contrast, denitrification requires the absence of dissolved oxygen. Thus, many plug flow facilities use a "tapered aeration" approach by stepping down the dissolved oxygen concentration along the length of the basin to maintain low dissolved oxygen concentration at the point of withdrawal for the internal recycle pumps. Newer research and operating experience by facilities using low dissolved oxygen concentrations have demonstrated that some micro-organisms can adapt to provide effective nitrification at suboptimal growth rates.

The effect of dissolved oxygen on the uptake of phosphorus has not been studied as extensively. Polyphosphate-accumulating organisms need a positive dissolved oxygen concentration for phosphorus uptake. In addition, some of the PAOs are also capable of using nitrate for phosphorus uptake.

Reactions in the aerobic zone determine system phosphorus removal. There must be sufficient dissolved oxygen for PAOs to completely metabolize stored organics to gain the energy for phosphorus uptake. The dissolved oxygen concentration should not limit the rate of oxygen transfer to the cells. There should also be enough dissolved oxygen or nitrate in bioreactor effluent to prevent phosphorus release in the secondary clarifier; however, the amount contained in the RAS should be near zero. High SRT oxidation-ditch systems have shown that EBPR performance is determined by the mass of oxygen transferred rather than the concentration maintained at any point in the tank.

Given the criticality of providing the right amount of dissolved oxygen at the right locations—as well as the challenges of maintaining dissolved oxygen under varying diurnal load conditions—more automation at the supervisory control and data acquisition system is getting incorporated into BNR facility designs. This may include a relatively simple control system using a single dissolved oxygen probe in a basin, and operations staff use judgment and trial-and-error to modify the setpoint to optimize the system. Others may use advanced aeration control systems including flow control valves for each dropleg, which provide a fine-tuned level of dissolved oxygen control that some facilities need to optimize combined nitrogen removal and EBPR systems. Some facilities may also use additional instruments to optimize the aeration system operation, including ammonia probes to monitor nitrification performance in real-time, or ORP probes which reflect an increase in ORP as dissolved oxygen increases and a decrease in ORP with decreasing dissolved oxygen and nitrate. With ORP probes, when the mixed liquor runs out of nitrate, there is a discernible bend in the downward line as a result of the sharp drop in redox potential, which is an indication that oxygen supply must be resumed.

## 4.6 Alkalinity and pH

Alkalinity and pH measurements are both needed for process control of a combined nitrogen and phosphorus removal WRRF. The pH in influent water to WRRFs typically varies from 6.6 to 7.4; however, indications are that there is a decline in efficiency of both nitrification and phosphorus removal when the pH drops to below approximately 6.8. There is some evidence that GAOs may also compete at pH values below 7, reducing the VFAs available for PAOs (Filipe, 1999). The effects of higher pH on BNR processes have not been as widely studied. However, the factors influencing the optimal target pH for a given WRRF will depend on site-specific factors, and it is important to keep in mind that activated sludge populations can adapt to operating at lower than "textbook" pH values as long as the process stably operates in that range without sudden changes. An important factor for maintaining stable mixed liquor pH is to have enough alkalinity to provide buffering capacity. It is recommended to maintain a bioreactor mixed liquor effluent alkalinity (or secondary effluent alkalinity) exceeding 80 mg/L as $CaCO_3$; this amount of alkalinity should be able to provide adequate buffering capacity in response to variations in influent pH, influent nitrogen loading, and variations in reactor performance for most WRRFs.

The effect of various operations on alkalinity in mixed liquor is shown in Table 9.1. Hydrolysis of organic nitrogen will supply 3.6 g of alkalinity/g of nitrogen hydrolyzed. Nitrification consumes 7.14 g of alkalinity for every 1 g of ammonia-nitrogen converted to nitrate-nitrogen. Approximately 3.5 g is regained for every 1 g of nitrate-nitrogen reduced to nitrogen gas by denitrification. When adding chemicals for phosphorus removal, more alkalinity is consumed. This may affect nitrification and require the addition of buffering chemicals (i.e., supplemental alkalinity chemicals) to prevent the pH from dropping too far.

**TABLE 9.1** Alkalinity Consumed or Produced by Certain Processes

| Process | Alkalinity Change, mg/L | Per mg/L of |
|---|---|---|
| Ammonification of organic nitrogen | +3.6 | Organic nitrogen hydrolyzed |
| Nitrification | −7.1 | Ammonia-nitrogen oxidized |
| Denitrification | +3.6 | Nitrate-nitrogen reduced |
| Phosphorus removal with aluminum salt addition | −5.6* | Aluminum added |
| Phosphorus removal with iron salt addition | −2.7* | Iron added |

*Values shown for phosphorus are approximate and depend on the quantity of chemical addition required. The values shown above are the alkalinity consumed caused by metal salt addition for reducing phosphorus to 1 mg/L from a medium-strength wastewater. For lower effluent phosphorus targets, proportionally more chemicals will be required and more alkalinity will be consumed.

The most probable cause of low alkalinity and pH is when a facility is fully nitrifying without denitrification while treating a low-alkalinity wastewater. The pH can generally be adjusted back toward the optimum range by promoting denitrification under anoxic conditions in the single sludge system. If the target effluent alkalinity value is difficult to maintain via optimization of denitrification to recover alkalinity after nitrification, then supplemental alkalinity dosing is required. Which chemical is used at a particular WRRF depends on a number of criteria, including availability, cost, storage and feed requirements, safety considerations, and operator preference. Table 9.2 summarizes some example of supplemental alkalinity chemicals that are commonly used and their typical alkalinity contents.

## 4.7 Return Activated Sludge

Return activated sludge brings settled solids from secondary clarifiers to process zones, thereby maintaining proper MLSS concentration and SRT in these zones. In BNR facilities, one way to determine the optimum RAS flowrate is to adjust the RAS flowrate to maintain about 30 cm of sludge blanket in the secondary clarifiers. Another way to determine optimum RAS flow is based on a mass balance around a secondary clarifier using influent flow, MLSS, and total suspended solids (TSS) concentration of the RAS. The optimum RAS flowrate can be determined using the following equation, which is simplified by assuming that the effluent TSS is negligible:

$$Q_{RAS} = (Q_{INF} \times MLSS)/(TSS_{RAS} - MLSS) \qquad (9.1)$$

Where

$Q_{RAS}$ = RAS flow rate ($m^3/d$),
$Q_{INF}$ = facility influent flow rate ($m^3/d$),
MLSS = mixed liquor suspended solids concentration ($g/m^3$), and
$TSS_{RAS}$ = TSS concentration in RAS ($g/m^3$).

Although optimum RAS flow rate can be calculated from Equation 9.1, the actual RAS flowrate required to maintain a 30-cm clarifier sludge blanket will change depending on the actual conditions and sludge settleability. The optimum RAS flowrate should be determined as a fraction of the influent flowrate, and the RAS pump rate should be modulated based on this ratio. To keep the sludge blanket variations to an acceptable range, $Q_{RAS}/Q_{INF}$ may be adjusted based on the observed sludge blanket.

Applying rates lower or higher than the optimum RAS flowrates will affect BNR facility performance. Lower RAS flow will improve EBPR efficiency because the mass of nitrate load to the anoxic and

**TABLE 9.2** Chemicals Commonly Used to Increase Alkalinity

| Chemical | Common Name | Chemical Formula | mg $CaCO_3$ Alkalinity Added per mg Chemical |
|---|---|---|---|
| Sodium bicarbonate | Baking soda | $NaHCO_3$ | 0.6 |
| Sodium carbonate | Soda ash | $Na_2CO_3$ | 0.9 |
| Calcium magnesium dicarbonate | Dolomite | $CaMg(CO_3)_2$ | 1.1 |
| Sodium hydroxide | Caustic/caustic soda | NaOH | 1.3 |
| Calcium hydroxide | Slaked lime | $Ca(OH)_2$ | 1.4 |
| Magnesium hydroxide | Milk of magnesia/brucite | $Mg(OH)_2$ | 1.7 |
| Calcium oxide | Quicklime | CaO | 1.8 |

anaerobic zone is reduced and more VFAs will be available for PAOs. On the other hand, lower RAS flow increases the sludge blanket and solids inventory in the secondary clarifier, perhaps allowing denitrification to occur via endogenous respiration. Nitrogen gas bubbles formed as result of denitrification in clarifiers can cause rising sludge and increased effluent TSS. In addition, deeper blankets also increase residence time of PAOs in the secondary clarifiers, perhaps resulting in secondary release of phosphorus.

Higher RAS rates, which can be as much as 75% to 100% of influent flow, may improve nitrogen removal as more nitrate is returned to the anoxic or anaerobic zones. However, EBPR will be impaired because denitrifiers will use more of the available VFAs.

In MBRs, RAS flow is returned from the membrane basins; the RAS will have high dissolved oxygen because of the scour air. When the oxygen-rich mixed liquor is returned to the anaerobic or anoxic zones, more VFAs will be used aerobically, impairing the denitrification and EBPR efficiency. The minimum possible RAS flow should be used in MBR facilities to maintain sufficient MLSS distribution between the membrane and process basins. Even then, the oxygen load in RAS can impair the BNR efficiency of the facility. For this reason, modern MBR designs use membrane tank aeration systems that reduce the dissolved oxygen in the RAS significantly.

## 4.8 Internal Recycle

There are a few significant types of internal recycle streams used in combined BNR processes. The first type of internal recycle returns nitrified mixed liquor from the aeration zone to the anoxic zone. Nitrified mixed liquor is high in nitrate, and therefore, higher internal recycle flow rate will increase the nitrate load to the anoxic basin, increasing denitrification efficiency.

Lower internal recycle flow (as low as 200% of influent flow rate) may be applied for low-efficiency denitrification processes consisting of single-stage anoxic zones, such as the A2/O process. On the other hand, much higher internal recycle flowrates (up to an even greater than 400% of the influent flowrate) may be applied for high-efficiency nitrogen removal processes consisting of preanoxic and postanoxic zones, such as the five-stage Bardenpho process. The internal recycle flowrate should be determined with respect to the residual nitrate level at the end of the anoxic zones. If the process consists of at least two or three anoxic zones in series, the internal recycle flowrate should be optimized to drive nitrate down to low levels in the last anoxic zone. Lower internal recycle flowrates may drive nitrate concentration down to low levels in most anoxic zones. This can lead to secondary release of phosphorus.

Because internal recycle is returned from the aeration zone, it may also have high dissolved oxygen that will reduce denitrification in anoxic zones. Therefore, if possible, low dissolved oxygen should be maintained at the point of internal recycle withdrawal. However, the minimum dissolved oxygen that can be maintained is typically determined by the minimum airflow requirement for mixing at that point in the aeration zone, unless equipment is provided to decouple process aeration from mixing (such as using mixer/aerators, which can provide mixing energy independently from the air provided for process oxygen). The internal recycle flowrate can be increased up to 400% or higher of the influent flow as long as dissolved oxygen in internal recycle is sufficiently low (such as below 1.0 mg/L) (WEF, 2010).

Another type of internal recycle is used for returning denitrified mixed liquor from the end of the anoxic zone to the anaerobic zone, such as in the UCT process (Figure 9.3). The purpose of this internal recycle flow is to minimize nitrate return to the anaerobic zone. This internal recycle flow can be 100% to 300% of the influent flow.

## 4.9 Sidestreams

Sidestreams can affect combined nitrogen and phosphorus removal processes just like any other BNR process. Sidestreams containing nitrogen and phosphorus are typically returned to process basins after being mixed with facility influent, but some facilities have the capability to provide sidestream treatment. Sidestreams from solids handling processes may contain ammonia, nitrate, and phosphorus depending on the sludge treatment and handling processes that are used and the sidestream treatment methods available.

As previously discussed, commonly used measures of influent amenability to BNR are based on the amount of readily biodegradable substrate available to sustain EBPR and denitrification. This determination should be made on the influent to the BNR bioreactor and should include all significant recycle loads. Primarily domestic municipal wastewater typically contains sufficient carbon substrate for BNR if excess BOD removal does not occur in the primary clarifiers. Recycle streams, which are characteristically low in BOD and relatively high in nitrogen and phosphorus, can decrease the ratios of BOD to total phosphorus and BOD to TKN in the bioreactor influent. In addition, if solids capture is not optimized in EBPR sludge thickening and dewatering operations, significant phosphorus-rich solids would be returned to the primary clarifier and mixed with substrate-rich primary sludge. This can potentially result in secondary phosphorus release if anaerobic conditions prevail in the sludge blanket. This released phosphorus will lower the bioreactor ratio of influent BOD to total phosphorus.

Some facilities have the capability to provide some sidestream treatment either of the solids handling sidestreams or of the RAS. Treatment of the solids handling sidestreams typically is used to reduce recycling of certain constituents, whereas RAS sidestream treatment is typically used to generate carbon to enhance the mainstream BNR process. Sidestream management and treatment is discussed in detail in Chapter 12.

## 4.10 Operation of Other Unit Treatment Processes for Nutrient Removal

### *4.10.1 Primary Clarifiers*

The influent ratios of BOD to TKN and BOD to total phosphorus may be reduced substantially when primary clarifiers are used. The majority of influent TKN in facility influent is typically in the form of ammonia. Ammonia and orthophosphate are soluble species and pass through primary clarifiers. On the other hand, a substantial amount of influent BOD is associated with influent TSS, and primary clarification can commonly remove 25% to 40% of influent BOD. Therefore, primary clarifier effluent will have lower ratios of BOD to TKN and BOD to total phosphorus compared to facilities without primary clarification. Lower influent BOD will negatively affect denitrification. It can also affect EBPR if the anaerobic zones are designed to be large enough to ferment the VFAs needed for EBPR.

Maintaining shallow blankets in primary clarifiers can reduce hydrolysis of organic matter, which is the mechanism that converts particulate nitrogen and phosphorus to soluble forms. Therefore, soluble nutrient loading to process basins will increase. It is also important to maintain a reasonable blanket depth in primary clarifiers to prevent dilute concentrations of primary sludge.

### *4.10.2 Tertiary Filters*

Tertiary filters are used for suspended solids removal from effluent. Effluent suspended solids contain particulate nitrogen and phosphorus at varying fractions depending on the upstream biosolids and chemical treatment process. Suspended solids removal with tertiary filters is more critical when low-effluent nitrogen and phosphorus levels are required.

Deep-bed filters combine suspended solids removal and additional nitrogen removal with external carbon source addition. Deep-bed denitrifying filters provide effective suspended solids removal while reducing effluent nitrate to a 1.0- to 2.0-mg/L level.

# 5.0 OTHER CONSIDERATIONS

## 5.1 Competition for Available Carbon in Combined Biological Nutrient Removal Systems

Operators of combined BNR facilities must be aware of the competition for soluble COD between PAOs and denitrifying organisms. For facilities with anaerobic zones ahead of anoxic zones, the uptake of soluble COD by PAOs in the anaerobic zone reduces denitrification efficiency in the anoxic zone, possibly requiring the use of supplemental carbon to meet low nitrogen limits.

Another critical consideration for EBPR facilities is the importance of preserving VFAs for use by PAOs, as they are more selective in the type of carbon they can use compared to denitrifying organisms. Excessive nitrate in the RAS return to the anaerobic zone results in VFAs being taken up by denitrifying organisms in the anaerobic zone rather than by PAOs, thereby reducing EBPR efficiency.

## 5.2 Secondary Release

*Secondary release* refers to the release of polyphosphate without the uptake of VFAs. Although this can occur at any EBPR facility, it is particularly common at combined NDN and EBPR facilities where both anoxic and anaerobic zones are present. When there is a sufficient supply of VFAs in the feed or from a fermenter, EBPR is forgiving. However, in many facilities, the VFA supply is just sufficient or may have to be augmented. When secondary release of phosphorus takes place, more VFAs are needed to take up the excess phosphorus that is released. Secondary release takes place in any zone where there are unaerated conditions without a supply of VFAs or nitrate.

Secondary release can take place in many different locations and for different reasons, as summarized in Table 9.3.

Figure 9.10 shows the effect of insufficient VFAs on phosphorus release in anaerobic zones that are too large. The top curve represents combined primary and secondary release. The primary release stops when VFAs are all absorbed and there is only secondary release. Ideally, the anaerobic zone should be just large enough for completion of the primary release to minimize the secondary release. The RAS recycle rate or the amount of flow going to the anaerobic zone can be adjusted for optimization.

Secondary release can also take place in either the first anoxic zone of a three-stage A2/O facility or a post-anoxic zone of a five-stage Bardenpho facility. If the anoxic zones are too large or run out of nitrate due to insufficient supply of recycled nitrified mixed liquor, secondary release will take place. As

**TABLE 9.3** Location and Cause of Secondary Phosphorus Release

| Environment/Operation | Reason for Secondary Phosphorus Release |
|---|---|
| Anaerobic zone | VFA depletion because of oversized anaerobic zone |
| Anoxic zone | Nitrate depletion because of oversized anoxic zone (becomes anaerobic) |
| Second anoxic zone | Lack of nitrates resulting in anaerobic conditions |
| Aerobic zone | Long SRT leading to cell death |
| Final clarifier | Septic conditions caused by deep sludge blanket and low RAS flowrate |
| RAS piping | Septic conditions from long detention time in pipe |
| Sludge storage | Septic conditions from poorly or unaerated sludge storage |
| Aerobic digestion | Cell death and phosphorus release |
| Anaerobic digestion | Cell death and phosphorus release |

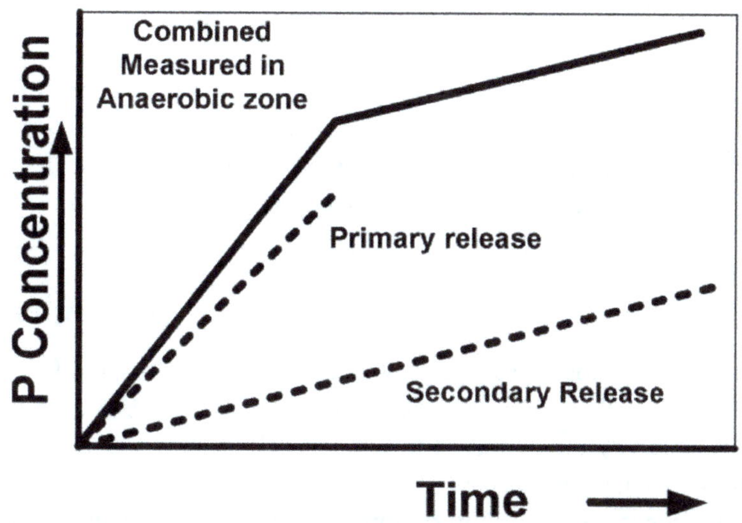


FIGURE 9.10 Primary and Secondary Release in the Anaerobic Zone

long as nitrate is present, PAOs, which can use nitrate as the electron acceptor, will take up any released phosphorus. On the other hand, without nitrate and oxygen, there is no means of taking up phosphorus that is released by other organisms, and too much secondary release takes place. This can be verified by measuring soluble phosphorus profiles through the anoxic zone. Figure 9.11 illustrates an example of a soluble phosphorus profile in a five-stage Bardenpho facility that used SND in the aeration basin using draught tubes with surface aerators. The SND was so efficient that there was no nitrate in the mixed liquor to the second anoxic zone. Phosphorus was released and could not be taken up again with aeration alone. If VFAs were added to the second anoxic zone, phosphorus would be taken up in the reaeration zone. In this instance, the solution was to aerate the second anoxic zone, and as a result, no further release took place.

Phosphorus may also be released when there is a low nitrate concentration in the mixed liquor flow to final clarifiers. When a deep sludge blanket develops in the final clarifiers, nitrate will be denitrified. When the nitrate is low, phosphorus will be released. This release may not affect effluent phosphorus directly but may return a large portion of the released phosphorus back to the anaerobic zone, where

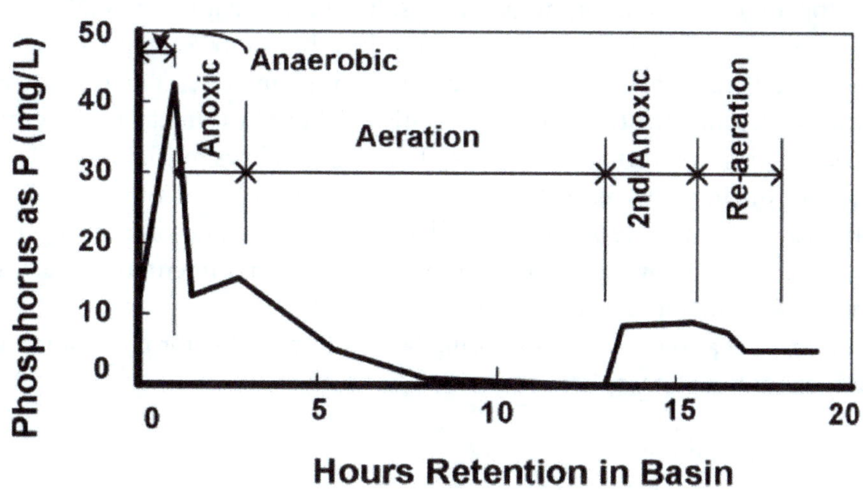


FIGURE 9.11 Secondary Release of Phosphorus in the Second Anoxic Zone

there may not be enough VFAs for uptake of this additional released phosphorus. At one facility with a relaxed requirement for nitrogen removal but a total phosphorus limit of less than 0.15 mg/L, the operator tried to reduce nitrate in RAS by reducing the recycle rate. A sludge blanket formed in the clarifiers and 10 mg/L of phosphorus was released in the RAS. Increasing the RAS rate reduced the secondary release of phosphorus, resulting in effluent soluble phosphorus dropping from 1.5 to 0.08 mg/L. Therefore, when the facility is operated for low-effluent nitrate in addition to low phosphorus, the sludge blanket should not be deep.

## 5.3 Considerations for Low-Effluent Nitrogen and Phosphorus Limits

For high levels of nitrogen and phosphorus removal in combined systems, the five-stage Bardenpho process typically is required followed by metal salt addition with filtration. Alternatively, processes that do not remove nitrate completely, such as the UCT and Westbank processes, could be used with attached-growth denitrification as a tertiary process.

A 2011 Water Environment Research Foundation (WERF) report concluded that, even though several flow sheets for combined nutrient removal statistically perform well, their effluent quality with respect to total nitrogen and total phosphorus varies somewhat depending on factors such as geographic location of the facility, climate, and other factors. The following are other important findings of this report:

- Separate-stage nitrogen removal facilities outperform combined nitrogen removal facilities because of a higher degree of denitrification control possible with a separate-stage process regarding total nitrogen.
- Bardenpho (four- or five-stage) facilities come close to meeting monthly total nitrogen goals of 3.0 mg/L 95% of the time.
- Single-stage chemical addition processes for total phosphorus removal outperformed multiple-stage processes, although often at the expense of higher chemical dosages.
- Tertiary chemical addition and effective filtration (gravity, media, or membrane) are required to achieve low phosphorus. Facilities with some form of tertiary chemical addition, clarification, and filtration slightly outperform those that have only effluent filters.

When a final effluent total nitrogen of less than 3 mg/L is required, the effluent nitrate must not exceed 1.5 mg/L to allow for some ammonia and recalcitrant soluble organic nitrogen that cannot be removed.

Thus far, most instances of consistently good removals of both nitrogen and phosphorus have been performed by five-stage Bardenpho-type facilities having point-source aerators because of the effects of SND. The Kelowna Wastewater Treatment Plant in British Columbia, Canada, is an example of such a facility. Aeration is supplied by a turbine aeration system, which allows for turning down the dissolved oxygen without loss of mixing. This, in turn, allows for a considerable amount of SND in the aeration basin. The facility was able to achieve an effluent of less than 1 mg/L ammonia-nitrogen and nitrate-nitrogen, which was maintained over months. Figure 9.12 presents representative data regarding total nitrogen performance for different BNR systems.

The 2011 WERF report also found the following factors responsible for increasing the variability of effluent total nitrogen and total phosphorus concentrations:

- Infrequent toxic event upsets;
- Unexpected interruptions in chemical supply—the majority of facilities in the survey used chemicals;

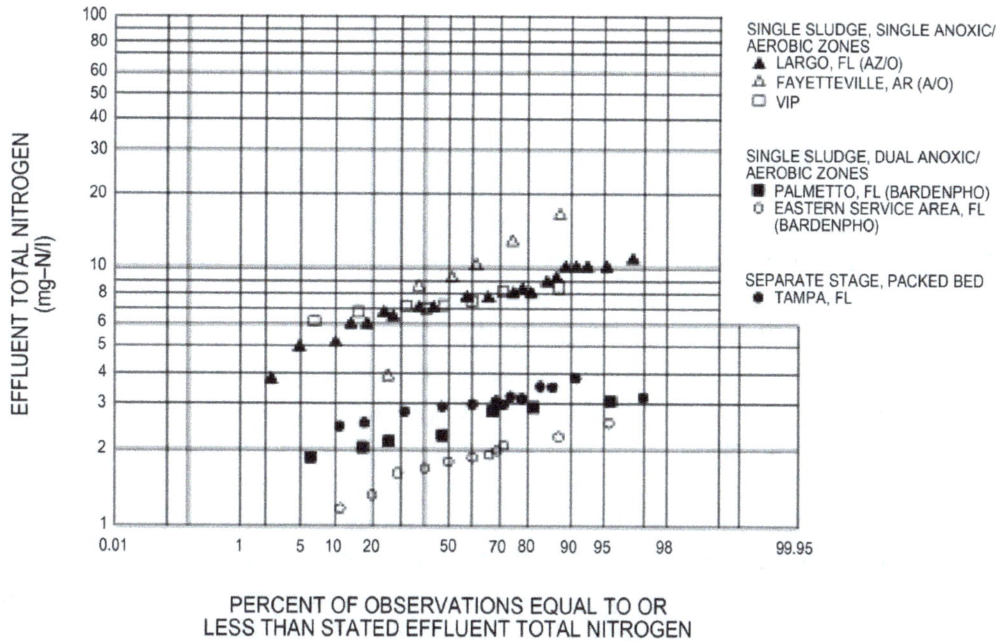


FIGURE 9.12 Removal of Nitrogen in BNR Facilities in the United States

- Facility upgrading projects and the effects of construction on effluent reliability;
- Peak flow events as the most difficult operating issues along with seasonal variations in flows and loads;
- Biological treatment capacity issues affecting performance during more stressed periods;
- Sludge supernatant recycle streams containing ammonia;
- Chemical feed control issues for phosphorus removal; and
- Fermenter control issues as the most difficult aspect of operations in facilities that rely solely on EBPR.

## 5.4 Effect of Chemical Phosphorus Removal on Biological Nutrient Removal Systems

The addition of chemicals to different points in the BNR system and how this affects nitrogen and phosphorus removal are discussed in the following subsections.

### *5.4.1 Primary Clarifiers*

In addition to precipitation of phosphorus compounds, metal salt addition upstream of primary clarifiers enhances suspended solids and BOD removal in primary clarifiers as a result of coagulation of suspended organic matter. Therefore, primary sludge will contain a greater amount of organic matter because it is captured with inorganic flocs. Removal of organic material in the primary clarifier reduces loading to secondary treatment facilities, resulting in capital and operation and maintenance cost savings for secondary treatment. However, removal of BOD and phosphorus from wastewaters could cause a low ratio of BOD to TKN, with many effects on the process including reduced nitrogen removal in the BNR process, phosphorus deficiency to the biological process if too much phosphorus is removed by chemical precipitation, and loss of alkalinity needed in the biological process. Chemical enhanced primary clarifiers will not remove soluble COD, and some rbCOD could be generated by fermentation of some of the primary sludge for denitrification and phosphorus removal.

### 5.4.2 Secondary Clarifiers

Addition of metal salts upstream of the secondary clarifiers provides a high level of phosphorus removal when used in conjunction with EBPR. At this point in the treatment process, phosphorus is typically in the orthophosphate form, which can be precipitated with metal salts. The metal salt and phosphorus precipitate can be removed with the flocculent biomass in the secondary clarifier. Chemical addition could also enhance nitrogen removal because colloidal nitrogen is flocculated as a result of chemical addition, thereby reducing total nitrogen in the effluent. The addition of chemicals to secondary clarifiers increases the amount of inerts carried in the mixed liquor because RAS returns inorganic precipitates to the aeration basin. This effectively means that the nitrification capacity of the aeration basin is reduced because of the lower ratio of mixed liquor volatile suspended solids to MLSS. This phenomenon is rarely considered when designing or operating a combined nutrient removal reactor, and depending on the amount of salts added to the clarifier, the effect on the nitrification system could be severe. This effect should be studied during the design phase because it may restrict the ability of a secondary clarifier to accept metal salts and subsequent provisions for polishing to lower total phosphorus levels may be required. To minimize the effects on the nitrification system, a two-stage chemical addition approach may be better suited for combined nutrient removal facilities performing chemical phosphorus removal in secondary or tertiary systems.

## 5.5 External Carbon Source Addition

In combined nitrogen and phosphorus removal processes, external carbon sources may be added to anaerobic and anoxic basins to enhance EBPR and denitrification, respectively. Use of an external carbon source for denitrification is a more common practice. Methanol has been a commonly used external carbon source because it provides lower sludge yield and higher denitrification efficiency indicated by relatively low ratios of gram of methanol per gram of nitrate and lower costs (Gu & Onnis-Hayden, 2010). On the other hand, the main concern regarding methanol is the safety issue related to storage and transportation because methanol is a highly flammable chemical. In addition, specialized methanol-oxidizing bacteria must be enriched in the mixed liquor, and therefore, methanol cannot be added on an "as needed basis."

Carbon sources can be directly added to the pre-anoxic basins of BNR processes with only pre-anoxic basins (i.e., A2/O) to lower effluent total nitrogen to the 5- to 8-mg/L range. The level of effluent that can be achieved is limited by internal recycle flow in this instance. The A2/O process can be coupled with denitrifying filters with carbon source addition to reduce effluent total nitrogen to 3.0 mg/L or lower.

Carbon sources can also be added to the post-anoxic basins of BNR processes (i.e., five-stage Bardenpho) with both pre-anoxic and post-anoxic basins. These BNR processes can lower effluent total nitrogen levels to 3.0 mg/L with carbon source addition to the post-anoxic basins. The carbon source dose should be optimized to achieve target effluent total nitrogen.

Carbon source addition to anoxic basins rather than to denitrification filters can reduce nitrate load in the RAS, thereby improving EBPR as well. On the other hand, carbon source addition to the filters can be controlled better, thereby allowing better nitrate removal than carbon source addition to the anoxic basins.

Depending on the BNR configuration and influent VFA levels, moderate effluent total phosphorus levels of 0.5 to 1.0 mg/L may be achieved. If there is insufficient VFA present in the influent, external

VFA sources may be added to anaerobic basins. Fermentation of primary solids is also commonly used at BNR facilities in Western Canada, the United States, and Europe to generate additional VFAs, thereby improving nutrient removal efficiency (Oleszkiewicz & Barnard, 2006).

Research shows that a mixture of acetic and propionic acids is optimal for improving EBPR. Full-scale observed supplemental VFA requirements are typically on the order of 5 to 10 mg/mg of phosphorus removed. The most common choice of VFA added for EBPR is acetic acid. Although the acetic acid unit cost may be high compared to the addition of metal salts (i.e., alum) for the removal of phosphorus by chemical precipitation, the cost of chemical sludge treatment and disposal should also be considered when making a decision. In some instances, use of metal salts for phosphorus removal may be needed when low-effluent phosphorus levels (i.e., 0.3 mg/L or lower) are required.

## 6.0 OPERATIONS SUMMARY

The key operating takeaways for facilities operating combined nitrogen and phosphorus removal are similar to those provided in Chapter 7 for nitrification and denitrification, but there are some additional considerations when combining a NDN process with EBPR:

- Waste sludge can provide sufficient aerobic SRT for reliable NDN performance, but avoid using excessively long SRTs that degrade EBPR performance.
- Pay careful attention to maximizing nitrate removal to limit the amount of nitrate into the anaerobic zones.
- Target dissolved oxygen that is just enough to meet nitrification needs without providing excessive dissolved oxygen that will inhibit denitrification or anaerobic zone activity.
- There should be enough dissolved oxygen or nitrate in bioreactor effluent to prevent phosphorus release in the secondary clarifier; however, the amount contained in the RAS should be near zero.
- Keep clarifier sludge blankets low to avoid secondary release of phosphorus in the clarifiers.

Operating with chemical phosphorus removal typically requires the consideration of fewer competing interests, but it must be practiced with care to avoid upsets to the biological populations, and operating personnel need to understand how operating expenses increase not only for chemical purchases but also for increase sludge production, handling, and disposal.

For detailed guidance on troubleshooting specific issues related to either nitrification, denitrification, or EBPR performance, refer to Appendix A.

## 7.0 REFERENCES

Avila, I., Freedman, D., Johnston, J., Wisdom, B., & McQuarrie, J. (2021). Inducing granulation within a full-scale activated sludge system to improve settling. *Water Science & Technology, 84*(2), 302–313.

Barnard, J. L. (1976). A review of biological phosphorous removal in activated sludge process. *Water SA, 2*(3), 136–144.

Cao, Y., Kwok, B. H., van Loosdrecht, M. C. M., Daigger, G. T., Png, H. Y., Long, W. Y., Chye, C. S., & Ghani, Y. A. (2017). The occurrence of enhanced biological phosphorus removal in a 200,000 $m^3$/day partial nitration and Anammox activated sludge process at the Changi water reclamation plant, Singapore. *Water Science & Technology, 75*(3–4), 741–751.

Daigger, G. T., Waltrip, G. D., Roman, E. D., & Morales, L. M. (1988). Enhanced secondary treatment incorporating biological nutrient removal. *Journal Water Pollution Control Federation, 60*, 1833.

Downing, L. (2023). *Water research foundation project #4975 final report: Practical considerations for the incorporation of biomass fermentation into enhanced biological phosphorus removal.* Water Research Foundation.

Ekama, G. A., & Marais, G. v. R. (1984). Biological nitrogen removal. In *Theory, design and operation of nutrient removal activated sludge processes.* Water Research Commission.

Filipe, C. D. M. (1999). *Competition between phosphate and glycogen accumulating bacteria: Stoichiometry, kinetics and the effects of pH* [PhD dissertation]. Clemson University.

Grady, C. P. L., Daigger, G. T., Love, N. G., & Filipe, C. D. M. (2011). *Biological wastewater treatment* (3rd ed.). IWA Publishing and CRC Press.

Gu, A., & Onnis-Hayden, A. (2010). *Protocol to evaluate alternative external carbon source for denitrification at full-scale wastewater treatment plants.* Water Environment Research Foundation.

Hamza, R., Rabii, A., Ezzahraoui, F., Morgan, G., & Terna Iorhemen, O. (2022, May). A review of the state of development of aerobic granular sludge technology over the last 20 years: Full-scale applications and resource recovery. *Studies in Chemical and Environmental Engineering, 5*, 100173. https://doi.org/10.1016/j.cscee.2021.100173

Kazmi, A. A., & Furumai, H. (2000). Field investigations on reactive settling in an intermittent aeration sequencing batch reactor activated sludge process. *Water Science & Technology, 41*(1), 127–135.

Keller, J., Watts, S., Battye-Smith, W., & Chong, R. (2001). Full-scale demonstration of biological nutrient removal in a single tank SBR process. *Water Science & Technology, 43*(3), 355–362.

Kobylinski, E., Durme, G. V., Barnard, J, Massart, N., & Koh, S. (2008, October 18–22). How biological phosphorus removal is inhibited by collection system corrosion and odor control practices. *Proceedings of the 81st Annual Water Environment Federation Technical Exhibition and Conference* [CD-ROM], Chicago, IL, United States. Water Environment Federation.

Neville, M., Wagoner, D., Doody, A., & Sangrey, K. (2015, September 24–30). Expanding the toolbox—operational strategies for optimizing the A2/O process at Upper Blackstone to push the limit of technology. *Proceedings of the 88th Annual Water Environment Federation Technical Exhibition and Conference*, Chicago, IL, United States. Water Environment Federation.

Oleszkiewicz, J. A., & Barnard, J. L. (2006). Nutrient removal in North America and European Union: A review. *Water Quality Research Journal, 41*(4), 449–462.

Pitman, A. R. (1992). *Process design of new BNR extensions to Northern Works, Johannesburg* (Paper presented at the Biological Nutrient Removal Conference). Leeds University.

Regmi, P., Sturm, B., Hiripitiyage, D., Keller, N., Murthy, S., & Jimenez, J. (2022). Combining continuous flow aerobic granulation using an external selector and carbon-efficient nutrient removal with AvN control in a full-scale simultaneous nitrification-denitrification process. *Water Research, 210*, 117991.

Sturm, B. (2020). Balancing flocs and granules for activated sludge process intensification in plug flow configurations (Water Research Foundation Project No. U1R14/4870 Final Report). The Water Research Foundation.

Water Environment Federation. (2010). *Nutrient removal* (Manual of Practice No. 34).

Water Environment Federation. (2023). *Wastewater engineering: Design of water resource recovery facilities* (MOP 8 series).

Water Environment Research Foundation. (2011). *Nutrient management volume II: Removal technology performance & reliability.*

Ydstebø, L., Bilstad, T., & Kommedal, R. (2000). Full-scale experience with enhanced biological phosphorus removal (EBPR) in cold climates. *Journal of Environmental Science & Health Part A, 35*(8), 1493–1502. DOI: 10.1080/10934520009377049

## 8.0 SUGGESTED READINGS

Water Environment Federation & Association of Boards of Certification. (2018). *Wastewater treatment fundamentals I, liquid treatment.* Water Environment Federation.

Water Environment Federation. (2023). *Standards for operation of water resource recovery facilities* (WEF 11-23).

# Chapter 10

# Chemical Precipitation of Phosphorus



## 1.0 INTRODUCTION

Chemical precipitation of phosphorus is a method frequently used as a primary phosphorus removal process or in conjunction with biological removal processes as a supplemental, polishing, or backup process. Chemical addition is often the phosphorus removal method of choice for small- and medium-sized facilities, where implementation of a dedicated, enhanced biological phosphorus removal (EBPR) process is not practical or economical. This is particularly the case when an existing water resource recovery facility (WRRF) has to meet new phosphorus limits and the configuration of the existing biological treatment process does not lend itself easily to an EBPR conversion. Because the EBPR process can be subject to periods of inferior performance or upsets, it is almost always supported by standby or polishing chemical addition facilities. In practice, whenever an effluent phosphorus limit has to be consistently met, chemical addition is almost always practiced, at least in a standby mode.

An advantage of chemical precipitation is that the phosphorus removed in this fashion remains fixed in the waste sludge as inorganic precipitate and is not readily released during sludge storage, thickening, or anaerobic digestion. At facilities with EBPR, phosphorus release from the sludge is common, particularly in oxygen-deficient conditions. Consequently, recycle streams from sludge-processing operations can return a significant phosphorus load to the head of the facility, particularly when anaerobic digestion is practiced. At such facilities, the sludge-processing train and return streams must be carefully designed and operated because uncontrolled releases of phosphorus can offset the purpose of the EBPR process. When chemical phosphorus precipitation is practiced, chemically bound phosphorus in sludge is not subject to such release.

The use of chemicals for phosphorus removal has a number of disadvantages, which are discussed in more detail in the subsequent sections. Some of these are increased sludge generation (Section 6.1), alkalinity depletion (Section 6.3), an increase in total dissolved solids (TDS) (Section 6.4), increased inert concentration in mixed liquor with related potential effects on ultimate biosolids disposal (Section 3.2), and potential effects on other processes such as ultraviolet (UV) disinfection (Section 6.7).

## 2.0 PHOSPHORUS FORMS IN RAW WASTEWATER AND IN THE EFFLUENT

The principal form of phosphorus removed by chemical addition is soluble orthophosphate ions. These ions represent the significant form of phosphorus in raw municipal wastewater. Biological treatment processes will convert most of the remaining phosphorus forms (polyphosphates, or condensed phosphates, and organic phosphorus) into orthophosphates while assimilating and incorporating some of the available phosphorus to the biomass (approximately 1 mg/100 mg of 5-day biochemical oxygen demand [$BOD_5$] removed in the biological process). Effluent from the activated sludge, or from any other biological process, will contain residual, soluble phosphorus primarily in the orthophosphate form, with a relatively minor contribution from refractory or recalcitrant dissolved organic phosphates (rDOP). For practical purposes, the contributions from rDOP are only of concern where stringent limits (at the level of 0.1 mg/L total phosphorus or less) are required (Section 5.1). Consequently, soluble phosphorus in the effluent from biological processes, including in the mixed liquor as it exits the aeration basin, is amenable to almost complete removal by chemical precipitation. In addition to soluble phosphorus, effluent from any treatment process will contain phosphorus in particulate form, either as inorganic precipitate or as part of biological cells. In a typical situation in which chemical is added to the activated sludge system, the effluent suspended solids will contain phosphorus in both forms enmeshed together into the floc (see Section 3.3 for a more detailed discussion of this topic).

Facility recycle streams from solids processing operations, particularly when anaerobic digestion is used and particularly for facilities practicing EBPR, could be rich in phosphorus released from sludge, as discussed in more detail in Section 4.0 of Chapter 2. The effect of the phosphorus load from recycle streams should be considered when designing and operating chemical phosphorus removal processes.

Chemical addition to the primary clarifier, which is known as *chemically enhanced primary treatment* (CEPT), or *pre-precipitation*, is frequently used for enhanced particulate and colloidal $BOD_5$ and total suspended solids (TSS) removal in the primary clarifiers to lower organic and solids loading on the downstream biological process. Chemicals commonly used in CEPT treatment (aluminum and iron salts) are the same ones used for orthophosphate precipitation. Consequently, most of the phosphorus present in raw wastewater in soluble, particulate, or colloidal form can be removed by chemical precipitation in the primary clarifier, potentially leading to phosphorus deficiency in downstream biological treatment processes. (For a detailed discussion of phosphorus sources and forms of phosphorus present in the influent and effluent, refer to Chapter 2 of these standards or Section 2.0 of Chapter 7 of the Water Environment Federation's [WEF's] *Nutrient Removal* [2010]).

# 3.0 PRINCIPLES OF CHEMICAL PHOSPHORUS REMOVAL

## 3.1 Mechanism of Phosphorus Removal

The basic mechanism of chemical phosphorus removal is precipitation followed by a solids separation step such as sedimentation and/or filtration. Phosphorus precipitation is the transformation of the soluble phosphorus present in wastewater as orthophosphate anion ($PO_4^{3-}$) to an insoluble chemical compound (salt) and the removal of these insoluble precipitates by sedimentation or filtration. Such a transformation occurs when a chemical agent bearing a proper cation, such as aluminum ($Al^{3+}$) or ferric ($Fe^{3+}$), is added to the wastewater and reacts with the soluble phosphorus to form an insoluble orthophosphate salt. In practice, more complex reactions are taking place, which involve co-precipitation and surface adsorption. These mechanisms are discussed in more detail in *Nutrient Removal* (WEF, 2010).

Precipitate formed in this way, together with other forms of particulate phosphorus present in the wastewater, can be separated from the wastewater by clarification or filtration. The phosphorus precipitated using chemicals is removed from the wastewater with sludge (i.e., either combined with waste primary or biological sludge or as a separate chemical sludge).

## 3.2 Fate of Chemicals Added During Treatment

The application dosages of chemicals needed to remove phosphorus to low levels are typically in excess of a stoichiometric requirement, as discussed in more detail in Section 4.0. When chemical is added to the mixed liquor at the effluent from the aeration basin (or influent to the final clarifier), which is the most common application point, any excess chemical that does not bind with orthophosphates will precipitate nevertheless. This precipitate will be in the form of aluminum hydroxide (for aluminum-based chemicals) and ferric hydroxide (for ferric-based salts), although, in reality, more complex hydrated-oxides mixtures will also be formed. These inorganic precipitates, only partially saturated with orthophosphates, will be enmeshed with biological floc and settle in the final clarifier. Thus, the excess, unused chemicals will be incorporated to the mixed liquor and returned to the head of the facility or upstream of secondary process with the return activated sludge (RAS). Here, these chemicals will be available to react with phosphorus entering the aeration basin with raw wastewater, reducing overall chemical use.

The precipitate accumulating in aeration basins will result in an increase in the basin inert solids concentration. The effect of an increased inert content of mixed liquor should be taken into account when designing and operating the activated sludge system. In particular, when operating the system at a given sludge age (such as that required for nitrification), inert solids from chemical addition will increase the equilibrium mixed liquor suspended solids (MLSS) concentration necessary to maintain the target sludge age. Conversely, if the same MLSS concentration is being maintained, the resulting sludge age will be lower after initiation of chemical addition. Refer to Section 6.1 for calculations of different chemicals' contribution to the inert content of MLSS.

## 3.3 Role of Solids Separation

Because effluent permit limits for phosphorus are commonly expressed as total phosphorus, precipitation of soluble phosphorus (orthophosphates) into a particulate form is only part of the job of phosphorus removal. Indeed, the solids separation step (clarification and/or filtration) must be capable of removing effluent TSS with associated particulate phosphorus forms to levels consistent with the effluent phosphorus limit.

The contribution of particulate phosphorus is particularly important for facilities with low effluent limits because even seemingly low effluent TSS can carry a significant phosphorus contribution (Figure 10.1). An additional consideration is that, in systems with a dedicated phosphorus removal process either by chemical addition to the activated sludge or by EBPR, the phosphorus content in MLSS and effluent TSS are increased. Whereas the phosphorus content of MLSS is approximately 2% in a municipal conventional activated sludge facility, the content increases to 4% to 5% for a facility with a stringent phosphorus limit, regardless of whether the phosphorus is incorporated into the sludge by precipitation or by EBPR. For facilities operating at a low sludge age and with an influent phosphorus concentration higher than that for typical municipal wastewater, this could even be much higher.

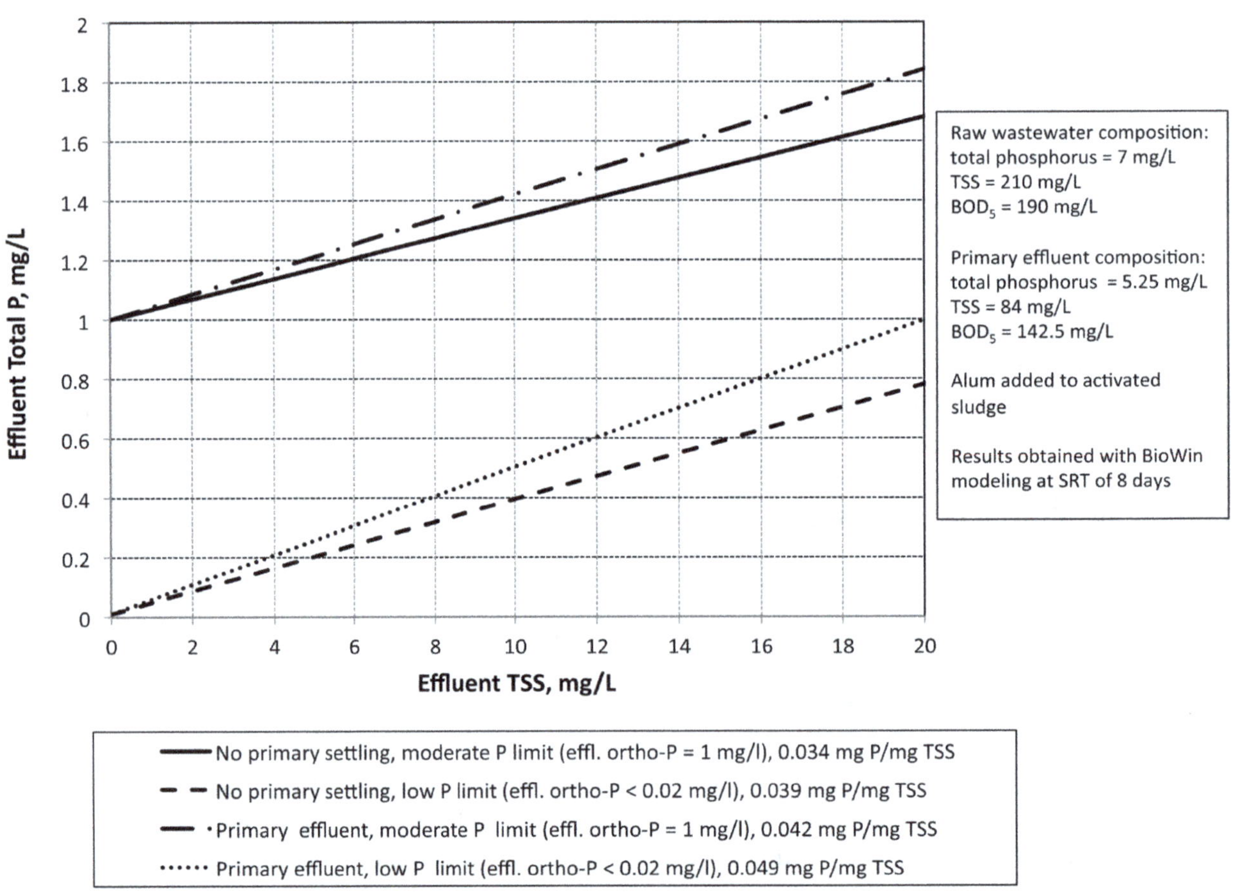


FIGURE 10.1 Contribution of Effluent TSS to Total Phosphorus in the Effluent for Different Treatment Scenarios

Figure 10.1 shows that, for total phosphorus limits of 0.1 mg/L, effluent TSS should be no more than 1 mg/L. Presently, this can only be consistently achieved by membrane filtration (as in a membrane bioreactor [MBR]), multistage filtration, or well-performing conventional filtration supported by polymer use. For limits of 0.5 to 1 mg/L total phosphorus, conventional filtration or even well-performing clarification (possibly supported by polymer addition) will be adequate, assuming that soluble orthophosphates are kept to a minimum by adequate chemical dose (or a well-functioning EBPR). As Figure 10.1 illustrates, for systems with primary clarification, which increases the mixed liquor volatile suspended solids (VSS) fraction of the MLSS, the phosphorus content in the MLSS and effluent TSS is even higher.

## 4.0 CHEMICALS USED FOR PHOSPHORUS PRECIPITATION

### 4.1 Introduction

In general, cations of the following metals can be used for the precipitation of phosphorus (orthophosphates) from wastewater:

- Aluminum;
- Iron;
- Calcium; and
- Magnesium.

Most commonly used are various chemicals containing aluminum (mainly alum) and iron (mainly ferric chloride), with calcium and magnesium being much less common. The following sections will discuss the different chemicals available, the precipitation process, and sample calculations.

### 4.2 Aluminum Salts

#### *4.2.1 Overview of Aluminum-Based Chemicals*

There are several aluminum compounds that are used in the wastewater industry for phosphorus removal, including aluminum sulfate, polyaluminum chloride (PACl), and sodium aluminate. The most widely used chemical is aluminum sulfate, commonly known as *alum*, although PACl is gaining popularity because of its ease of handling and lowering effect on pH. Polyaluminum chloride was originally used in water treatment when enhanced solids removal was a treatment objective. Because the strength and composition of sodium aluminate and PACl products vary considerably, efficacy of a specific product considered for an application should be established and compared to alternative products in jar tests, as discussed in Section 4.6. Sodium aluminate in a dry form or in solutions of various strengths was also originally developed for water treatment applications and is typically used when additional alkalinity is required.

Solid alum could be supplied as a dry, powdery chemical ("filter alum"), with an average molecular composition of $A1_2(SO_4)_3{\cdot}14H_2O$. This is a hydrated aluminum sulfate salt that indicates the presence of a defined number of water molecules in the structure of the dry salt crystal under normal conditions. However, alum for wastewater treatment is commonly delivered as a 49% solution (indicating that 49 parts of dry alum are combined with 51 parts of water, by weight). By convention, the alum dose applied in the treatment process is expressed in equivalent units of weight of dry alum (including water of crystallization) per volume of water (wastewater). Properties of alum and other aluminum-based chemicals are summarized in Table 10.1. The effect of aluminum-based chemicals on sludge generation, alkalinity consumption, and the increase in TDS are discussed in Sections 6.1, 6.3, and 6.4, respectively.

Regardless of the chemical, the active precipitating agent is aluminum ion, which combines with orthophosphate ions to form aluminum phosphate, as follows:

$$Al^{3+} + (PO_4)^{3-} \rightarrow AlPO_4 \qquad (10.1)$$

In reality, precipitation of phosphorus with aluminum and iron salts is a more complex process involving co-precipitation with aluminum hydroxide, hydrated aluminum oxides, and surface adsorption. This makes stoichiometric (based on chemistry of the reaction) calculations approximate; however, Equation 10.1 will be used as the basis of dose calculations for illustrative purposes.

Equation 10.1 indicates that it will take 1 mol of aluminum ion to react with 1 mol of phosphate, and thus, a 1-to-1-mol ratio of aluminum to phosphorus is required for this reaction. Because the molecular weight of aluminum is 27 and the molecular weight of phosphorus is 31, the weight ratio of aluminum to phosphorus in Equation 10.1 is 0.87 to 1, as shown in the following calculation:

$$(27 \text{ g Al}/1 \text{ mol A1})/(31 \text{ g P}/1 \text{ mol P}) = 0.87/1 \text{ or } 0.87:1$$

With aluminum sulfate, there are two molecules of aluminum per molecule of alum; therefore, the stoichiometric weight ratio is

$$594 \text{ g } Al_2\,(SO_4)_3 \times 14\,H_2O/2/31 \text{ g P} = 9.6:1.$$

A larger amount of aluminum salt is required for actual operation than the chemistry of the reaction (stoichiometry) predicts. The amount of excess chemical required increases as the target residual

**TABLE 10.1** Properties of Aluminum-Based Chemicals Used for Phosphorus Precipitation

| Name | Chemical formula | Molecular weight | Aluminum metal contents, % by weight | Specific density | | Weight ratio of dry chemical ($Al_2(SO_4)_3 \times 14H_2O$) for stoichiometric phosphorus precipitation (g of chemical/g of P) |
|---|---|---|---|---|---|---|
| | | | | kg/L | lb/gal | |
| Alum, dry | $A1_2(SO_4)_3{\cdot}14H_2O$ | 594 | 9.1 | 0.6–1.1 | 5–9.5 | 9.6 |
| Alum, 49% solution* | $A1_2(SO_4)_3{\cdot}14H_2O$ | 594 | 4.4 | 1.33 | 11.1 | 9.6 |
| Sodium aluminate, anhydrous (powder) | $NaAlO_2$ | 82 | 33 | 0.72 | 6.0 | 2.64 |
| Sodium aluminate, trihydrate (granular) | $Na_2O{\cdot}Al_2O_3{\cdot}3H_2O$ | 218 | 25 | 1.02 | 8.51 | 3.52 |
| Sodium aluminate, 20%–45% solution (could vary significantly) | $NaAlO_2$ | 82 | Varies | Varies | Varies | |
| Polyaluminum chloride, solutions of various strengths | $Al_nCl_{(3n-m)}(OH)_m$ example: $Al_{12}Cl_{12}(OH)_{24}$ | Varies | Varies | Varies | Varies | |

*49% alum solution has a dry alum [$Al_2(SO_4)_3{\cdot}14H_2O$] content of 0.647 kg/L (5.4 lb/gal) and an aluminum metal content of 0.059 kg/L (0.492 lb/gal).

phosphorus concentration decreases. The excess aluminum salt will precipitate as aluminum hydroxide, with a simplified reaction as follows:

$$Al^{3+} + 3OH^{-} \rightarrow Al(OH)_3 \tag{10.2}$$

Therefore, sludge that is generated at the point of chemical application will include a mixture of aluminum hydroxide and aluminum phosphate generated according to Equations 10.1 and 10.2. Equation 10.2 also illustrates the consumption of alkalinity (pH reduction) encountered when applying alum. These effects are discussed in more detail in Section 6.3.

An alternative aluminum-based chemical available for phosphorus removal is PACl, with a general composition of $Al_nCl_{(3n-m)}(OH)_m$. Polyaluminum chloride products also include an admixture of polyaluminum chlorohydrate and vary in the degree of acid neutralization (amount of caustic added, or basicity) and polymerization and aluminum content. In the water treatment industry, PACl is used as a more effective coagulant for TSS removal than alum and is also being used for phosphorus removal from wastewater. Polyaluminum chloride use generally does not depress wastewater pH, although its effects on pH depend on the exact formulation. Polyaluminum chloride use for phosphorus removal is gaining popularity, particularly at smaller facilities, because it could eliminate the need to use caustic to increase pH when straight alum is used.

Sodium aluminate can also serve as a source of aluminum for the precipitation of phosphorus. It is generated by reacting aluminum hydroxide with caustic soda. The chemical formula for sodium aluminate is $Na_2Al_2O_4$ or $NaAlO_2$. One commercial form of sodium aluminate is granular trihydrate, which may be written as $Na_2O{\cdot}Al_2O_3{\cdot}3H_2O$ and contains approximately 46% $Al_2O_3$ or 25% Al. It is also being sold as a solution in various strengths. In contrast to alum, which reduces wastewater pH, a rise in pH may be expected upon addition of sodium aluminate to wastewater. Consequently, in situations in which caustic addition would be required to counter pH depression caused by alum addition, application of sodium aluminate may eliminate the need for handling and dosing two different chemicals.

### 4.2.2 Determination of Dose for Aluminum-Based Chemicals

An approximate dose of alum (or other aluminum-based chemicals) required for a particular application can be calculated based on calculations presented in this section (including Equations 10.4 and 10.5 in Section 4.2.3). Refinement of the chemical application rate could then be performed in full scale based on the results obtained because the system response is fairly quick (Section 5.6).

Jar tests may also be performed to determine the chemical dosage for phosphorus removal in atypical applications (e.g., unusual wastewater composition) and to assess alkalinity destruction and pH depression from dosing of aluminum-based chemicals (and iron-based chemicals). Jar testing will also aid in determining requirements for supplemental alkalinity addition that may be necessary to maintain the required pH for nitrification (see also Section 6.3 for a discussion of the effects of pH and alkalinity). Jar test results may only be an approximation of the dose requirements and the chemical's effect on pH because many aspects of full-scale, continuous chemical addition may not be well duplicated in jar tests. However, jar tests are well suited for comparative testing of efficiency and costs of alternative chemicals (see also Section 4.6). Because of the inherent variability of wastewater, jar tests should be repeated on several different samples.

A design dose curve for chemical phosphorus removal using aluminum ions was developed using literature and pilot-facility data (Figure 10.2). Data in the literature include laboratory-scale test data and data from sites that are operating with chemical phosphorus removal (Gates et al., 1990). The full data set includes results for a range of pH values (mostly 6.5 to 7.5), temperatures, and wastewater

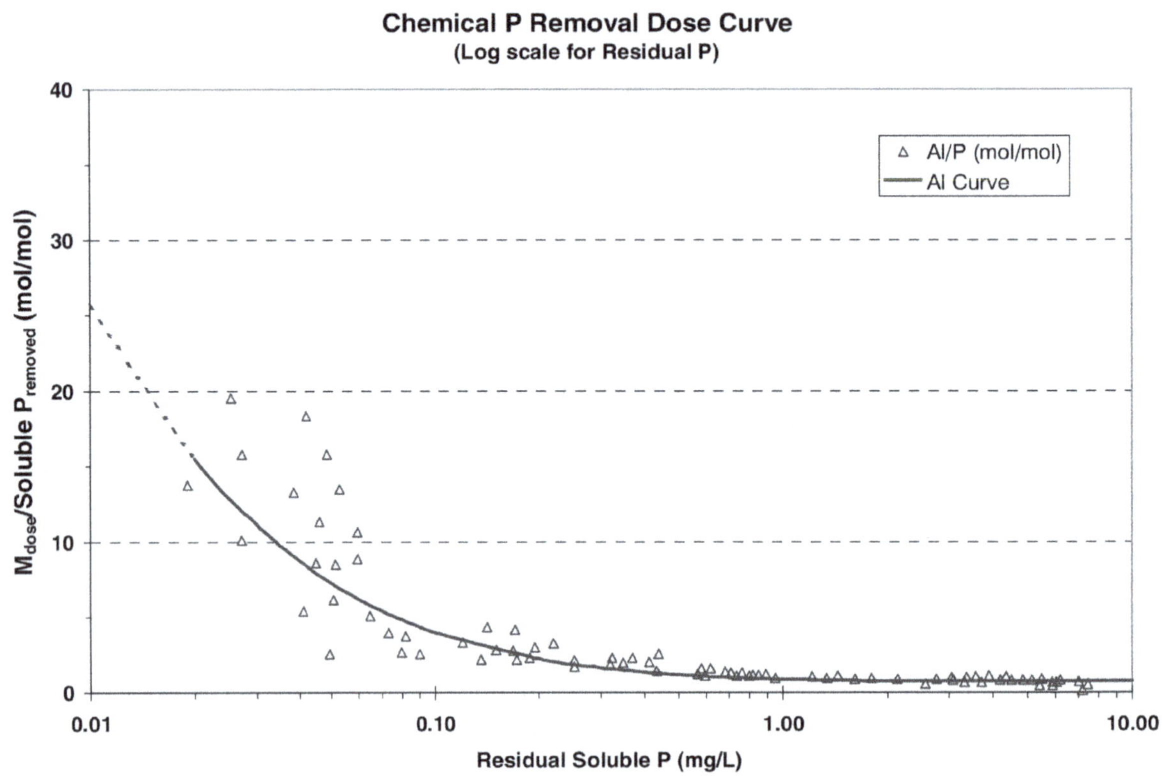


FIGURE 10.2 Ratio of Aluminum ($Al^{3+}$) Dose to Phosphorus Removed as a Function of Residual Orthophosphate Concentration

characteristics. The effluent soluble phosphorus concentration is labeled "residual soluble P" on the logarithmic x-axis. The molar ratio for metal ion dose to soluble phosphorus removed is labeled "$M_{dose}$/ soluble $P_{removed}$ (mol/mol)" on the y-axis. It is important to note that the curves apply to the soluble portion of the phosphorus only. The curve used to fit the data is based on the following equation:

$$y = a/(1 + be^{-cXe}) \tag{10.3}$$

Where

$y$ = moles aluminum required per mole soluble phosphate removed,
$a$ = 0.8,
$b$ = −0.95,
$c$ = 1.9, and
$Xe$ = target effluent soluble phosphorus concentration (mg/L).

Because Equation 10.3 was derived as a best-fit formula for empirical data, it is valid only for a limited range of residual soluble phosphorus concentrations (approximately 0.1 to 0.8 mg/L). For residual phosphorus concentrations above 0.8 mg/L, the value of $y$ should be assumed to be 1.0 (i.e., no excess over the stoichiometric amount indicated).

In general, the dose response found in the literature is quite variable and the data scatter increases at residual soluble phosphorus concentrations below 0.1 mg/L. Because of this variability, pilot-scale testing is recommended to determine the actual chemical dose required to reach a low-targeted effluent soluble phosphorus concentration. Whereas jar tests will provide the approximate dose required and are effective for comparison of efficacy of alternative chemicals, the actual dose required in the full-scale application, particularly when applied to an activated sludge process, could be different. This is attributable to the full-scale effects and recirculation of the excess chemicals, as discussed in Section 3.2, which are not duplicated in jar tests. The solids separation step, which plays a critical role in achieving a low effluent phosphorus concentration (Section 3.3), cannot be well simulated in jar tests, emphasizing the need for pilot (or full-scale) testing.

Table 10.1 includes the molecular weight for the different chemicals mentioned in this section. A sample calculation is provided in this section to determine the dose required to precipitate soluble phosphate. Following the detailed, step-by-step derivation, simplified formulas for both International System of Units and U.S. customary units are provided. The following conditions are assumed:

- Influent facility flow rate is 37 850 $m^3/d$ (10 mgd);
- A residual (influent) soluble phosphorus concentration at the point of chemical application is 3.0 mg/L; and
- A residual (effluent) soluble phosphorus concentration of 0.4 mg/L is required.

The amount of soluble phosphorus to be removed in kilograms (pounds) per day is as follows:

$$\text{Phosphorus} = (3 \text{ to } 0.4 \text{ mg/L}) \times 37\,850 \text{ m}^3\text{/d} \times 1\,000 \text{ L/m}^3 \times 1 \text{ kg/}1\,000\,000 \text{ mg} = 98.4 \text{ kg/d}$$

or

$$(3 \text{ to } 0.4) \text{ mg/L} \times 10 \text{ mgd} \times 8.34 \text{ (lb/mg} \times \text{L/mg)} = 217 \text{ lb/d of soluble phosphorus to be removed.}$$

This is equivalent to (98.4 kg/d)/(31 kg/kg-mol P) = 3.17 kg-mol or (217 lb/d)/(31 lb/lb-mol P) = 7.0 lb-mol phosphorus to be removed.

The amount of aluminum ions required is calculated from Equation 10.3 as follows:

$$Y = 0.8/(1 - 0.95e^{(-1.9)(0.4)}) = 1.44 \text{ kg-mol Al/kg-mol soluble phosphorus removed (or lb-mol Al/lb-mol P).}$$

Consequently, the amount of aluminum ions required per day is

$$3.17 \text{ kg-mol P} \times 1.44 \text{ kg-mol Al/kg-mol P} = 4.57 \text{ kg-mol Al}$$

or

$$7.0 \text{ lb-mol P} \times 1.44 \text{ lb-mol Al/lb-mol P} = 10.1 \text{ lb-mol Al.}$$

As the molecular weight of the Al ion is 27, this is converted to mass of aluminum ions per day, as follows:

$$4.57 \text{ kg-mol Al} \times 27 \text{ kg Al/kg-mol Al} = 123 \text{ kg of aluminum ions per day}$$

or

$$10.1 \text{ lb-mol Al} \times 27 \text{ lb Al/lb-mol Al} = 272 \text{ lb of aluminum ions per day.}$$

As indicated in Table 10.1, dry alum contains 9.1% aluminum. Consequently, the amount of dry alum required for this application will be

$$123 \text{ kg Al}/0.091\text{kg Al/kg alum} = 1356 \text{ kg alum}$$

or

$$272 \text{ lb Al}/0.091 \text{ lb Al/lb alum} = 2988 \text{ lb of dry alum per day.}$$

If 49% alum solution is used, it contains 0.059 kg/L (or 0.492 lb/gal) of aluminum ions (Table 10.1). Thus, the amount of liquid alum required for this application will be

$$123 \text{ kg Al}/0.059 \text{ kg Al/L} = 2092 \text{ L}$$

or

$$273 \text{ lb Al}/0.492 \text{ lb Al/gal} = 553 \text{ gal of solution per day.}$$

The liquid alum required per day to reduce soluble phosphorus from 3.0 to 0.4 mg/L in the aforementioned example is 2092 L/d (553 gpd). The alum dose equivalent to this alum use rate may be calculated from a factor of 0.647 kg/L (5.4 lb/gal) of alum in the 49% solution, as listed in Table 10.1, as follows:

$$2092 \text{ L/d} \times 0.647 \text{ kg/L}/37\,850 \text{ m}^3\text{/d} = 0.0358 \text{ kg/m}^3$$

$$0.0358 \text{ kg/m}^3 \times 1\,000\,000 \text{ mg/kg}/1000 \text{ L/m}^3 = 35.8 \text{ mg/L alum dose.}$$

In U.S. customary units, calculation for this 10-mgd facility will be as follows:

$$553 \text{ gpd} \times 5.4 \text{ lb/gal}/8.34 \text{ (lb/mg} \times \text{L/mg)}/10 \text{ mgd} = 35.8 \text{ mg/L alum dose.}$$

### 4.2.3 Summary Dose Formulas for Alum

The aforementioned calculations can be abbreviated to the following overall formula for calculation of the application rate of 49% alum solution in International System of Units:

$$A = (0.0118)(Xi - Xe)(Q)/[1 - 0.95 \times \exp(-1.9 \times Xe)] \tag{10.4}$$

Where

$A$ = 49% alum solution application rate (L/d),
$Xi$ = soluble phosphorus concentration at the application point (mg/L),
$Xe$ = target effluent soluble phosphorus concentration (mg/L), and
$Q$ = facility flow ($m^3$/d).

In U.S. customary units, the formula is as follows:

$$A = (11.8)(Xi - Xe)(Q)/[1 - 0.95 \times \exp(-1.9 \times Xe)] \tag{10.5}$$

Where

$A$ = 49% alum solution application rate (gpd),
$Xi$ = soluble phosphorus concentration at the application point (mg/L),
$Xe$ = target effluent soluble phosphorus concentration (mg/L), and
$Q$ = facility flow (mgd).

The aforementioned dose calculations are approximate. Many site-specific factors such as mixing conditions (Section 5.3), application point (Section 5.4), and wastewater chemistry and temperature will affect the actual dose of chemical required to accomplish treatment objectives. This is particularly true at residual soluble phosphorus concentrations below 0.1 mg/L.

The aforementioned calculations were based on a single application point; however, the optimum operating mode may involve dual application points (i.e., chemical addition to the primary clarifiers and to the secondary treatment process), which could result in overall savings in chemical consumption and cost (Section 5.4).

### 4.2.4 *Other Considerations for Aluminum-Based Chemicals*

A significant amount of sludge is produced when aluminum salts are added to the process to remove phosphorus. The quantity of sludge and handling considerations are discussed in Sections 6.1 and 6.2. Alkalinity and pH effects of aluminum sulfate addition are discussed in Section 6.3. Aluminum sulfate and, to a lesser extent, sodium aluminate and PACl will increase TDS in the system. These effects are discussed in more detail in Section 6.4.

Aluminum compounds (and ferric compounds, discussed later) are mildly acidic, and therefore, storage and handling issues are of concern. Fiber-glass-reinforced plastic or polyethylene tanks can be used to store any of the aluminum compounds. Recommended metering pumps include solenoid, peristaltic, and diaphragm types. Carrier water should be avoided, if possible, because it will result in a more neutral pH; in addition, aluminum hydroxide may precipitate, causing plating in the chemical feed lines. If it is necessary to add carrier water for mixing or dilution, then it should be added as close to the injection point as possible to minimize plating effects. The point of addition of chemical should be in an accessible location above the water level so the delivery rate can be verified with a "bucket and stopwatch" method. For mixing recommendations, refer to Section 5.3. Pump heads should be constructed of polyvinyl chloride (PVC). Piping, valves, and fittings should be PVC or chlorinated PVC. Where feasible, chemical storage and handling equipment should be designed to be compatible with alternative chemicals to provide flexibility when availability and prices change. Storage and delivery piping should be heat-traced, where necessary, to prevent crystallization (49% alum will start to crystallize at −1 °C [30 °F]).

Facility personnel should wear personal protection equipment (PPE) when handling chemicals. The PPE should include, but not be limited to, gloves, respirators, goggles, aprons, and face shields and should be worn when working or handling any aluminum salt solutions. The dangers and handling requirements of chemicals can vary significantly. Consequently, it is essential to remember that each chemical is unique and can interact differently with different materials, including PPE. Consult the safety data sheet provided by the manufacturer of the chemical for specific instructions on handling, storage, and required protective equipment. Consult a chemical safety expert or a professional who is familiar with the specific chemical you are working with if you are uncertain about the proper PPE.

## 4.3 Iron Salts

### 4.3.1 *Overview of Iron-Based Chemicals*

The most common iron compounds used for phosphorus precipitation from wastewater are trivalent ferric ($Fe^{3+}$) salts, chiefly ferric chloride ($FeCl_3$), and, sometimes, ferric sulfate [$Fe_2$ $(SO4)_3$]. Because ferric salts addition to preliminary treatment facilities or to primary clarifiers is sometimes used for odor control, ferric may be a chemical of choice for phosphorus removal as it could serve dual purposes at facilities with a need to control odor in the liquid train. Table 10.2 includes general information for the different chemicals mentioned in this section. The chemistry of phosphorus precipitation with ferric salts is similar to that of aluminum compounds. Bivalent ferrous ($Fe^{2+}$) salts, primarily in the form of

**TABLE 10.2** Properties of Iron-Based Chemicals Used for Phosphorus Precipitation

| Name | Chemical formula | Molecular weight | Iron metal contents, % by weight | Weight ratio of dry chemical for stoichiometric phosphorus precipitation (g of chemical/g of P) |
|---|---|---|---|---|
| Ferric chloride, dry | $FeCl_3$ | 162.5 | 34.5 | 5.24 |
| Ferric chloride, 37% solution* | $FeCl_3$ | 162.5 | 12.8 | 5.24 |
| Ferric sulfate, dry | $Fe_2(SO_4)_3$ | 400 | 28 | 6.45 |
| Ferrous chloride, dry | $FeCl_2$ | 127 | 79 | 6.14 |
| Ferrous sulfate, dry | $Fe(SO_4)_2$ | 152 | 37 | 7.36 |

*37% ferric chloride solution has a specific density of 1.36 kg/L (11.4 lb/gal), a dry $FeCl_3$ content of 0.504 kg/L (4.2 lb/gal), and an iron metal content of 0.173 kg/L (1.44 lb/gal).

spent pickle liquor from metal surface cleaning with acids, are sometimes used as an inexpensive alternative, when locally available. Because pickle liquor can contain metal contaminants, it is important to ensure that these contaminants do not have an adverse effect on the sludge generated or facility effluent quality. It is not clear if the use of spent pickle liquor will continue to be a common practice in wastewater treatment because alternative methods of metal surface cleaning have recently been developed.

Ferrous salts (such as those found in pickle liquor) react with phosphates and precipitate rather poorly (WEF, 2010); therefore, they should only be added to the aeration basin, where they will first be oxidized to ferric salts. It should be noted that some additional oxygen demand will be created in the aeration basin to oxidize ferrous ions to ferric ions. Ferrous ions should never be added to the final clarifier for phosphorus removal because excess or unreacted ferrous ions will carry over into the disinfection system to consume chlorine and form a precipitate, contributing to effluent TSS. Furthermore, if a UV disinfection system is used, iron added to the final clarifier may interfere with UV absorbance and foul the lamp sleeves, increasing the frequency of lamp cleaning (Section 6.7).

Effects of iron-based chemicals on sludge generation, alkalinity consumption, and TDS increase are discussed in Sections 6.1, 6.3, and 6.4, respectively. Ferric salts can be added at various WRRFs, similar to aluminum-based chemicals. Ferric ion combines with orthophosphate ions to form ferric phosphate, as follows:

$$Fe^{3+} + PO_4^{3-} \rightarrow FePO_4 \tag{10.6}$$

The reaction between ferrous ions and phosphate ions can be written as follows:

$$3\ Fe^{2+} + 2\ PO_4^{3-} \rightarrow Fe_3(PO_4)_2 \tag{10.7}$$

In reality, precipitation of phosphorus with ferric salts is a more complex process (as is the case with aluminum compounds) involving co-precipitation with ferric hydroxide and hydrated ferric oxides and surface adsorption. This makes stoichiometric calculations approximate; however, Equation 10.6 will be used as the basis of dose calculations for illustrative purposes.

Equation 10.6 indicates that it will take 1 mol of ferric ion to react with 1 mol of phosphate, and thus, a 1-to-1-mol ratio of iron to phosphorus is required for this reaction. Because the molecular weight of iron is 56 and the molecular weight of phosphorus is 31, the weight ratio of iron to phosphorus in Equation 10.6 is 1.8 to 1, as shown in the following calculation:

$$(56\ \text{g Fe/1 mol Fe})/(31\ \text{g P /1 mol P}) = 1.81/1 \text{ or } 1.8:1$$

With ferric chloride, there is one molecule of ferric per molecule of ferric chloride; therefore, the stoichiometric weight ratio of ferric chloride to phosphorus is

$$(162.5 \text{ g FeCl}_3)/31 \text{ g P} = 5.24:1.$$

Similarly calculated chemical stoichiometric weight ratios for other ferric chemicals are provided in Table 10.2.

Bench-, pilot-, and full-scale studies have shown that considerably higher stoichiometric quantities of chemicals are typically necessary to meet phosphorus-removal objectives as a result of competing hydroxide and sulfide reactions. If iron salts are added to the primary clarifier, any sulfides present will first react with the ferric or ferrous ion. Consequently, if a facility influent has significant levels of sulfide, then the iron dosing must be higher for the same amount of phosphorus removed in the primary clarifier.

The excess chemical required increases as the target residual phosphorus concentration decreases. The excess ferric salt will precipitate as ferric hydroxide, with a simplified reaction as follows:

$$Fe^{3+} + 3\ OH^- \rightarrow Fe(OH)_3 \qquad (10.8)$$

Therefore, sludge that is generated at the point of ferric application will include a mixture of ferric hydroxide and ferric phosphate generated according to Equations 10.6 and 10.8. Equation 10.8 also illustrates the consumption of alkalinity (pH reduction) encountered when applying ferric. These effects are discussed in more detail in Section 6.3.

### 4.3.2 Determination of Dose for Iron-Based Chemicals

An approximate dose of ferric salt required for the particular application can be determined based on calculations presented in this section, including Equations 10.10 and 10.11 presented in Section 4.3.3). Refinement of the chemical application rate can then be performed in full scale based on the results obtained because the system response is fairly quick (Section 5.6).

For comments on performing jar tests to determine chemical dose, refer to the beginning of Section 4.2.2 and to Section 4.6. A design dose curve as illustrated in Figure 10.3 for chemical phosphorus removal using ferric ions was developed using literature and pilot-facility data (including but not limited from Blue Plains Wastewater Treatment Plants, Washington, D.C.). Literature data include laboratory-scale test data and data from sites that are operating with chemical phosphorus removal (Luedecke et al., 1988). The dataset includes results for a range of pH values (mostly 6.5 to 7.5), temperatures, and wastewater characteristics. The effluent soluble phosphorus concentration is labeled "residual soluble P" on the logarithmic x-axis. The molar ratio for metal ion dose to soluble phosphorus removed is labeled "$M_{dose}$/soluble $P_{removed}$ (mol/mol)" on the y-axis. It is important to note that the curves apply to the soluble portion of the phosphorus only. The curve used to fit the data is based on the following equation:

$$y = a/(1 + b \times e^{-cXe}) \qquad (10.9)$$

Where

$y$ = mole iron required per mole soluble phosphate removed,
$a$ = 1.48,
$b$ = −1.07,
$c$ = 2.25, and
$Xe$ = target effluent soluble phosphorus concentration at the chemical application point (mg/L).

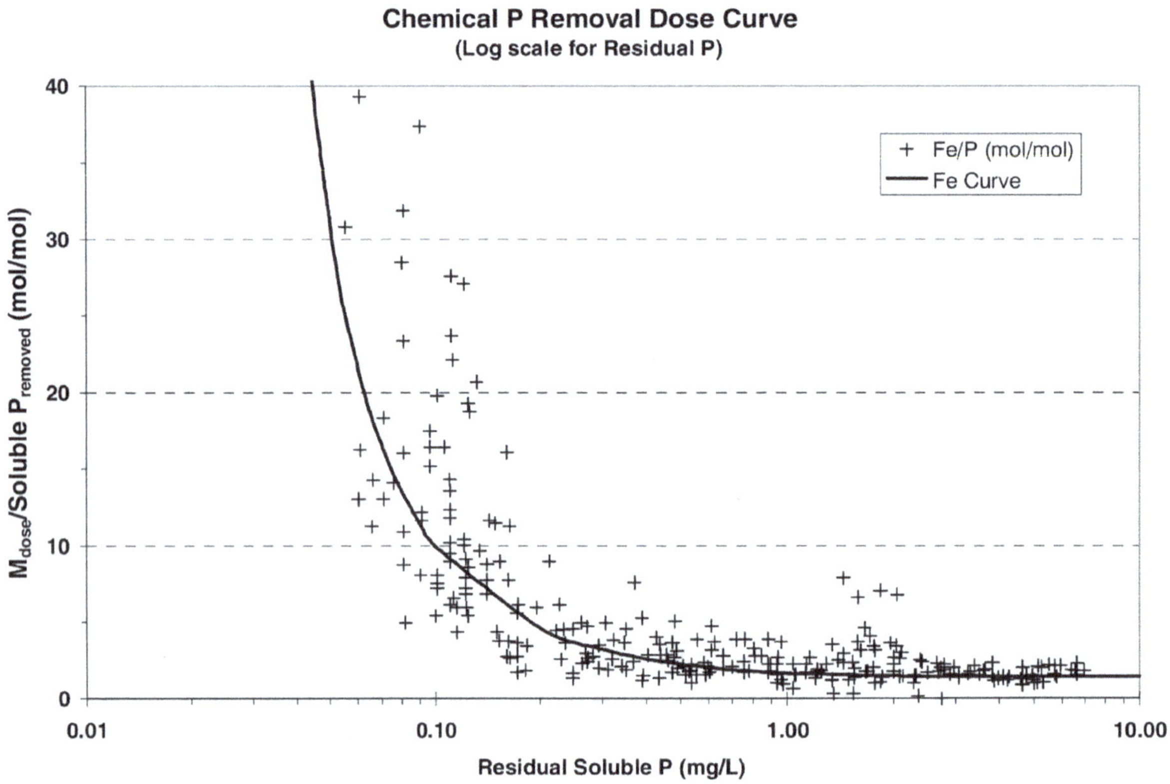


FIGURE 10.3 Ratio of Iron ($Fe^{3+}$) Dose to Phosphorus Removed as a Function of Residual Soluble Orthophosphate Concentration

In general, the data at lower residual soluble phosphorus concentrations are more scattered. Because of this variability, bench- or pilot-scale testing is recommended at each facility to determine the actual molar dose required to reach the targeted effluent soluble phosphorus concentration.

The molar dose for phosphorus precipitation is based on the desired final effluent soluble phosphorus concentration rather than the starting phosphorus concentration. For example, to meet a 0.5-mg/L soluble phosphorus concentration, a 2.7-mol ratio of ferric ion to phosphorus or a weight ratio of 4.1 g $Fe^{3+}$/g P is required. To remove 2.5 mg/L P (from 3 to 0.5 mg/L), an iron dose of 10.25 mg/L $Fe^{3+}$ is required. It is important to determine if a value greater than the influent soluble phosphorus concentration should be used because of the potential for solubilization of the particulate phosphorus, which would increase the soluble phosphorus concentration above measured influent concentrations.

As noted previously, using dual application points may yield the optimum operating point with respect to chemical dose and sludge production. Using the same concentrations as those previously cited, if the phosphorus concentration were to be reduced to 1 mg/L in the primary clarifier, with additional iron added to the aeration basin to achieve a final effluent of 0.5 mg/L soluble phosphorus, less iron will be used in total. To achieve 1 mg/L of soluble phosphorus out of the primary clarifier requires a molar ratio of 1.67 to 1 or a weight ratio of 3 g $Fe^{3+}$/g P. Therefore, to remove 2 mg/L of phosphorus across the primary clarifier, an iron dose of 6 mg/L is required. To remove the remaining 0.5 mg/L of soluble phosphorus across the secondary treatment system requires a molar ratio of 2.27 to 1 or a weight ratio of 4.1 g $Fe^{3+}$/g P, which equates to an iron dose of 2.05 mg/L $Fe^{3+}$. The total iron dose to meet a residual soluble phosphorus concentration of 0.5 mg/L is, therefore, $6 + 2.05 = 8.05$ mg/L $Fe^{3+}$, as opposed to 10.25 mg/L $Fe^{3+}$ if all of the phosphorus is removed at one time. There is a 20% savings in chemical use and a reduction in the overall chemical sludge production. It should be noted that this example neither has taken any credit for phosphorus that would be removed biologically across the secondary treatment system nor has it accounted for potential solubilization of particulate phosphorus. Actual dosages must be fine-tuned in the field to account for these issues. A similar relationship exists when aluminum is used for phosphorus precipitation.

A sample calculation is provided in this section to determine the ferric chloride dose required to precipitate soluble phosphorus. Following the detailed, step-by-step derivation, simplified formulas for both International System of Units and U.S. customary units are provided. The following conditions are assumed:

- Influent facility flow rate is 37 850 $m^3/d$ (10 mgd),
- Soluble phosphorus influent concentration to the facility is 3.0 mg/L, and
- A residual primary effluent phosphorus concentration of 1 mg/L is required.

The amount of soluble phosphorus to be removed in kilograms per day (pounds per day) is as follows:

$$\text{Phosphorus} = (3 - 1\ \text{mg/L}) \times 37\,850\ \text{m}^3\text{/d} \times 1000\ \text{L/m}^3 \times 1\ \text{kg}/1\,000\,000\ \text{mg} = 75.7\ \text{kg/d}$$

or

$$(3 - 1)\ \text{mg/L} \times 10\ \text{mgd} \times 8.34\ (\text{lb/mg} \times \text{L/mg}) = 167\ \text{lb/d of soluble phosphorus to be removed.}$$

This is equivalent to $(75.7\ \text{kg/d})/(31\ \text{kg/kg-mol P}) = 2.44\ \text{kg-mol P}$

or

$(167\ \text{lb/d})/(31\ \text{lb/lb-mol P}) = 5.38\ \text{lb-mol of P to be removed.}$

The amount of ferric ions required can be calculated from Equation 10.9 as follows:

$$y = 1.48/(1 - 1.07e(-2.25)(1)) = 1.67\ \text{kg-mol Fe/kg-mol}$$

or

$$(1.67\ \text{lb-mol Fe/lb mol})\ \text{of soluble phosphorus removed.}$$

Consequently, the amount of ferric ions required per day is

$$2.44\ \text{kg-mol P} \times 1.67\ \text{kg-mol Fe/kg-mol P} = 4.07\ \text{kg-mol Fe}$$

or

$$5.38\ \text{lb-mol P} \times 1.67\ \text{lb-mol Fe/lb-mol P} = 8.99\ \text{lb-mol Fe.}$$

The molecular weight of the Fe ion is 56, which can be converted to mass of ferric ions per day, as follows:

$$4.07\ \text{kg-mol Fe} \times 56\ \text{kg Fe/kg-mol Fe} = 228\ \text{kg}$$

or

$$8.99 \text{ lb-mol Fe} \times 56 \text{ lb Fe/lb-mol Fe} = 503 \text{ lb/d of ferric ions.}$$

As indicated in Table 10.2, dry ferric chloride contains 34.5% ferric. Consequently, the amount of dry ferric required for this application will be

$$228 \text{ kg Fe}/0.345 = 661 \text{ kg,}$$

or

$$503 \text{ lb Fe}/0.345 = 1459 \text{ lb/d of dry ferric chloride.}$$

If 37% ferric chloride solution is used, it contains 0.173 kg/L of ferric ions (Table 10.2). Thus, the amount of ferric solution required for this application will be

$$228 \text{ kg Fe}/0.173 \text{ kg/L} = 1318 \text{ L}$$

or

$$503 \text{ lb Fe}/1.44 = 349 \text{ gpd of solution.}$$

The 37% ferric chloride solution required per day to reduce the soluble phosphorus from 3.0 to 1.0 mg/L in the aforementioned example is 1318 L/d (349 gpd). The ferric dose equivalent to this ferric chloride use rate can be calculated from the factor listed in Table 10.2 of 0.504 kg/L (4.2 lb/gal) of ferric chloride in 37% solution, as follows:

$$1318 \text{ L/d} \times 0.504 \text{ kg/L}/37\,850 \text{ m}^3\text{/d} = 0.0176 \text{ kg/m}^3$$

$$0.0175 \text{ kg/m}^3 \times 1\,000\,000 \text{ mg/kg}/1000 \text{ L/m}^3 = 17.6 \text{ mg/L}$$

In U.S. customary units, calculation for this 10-mgd facility will be as follows:

$$349 \text{ gpd} \times 4.2 \text{ lb/gal}/8.34 \text{ (lb/mg} \times \text{L/mgd)}/10 \text{ mgd} = 17.6 \text{ mg/L}$$

### 4.3.3 Summary Dose Formulas for Ferric Chloride

The aforementioned calculations can be abbreviated to the following overall formula for calculation of the application rate of 37% ferric chloride solution in International System of Units:

$$A = (0.0155)(Xi - Xe)(Q)/[1 - 1.07 \times \exp(-2.25 \times Xe)] \tag{10.10}$$

Where

$A$ = 37% ferric chloride solution application rate (L/d),
$Xi$ = soluble phosphorus concentration at the application point (mg/L),
$Xe$ = target effluent soluble phosphorus concentration (mg/L), and
$Q$ = facility flow ($m^3$/d).

In U.S. customary units, the formula is as follows:

$$A = (15.5)(Xi - Xe)(Q)/[1 - 1.07 \times \exp(-2.25 \times Xe)] \quad (10.11)$$

Where

$A$ = 37% ferric solution application rate (gpd),
$Xi$ = soluble phosphorus concentration at the application point (mg/L),
$Xe$ = target effluent soluble phosphorus concentration (mg/L), and
$Q$ = facility flow (mgd).

It should be recognized that the aforementioned dose calculations are approximate. Many site-specific factors such as mixing conditions (Section 5.3), application point (Section 5.4), wastewater chemistry, and temperature, discussed in more detail later in this chapter, will affect the actual dose of chemical required to accomplish treatment objectives. This is particularly true at low residual soluble phosphorus concentrations (below 0.1 mg/L).

The aforementioned calculations were based on a single application point; however, the optimum operating mode may involve dual application points (i.e., chemical addition to the primary clarifiers and to the secondary treatment process), which could result in savings in chemical consumption and cost (Section 5.4).

### *4.3.4 Other Considerations for Iron-Based Chemicals*

As with aluminum-based compounds, there will be a significant amount of sludge produced when iron salts are added to the process to remove phosphorus. The quantity of sludge and handling considerations are discussed in Sections 6.1 and 6.2, respectively. Alkalinity and pH effects of ferric salts are discussed in Section 6.3. Metal salts will also increase TDS in the system. These effects are discussed in more detail in Section 6.4. A potentially significant drawback of using ferric is its potential effect on UV disinfection, as discussed in Section 6.7. Tank and pipe material, pump type, and safety considerations are all similar to the aluminum-based options discussed in Section 4.2.4.

## 4.4 Lime

Lime has historically been used to increase alkalinity, remove phosphorus, and improve removal efficiencies across primary clarifiers. More recently, the use of lime at WRRFs for phosphorus removal has become less common, presumably because of lime-handling difficulties and the availability of more effective chemicals.

Because high pH ($>10$) is needed for phosphorus precipitation with lime, lime addition for the purpose of phosphorus removal is not compatible with biological treatment as the required pH is too high for microorganisms. However, lime-treated primary effluent may potentially be accepted by the subsequent biological treatment step without neutralization (or with minimal neutralization) because carbon dioxide generated during aerobic treatment will lower the pH. Nitrification, if practiced, could consume alkalinity (i.e., lower the pH) as well.

In addition to the aforementioned use in primary clarifiers, lime addition for phosphorus removal may potentially be used in sidestream treatment (such as in the Phostrip process) or as a post-treatment step with a subsequent pH adjustment. However, the use of lime for phosphorus removal is rare. The residual phosphorus concentration typically achieved with lime treatment is approximately 1 mg/L. The dose of lime required for phosphorus removal is typically governed by the alkalinity of the wastewater because lime will first react with bicarbonate before precipitating as hydroxyapatite. That dose is approximately 1.5 times that of total alkalinity in milligrams per liter as $CaCO_3$ (Sedlak, 1991). The hydroxyapatite has a variable composition; however, an approximate equation for its formation can be written as follows, assuming, in this instance, that the phosphate present is the hydrogen phosphate ion ($HPO_4^{2-}$):

$$3\ HPO_4^{2-} + 5\ Ca^{2+} + 4\ OH^- \rightarrow Ca_5\ (OH)(PO_4)_3 + 3\ H_2O \qquad (10.12)$$

The use of lime for phosphorus removal results in generation of inert solids, the quantity of which is governed mainly by wastewater alkalinity. The following two issues should also be considered when lime is added to the treatment process: (1) relatively high calcium concentrations in the process water can inhibit VSS destruction in digesters, and (2) high calcium content in the final sludge may not be advantageous to certain soils if the sludge is ultimately used for soil amendments.

Lime is either gravity fed or pumped to the point of application. Materials of construction for lime systems are carbon steel or PVC. Personnel should wear PPE when handling chemicals; the PPE should include, but not be limited to, gloves, respirators, goggles, aprons, and face shields and should be worn when working or handling chemical solids or slurries.

## 4.5 Magnesium Hydroxide

Another alternative chemical available to precipitate phosphorus is magnesium hydroxide, although its use in the main liquid train is presently limited because of its high cost. Magnesium hydroxide raises the pH to precipitate phosphorus and, therefore, yields similar results as lime addition, although chemical handling issues are not as significant as those for lime.

In anaerobic digesters, magnesium in the presence of ammonia and orthophosphate promotes formation of magnesium ammonium phosphate (struvite), which is known to cause severe clogging of digester piping. Magnesium addition is used in dedicated treatment processes recently developed for removal and recovery of phosphorus and ammonia from sludge recycle streams (e.g., the Pearl® process by Ostara). This is discussed in more detail in Chapter 12.

Magnesium hydroxide is received in liquid form and can be used similarly to ferric or aluminum addition. Tank materials are typically constructed of fiberglass reinforced plastic (FRP) or polyethylene. Piping, valves, and fittings are typically constructed of PVC or other compatible plastic material.

## 4.6 Proprietary Formulations

A number of proprietary formulations for phosphorus removal have been marketed in recent years. Whereas the exact composition of these products is typically not known, they are likely based on a combination of some of the chemicals previously discussed in this chapter. The products variously claim to be easier to handle, generate less sludge, be less sensitive to lower wastewater temperatures, and/or have a lesser effect on facility pH. Whereas claims of lower sludge generation should be treated with skepticism, the other claims may be valid. This is attributable to the fact that these formulations typically contain some form of neutralizing alkali agent, which increases pH and alkalinity and makes the chemicals less corrosive, as in the case of sodium aluminate and PACl discussed earlier in this chapter.

However, this convenience typically comes at a higher chemical cost. The recommended method of evaluating the expected benefits of using alternative chemicals is to determine the dose of each chemical needed to achieve the required phosphorus removal in side-by-side jar tests. Such jar tests should include evaluation of the required dose of a neutralizing agent. The dose requirements established in the comparative jar tests can then be combined with unit costs of alternative chemicals to determine overall application costs. Only then can the relative costs of using different chemicals be compared to other factors, such as ease of handling and need for any additional neutralizing chemicals, and an informed decision be made. For small facilities, the use of specialized or proprietary formulations may sometimes be justified; however, at larger facilities, chemical costs will likely be an overriding consideration.

## 4.7 Water Treatment Sludge

Water treatment residuals typically contain a large fraction of aluminum or ferric salts, which may have residual capacity to bound and adsorb phosphorus. Such sludges can be used as cost-effective materials to reduce soluble phosphorus in wastewater. Although fresh alum was found to be more efficient at phosphorus removal than spent alum sludge (Georgantas et al., 2006), alum sludge may be a reasonable substitute for alum because of its low cost and high availability. It should also be noted that the efficiency of alum sludge to remove phosphorus decreases through aging. More than 90% phosphorus removal from wastewater has been observed at sufficiently high spent-water-treatment sludge application doses (Asada et al., 2010; Ippolito et al., 2011; Mortula & Gagnon, 2007). Naturally, the use of spent water treatment facility sludge for wastewater treatment transfers the burden of water treatment residual handling to the wastewater facility and could result in a large increase of waste sludge generated at the facility.

## 4.8 Role of Polymers

Organic polymers do not remove or precipitate soluble phosphorus on their own to any significant extent. However, their use can significantly improve the solids separation process by coagulating colloids and increasing the size and compactness of the chemical and biological solids, thus increasing their settling velocity.

The use of metal salts for phosphorus removal can create, at least initially, fine precipitate (floc) or even colloids. Flocculation and settling of such particulates can be greatly improved by the application of a polymer. The use of polymer can be particularly effective when low-effluent phosphorus levels are required. In some high-rate applications used for phosphorus removal, such as several patented sand-ballasted flocculation and sedimentation processes, the use of polymer at a proper dose is critical to process performance.

When used, the polymer addition point should be as far downstream from the point of addition of the metal salts as practical and should be located in a place where adequate mixing is available or can be created. In many typical configurations of the activated sludge process, metal salts are added at the outlet of the aeration basin, with polymer addition at the final clarifier splitter box. Adequate mixing, which is critical to polymer effectiveness, can be provided by simple air agitation.

When added to mixed liquor, the polymer dose required is typically in the range of 0.5 to 1 mg/L for dry polymers, which typically have approximately 90% active content. For emulsion polymers, with only 30% to 40% active content, the required dose will be about 2.5 to 3 times higher (as expressed in the weight of the emulsion, as supplied).

It should be noted that, at some activated sludge facilities, polymer addition is continually added to final clarifiers to improve their performance. Other facilities use polymer addition to final clarifiers in a standby mode during storm flows.

When metal salts are added to the primary clarifier (CEPT), polymer addition can significantly increase phosphorus, TSS, and particulate BOD removal compared to that obtained by coagulant alone. The optimal polymer dose in CEPT treatment, or when applied in a tertiary phosphorus removal process, will typically be lower for a mixed liquor application than mentioned previously.

Typical polymer systems require stainless steel or FRP storage or aging tanks. Polyvinyl chloride piping, valves, and fittings are typically used for polymer service. Because polymers are sensitive to shear and have higher viscosity than most metal salts, progressive cavity pumps are often recommended for polymer service.

Polymer does not have the chemical handling issues associated with most other chemicals. Generally, slips and falls are the most common hazard when handling a polymer solution. Care should be taken when changing the type of the polymer from anionic to cationic based (or vice versa) because a thorough cleaning of polymer storing and handling equipment with solvents may be needed to prevent formation of scale-type deposits.

# 5.0 PROCESS CONSIDERATIONS

## 5.1 Limits of Technology

The minimum achievable effluent phosphorus concentration depends on a number of factors, including efficiency of the solids separation process, as discussed in Section 3.3. In general, soluble orthophosphate concentrations in wastewater can be reduced to below 0.01 mg/L with a sufficiently high chemical dose (Smith et al., 2008; U.S. Environmental Protection Agency [U.S. EPA], 2007). Dual-step filtration processes are particularly suited to achieving low concentrations because they provide extended contact time for precipitation, coagulation, and separation of fine particulates initially generated after chemical addition. Microfiltration or ultrafiltration (e.g., in an MBR) while achieving a reliable separation of suspended matter may not be as effective in lowering soluble orthophosphate concentration to the limits of solubility because contact time in the filtration step is limited (WEF, 2010). In some effluents, a measurable concentration (as much as 0.05 mg/L) of non-reactive rDOP may be present (refer to Chapter 11).

In practice, taking into account process and wastewater chemistry variability, the potential presence of measurable rDOP, the effect of any residual particulate phosphorus, and any imperfections in the solids separation step, a monthly average effluent total phosphorus concentration below 0.05 mg/L is generally achievable at a well-designed and -operated full-scale facility. In a survey of advanced WRRFs (U.S. EPA, 2007), the range of monthly average phosphorus concentrations for facilities having a phosphorus limit of 0.1 mg/L or less typically was approximately 0.03 to 0.09 mg/L.

It should be noted that analyzing residual phosphorus at low levels by the standard colorimetrical method is uncertain. Concentration readings from a spectrophotometer calibration curve should be verified manually (as opposed to relying on a regression equation), with calibration data developed at the low end of the concentration range.

## 5.2 Effect of pH

The minimum solubility of ferric and aluminum orthophosphates in wastewater is in the acidic pH range of 3 to 5 (WEF, 2010), with the exact optimum pH likely being site specific depending on a number of factors, including wastewater chemistry. At a pH greater than 7, the residual dissolved orthophosphate concentration starts to increase for the same chemical dose applied. Considering that most facilities must maintain their effluent pH above 6, the optimal pH for minimizing residual soluble phosphorus concentration will be approximately 6.5 to maintain a safety margin. However, maintaining a particular target pH may not be practical or necessary because low effluent phosphorus concentrations were achieved at facilities operating at a pH as high as 7.5 (Takacs et al., 2006).

The target pH of the precipitation reaction, as discussed in this section, should be measured at the point of the precipitate formation, that is, downstream of the chemical application point or at the solids separation point (clarifier effluent). The final effluent pH or aeration basin pH could be somewhat different.

An important consideration is that some commonly used chemicals, including alum and all iron salts, are acidic and will lower the wastewater pH to a degree depending on the alkalinity of the wastewater and the dose of chemical. Typically, access metal salts applied can form acid that can attack and corrode metal components in the wastewater treatment, such as pipes, valves, and gates. To prevent damage to metal components in the wastewater, it is essential to monitor the water's pH and control the dosage of metal salts. In addition, the selection of corrosion-resistant materials for the construction of metal system components can aid in mitigating the harmful effects of ferric chloride. For larger chemical doses and/or for wastewater with low alkalinity, the use of neutralizing agents such as caustic may be necessary. In some instances, the use of ferric or alum may actually lower the pH of the wastewater to around 6.5, at which point the use of chemicals for phosphorus removal is likely to be optimal.

## 5.3 Effect of Mixing and Contact Time

Orthophosphate removal with metal salts significantly improves if vigorous mixing at the point of chemical addition is provided. The benefits of mixing seem to be greater at a higher pH and are more pronounced for alum than for ferric (Sagberg et al., 2006; Smith et al., 2008). Extending the contact time between the precipitated matter and wastewater allows for additional adsorption of soluble orthophosphates and increases the removal efficiency. Consequently, the recommended chemical addition location should be at a turbulent point as far upstream from the solids separation point (i.e., clarifier or filter) as practical. A convenient location for coagulant addition to an activated sludge train is to a turbulent area at the tail end of the aeration basin (e.g., effluent weir).

When chemical addition to the primary clarifier is contemplated, a suitable addition point may be a turbulent location at a preliminary treatment facility, such as at the screens, or to grit tanks (if available), with conveyance and flow distribution structures upstream of the primary clarifiers providing a good flocculating environment. Vigorous mixing desired at the point of chemical addition could also be enhanced by an additional mixing device, such as an air sparger. G-value (velocity gradient) can be used to quantify the mixing intensity, which is typically between 20 and 200 s-1 (seconds inverse). However, these values should serve as a starting point, and a site-specific bench or pilot study should be conducted to corroborate the optimal mixing conditions for coagulation.

## 5.4 Points of Addition

Metal salts addition points should be upstream of solids separation steps such as clarification of filtration. Figure 10.4 depicts typical available options. The use of chemicals is most efficient at a tertiary application point, where most primary or biological solids or other constituents that can bind with metal salts are already removed. However, most existing facilities are not equipped with a suitable tertiary treatment process.

Chemicals can be added in front of conventional sand filters; however, this decreases the filter run time (the higher the dose, the shorter the filter run), and the acceptable chemical dose in such an application can be too small for adequate phosphorus removal. In such instances, split-point addition may be most effective, with part of the chemical going to secondary clarifier influent (aeration basin effluent) and part going to secondary clarifier effluent (filter influent).

At conventional activated sludge facilities, and particularly at smaller, extended aeration facilities, common practice is to add chemicals to a turbulent location in front of final clarifiers such as the effluent weir of the aeration basin or to a clarifier's distribution box.

Application of chemicals for phosphorus removal to the primary clarifier (CEPT) is effective, particularly when biological processes could benefit from lower loading of organics, nitrogen, and TSS. Such practice also minimizes the inert fraction of the mixed liquor and, in case of ferric addition,

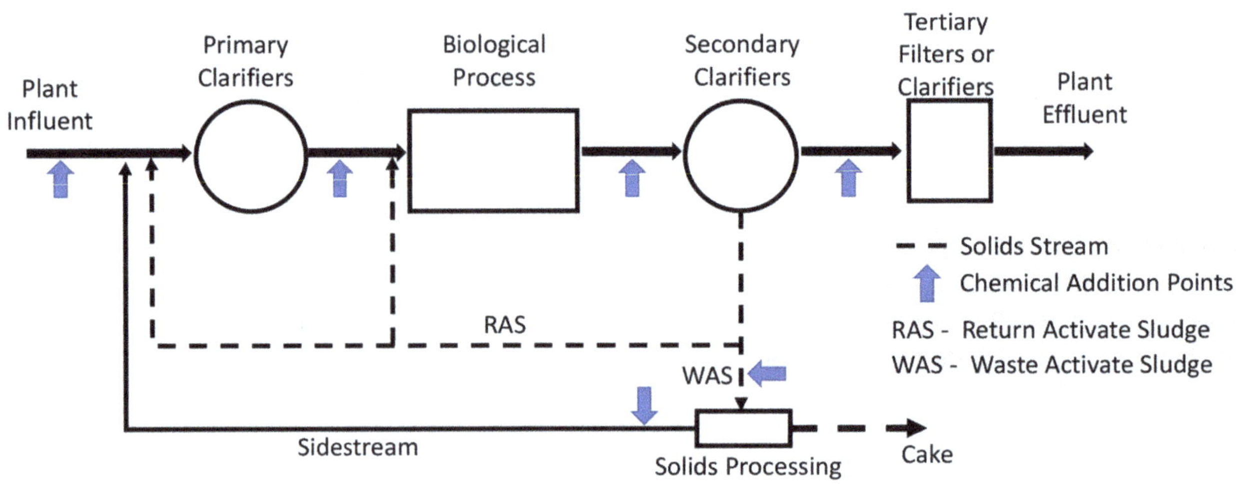


FIGURE 10.4 Chemical Phosphorus Removal Dosing Locations (Courtesy Raj Chavan, AtkinsRealis)

minimizes the potential effects of chemicals on UV disinfection. Selection of the chemical application point at existing facilities may be limited by the need for good mixing, as previously discussed, which may not be easy to achieve at all potential locations.

A potentially significant reduction in chemical use could be accomplished by multipoint (or split-point) chemical addition. A commensurable reduction in waste sludge generation and the need for neutralizing chemicals and other process effects can also be realized. Multipoint addition is typically achieved by splitting chemical addition between primary clarifier and aeration basin effluent or between aeration basin effluent and final clarifier effluent (i.e., filter influent).

When phosphorus is removed by chemical addition to tertiary facilities (tertiary clarifiers and/or polishing filters), recycling the resulting chemical sludge to the primary treatment process should be considered, if feasible. Such practice will facilitate the use of the residual phosphate bonding capacity of tertiary sludge and could substantially reduce the overall use of chemicals (Takacs et al., 2006).

## 5.5 Chemical Feed Control

Many smaller and medium-sized facilities adding chemicals to the aeration basin operate at a constant chemical flowrate that is manually adjusted periodically, as needed. Diurnal variations in influent phosphorus loading are partially absorbed by large quantities of partially active chemical precipitate present in the aeration basin. Such facilities add chemicals in excess of what is required to meet the permit limit.

Because of the increasing availability and reliability of online probes measuring residual orthophosphate, real-time dose control is becoming feasible and cost effective, even at smaller facilities. A less efficient but simpler option would be to flow-pace the chemical addition with periodic adjustment of the feed ratio based on observed results. Implementation of phosphorus removal process monitoring and dose control is recommended because it lowers chemical costs and mitigates other negative aspects of chemical addition. Chapter 14 provides more details on online analyzers and dose control.

## 5.6 Response Time and Startup of Chemical Addition

Phosphorus removal by chemical addition can be initiated on demand without the prolonged acclimation period required for some biological processes, such as EBPR. This is typically the case for chemical processes. Some dedicated tertiary processes with a minimal hydraulic retention time (HRT) can lower the effluent phosphorus concentration within minutes of startup. For more conventional applications, the response is more measured and predictable because it depends on HRT in downstream treatment facilities.

Figure 10.5 illustrates results from a test performed at an oxidation-ditch facility without primary clarifiers or tertiary filters (Patoczka, 2009). The alum dose required for achieving the desired effluent total phosphorus level (below 0.5 mg/L) was determined before initiating the test; however, no chemical was added for more than 3 weeks before initiation of the test. As Figure 10.5 shows, the clarifier effluent total phosphorus, originally at 3.7 mg/L, started to decrease almost immediately upon activation of alum addition and dropped below 1 mg/L within 24 hours. In 3 days, the concentration stabilized at 0.35 mg/L.

The delay in system response to initiation of chemical addition to activated sludge is likely attributable to a combination of the followings factors:

- Travel time between the chemical application point and the effluent sampling location: If only a final clarifier is present, this delay will be marginal. Any tertiary facilities, chlorine contact tanks, and, particularly, a stabilization lagoon, will add delay commeasurable with HRT of these facilities.
- As discussed in Section 3.2, the excess, unused chemicals applied at the secondary clarifier influent will be returned with RAS to the aeration basin and will bind with phosphorus in the raw wastewater. The time required to reach an accumulation of these chemicals close to a saturation point (steady state) corresponds to several multiples (3 is a good number) of HRT. For an aeration basin with a nominal HRT of 1 day, the time to approach equilibrium conditions will thus be approximately 3 days.
- Metal salts are coagulating agents and will be partially used by biomass and colloids in the aeration basin for coagulation of mixed liquor flocs until a degree of saturation of the biomass inventory is

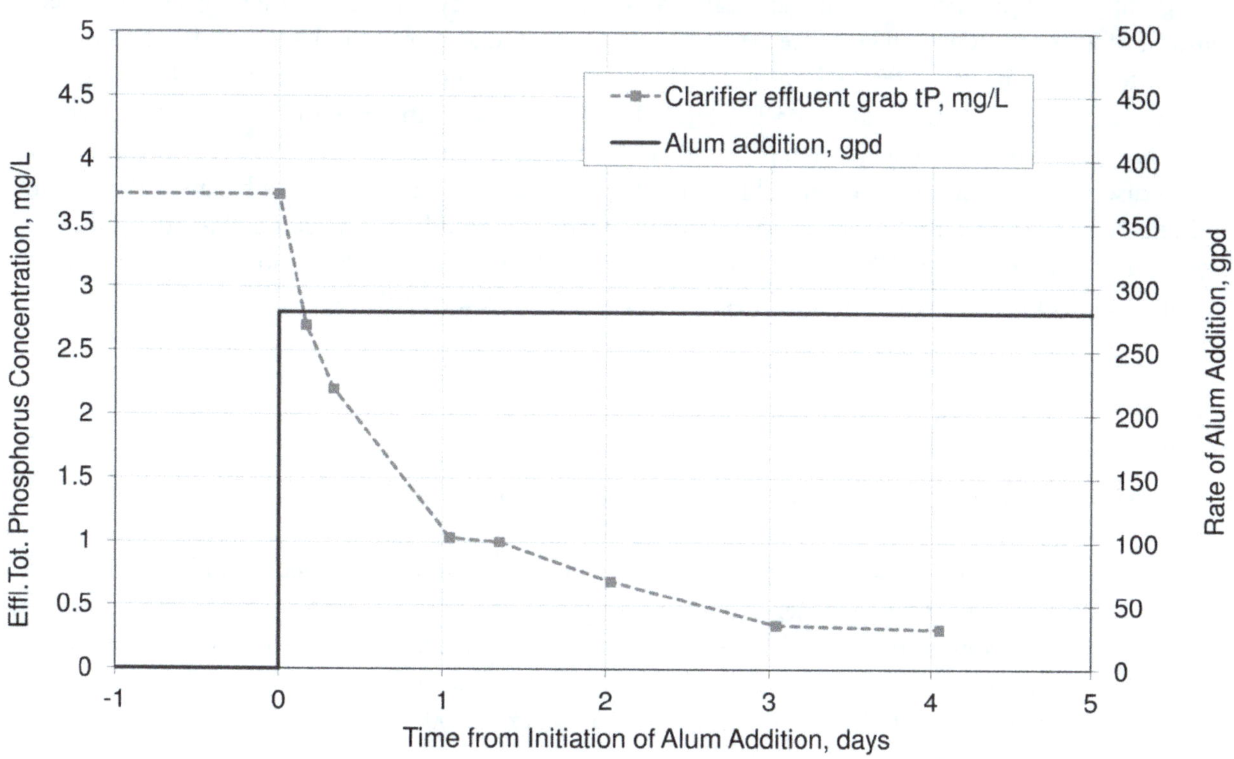


FIGURE 10.5 Response Time to Initiation of Chemical Addition

achieved and the mixed liquor at the point of the chemical application is presaturated with chemicals and pre-coagulated.

- Precipitates resulting from metal salt addition continue to adsorb orthophosphate for several hours or even days, delaying the time to achieve equilibrium.

## 6.0 EFFECT OF CHEMICAL ADDITION ON FACILITY OPERATIONS

### 6.1 Sludge Generation

The addition of chemicals for precipitation of phosphorus will generate additional inert chemical sludge at the point of addition, be it in a primary clarifier, activated sludge system, or tertiary application. The amount of additional sludge (on a dry-weight basis) generated by the addition of selected chemicals can be calculated based on the conversion factor 0.411 (Patoczka, 2006). For example, for a 3785-ML/d (1-mgd) facility adding 10 mg/L of alum, the expected additional sludge generation will be 4.11 mg/L, or 15.6 kg/d (33.9 lb/d), based on a 0.411-conversion factor. In many applications, particularly when levels of phosphorus below 1 mg/L are targeted with chemical addition, an excess of chemical over that required by the stoichiometric equation is needed. Because the excess chemical will precipitate in a different form, a different conversion factor will result. For example, the conversion factor for excess alum without neuralization is 0.263 and typical alum application for phosphorus removal with and without neutralization is 0.312 (Patoczka, 2006). As discussed in Section 4.0, the stoichiometric amount of chemical for the removal of 1 mg/L of phosphorus is 9.6 mg/L for alum and 5.2 mg/L for ferric chloride.

Conversion factors mentioned above are approximate because simplifying assumptions of precipitate composition were used in their derivation. The addition of coagulating chemicals will increase the capture of colloidal and particulate matter, which may further increase additional sludge generation. This is particularly applicable when chemicals are added to primary clarifiers (CEPT), where enhanced removal of TSS and BOD will generate even more primary sludge. However, decreased TSS and organic loading to the secondary system will result in a lower waste biological sludge generation. Typical primary clarifier removal rates of 50% for TSS and 30% for $BOD_5$ may increase to 70% and 50%, respectively, or even more, particularly when polymer addition is also provided.

As discussed in Section 3.2, chemical precipitates accumulating in the aeration basins will result in an increase in the basin inert solids concentration. This effect should be considered when designing a biological treatment system. The equilibrium concentration of inert chemical residue resulting from chemical addition to the activated sludge system can be calculated as follows:

$$\text{MLSSci} = D \times F \times \text{SRT/HRT} \tag{10.13}$$

Where

MLSSci = concentration of chemical inerts in the mixed liquor arising from chemical addition, mg/L,

D = chemical dose applied to the activated sludge process, mg/L (based on the nominal, forward flow of wastewater),

$F$ = TSS conversion factor for the chemical applied (conversion factor for alum for stoichiometric P precipitation and excess alum without neutralization is 0.165 & 0.485, respectively), for detailed list of values refer to Patoczka, 2006,

SRT = solids retention time, days, and

HRT = hydraulic retention time, days.

The following conditions are assumed for the sample calculations described herein:

- Flowrate, 3785 $m^3$/d (1 mgd);
- Aeration basin volume, 1892 $m^3$ (500 000 gal);
- Sludge age of the process, 12 days; and
- Ferric (37% solution) application rate, 300 L/d (79.3 gpd).

The HRT of this activated sludge system is equal to the aeration basin volume divided by flow, that is, 1892 $m^3$/3785 $m^3$/d = 0.5 days (or 12 hours). The ferric solution application rate is equal to the ferric chloride dose of 40 mg/L (refer to Section 4.3.2 for a calculation example). With 3-mg/L phosphorus concentration in the influent, this represents a stoichiometric ratio of iron to phosphorus of approximately 3 to 1 (again, refer to Sections 4.2.2 and 4.3.2 for exemplary calculations). The TSS increase conversion factor (*F*) for $FeCl_3$ applied at a 3-to-1 ratio is 0.748 (Patoczka, 2006). Consequently, the inert MLSS concentration is

$$40 \text{ mg/L} \times 0.748 \times 12 \text{ d}/0.5 \text{ d} = 718 \text{ mg/L}$$

The MLSSci calculated from Equation 10.13 only accounts for the component of the inert fraction of the mixed liquor resulting from chemical dosing in addition, and above, the typically expected inert MLSS fraction results from inert TSS present in the influent or generated during the treatment.

Equation 10.13 can also be used to calculate the amount of additional waste sludge generated by chemical addition. If the facility operated without ferric addition at 3000 mg/L MLSS, the equilibrium MLSS concentration with ferric addition will be 3718 mg/L for the conditions in the aforementioned example. This corresponds to a 24% (3718 mg/L ÷ 3000 mg/L) increase in the dry mass of waste activated sludge generation.

An even more straightforward way of calculating the additional mass of dry-waste inert chemical sludge that is generated is to multiply the chemical use rate expressed in kilograms per day (pounds per day) (see Tables 10.1 and 10.2 for chemical content of typical technical solutions) by the applicable TSS conversion factor mentioned above or refer to Patoczka (2006) for detailed list. For example, 49% alum solution has 0.647 kg/L (5.4 lb/gal) of alum (Table 10.1). If a facility is using 200 L/d (52.8 gpd) of the 49% solution, the mass of alum used is 200 L/d × 0.647 kg/L = 129 kg/d (285 lb/d). Assuming a typical application with a 3-to-1 stoichiometric rate, the appropriate TSS conversion factor is 0.312 (Patoczka, 2006). Consequently, this alum addition will result in generation of 129 kg/d × 0.312 = 40.3 kg/d (88.7 lb/d) of inert alum sludge.

When using iron-based coagulants, the presence of iron in the waste sludge disposed of off-site may be beneficial for using the sludge as a soil amendment. Aluminum has no advantage as a soil amendment; therefore, if the final sludge is blended with soil, there may be additional concerns with aluminum content in the sludge.

## 6.2 Sludge Settling, Thickening, and Dewatering

In addition to increasing the waste solids generation rate on a dry-mass basis, the addition of chemicals such as aluminum and ferric salts will affect sludge settling and dewatering properties. When added to activated sludge, these flocculating chemicals will generally improve solids separation in the secondary clarifier because of the increased capture of fine floc and colloids. Alum and, in particular, iron-based chemicals will increase activated sludge specific density, thereby improving its settling properties (lowering the sludge volume index). This could significantly improve performance of secondary clarifiers, allowing operation at a lower sludge blanket and/or higher MLSS concentration. Ferrous, ferric, and aluminum-based chemicals could also help control activated sludge foaming and possibly inhibit filamentous bulking.

The effect of chemicals is more difficult to predict in sludge thickening and dewatering operations. On several occasions, the Woodland Plant in Morris Township, New Jersey, has initiated and stopped adding alum to remove total phosphorus to approximately 0.7 mg/L. Each time, the facility experienced a decrease in dry solids concentration during gravity belt thickening from approximately 5.5% to 4.5% solids when chemical was being added. Another New Jersey facility, the East Windsor Municipal Utilities Authority, observed a decrease from 3.2% to 2.8% solids concentration from gravity thickeners when alum was used for phosphorus removal.

## 6.3 Alkalinity and pH

Addition of alum and, particularly, iron compounds will lower the pH of wastewater because these agents are acidic and alkalinity is consumed during precipitation reactions (Section 4.0). As discussed in Section 4.2, some alternative aluminum-based products such as sodium aluminate and PACl will not depress pH or consume alkalinity.

The extent of pH reduction mainly depends on the alkalinity of the wastewater; the higher the alkalinity, the lower the reduction in pH for a given chemical dosage. In some instances of low wastewater alkalinity, addition of an alkaline substance such as sodium hydroxide, soda ash, or lime may be required to maintain acceptable pH. If the facility also nitrifies, alkalinity consumption by phosphorus precipitation must be added to the nitrification alkalinity demand to evaluate the overall effect on the system.

Alum addition consumes alkalinity at a rate of 0.505 mg (as $CaCO_3$) per milligram of alum added, which corresponds to a need for 0.404 mg of caustic (NaOH). Thus, a 3.785-ML/d (1-mgd) facility that applies a 60-mg/L alum dose (227 kg/d or 500 lb/d alum application rate, corresponding to 351 L/d or 92.7 gpd of 49% alum solution, respectively) would require 91.7 kg/d (202 lb/d) of dry caustic to completely recover alkalinity loss induced by alum. This amount of dry caustic corresponds to 120 L/d (31.8 gpd) of 50% caustic solution.

Similarly, ferric chloride addition consumes alkalinity at a rate of 0.923 mg (as $CaCO_3$) per milligram of ferric chloride added, which corresponds to a need for 0.738 mg of caustic (NaOH). Thus, a 3.785-ML/d (1-mgd) facility that applies 30 mg/L of ferric chloride dose (114 kg/d or 250 lb/d dry ferric chloride application rate, corresponding to 226 L/d or 59.5 gpd of 27% ferric chloride solution, respectively) would require 84.1 kg/d (184 lb/d) of dry caustic to completely recover alkalinity loss induced by ferric chloride. This amount of dry caustic corresponds to 111 L/d (29 gpd) of 50% caustic solution.

It is important to note that complete neutralization may or may not be necessary or desired, depending on wastewater alkalinity, effluent pH requirements, and other process considerations such as maintenance of optimum pH for phosphorus precipitation (typically below pH 7) or optimum pH for nitrification (above 7). All technical iron solutions and, in particular, pickle liquor, contain substantial amounts of free sulfuric acid or hydrochloric acid, which will additionally consume alkalinity and suppress the pH of the wastewater, depending on the acid content. As the effectiveness of chemical phosphorus precipitation increases with the lowering of pH (in the typical range of WRRF operations), some depression of the pH by chemicals may be desirable, providing that the nitrification process is not affected and the effluent remains within the acceptable range of pH.

## 6.4 Total Dissolved Solids Increase

Addition of chemicals will, in most instances, result in an increase in TDS concentration. *Total dissolved solids* represents a parameter of growing regulatory concern in many regions. In many areas, regulators are enforcing water quality standards for TDS, which, for fresh waters, typically are 500 mg/L (e.g., New

Jersey Water Quality Standards). Discharging high TDS can also be of concern in instances in which reclaimed water is used for irrigation purposes or as a cooling water makeup.

However, in most typical applications, an increase in TDS because of chemical addition will be relatively modest. Total dissolved solids increase factors for selected chemicals and processes commonly used in wastewater treatment are provided in Patoczka (2006). For example, the addition of a relatively high (100-mg/L) dose of alum, which could be sufficient in producing 0.1 mg/L of effluent total phosphorus concentration according to Equation 10.2, will increase TDS by 53.3 mg/L based on a factor of 0.533 (Patoczka, 2006) (assuming a 3-to-1 stoichiometric rate). In most situations in which chemical addition is lower, TDS effects from using chemicals for phosphorus removal will be relatively small.

## 6.5 Enhanced Biological Phosphorus Removal

Whereas the EBPR process can sometimes be effective in producing low-effluent phosphorus, consistently meeting low-effluent limits could be impaired by periods of substandard performance or upsets. For these reasons, chemical addition is commonly practiced at EBPR facilities as a tertiary polishing step, a backup process, or, more commonly, in a form of simultaneous biological and chemical phosphorus removal by coagulant addition to the activated sludge train (Gebremariam et al., 2011; Neethling et al., 2005). Although the addition of metal salts to an EBPR process has been shown to improve the overall phosphorus removal efficiency, there are concerns that continuous dosing of coagulants to an activated sludge facility may lower efficiency of the phosphorus release and uptake cycle by competition or inhibition. As chemicals added to activated sludge remain in the system for a period commensurable with sludge age, the effect of a chemical added at a high dose during a period of inferior EBPR performance could linger for a considerable time, delaying recovery of the EBPR system.

Consequently, metals addition to the EBPR process should be practiced carefully, at the lowest necessary dose, with the point of addition being at the effluent from the aeration basin. Where possible, chemical addition to the EBPR activated sludge train should be avoided, and if necessary, chemicals should be applied to the primary clarifier and/or to the tertiary process.

## 6.6 Anaerobic Digestion

Because of the low solubility of ferrous sulfide (FeS), ferric salt addition directly to anaerobic digesters has been used to control $H_2S$ in the digester gas and for odor control. Lee et al. (2009), Novak and Park (2010), and Yuan and Bandosz (2007) suggested that the addition of iron salts for phosphorus removal in the liquid train will have a similar effect on digester gas quality. Aluminum-based salts are not expected to have this effect.

When metal salts are used in the liquid train for phosphorus removal or ferric is added directly to the digester for odor control, struvite (magnesium ammonium phosphate) formation in the anaerobic digesters is typically prevented. This is because most of the orthophosphate in the waste sludge is bound in a stable precipitate, making it unavailable for struvite formation. There are some reports (e.g., Chen et al., 2008; Dentel & Gosset, 1982; Monteith & Atkinson, 2001) that at certain aluminum concentrations anaerobic digestion could become inhibited. However, this does not appear to be a problem at metal addition rates typically used for phosphorus control because no instances of full-scale anaerobic digestion inhibition were found at a number of facilities using alum for phosphorus removal.

In summary, when anaerobic digestion is part of the treatment process, the effect of chemicals added for phosphorus removal should be carefully considered. Ferric salts may be preferred over aluminum-based products in such applications, mainly because of their side benefit in odor control. However, under some conditions, ferric addition at a high dose could cause formation of iron scales in digester piping. Separation of biological and chemical sludges should be considered, if practical, to avoid increased inert load to the digester. Additionally, when phosphorus is bound with aluminum or ferric in the waste (and digested) sludge, it becomes largely unavailable as fertilizer, potentially diminishing the value of biosolids derived from sludge. It is important to note that magnesium ammonium phosphate (struvite) is more soluble than aluminum or ferric sulfates. Consequently, this compound recovered from sidestreams by processes such as Pearl (see Chapter 12) may be more suitable as a long-release fertilizer.

## 6.7 Ultraviolet Disinfection

Effectiveness of UV disinfection is dependent on the delivered UV dose, contact time, liquid film thickness, wastewater absorbance, wastewater turbidity, wastewater chemistry (including hardness, alkalinity, pH, and oxidation-reduction potential), system configuration, and temperature. Fouling of UV lamp sleeves decreases the effective UV dose reaching the wastewater and is the main cause of decreased UV-system disinfection efficiency.

Lu et al. (2012) observed that iron salts have a greater negative effect on UV transmittance in wastewater than other coagulants. The presence of iron in UV-system influent has been documented to be a significant factor in the overall decrease of the disinfection efficacy through the following mechanisms:

- Dissolved iron molecules can absorb UV radiation in critical wavelengths, preventing UV light from reaching target organisms. This may include cationic free iron or iron complexes adsorbed into the residual suspended solids and bacteria flocs (Kozak et al., 2011).
- Iron precipitates can add to the resulting suspended solids of a treated effluent, thereby causing enhanced shielding effects, blocking the transmittance of UV light. Ultraviolet radiation is believed to promote precipitation processes of residual dissolved iron as $Fe(OH)_3$ and other compounds (Kozak et al., 2011; Nessim & Gehr, 2006).
- Residual iron, particularly at concentrations in excess of 0.5 mg/L, has been found to be the main constituent associated with fouling of UV lamps, although the presence of calcium, magnesium, phosphorus, and organic matter is also important for scale formation. Formation and deposition of precipitates on the lamp's quartz sleeves is promoted by the lamp's high temperature. The precipitate, in addition to ferric salts, could include $CaCO_3$, $CaSO_4$, $MgCO_3$, $MgSO_4$, $Al(OH)_3$, and $Al_2(SO_4)_3$ (Black & Veatch Corporation, 2010; Kozak et al., 2011; Nessim & Gehr, 2006; Peng et al., 2005; Sehnaoui & Gehr, 2001; Sheriff & Gehr, 2001).

Operation of the UV system could also be affected by fouling of UV light sensors, causing an unnecessary increase in power to compensate for the errant light-intensity measurements.

With increasing iron concentrations above 0.5 mg/L, particularly in the presence of other scale-forming constituents, scaling may be quite rapid, resulting in decreased effectiveness of UV disinfection in a matter of days or even hours, depending on the concentration of iron and other constituents (Sheriff & Gehr, 2001). It is important to note that the presence of residual ferric at such elevated concentrations is not typical (and not necessary), and the ferric application rate should be reduced. Chemical and mechanical cleaning of the sleeves could cause permanent fouling because of the scratches that trap foulants (Peng et al., 2005). On the other hand, addition of coagulants and/or polymer to the upstream treatment processes could significantly reduce the concentration of dispersed solids and colloidal matter in the UV influent, thus reducing turbidity and improving disinfection performance.

## 6.8 Potential Phosphorus Deficiency

It is well established that activated sludge bacteria need both nitrogen and phosphorus (sometimes called *macronutrients*) in addition to other microconstituents. These are typically available in adequate concentrations in municipal wastewater. An aggressive phosphorus removal approach using chemical addition ahead of the biological process, like CEPT, could result in inadequate phosphorus supply for the bacteria.

As discussed in Section 3.3, a conventional activated sludge biomass has a phosphorus content of about 2% (and much more if EBPR is practiced). As a first approximation, a general rule can be used that for each 100 mg/L of $BOD_5$ consumed by the biological process, 1 mg/L of phosphorus will be needed for biological growth (in addition to 5 mg/L of nitrogen). Consequently, to avoid phosphorus

**TABLE 10.3** Summary of Chemical Phosphorus Removal's Operations Philosophy

| Process Aspects | Description |
|---|---|
| Pretreatment and monitoring | Before beginning the chemical phosphorus removal procedure, the influent wastewater must be screened and pre-treated to remove larger solids and debris. Continuous monitoring of influent and effluent phosphorus levels is necessary for determining the performance of a process and treatment based on phosphorus speciation. |
| Chemical selection | The selection of chemicals for phosphorus removal is crucial. Metal compounds such as alum (aluminum sulfate) and ferric chloride are common chemicals. These chemicals react with phosphorus to produce water-removable, insoluble precipitates. |
| Dosage optimization | The quantity of chemicals added (the dosage) must be optimized. Overdosing can result in excessive chemical costs, increased solids production, and potential negative effects on subsequent processes including corrosion of structures and equipment. |
| Mixing and flocculation | Effective mixing and flocculation are required to ensure proper contact between the chemicals and phosphorus, allowing the formation of larger particles (flocs) that settle more easily. |
| Settling and clarification | The formed flocs then settle in a sedimentation basin or clarifier or can be further filtered. Proper design and operation of these units ensure effective separation of the flocs from the treated water. |
| Sludge management | The sludge collected in the settling basin must be managed correctly. Before disposal, this may involve thickening, dewatering, and possibly additional treatment. |
| pH adjustment | The pH of the wastewater can influence the effectiveness of the chemical reaction. Monitoring and adjusting pH can enhance the phosphorus removal efficiency. |
| Operator training and monitoring | The operators of the chemical phosphorus removal process should be well-trained in the operation of the process. Monitoring key process parameters—such as chemical dosages, pH, settling rates, and process influent and effluent phosphorus concentrations—on a regular basis is essential for ensuring consistent and dependable performance. |
| Emergency response plan | It is essential to have a well-defined plan to resolve potential process upsets, chemical spills, and equipment failures to minimize disruptions and mitigate potential environmental impacts. |
| Process optimization | To ensure long-term effectiveness, it is essential to continuously optimize the process through data analysis, adjustments, and modifications based on changing influent characteristics or regulatory requirements. |
| Regulatory compliance | The operations philosophy should adhere to local, state, and federal phosphorus discharge limits and water quality standards. |

deficiency in the downstream biological treatment process, effluent from a primary clarifier with 50 mg/L of $BOD_5$ should have residual phosphorus of no less than 0.5 mg/L in soluble orthophosphate form because other forms of phosphorus may not be bioavailable.

Another treatment process that could experience phosphorus deficiency is a separate-stage tertiary denitrification process at facilities required to also meet a tight effluent phosphorus limit. Some facilities, particularly in the Chesapeake Bay watershed, are required to simultaneously meet total nitrogen limits of 3 mg/L and total phosphorus limits as low as 0.18 mg/L. To ensure that denitrifying bacteria have an adequate supply of phosphorus, deBarbadillo et al. (2006) recommended maintaining an orthophosphorus-P (i.e., expressed as phosphorus to $NO_x$-N—nitrates and nitrites expressed as nitrogen) ratio of 0.02 or more.

### 6.9 Safety and Operational Considerations

All chemicals, whether they are a gas, solid, or liquid, require a feeding system to accurately and repeatedly control the amount applied. Effective use of chemicals depends on accurate dosages and proper mixing. The effectiveness of certain chemicals is more sensitive to dosage rates and mixing than that of others. Design of a chemical feed system must consider the physical and chemical characteristics of each chemical used for feeding; minimum and maximum ambient or room temperatures; minimum, average, and maximum wastewater flows; minimum, average, and maximum anticipated dosages required; and the reliability of feeding devices.

Operators and maintenance personnel should be aware of the hazards and characteristics of the chemicals that are used at a facility. Material safety data sheets and technical specifications provided by suppliers are a good source of this information. Additional resources for design and operation of chemical feed systems are cited in the Suggested Readings section of this chapter.

## 7.0 OPERATIONS SUMMARY

The operations philosophy of a chemical phosphorus removal process comprises a systematic approach and guidelines for removing phosphorus from effluent using chemical agents efficiently and effectively. A summary of the most important aspects of this process's operations philosophy is presented in Table 10.3.

Chemical phosphorus removal processes emphasize efficient chemical dosing, proper mixing and flocculation, effective settling, and cautious sludge management. Consistent and dependable phosphorus removal results require regular monitoring, expert operator knowledge, and a commitment to process optimization.

## 8.0 REFERENCES

Asada, L. N., Sundefeld, G. C., Jr., Alvarez, C. R., Filho, S. S., & Piveli, R. P. (2010). Water treatment plant sludge discharge to wastewater treatment works: Effects on the operation of upflow anaerobic sludge blanket reactor and activated sludge systems. *Water Environment Research*, *82*(5), 392–400.

Black and Veatch Corporation. (2010). *White's handbook of chlorination and alternative disinfectants* (5th ed.). Wiley & Sons.

Chen, Y., Cheng, J. J., & Creamer, K. S. (2008). Inhibition of anaerobic digestion process: A review. *Bioresource Technology*, *99*(10), 4044–4064.

deBarbadillo, C., Rectanus, R., Canham, R., & Schauer, P. (2006, October 21–25). Tertiary denitrification and very low phosphorus limits: A practical look at phosphorus limitations on denitrification filters. *Proceedings of the 79th Annual Water Environment Federation Technical Exhibition and Conference* [CD-ROM], Dallas, TX, United States. Water Environment Federation.

Dentel, S. K., & Gosset, J. M. (1982). Effect of chemical coagulation on anaerobic digestibility of organic materials. *Water Research*, *16*(5), 707–718.

Gates, D. D., Luedecke, C., Hermanowicz, S. W., & Jenkins, D. (1990). Mechanisms of chemical phosphorus removal in activated sludge with Al(III) and Fe(III). *Proceedings of the 1990 Specialty Conference on Environmental Engineering* (p. 322). American Society of Civil Engineers.

Gebremariam, S. Y., Beutel, M. W., Christian, D., & Hess, T. F. (2011). Research advances and challenges in the microbiology of enhanced biological phosphorus removal—A critical review. *Water Environment Research*, *83*(3), 195–219.

Georgantas, D. A., Matsis, V. M., & Grigoropoulou, H. P. (2006). Soluble phosphorus removal through adsorption on spent alum sludge. *Environmental Technology*, *27*(10), 1081–1088.

Ippolito, J. A., Barbarick, K. A., & Elliot, H. A. (2011). Drinking water treatment residuals: A review of recent uses. *Journal of Environmental Quality*, *40*(1), 1–12.

Kozak, J. A., Lordi, D. T., Abedin, Z., O'Connor, C., Granato, T., & Kollias, L. (2011). The effect of ferric chloride addition for phosphorus removal on ultraviolet radiation disinfection of wastewater. *Environmental Practice*, *12*(4), 275–284.

Lee, Y., Baker, S., Wang L., Gibson, D., Amos, J., & Zaleski, A. (2009, June 28–July 1). Operational impact of phosphorus removal using ferric chloride on anaerobic digesters and dewatering. *Proceedings of the Water Environment Federation Nutrient Removal Specialty Conference* (pp. 7411–7418), Washington, DC, United States. Water Environment Federation.

Lu, G., Li, C., Zheng, Y., & Deng, A. (2012). Effect of different coagulants on the ultraviolet light intensity attenuation. *Desalination Water Treatment*, *37*(1–3), 302–307.

Luedecke, C., Hermanowicz, S. W., & Jenkins, D. (1988). Precipitation of ferric phosphate in activated sludge: A chemical model and its verification. *Water Science & Technology*, *21*, 325–338.

Monteith, H., & Atkinson, D. (2001, October 13–17). By the numbers: Process data review reveals the causes of digester inhibition. *Proceedings of the 74th Annual Water Environment Federation Technical Exhibition and Conference* [CD-ROM] (pp. 389–400), Atlanta, GA, United States. Water Environment Federation.

Mortula, M. M., & Gagnon, G. A. (2007). Phosphorus treatment of secondary municipal effluent using oven-dried alum residual. *Journal of Environmental Science and Health, 42*, 1685–1691.

Neethling, J. B., Bakke, B., Benish, M., Gu, A., Stephens, H., Stensel, H. D., & Moore, R. (2005). *Factors influencing the reliability of enhanced biological phosphorus removal* (Report No. 01-CTS-3). Water Environment Research Foundation.

Nessim, Y., & Gehr, R. (2006). Fouling mechanisms in a laboratory-scale UV disinfection system. *Water Environment Research*, *78*(12), 2311–2323.

Novak, J.T., & Park, C. M. (2010). *Effect of aluminum and iron on odors, digestion efficiency, and dewatering properties* (WERF Research Report). International Water Association.

Patoczka, J. (2006, October 21–25). TDS and sludge generation impacts from use of chemicals in wastewater treatment. *Proceedings of the 79th Annual Water Environment Federation Technical Exhibition and Conference* [CD-ROM], Dallas, TX, United States. Water Environment Federation.

Patoczka, J. (2009). Unpublished report prepared for Bernards Township Sewerage Authority, New Jersey, by Hatch Mott MacDonald, Millburn, New Jersey.

Peng, J., Qiu, Y., & Gehr, R. (2005). Characterization of permanent fouling on the surfaces of UV lamps used for wastewater disinfection. *Water Environment Research, 77*(4), 309–322.

Sagberg, P., Ryrfors, P., & Berg, K. G. (2006). 10 years of operation of an integrated nutrient removal treatment plant: Ups and downs. *Water Science & Technology, 53*(12), 83-90.

Sedlak, R. (1991). *Phosphorus and nitrogen removal from municipal wastewater.* Lewis Publishers.

Sehnaoui, K., & Gehr, R. (2001, June 14–16). Fouling of UV lamp sleeves: Exploring inconsistencies in the role of iron. *Proceedings of the First International Congress on UV Technologies.* Washington, DC, United States. International UV Association.

Sheriff, M., & Gehr, R. (2001). Laboratory investigation of inorganic fouling of low pressure UV disinfection lamps. *Water Quality Research, 36*(1), 71–92.

Smith, S., Takacs, I., Murthy., Daigger, G. T., & Szabo, A. (2008). Phosphate complexation model and its implications for chemical phosphorus removal. *Water Environment Research, 80*(5), 428–438.

Takacs, I., Murthy, S., Smith, S, & McGrath, M. (2006). Chemical phosphorus removal to extremely low levels: Experience at two plants in the Washington, D.C. area. *Water Science & Technology, 53,* 21.

U.S. Environmental Protection Agency. (2007). *Advanced wastewater treatment to achieve low concentration of phosphorus* (EPA 910-R/07-002).

Water Environment Federation. (2010). *Nutrient removal* (Manual of Practice No. 34).

Yuan, W., & Bandosz, T. J. (2007). Removal of hydrogen sulfide from biogas on sludge-derived adsorbents. *Fuel, 86*(17–18), 2736–2746.

## 9.0 SUGGESTED READINGS

Cabirol, N., Barraga'n, E. J., Dura'n, A., & Nayola, A. (2003). Effect of aluminum and sulphate on anaerobic digestion of sludge from wastewater enhanced primary treatment. *Water Science & Technology, 48*(6), 235–240.

Gehr, R., & Wright, H. (1998). UV disinfection of wastewater coagulated with ferric chloride: Recalcitrance and fouling problems. *Water Science & Technology, 38*(3), 15–23.

Stumm, W., & Morgan, J. J. (1970). *Aquatic chemistry.* Wiley & Sons.

Water Environment Federation. (2023a). *Standards for design of water resource recovery facilities* (WEF 8-23).

Water Environment Federation. (2023b). *Standards for operation of water resource recovery facilities* (WEF 11-23).

# Chapter 11

# Optimization of Nutrient Removal Systems



## 1.0 INTRODUCTION

Operators should strive to optimize the performance of the nutrient removal process in terms of effluent quality and/or reducing ongoing operational costs associated with power and chemical consumption. The complexity of nutrient removal processes and the many interrelated and often-competing operational parameters dictate that a deliberate, integrated, orderly process occur to achieve process optimization.

# 2.0 PHASE 1 PROCESS EVALUATION

The first step in optimizing process performance is to evaluate the current operation and set process improvement goals. This first phase of process evaluation may include review of facility influent and effluent discharge data, evaluation of each unit process performance, and a review of operations efficiency in terms of chemical dosages, electricity consumption, and sludge production. Optimization goals may include improving effluent performance and/or reducing the cost of operation. Based on this initial evaluation, it is likely that additional data will be required to meet optimization goals.

## 2.1 Data Review

Review of facility data and process optimization should be an ongoing task. Interpreting data requires technical knowledge, experience, and a strong familiarity with the facility. Depending on the specific facility and the particular goals of process optimization, there are numerous areas on which to focus attention. The factors that are discussed in Sections 2.1.1 through 2.1.4 of this chapter were identified as those that will typically result in the highest effect on a facility's effluent performance and operating efficiency.

### *2.1.1 Nutrient Loading and Loading Variability*

Nutrient mass loads that are received by a water resource recovery facility (WRRF) are products of the flow rate and pollutant concentration. Loading rates may vary considerably during a typical day and can result in a biological nutrient removal (BNR) process experiencing loads both significantly higher and lower than the average design. Figure 2.2 in Chapter 2 illustrates a typical diurnal variation of flow and load. This variation can be more pronounced in smaller collection systems, where there is less storage capacity and transit time to dampen the effects. Typical effects of high nutrient loading may include deteriorated effluent quality (i.e., higher effluent concentrations of effluent total nitrogen and total phosphorus), an inability to maintain target dissolved oxygen concentrations, and reduced sludge settleability. A common effect of a lowly loaded biological process is higher dissolved oxygen concentrations and the potential for interference with denitrification and enhanced biological phosphorus removal (EBPR).

Table 11.1 presents typical ranges of peaking factors for nutrient loading in raw wastewater (Water Environment Federation [WEF], 2023). Note that there can be considerable variability in these peaking factors, depending on the characteristics of the collection system. Peaking factors tend to be lower when the WRRF's design flowrate is larger and somewhat higher as the facility's design flow decreases.

### *2.1.2 Influent Fractions and Constituent Ratios*

Commonly used measures of influent amenability to BNR are based on the amount of readily biodegradable substrate available to sustain EBPR and denitrification. As a first approximation, ratios of minimum 5-day $BOD_5$ ($BOD_5$) to total Kjeldahl nitrogen (TKN) and $BOD_5$ to total phosphorus of 3 to 1 and 25 to 1, respectively, may be used to assess the site-specific availability of an adequate carbon source for BNR. This determination should be made on the influent to the BNR bioreactor and should include all significant sidestream loads. Municipal wastewater that is primarily domestic in origin typically contains sufficient carbon substrate for BNR if excessive $BOD_5$ removal does not occur in the primary clarifiers. Sidestreams, which are characteristically low in $BOD_5$ and relatively high in nitrogen and phosphorus, can depress the ratios of $BOD_5$ to total phosphorus and $BOD_5$ to TKN in bioreactor influent. It is important to note that the aforementioned minimum ratios are a first approximation of

**TABLE 11.1** Design Peaking Factor Summary (Adapted from WEF, 2023)

| | Flow | $BOD_5$ | TSS | TKN | Ammonia-Nitrogen | Phosphorus |
|---|---|---|---|---|---|---|
| **Separate collection systems** | | | | | | |
| Number of valid data sets | 40 | 33 | 32 | 6 | 15 | 6 |
| **Peaking factors*** | | | | | | |
| Minimum daily | 0.73 | 0.60 | 0.55 | 0.74 | 0.70 | 0.73 |
| Maximum monthly | 1.30–1.45 | 1.30–1.45 | 1.30–1.53 | 1.15–1.36 | 1.21–1.30 | 1.12–1.27 |
| Maximum daily | 1.40–2.60 | 1.30–1.45 | 1.53–2.86 | 1.27–3.57 | 1.32–2.09 | 1.25–1.90 |
| **Combined collection systems** | | | | | | |
| Number of valid data sets | 20 | 19 | 18 | 2 | 11 | 3 |
| **Peaking factors*** | | | | | | |
| Minimum daily | 0.68 | 0.60 | 0.53 | 0.67 | 0.66 | 0.73 |
| Maximum monthly | 1.32–1.64 | 1.26–1.51 | 1.31–1.81 | 1.24–1.48 | 1.21–1.31 | 1.35–2.20 |
| Maximum daily | 1.62–3.58 | 1.60–3.60 | 1.87–6.22 | 1.40–3.07 | 1.35–2.87 | 1.35–2.20 |

nutrient removal capability. Even if these ratios indicate that adequate readily biodegradable substrate is available, other influent characteristics and operating factors could compromise the ability of the system to achieve reliable BNR. For a detailed discussion of this topic, refer to other chapters in these standards.

### *2.1.3 Effluent Quality and Fractions*

Biological nutrient removal facilities are designed and operated to achieve effluent criteria for total nitrogen and/or total phosphorus. As performance requirements become increasingly stringent and required effluent concentrations of total nitrogen and total phosphorus are reduced to levels that challenge the capabilities of feasible treatment processes, the fractions of effluent total nitrogen and total phosphorus become an important aspect in the evaluation of process performance. The parameter of concern for compliance with total nitrogen limits is refractory dissolved organic nitrogen (rDON), and for a low total phosphorus limit, the parameter of concern is recalcitrant dissolved organic phosphorus (rDOP). Refractory dissolved organic nitrogen passes through BNR processes and typically ranges from 0.5 to 2.0 mg/L in BNR process effluent. Likewise, rDOP passes through both EBPR processes and chemical phosphorus removal and may typically comprise about 0.05 mg/L in BNR process effluent. These parameters are discussed in detail in Section 3.0.

### *2.1.4 In-Facility Sidestreams*

Internal facility sidestreams—such as those from solids dewatering processes, supernatant from sludge digestion processes, thickening systems, and filter backwashes—can have a significant effect on the nutrient removal process and can introduce tremendous variability in WRRF flows and loads. Chapter 12 describes alternatives for addressing these effects.

Sidestreams containing nitrogen and phosphorus are typically blended with facility influent and recycled through the bioreactor. The sidestream loads may be intermittent and highly variable. Biological nutrient removal process reactions are sensitive to influent load variations. Whereas the average sidestream loading may or may not be significant, the short-term peak loads imposed by return streams can overwhelm the BNR system. For example, a peak load of ammonia introduced back to the head of a treatment process may exceed the nitrification capacity of the biological process or aeration capacity, thereby resulting in a spike of ammonia in the effluent.

Sidestream flows also have the potential to upset the alkalinity-to-nitrogen ratio (as measured by total alkalinity-to-TKN ratio), the carbon-to-nitrogen ratio (as measured by the $BOD_5$-to-TKN ratio), or the carbon-to-phosphorus ratio (as measured by the $BOD_5$-to-total phosphorus ratio) described earlier in this chapter. If operational adjustments cannot be made to maintain acceptable ratios, then supplementation of alkalinity or carbon may be needed.

Facilities that provide EBPR need to be careful in their handling of waste sludge, particularly waste activated sludge. If the sludge is held in an anaerobic condition, such as in a holding tank, gravity thickener, or anaerobic digester, then sidestream flows from these processes can have high concentrations of soluble reactive phosphorus, generally measured as orthophosphate. These heavy loads can overwhelm the capacity of the process to adequately remove the orthophosphate biologically, especially if there is not enough readily degradable carbon available. If operational adjustments cannot be made to maintain acceptable ratios, then carbon supplementation to support EBPR or chemical addition in the dewatering process to reduce the orthophosphate in the sidestream may be needed.

Facilities that incinerate their sludge may be introducing cyanide from the scrubber blow-down water to their process through sidestream flows. Depending on the concentration, the cyanide may be toxic or at least inhibitory to nitrifying organisms.

## 2.2 Establishing Optimization Goals and Metrics

Successful optimization of BNR process operation first requires setting optimization goals and a system by which optimization will be measured.

### *2.2.1 Consistent Effluent Quality*

Biological nutrient removal optimization is perhaps most often targeted to improve process effluent quality to reliably comply with regulatory requirements. The performance goals and means of measuring progress can, therefore, be easily established.

### *2.2.2 Chemical Doses*

Operation of nutrient removal processes typically requires dosing of chemicals to increase available alkalinity, adjust pH, provide supplemental carbon to improve denitrification and/or biological phosphorus removal, precipitate orthophosphate, and improve solids separation. The dose required for these chemicals is dependent on effluent quality requirements, although there may be opportunities to reduce dose by modifying operations to reduce orthophosphate load and/or alternative dosing locations or using alternate chemicals.

### *2.2.3 Energy Use*

Biological nutrient removal processes are typically energy intensive. Therefore, finding a means to reduce power consumption should be a high priority. Setting goals for energy consumption by process optimization should follow a step-by-step procedure as follows: (1) assess current energy use at the facility, setting a baseline from which to measure progress; (2) take an inventory of facility assets, equipment, and systems to identify areas of significant energy consumption; and (3) prioritize areas of potential optimization. The U.S. Environmental Protection Agency's [U.S. EPA's] (2008) *Energy Management Guidebook for Wastewater and Water Utilities* and WEF (2021) MOP-32, *Energy in Water Resource Recovery Facilities*, contain detailed guidance procedures for energy management.

### 2.2.4 Sludge Production

Sludge produced through operation of a BNR process is primarily a function of influent load, which is beyond an operator's control. However, the net solids yield is also affected by the operating solids retention time (SRT), the process configuration, and chemical addition.

### 2.2.5 Operation and Maintenance Requirements

Efficient operation of a BNR process requires qualified, diligent attention by operators and maintenance staff. The level of effort and associated costs may be a significant drain on resources, and optimization may focus on ways of reducing the labor and resources required to operate the process such as use of online analyzers (online sampling and analysis), enhanced supervisory control and data acquisition systems, remote observation and notification systems, and consistency of maintenance-friendly use of mechanical equipment.

## 2.3 Identification of Additional Data Needs

Once the goals of the BNR optimization process are established, it will be necessary to collect and review relevant information such as facility historical performance data, maintenance records, energy use data, and operations budgets. Typically, available facility data do not provide sufficient information on which to complete an evaluation (e.g., facility data may include influent and effluent $BOD_5$, but no carbonaceous biological oxygen demand [$CBOD_5$] or chemical oxygen demand [COD] data when needed). In this situation, it will be necessary to develop a supplemental process sampling and/or monitoring plan to collect the necessary data. Section 3.0 provides detailed guidance on preparing such a plan and describes key parameters that should be considered.

# 3.0 PROCESS SAMPLING, TESTING, AND MONITORING

## 3.1 Sampling and Testing

Sampling and testing include determining sampling locations, frequency, techniques, and appropriate tests that will allow evaluation of process performance and identifying operational improvements needed to meet optimization objectives.

### 3.1.1 Sampling Plan

Every WRRF is unique and, therefore, will require a site-specific sampling plan to monitor performance. Operations personnel and facility administrators should work together to develop a sampling plan that allows for a complete understanding of key operating parameters without creating unneeded effort and excessive expense. During this exercise, everyone involved should ask the following questions:

- What are our regulatory reporting requirements?
- What reactions occur at each point in the process, and how do they affect the facility's ability to meet regulatory requirements?
- What environmental conditions are needed for the reactions to occur?
- What parameters must be sampled to evaluate if these reactions are occurring?

When creating a sampling plan, it is helpful to start with a copy of the facility's flow schematic(s) to illustrate the reactions in each process and various sample locations of interest at the facility. Figure 11.1

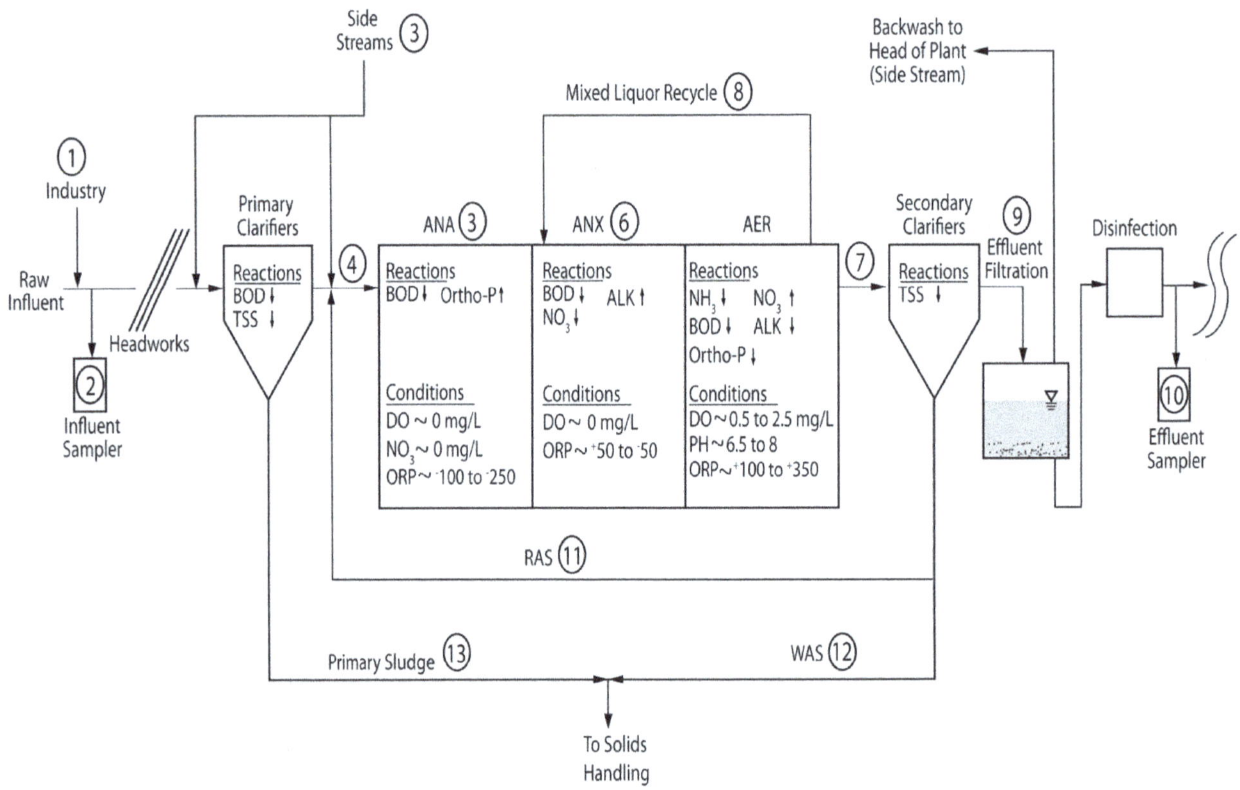


FIGURE 11.1 Flow Schematic and Sampling Plan Example

presents a sample flow schematic of a typical BNR facility. It is important to record the general reactions that occur in each process and the required environmental conditions to understand the parameters of interest. Operators should identify the location and type of samples that are required for permit compliance and the location of key parameters that indicate reactor performance for a specific step in a treatment process. Operators should then build a table summarizing the data that should be collected and frequency of collection for each of the sample locations identified on the schematic. Each sample location and type are chosen to provide specific information or an indication of the process and reaction that is meant to occur. Table 11.2 contains an example of a sampling plan that could be prepared for the facility characterized in Figure 11.1. Refer to Chapter 14 for a detailed discussion of these parameters.

### 3.1.2 Sample Handling and Collection

Sample handling should occur in accordance with regulatory requirements that govern the facility and/or as suggested by a generally accepted resource, such as 40 CFR 136 (U.S. EPA, 2005) or *Standard Methods* (American Public Health Association [APHA] et al., 2023). Once a facility's staff has determined the information needed and marked it on the sampling plan, a decision must be made regarding how the raw data will be collected. The raw data will generally be collected in one of the following three forms:

- Online in situ sampling: Typically, samples are measured with probes or automated devices to achieve a real-time indication of the reactor environment and performance. Examples include dissolved oxygen, temperature, oxidation-reduction potential (ORP), ammonia, orthophosphate, nitrate, alkalinity, and pH.
- Grab sampling: A sample is taken as a "snapshot" in time and used to measure real-time reactor performance. Grab samples differ from online in situ sampling in that they are typically used for

TABLE 11.2 Sampling Plan Example of a Typical BNR Facility

| Sample Location | Sample Description | Flow | Temperature | pH | ORP[a] | BOD/$BOD_5$, $CBOD_5$, and/or COD | TSS[b] | TKN | $NH_3$[c] | $NO_3$[d] | Total Phosphorus | Orthophosphate Orthophosphorus | Dissolved Oxygen | Alkalinity |
|---|---|---|---|---|---|---|---|---|---|---|---|---|---|---|
| | Sample type (grab [G]/composite [C]) | — | G | G | G | C | C | C | C | C | C | C | G | C |
| 1 | Industry effluent | Daily | | | | M/W/F[e] | | M/W/F | | | M/W/F | | | |
| 2 | Raw influent | Daily | Daily | Daily | | Daily | Daily | Daily | Daily | Daily | Daily | Daily | Daily | Daily |
| 3 | Sidestreams | | | | | | | | | | | | | |
| | a. Filtrate | Daily | | | | M/W/F | M/W/F | | M/W/F | | | M/W/F | | |
| | b. Thickening decant | Daily | | | | M/W/F | M/W/F | | M/W/F | | | M/W/F | | |
| 4 | Primary effluent | | | Daily | | Daily | Daily | | Daily | | | M/W/F | | |
| 5 | Anaerobic basin effluent (supernatant, other than TSS) | | | Daily | Daily | Weekly | Weekly | | | M/W/F | Daily | Daily | Daily | |
| 6 | Anoxic basin effluent (supernatant, other than TSS) | | | Daily | Daily | Weekly | | | M/W/F | M/W/F | | M/W/F | Daily | M/W/F |
| 7 | Aerobic basin effluent (supernatant, other than TSS) | | | Daily | Daily | Daily | Daily | Daily | Daily | M/W/F | M/W/F | Daily | Daily | M/W/F |
| 8 | Mixed liquor recycle | Daily | | | | | | | | | | | | |
| 9 | Secondary effluent | | | | | | Daily | | | | M/W/F | | | |
| 10 | Final effluent | Daily | | | | Weekly | Weekly | Weekly | Weekly | Weekly | Weekly | Weekly | Weekly | Weekly |
| 11 | RAS[f] | Daily | | | | | Daily | | | | | | | |
| 12 | WAS[g] | Daily | | | | | Daily | M/W/F | | | M/W/F | | | |
| 13 | Primary sludge | Daily | | | | | Daily | M/W/F | | | M/W/F | | | |

[a]ORP = oxidation-reduction potential
[b]TSS = total suspended solids
[c]$NH_3$ = ammonia
[d]$NO_3$ = nitrate
[e]M/W/F = Monday, Wednesday, and Friday
[f]RAS = return activated sludge
[g]WAS = waste activated sludge

parameters that require laboratory analysis. Multiple grab samples during the course of a day may be necessary to identify and help understand variabilities in performance.

- Composite sampling: Small volumes of sample are taken throughout the day at the same location and combined to create a single sample to represent an average for that day. Samples can be collected based on a time interval (e.g., 5 mL of sample per hour) or flow paced (e.g., 5 mL of sample per 4000 L of flow).

Chapter 15 provides additional information on sample collection and storage.

### 3.1.3 Sample Locations

After a sample location is selected, it is important that all personnel collect the sample at precisely the same place and in the same manner. The sample location should be clearly marked. Sample locations should be selected that allow representative samples to be collected that provide the desired information related to process performance. Additional samples should be collected if there is a possible change in conditions as the flow passes through the tanks. The following are characteristics of a good sample location:

- The contents are mixed well enough at that location to collect a representative sample.
- There is accessibility, allowing safe collection of the sample.

## 3.2 Online Process Monitoring and Automation

The use of online monitoring and automation has been more prevalent in recent years because of low effluent nutrient requirements and an increasing focus on energy and chemical costs at many facilities. Advancements in online analyzers allow operators to monitor virtually every step in a BNR process in real time. When considering online instrumentation, the benefits of each parameter and monitoring capability must be compared to the relative need based on regulatory requirements and the capital and maintenance costs of the instrumentation. The following subsections briefly discuss key parameters to consider when using online analyzers for process monitoring and control. Online analyzers are discussed in further detail in Chapters 13 and 14.

### 3.2.1 Monitoring Plan

Similar to the sampling plan discussed in Section 3.1.1, a process flow schematic should be the starting point of any online monitoring plan. It is important to understand the function of each unit process to select the appropriate parameter and location to monitor.

### 3.2.2 Monitoring Locations

Table 11.3 lists several common parameters and corresponding locations that can be monitored continuously as an indication of treatment performance.

### 3.2.3 Use of Monitoring for Automation

Data from online monitors can also be used to control chemical feed rates, internal recycle flow rates, influent loadings, and blower output, among other operations. Examples of some common process monitoring and corresponding control parameters are given in Table 11.4.

**TABLE 11.3** Potential Monitoring Locations of Common On-Line Analyzers and Process Indication

| Parameter | Monitoring Location | Purpose and Process Indication |
|---|---|---|
| Ammonia ($NH_3$) | • Raw influent<br>• Aerobic zone effluent and mid-point if plug flow reactors<br>• Final effluent | Monitor influent ammonia loading, nitrification performance, and total nitrogen in the final effluent. |
| Alkalinity | • Raw influent<br>• Aerobic zone influent (upstream of the aerobic zone)<br>• Aerobic zone effluent | Monitor alkalinity consumption and recovery to ensure sufficient alkalinity levels required for nitrification. |
| pH | • Aerobic zone effluent | Indication that pH level is sufficient for nitrification |
| Dissolved oxygen | • Various locations in the aerobic zone | Indication of sufficient dissolved oxygen for nitrification and biological phosphorus removal in upstream portions of the aerobic zone; monitor for low enough dissolved oxygen in aerobic zone effluent and internal recycle to minimize oxygenation of preanoxic zones. |
| Orthophosphorus ($PO_4^{-3}$) | • Raw influent<br>• Anaerobic zone<br>• Aerobic zone effluent<br>• Final effluent | Monitor biological and chemical phosphorus removal process performance. |
| Oxidation-reduction potential | • Anaerobic, anoxic, and aerobic zones | Monitor the environmental conditions in each zone. For example, ORP levels can indicate elevated/high levels of nitrate in an anaerobic zone, elevated/high dissolved oxygen in anaerobic or anoxic zones, or other environmental factors that reduce process performance. |
| Nitrate ($NO_3^-$) | • Anoxic zone effluent<br>• Aerobic zone effluent<br>• Secondary effluent<br>• Filtered effluent (denitrification filters) | Indication of nitrification and denitrification performance |
| Total suspended solids | • Various locations in BNR basins<br>• Secondary effluent<br>• RAS<br>• Final effluent | Monitor mixed liquor suspended solids (MLSS) concentrations.<br>Monitor filter performance. |

*Note.* Parameters given are relevant to a BNR treatment process to remove both total nitrogen and total phosphorus. Some parameters may not be applicable depending on the treatment process used and permit limits at each facility. Instrument reliability can vary depending on the application, wastewater characterization, and environment. All online instruments should be tested in the field during a trial period before purchase and installation at full scale. Sample preparation and location must be closely coordinated with vendor recommendations.

TABLE 11.4 Commonly Monitored Parameters Used to Control Unit Processes in a WRRF

| Parameter | Monitoring Location | Process and Equipment Controlled |
|---|---|---|
| Facility flow | • Influent (upstream of sidestream addition) | • Coagulant for phosphorus removal<br>• RAS and internal recycle flow rates<br>• Influent flow and load pacing<br>• Disinfection chemical dosing |
| Alkalinity | • Raw influent<br>• Aerobic zone influent (upstream of the aerobic zone)<br>• Aerobic zone effluent | • Alkalinity addition |
| Dissolved oxygen | • Various locations in the aerobic zone | • Aeration blowers' airflow and energy consumption<br>• Control dissolved oxygen in MLSS recirculation to anoxic zone |
| Orthophosphorus ($PO_4$) | • Raw influent<br>• Anaerobic zone<br>• Aerobic zone effluent<br>• Final effluent | • Coagulant feed rate for phosphorus removal<br>• Supplemental carbon feed in anaerobic zone for EBPR<br>• Orthophosphate release |
| Nitrate | • Anoxic zone effluent<br>• Aerobic zone effluent<br>• Secondary effluent<br>• Filtered effluent (denitrification filters) | • Carbon feed for denitrification<br>• MLSS recirculation rate for denitrification |

*Note.* Use and effectiveness of the instruments for process control are dependent on site-specific conditions including permit limits, chemical and energy costs, and the BNR process being implemented.

# 4.0 PHASE 2 PROCESS EVALUATION—DATA ANALYSIS AND INTERPRETATION

## 4.1 Nitrification

The parameters critical to nitrification are related to operating conditions that favor or indicate the activity of nitrifying organisms. These conditions include the biochemical environment, SRT, and performance indicators.

### *4.1.1 Environmental Conditions*

Biochemical environment conditions most affecting nitrification are

- alkalinity and pH;
- temperature;
- dissolved oxygen; and
- inhibitory compounds.

The pH in the bioreactor can greatly affect nitrification. Low pH can occur when the alkalinity concentration is insufficient to buffer the loss of alkalinity resulting from nitrification and tends to inhibit nitrification. The pH and alkalinity of the incoming wastewater affect the pH and alkalinity in the aeration basin, where nitrification occurs. An important factor to consider regarding the effects of pH on the nitrification process is the degree of acclimation that the particular system has achieved. Wide swings in pH have been demonstrated to be detrimental to nitrification performance, although acclimation generally allows satisfactorily performance with consistent pH control within the range of 6.5 to 8.0 standard pH units. It is generally recommended that sufficient alkalinity be present through the reactors by maintaining a minimum effluent alkalinity of at least 50, but higher minimum alkalinity may be necessary depending on the relationship of nitrogen to alkalinity loading consistency to the WRRF. By trending this relationship, the critical minimum alkalinity can be determined for a WRRF, and if necessary, alkalinity can be supplemented to maintain adequate buffering capacity. If there are periods of high nitrogen loading, equalizing that load to reduce the peak loads may allow more effective use of naturally occurring alkalinity and alkalinity supplementation may be reduced or avoided. Low pH inhibition of nitrification is most common in the following instances:

- Nitrification-only (no denitrification) facilities with low influent alkalinity relative to nitrogen loading; and
- Continuously aerated aerobic digesters operated at SRT exceeding 15 days.

The growth rates of nitrifying bacteria *Nitrosomonas* and *Nitrobacter* are particularly sensitive to the liquid temperature in which they live. Nitrification occurs in wastewater temperatures from 4 to 45 °C (39 to 113 °F), with an optimum growth rate occurring in the temperature range of 35 to 42 °C (95 to 108 °F) (U.S. EPA, 1993). However, most WRRFs operate with a liquid temperature between 10 and 25 °C (50 and 77 °F). It is generally recognized that the nitrification rate doubles for every 8 to 10 °C rise in temperature. The effect of temperature on the nitrification rate is shown in Figure 3.2 in Chapter 3. To compensate for lower nitrification rates in winter operating conditions, incremental increases in SRT may be necessary to provide additional nitrifiers to sustain adequate nitrification performance. Care must be taken, as this increase in SRT may exceed the level needed for nitrification and can lead to increases in filamentous bacteria that affect settleability and cause foaming/scum formation. Managing this balancing act to compensate for lower temperatures using SRT adjustments is critical for maintaining a healthy, properly settling biomass and optimum facility performance.

Nitrifiers are only able to function under aerobic conditions. Consequently, the dissolved oxygen concentration in the bulk liquid can have a significant effect on the nitrifier growth rate. The value at which the dissolved oxygen concentration reduces the rate of nitrification varies on a site-specific basis depending on temperature, organic loading rate, SRT, and diffusional limitations. It has generally been accepted that a bioreactor dissolved oxygen concentration of greater than 2 mg/L is required to ensure full nitrification. However, with greater emphasis on the reduction of energy use in WRRFs, many facilities are capable of achieving full nitrification while operating at dissolved oxygen levels lower than 2 mg/L. Careful monitoring of process performance parameters and nitrification performance trends is needed, as optimization of nitrification at dissolved oxygen less than 2.0 is considered as shifts in microbiology, and sludge volume index trends and effects to BNR may develop.

### *4.1.2 Solids Retention Time*

Nitrifying bacteria include species that do not grow as quickly as bacteria that oxidize organic material. A sufficiently long aerobic SRT is needed for a stable, adequate nitrifier population. Nitrification tends to be an "all or nothing" process, meaning that effluent ammonia concentrations will quickly rise once favorable conditions for nitrification are not met. *Washout SRT* is a term applied to the SRT at which no nitrifying organisms are present and no nitrification is occurring. For a more detailed discussion of this topic, refer to Chapter 3.

The target operating aerobic SRT for a BNR facility is determined by applying a peaking factor to the washout SRT, with a peaking factor of 2 to 3 being most common. Highly variable influent wastewater characteristics; intermittent inconsistent internal facility recycles/sidestreams; stringent permit limits; intermittent, manually controlled operation; and smaller process tanks warrant greater peaking factor values. Highly variable influent ammonia-nitrogen loading may also suggest that swing zones be used to quickly increase aerobic SRT when needed (refer to Chapter 6). Consistent influent and internal facility recycles; less frequent (i.e., moving annual average) permit requirements; well-automated processes; and facilities with larger, more forgiving process tanks or equalization allow lower peaking factor values. If a WRRF is prone to filamentous bacteria growth associated with elevated SRT conditions, efforts to reduce conditions forcing a higher peaking factor to enable operating at a lower SRT may prove beneficial for reducing filamentous growth.

### 4.1.3 Performance Indicators

The profile of nitrogen species ammonia, organic nitrogen, nitrite, and nitrate throughout the treatment process indicates nitrification performance. Raw wastewater will typically have about 80% ammonia and 20% organic nitrogen. It is typically assumed that all but 1 to 2 mg/L of organic nitrogen will be converted to ammonia in the activated sludge process. There is typically little to no nitrate or nitrite in raw wastewater. Ammonia is converted to nitrate in the nitrification process in the aeration basins and nitrate is converted to nitrogen gas in the denitrification process in the anoxic basins.

Nitrite typically does not accumulate in large concentrations in biological treatment systems under stable operation because the maximum growth rate of nitrite oxidizing bacteria, *Nitrobacter*, is significantly higher than that of *Nitrosomonas*. As a result, the growth rate of *Nitrosomonas* generally controls the overall rate of nitrification. However, facilities that transition into and out of nitrification may experience a condition wherein the growth rate of *Nitrobacter* is not able to keep up with that of *Nitrosomonas*, and effluent nitrite concentrations increase. This condition is known as "nitrite lock" and results in a significant increase in effluent chlorine demand (5.1 mg chlorine/mg nitrite). In situations wherein effluent permits contain both seasonal nitrification and seasonal disinfection requirements (using chlorine), it is advisable to establish complete nitrification before the start of chlorination.

For BNR systems, by setting up the operational conditions to achieve complete nitrification and adequate denitrification to meet permit conditions, the appropriate SRT for nitrification should be adequate for EBPR. However, EBPR phosphorus-accumulating organisms (PAOs) prefer a shorter SRT than that needed by nitrifiers, so care should be taken not to run excessive SRTs to maintain a healthy PAO population and overall balanced and well-settling mixture of microbiological populations.

Once adjustments to SRT are made, it takes time for the biological system to adjust to the new conditions. The full biological response to a change in SRT may take 2 to 3 SRTs. This presents why it is important to track and trend performance conditions to identify changes in performance as early as possible to make proactive adjustments before significant performance issues arise.

## 4.2 Denitrification

Parameters critical to denitrification are related to operating conditions that favor or indicate the activity of denitrifying organisms (heterotrophic organisms in an anoxic environment). These conditions include the chemical environment, SRT, and performance indicators.

#### 4.2.1 Environmental Conditions

The environment for denitrification is anoxic, that is, it lacks molecular oxygen (i.e., dissolved oxygen). The important parameters are pH, temperature, dissolved oxygen, ORP, nitrate-nitrogen, and COD or $BOD_5$. While optimum denitrification occurs within a narrow pH range, denitrification is significant unless the pH is lower than 6 or higher than 8. Denitrification can be affected by low temperature as a result of lowered activity. Because the denitrifiers are capable of using both dissolved oxygen and nitrate as the electron acceptor and the use of dissolved oxygen requires less energy than nitrate, denitrification is inhibited by the presence of dissolved oxygen. Oxidation-reduction potential can be used as an indicator of anoxic conditions. Anoxic ORP readings in the range of approximately –50 mV to 50 mV present anoxic conditions for denitrification. However, the optimum ORP conditions for each WRRF should be determined for optimum performance. The organic concentration, reported as COD or $BOD_5$, needs to be adequate for the desired amount of denitrification to occur. The minimum ratio of $BOD_5$ to TKN for good denitrification is 4.5 to 1, with 6 to 1 being preferred.

#### 4.2.2 Solids Retention Time

Denitrifying bacteria include many species, such as those that can use dissolved oxygen or nitrate for metabolism. Therefore, anoxic SRT need not be equal to aerobic SRT for nitrification.

#### 4.2.3 Performance Indicators

The nitrate profile throughout the treatment process will indicate denitrification performance. Using the modified Ludzack–Ettinger (MLE) process described in Chapter 7 as an example, a properly operated MLE process should have little to no nitrate in the anoxic reactor effluent despite measurable final effluent nitrate because a fraction of the nitrate will not be recycled from the aerobic reactor to the anoxic reactor.

Dissolved oxygen in MLSS that is recycled to anoxic zones for denitrification should be minimized. The dissolved oxygen levels at the recirculation pump intake should be low (2 mg/L or less) so excessive dissolved oxygen laden MLSS is not returned to the anoxic zone to impact anoxic zone performance. The higher the dissolved oxygen in the recirculation flow, the greater the carbon demand to reduce the dissolved oxygen before anoxic conditions are reached and the shorter the effective anoxic detention time for denitrification.

### 4.3 Enhanced Biological Phosphorus Removal

Parameters critical to EBPR are related to operating conditions that favor or indicate the activity of PAO organisms. These conditions include environmental conditions, SRT, and performance indicators.

#### 4.3.1 Environmental Conditions

The environment for EBPR is cyclic anaerobic and aerobic conditions that are fundamental for selection and growth of PAOs. Given these conditions, the population of the PAOs and EBPR performance is dependent upon conditions of key parameters: phosphorus load, SRT, temperature, dissolved oxygen, ORP, nitrate-nitrogen, the volatile fatty acid (VFA) fraction of COD or $BOD_5$ and initial dissolved oxygen in aerobic zone. Operation of EBPR is affected by temperature because the ability of PAOs, glycogen-accumulating organisms, and other heterotrophic microorganisms to out-compete each other for organics varies with temperature (refer to Chapters 6 and 8 for more information). Dissolved oxygen and nitrate input to anaerobic basins must be minimized to maintain adequate anaerobic detention time to optimize the availability and uptake of VFAs and release of phosphate by PAOs. Recycle streams are the main sources of dissolved oxygen (internal recycles) and nitrate-nitrogen return activated sludge (RAS) input to anaerobic basins. The greater the dissolved oxygen and/or nitrate load to the anaerobic zone, the less efficient EBPR process will become. Sufficient VFA should be available relative to phosphorus load in the WRRF influent to maximize EBPR performance. Inadequate VFA throttles anaerobic phosphate release and reduces selection of PAOs resulting in a decline of EBPR performance. If VFA is insufficient, VFA/carbon supplements may be required to maintain the proper carbon to phosphorus balance to promote and maintain optimum EBPR. Conversely, excessive production of VFAs should be avoided as this condition can lead to excessive growth of filamentous bacteria.

Oxidation-reduction potential can be used as a quick indicator of anaerobic basin conditions. For example, low ORP (approximately –100 mV) is an indicator of conditions that favor phosphate release. Higher (more positive) anaerobic ORP conditions would be a reason to look at potential causal factors such as dissolved oxygen and/or nitrate-nitrogen entrained in RAS and/or recycles. In the anaerobic stage, ORP can also go too low (generally lower than –250 mV), resulting in reduction of sulfate to sulfide and even methanogenesis. Chapter 13 contains a detailed description of operations based on ORP.

Initial dissolved oxygen levels in the initial stage of the aeration zone of at least 2 mg/L are important for optimizing phosphorus uptake by PAOs.

### 4.3.2 Solids Retention Time

Aerobic SRT and total system SRT affect EBPR rather than an "EBPR" SRT. Anaerobic SRT of 1 to 2 days is sufficient for EBPR. A long SRT results in lower mass removed in waste activated sludge, which lowers the amount of total phosphorus removed through biomass. A long SRT will also lower the EBPR rate as a result of more endogenous activity of the biomass and less storage capability within PAOs. Long anoxic SRT, particularly in post-anoxic basins, in five-stage Bardenpho or similar process configurations, can cause secondary release of phosphorus by PAOs when nitrate-nitrogen levels are low. Contact time under anaerobic conditions is important to produce VFAs, which is the preferred food source of EBPR organisms. This contact time is determined by the flowrate entering the process and the anaerobic volume that is provided. In some facilities, VFAs are formed in the collection system and no contact time or only a short contact time is needed to produce adequate VFAs for EBPR. A balanced SRT to maintain high enough SRT for healthy nitrifiers and low enough for healthy PAOs must be determined for each BNR WRRF. Proper control of environmental/selective conditions and performance indicators can provide a foundation to optimize the selection of adequate SRT for nitrification while not excessively impacting PAOs.

### 4.3.3 Performance Indicators

The orthophosphate profile throughout the treatment process will indicate EBPR conditions and provide indication of where conditions are occurring that may be affecting optimum performance. High orthophosphate concentrations in the anaerobic basin are a key goal and the indicator of the performance of the PAOs to release phosphate that they had taken up in the aeration basin. Depending upon the conditions at each EBPR WRRF, orthophosphate release concentration in the anaerobic zone may range from 5 to 50 mg/L. The relative level of orthophosphate release provides indication of the health of PAOs, availability of VFAs, and anaerobic conditions. Low orthophosphate release may indicate one or more of these conditions are affected or insufficient and results in incomplete/inefficient uptake of orthophosphate under aerobic conditions. These conditions may be transient, but if one or more continue, over time, a downtrend of PAO selection can lead to trends of decreasing EBPR capacity and performance. Orthophosphate concentration in the aeration basin indicates the efficiency of phosphorus uptake by PAOs. It is important to maintain at least 2 mg/L dissolved oxygen at the beginning of the aerobic zone to promote optimum phosphorus uptake by PAOs. It is in this initial aerobic zone that soluble phosphate released in the anaerobic zone is at its highest concentration and available for the most efficient PAO uptake. Increased phosphorus levels in post-aeration zone units—post-anoxic basins and secondary clarifier effluent—indicate secondary phosphorus release, which may occur when dissolved oxygen and nitrate are absent and bacterial decay is occurring and can be enhanced if operating with excessive SRT.

## 4.4 Chemical Phosphorus Removal

The basic principle of chemical phosphorus removal is commonly termed *precipitation* followed by *solids removal.* More accurately, the orthophosphate is complexed in metal hydroxide compounds/solids for removal. The critical parameters for chemical phosphorus removal are physical and chemical and are related to the type and dose of chemical being added and to sufficient solids removal or filtration. Parameters to monitor include

- total phosphorus and orthophosphorus influent design loadings and target removal rates;
- chemical injection location, dispersion, and mixing;
- chemical type, dosing, and usage;
- alkalinity and pH control; and
- sedimentation and filtration performance.

Coagulants (metal salts) containing aluminum, iron, or calcium are commonly used to form insoluble compounds with orthophosphate and other reactive phosphorus and to turn soluble phosphorus into a solid for removal. Metals salts used for chemical phosphorus removal consume alkalinity and can lower pH. It is important to monitor alkalinity consumption and influent alkalinity to determine if supplemental alkalinity is required to maintain adequate alkalinity for pH stability through the WRRF.

After phosphorus is removed from the liquid stream, it is important to monitor the phosphorus mass flow in other areas in the facility because sidestreams containing high concentrations of phosphorus that return to the facility can be detrimental to the treatment process. Chemically bound phosphorus is removed from the liquid phase in the following three primary forms:

- Primary sludge;
- Secondary sludge; and
- Filter backwash.

Sidestream flows and phosphorus loadings from the handling of the aforementioned processes need to be properly managed and equalized so that high loads of phosphorus are slowly added back to the secondary process.

To control sidestream phosphorus loading, metal salts may be added ahead of sludge dewatering operations. Phosphorus is captured in the dewatered solids and removed from the WRRF in biosolids disposal.

### *4.4.1 Environmental Conditions*

Environmental conditions for chemical phosphorus removal are related to the physical process configuration and include four key parameters: injection dose, injection location and mixing, alkalinity, and solids removal performance. The chemical dose is unique to each chemical and is dependent on the target total phosphorus to be removed. After the chemical and dose have been selected, the chemical injection point should be placed far enough upstream of the sedimentation (primary or secondary clarifiers) and/or effluent filtration process and/or dewatering device to allow enough contact time for the chemical to react with the reactive phosphorus. The final condition requires successful performance of the solids removal or filtration process to remove the phosphorus that has been chemically bound as a solid.

#### 4.4.2 Performance Indicators

The main performance indicator of chemical phosphorus removal is simply a measure of whether the phosphorus removed from the treatment process meets the permit objective. Alkalinity consumption should also be monitored in relation to the permit and process requirements of the WRRF. Chemical dosage of both the metal salt and alkalinity supplementation should be closely monitored to optimize performance and minimize operation costs.

## 5.0 USE OF PROCESS MODELING FOR OPTIMIZATION

### 5.1 Introduction

Operators are now using models and simulators to understand nutrient removal processes, evaluate new designs, and optimize and troubleshoot. Models and simulators are becoming easier to understand and more user friendly. This chapter briefly describes what they are and how they can be used. More detailed information on this topic can be found in *Wastewater Treatment Process Modeling* (WEF, 2013).

Simulators are computer- or spreadsheet-based programs used to solve the various established equations that have been developed to simulate the response of a given activated sludge system to changes in various parameters. Typically, this is done in two steps. First, the reactor configuration and the flow scheme must be specified, including the reactor sizes, influent characteristics, recycle flowrates, wastage rate, and other specifics. After this information is fixed, it is possible to perform mass balances over each reactor (or reactor zone) for each model. The mass balances form a set of simultaneous equations, which, when solved, characterize the system behavior. The simultaneous solution provides values of the concentrations in different reactors and time. In this way, the change in concentrations throughout the system is related to the input and output and conversion processes occurring within the system.

It is important to emphasize that models are not perfect. This is particularly true for nutrient removal systems, in which there are many unresolved technical aspects that are subject to ongoing research. Perhaps the biggest limitation is the expense and effort required for comprehensive influent wastewater characterization. Furthermore, the nitrifier growth rate is a parameter with high variability, and therefore, using the default value may not give accurate model predictions. The user must understand the kinetics of the process and how the various coefficients relate to the prediction of BNR performance.

### 5.2 Use of Simulators to Optimize Facility Operations

Engineers and operators use simulators to assist in the design and operation of nutrient removal systems. An important function of the simulator is to help troubleshoot the process. The following are some troubleshooting examples:

- Models can be used to evaluate changing operational parameters such as RAS rate, mixed liquor recycle rate, dissolved oxygen concentrations in the bioreactor, and distribution of influent and recycles at various points within the bioreactor.

- Influent wastewater characteristics often change from those used during design, as industry leaves or a new industry comes to town, for example. An operator can input the new characteristics to the simulator to determine how those changed characteristics have affected predicted effluent quality. Through the simulator, the operator can decide whether to change operational parameters, to take tanks offline or add tanks (if no remedies exist within the existing facilities), or to request a capital expenditure for improvements.
- The simulator can be used to understand the effect of taking tanks out of service for maintenance. An operator can evaluate what time of year is the best for preventive maintenance by entering in various temperatures to determine which temperatures favor the available volume. The operator can also determine if carbon addition would attenuate the effects of the reduced volume.
- The potential for an existing facility to meet stricter effluent discharge requirements can be evaluated by operating at a higher MLSS concentration or a longer SRT. It is important to note that models do not predict sludge settling characteristics and how they are affected as environmental conditions change. Thus, model results alone may not result in meaningful effluent quality predictions unless sludge settleability is also addressed.
- Simulators give tools to the operator to understand what the design engineer is proposing, to help meet permits under all conditions, to improve the economics of operating a BNR facility, and to understand more fully what is happening within these processes. Models are not easy to use and require not only training but also knowledge of the kinetics of the biological system. However, these simulators are improving continuously. Indeed, modeling and simulation are assuming a prominent role in nutrient removal system design and can be useful for operation and control. See "Suggested Readings" (Section 9.0) for more on the application of modeling and process simulators for process optimization.

## 6.0 TROUBLESHOOTING

Refer to Appendix A for an optimization and troubleshooting guide.

## 7.0 OPERATIONS SUMMARY

Operating a BNR WRRF is a complex, integrated challenge. Whereas there are many common designs and processes for BNR, each application is unique. Differences in waste load character, intensity, consistency, required level of treatment, and environmental conditions create significant variables in optimizing treatment performance. Operators must take the fundamental control mechanisms and features used to control and maintain biological systems and apply them appropriately in each WRRF to provide the performance required to comply with permit conditions.

Key performance indicators should be trended and compared to provide early indication of changes in performance so proactive adjustments can be determined and initiated and responses monitored.

## 8.0 REFERENCES

American Public Health Association, American Water Works Association, & Water Environment Federation. (2023). *Standard methods for the examination of water and wastewater* (24nd ed.). American Public Health Association.

U.S. Environmental Protection Agency. (1993). *Nitrogen control manual* (EPA-625/R-93-010).

U.S. Environmental Protection Agency. (2005). *Guidelines establishing test procedures for the analysis of pollutants code of federal regulations* (Part 136, Title 40).

U.S. Environmental Protection Agency. (2008). *Energy management guidebook for wastewater and water utilities.*

Water Environment Federation. (2013). *Wastewater treatment process modeling* (2nd ed.; Manual of Practice No. 31).

Water Environment Federation. (2021). *Energy in water resource recovery facilities* (2nd ed.; Manual of Practice No. 32).

Water Environment Federation. (2023). *Standards for design of water resource recovery facilities* (WEF 8-23).

## 9.0 SUGGESTED READINGS

Bernardelli, A., Marsili-Libelli, S., Manzini, A., Stancari, S., Tardini, G., Montanari, D., Anceschi, G., Gelli, P., & Venier, S. (2020). Real-time model predictive control of a wastewater treatment plant based on machine learning. *Water Science & Technology, 81*(11), 2391–2400. https://doi.org/10.2166/wst.2020.298

Jenkins, D., Richard, M., & Daigger, G. (2003). *Manual on the causes and control of activated sludge bulking, foaming and other solids separation problems* (3rd ed.). Lewis Publishers.

U.S. Environmental Protection Agency. (2015). *Case studies on implementing low-cost modifications to improve nutrient reduction at wastewater treatment plants* (EPA-841/R-15-004).

Wahlberg, M. (2018). *Activated sludge wastewater treatment: Process control & optimization for the operations professional.* Destech Publications.

Water Environment Federation & Association of Boards of Certification. (2018). *Wastewater treatment fundamentals I, liquid treatment.* Water Environment Federation.

Water Environment Federation & Association of Boards of Certification. (2022). *Wastewater treatment fundamentals III, advanced treatment.* Water Environment Federation.

Water Research Foundation. (2023). *Guidelines for optimizing nutrient removal plant performance* (Project No. 4973).

# Chapter 12

# Recycle Streams Management



## 1.0 INTRODUCTION

This section addresses concerns associated with recycle streams and provides potential management strategies for the recycle of solids, nutrients, and other constituents from solids management processes. These recycle streams can represent a significant nutrient load on a water resource recovery facility's (WRRF's) nutrient removal process and must be managed to ensure compliance with effluent quality goals and improve treatment efficiency.

In this chapter, *recycle streams* refers to flows typically generated from solids processing units such as thickening, stabilization, and dewatering units. *Sidestream flows* refers to the effluent of add-on treatment processes that modify the quality of the recycle stream, whether via physical and chemical processes, biological processes, or both.

# 2.0 ISSUES AND CONCERNS

## 2.1 Introduction

Key parameters in recycle streams that can affect biological nutrient removal (BNR) treatment processes are solids, nutrients, pH, and alkalinity. Internal facility recycles such as those from solids dewatering processes, supernatant from sludge digestion processes, and overflow and underflow from thickening systems can introduce tremendous variability in WRRF solids and nutrient loadings on the BNR treatment process.

## 2.2 Suspended Solids

Recycle solids loads from solids thickening and solids dewatering processes can significantly increase solids loading returned to the BNR treatment process. When they are discharged to the BNR treatment process (either directly discharged or indirectly discharged upstream of primary clarification), there can be a reduction in the solids retention time (SRT) of the BNR treatment process, which can subsequently reduce the population of microorganisms and their treatment effectiveness. Lower SRTs can compromise nitrification effectiveness in particular, especially during the winter when wastewater temperatures are colder.

## 2.3 Nutrients

### *2.3.1 Nitrogen*

#### *2.3.1.1 Recycle of Ammonia*

Anaerobic processes, including fermentation and thermal stabilization processes that cause cell rupturing, will result in significant ammonia concentrations in the associated recycle streams (such as those from thickening and dewatering). The following are concerns associated with higher ammonia loading to the BNR treatment process:

- Increase in short-term (slug) peak nitrogen loading caused by intermittent dewatering operations;
- Increase in aeration demands, especially if the recycle stream is returned during the daytime, when diurnal influent ammonia loads are also highest;
- Reduction in the ratio of carbon to nitrogen, which may stress the denitrification process in the BNR treatment process; and
- Increase in the ammonia concentration being discharged from the BNR treatment process (i.e., higher effluent nitrogen concentrations).

#### *2.3.1.2 Recycle of Oxidized Nitrogen*

Aerobic digesters that are operated without significant anoxic (nonaerated) periods for partial denitrification can result in a significant load of oxidized nitrogen recycled to the BNR treatment process. The following are concerns associated with higher oxidized nitrogen loading to the BNR treatment process:

- Reduction in the ratio of carbon to nitrogen, which may negatively affect the denitrification capacity of the BNR treatment process;
- Increase in the oxidation-reduction potential of anaerobic zones (i.e., less negative) of the BNR treatment process, due to higher nitrate concentrations; and

- Increase in the oxidized nitrogen concentration being discharged from the BNR treatment process (i.e., higher effluent nitrogen concentrations).

### *2.3.2 Phosphorus*

Phosphorus recycled from anaerobic digestion processes can significantly affect the performance and operation of biological and chemical phosphorus removal processes at BNR treatment facilities. Concerns with recycled phosphorus are:

- Slug loading of phosphorus because of intermittent dewatering operation;
- Increase in metal salt demand for chemical phosphorus removal;
- Decrease in the ratio of carbon to phosphorus for the enhanced biological phosphorus removal (EBPR) process; and
- Increase in the potential for struvite formation.

Struvite ($MgNH_4PO_4$) formation can occur because all of the required constituents (ammonia, phosphorus, and magnesium) are present following anaerobic digestion. Struvite formation occurs when the pH increases in the digester and/or subsequent piping and process handling equipment, due to the release of carbon dioxide. While struvite is the most common chemical precipitate formed in the digester environment, other compounds such as bushite ($CaHPO_4$ $2H_2O$) and vivianite [$Fe_2(PO_4)$ $8H_2O$] may also form if favorable conditions are encountered. These favorable conditions include high calcium or reduced iron concentrations.

# 3.0 RECYCLE STREAM CHARACTERISTICS

## 3.1 Potential Concentrations From Various Solids Management Processes

Recycle streams from solids processing units typically contain biochemical oxygen demand (BOD), total suspended solids (TSS), nitrogen, and phosphorus. The characteristics of recycle streams can vary significantly depending on the solids treatment process that is implemented. In general, recycle streams of concern are from thickening, stabilization, and dewatering activities. The characteristics of these recycle streams cannot be generalized because of the wide variability exhibited within an individual WRRF and the strong influence of many site-specific factors, including the following:

- Facility influent characteristics;
- Mainstream unit processes that are implemented and their corresponding performance;
- Solids stream unit processes that are implemented and their corresponding performance;
- Nitrogen and phosphorus content of the solids;
- Operating schedule for solids handling processes, particularly dewatering; and
- Extent of other unintentional internal removal mechanisms (struvite formation).

The recycled BOD and TSS loads resulting from most conventional sludge processing operations are modest and generally do not pose any treatment challenges. Ammonia and orthophosphate loads, however, can be quite significant. Table 12.1 provides typical BOD, TSS, ammonia, and orthophosphate concentration ranges for sludge processing recycle streams.

TABLE 12.1 Biochemical Oxygen Demand and TSS Levels in Sludge Processing Sidestreams (Adapted from U.S. Environmental Protection Agency, 1987; Water Environment Federation [WEF], 2023)

| Source | Recycle Stream | $BOD_5$, mg/L | TSS, mg/L | Ammonia (mg $NH_3$-N/L) | Orthophosphate (mg P/L) |
|---|---|---|---|---|---|
| **Thickening** | | | | | |
| Gravity thickening | Supernatant | 100–1200 | 200–2500 | 1–30 | 1–5+ |
| Dissolved air flotation | Subnatant | 50–1200 | 100–2500 | 1–30 | 1–5 |
| Centrifuge | Centrate | 170–3000 | 500–3000 | 1–30 | 1–5 |
| **Stabilization** | | | | | |
| Aerobic digestion | Decant | 100–2000 | 100–10 000 | 1–50 | 2–200 |
| Anaerobic digestion | Supernatant | 100–2000 | 100–10 000 | 300–2000+ | 50–200 |
| Incineration | Scrubber water | 30–80 | 600–10 000 | 10–50 | 1–5 |
| **Dewatering** | | | | | |
| Belt filter press | Filtrate | 50–500 | 100–2000 | 1–1500 | 2–200 |
| Centrifuge | Centrate | 100–2000 | 200–20 000 | 1–1500 | 2–200 |
| Sludge drying beds | Underdrain | 20–500 | 20–500 | 1–1500 | 2–200 |

Characteristics of recycle streams can vary significantly depending on the treatment process. Therefore, facility operators should review their facility operation and develop real-time data. Additional care must be taken if and when future modifications to the treatment process are being considered. For example, measuring existing phosphorus recycle concentrations before EBPR will not reflect the additional phosphorus that will be in the recycle stream following implementation of EBPR. The following guidelines can be used to estimate recycle nitrogen and phosphorus loads:

- Development of a mass balance of the wastewater treatment facility (liquid and solids) is critical to understanding the magnitude of the flows and loads contained in the recycle streams. Even small flows with high concentrations of a particular wastewater constituent can have a significant effect on the biological treatment process. Additional sampling of internal process and recycle streams may be required for a period of time to develop a complete mass balance.
- Mass balances should be developed for average conditions with adjustments to assess the effect of operating schedules and peak conditions. Both present and projected mass balances should be performed and examined. It is important to note that a mass balance provides a "snapshot" of operating conditions and, therefore, cannot be used to predict the dynamic behavior of the biological treatment system and the effects of the recycle streams.
- Nitrogen release during digestion is directly related to volatile solids reduction. Nitrogen that is released during the digestion process is equivalent to approximately 8% to 10% of the volatile suspended solids (VSS) removed via the digestion process.
- Phosphorus release from EBPR sludges represents approximately 4% to 8% of VSS removed during digestion, whereas the potential phosphorus release from non-EBPR sludges represents approximately 1% to 3% of the digester influent VSS removed during the digestion process.

### *3.1.1 Thickening*

Sludge thickening processes can significantly affect the BNR treatment process with elevated solids recycle loads because of poor capture efficiency. The recycle solids can increase the influent inert solids when directed to the BNR treatment process. These inert solids decrease the active fraction at a given mixed liquor suspended solids (MLSS) concentration, which can negatively affect treatment performance. Recycle of microorganisms from sludge thickening processes can also cause treatment process upsets, which depend on the discharge location and inert solids load of the recycle stream.

In general, primary sludge and waste activated sludge (WAS) should be gravity thickened separately to minimize biological activity in the sludge-thickening process and reduce the effect of nutrient recycle. Mechanical thickening, such as via gravity belt or dissolved air flotation, may be a viable thickening approach that may minimize biological activity when operated continuously and immediately after mixing primary sludge and WAS. Nutrient release in thickening processes is typically not significant when primary and WAS sludges are thickened separately while minimizing the anaerobic detention time. For EBPR facilities, co-mixing of primary and WAS must be avoided to prevent the release of phosphorus from WAS. Anaerobic conditions in the thickening device can be similar to those in the anaerobic zone of an EBPR treatment process, which causes a significant release of phosphorus under anaerobic conditions.

### 3.1.2 Stabilization

#### 3.1.2.1 Aerobic Digestion

Aerobic digestion processes have an extremely long SRT (typically greater than 40 days) to achieve biosolids designation as per 40 CFR Part 503. This extended SRT allows for complete nitrification of the ammonia released from the breakdown of organic nitrogen in the digestion process. Therefore, aerobic digestion processes that are operated continuously and fully aerobically can produce a significant loading of oxidized nitrogen (i.e., nitrite and/or nitrate). This oxidized nitrogen load will be present in the aerobic digester decant recycle stream or subsequent dewatering recycle stream.

As mentioned previously, these oxidized nitrogen recycle streams can overwhelm the anoxic zones used for denitrification or negatively affect the anaerobic zones required for EBPR. Some facilities with aerobic digestion processes have incorporated unaerated (anoxic) periods to reduce oxidized nitrogen recycle loadings and electrical power requirements associated with the aerobic digestion process. Additionally, this anoxic period allows potential alkalinity recovery because of denitrification of the oxidized nitrogen in the aerobic digestion process.

Another strategy is to recycle the elevated oxidized nitrogen recycle streams to the preliminary treatment processes. This provides odor control benefits because it suppresses hydrogen sulfide production and can also cause sulfide oxidation.

#### 3.1.2.2 Anaerobic Digestion

Anaerobic digestion, through fermentation processes, releases high concentrations of ammonia through the breakdown of organic nitrogen. Typically, the ammonia concentrations associated with anaerobic digestion process recycle streams will be greater than 85% of the total nitrogen present. High levels of phosphorus may also be released into the recycle streams, particularly at EBPR facilities. These elevated phosphorus concentrations are present in EBPR because WAS contains high levels of stored polyphosphates (4% to 10%, by dry weight). A portion of the stored polyphosphates will be released from WAS when it is subjected to anaerobic conditions in the digestion process. Ultimately, the actual ammonia and phosphorus concentrations in the anaerobic digester will depend on influent solids characteristics and the volatile solids destruction performance of the anaerobic digestion process.

### 3.1.3 Dewatering

Dewatering of unstabilized sludges will have insignificant nutrient recycles, with the possible exception of EBPR WAS that has been subjected to anaerobic conditions (in which significant phosphorus release can occur). Dewatering recycle streams from stabilized sludges (both aerobic and anaerobic digestion processes) represents more of a concern because of the potential significant nutrient concentrations associated with these recycle streams. The following two dewatering process examples reflect those dewatering processes without (or with minimal) dilution water versus dewatering processes with significant dilution water.

#### 3.1.3.1 Presses (With and Without Spray Water in Filtrate)

Presses, such as belt filter presses and screw presses, can have recycle streams (called *filtrate*) that are diluted by spray water that is used to clean the cloth or other screening material used for dewatering the stabilized solids. A number of belt filter presses now have the ability to isolate the filtrate discharged from the gravity and belt pressure zone from the belt cleaning operation. However, many treatment facilities direct all of this liquid (filtrate and spray water) to the internal drain system as a combined recycle stream. In many applications, the spray water can be equal in volume to the filtrate, thus diluting and cooling the filtrate recycle stream. As will be discussed in Section 4.0, add-on sidestream treatment processes can take advantage of the higher temperature and concentration of the filtrate, suggesting that separation of the makeup water from the filtrate is desirable.

#### 3.1.3.2 Centrifuges

Centrifuges can dewater influent feed solids to between approximately 20% to 35% total solids. The soluble nutrient concentration of the centrate is equivalent to the stabilization process soluble nutrient concentration because makeup water is not used in the dewatering centrifuge operation. Accordingly, the centrate will be similar in terms of soluble constituent characteristics of the upstream stabilization process, but the particulate constituent characteristics will vary from the upstream stabilization process.

## 3.2 Intermittent Versus Continuous Recycle Issues

Recycle streams containing released nitrogen and phosphorus are typically blended with the facility influent and recycled through the biological treatment process. These recycle streams are not always continuous and, in many facilities, occur intermittently. For example, sludge dewatering operations conducted 5 days per week, 8 hours per day, will increase the recycle nutrient loading rate to the biological treatment process by a factor of 4 to 5 versus a 7-day-per-week, 24-hour-per-day operation. Accordingly, significant nutrient concentration increases can occur during this intermittent discharge, which challenges the ability of the BNR treatment process to achieve effluent quality goals. Additionally, the time of day and discharge location of the recycle loadings to the BNR treatment process can significantly affect loading on the BNR treatment process.

Figure 12.1 shows an example of three different modes of recycle loadings discharging directly to the BNR treatment process. The dashed line shows a fully equalized recycle loading (24-hour-per-day recycle of dewatering recycle stream). The dotted line shows the effect of dewatering 5 days per week between 0800 and 1600 hours (i.e., mid-day), and the solid line shows the loading on the BNR treatment process when dewatering is performed 5 days per week between 2300 and 0700 hours (i.e., night). In this example, dewatering overnight provides the most equalized ammonia loading on the BNR treatment process. It also represents a cost savings (electricity) by operating during nonpeak periods of electrical demand. However, there are two other considerations. The first consideration is that the carbon available in the raw wastewater will be lower during morning hours; thus, the ratio of carbon to nitrogen may not be favorable for good denitrification. The second consideration is that, if the centrate for the overnight dewatering operation is sent to the facility headworks, then it will significantly increase

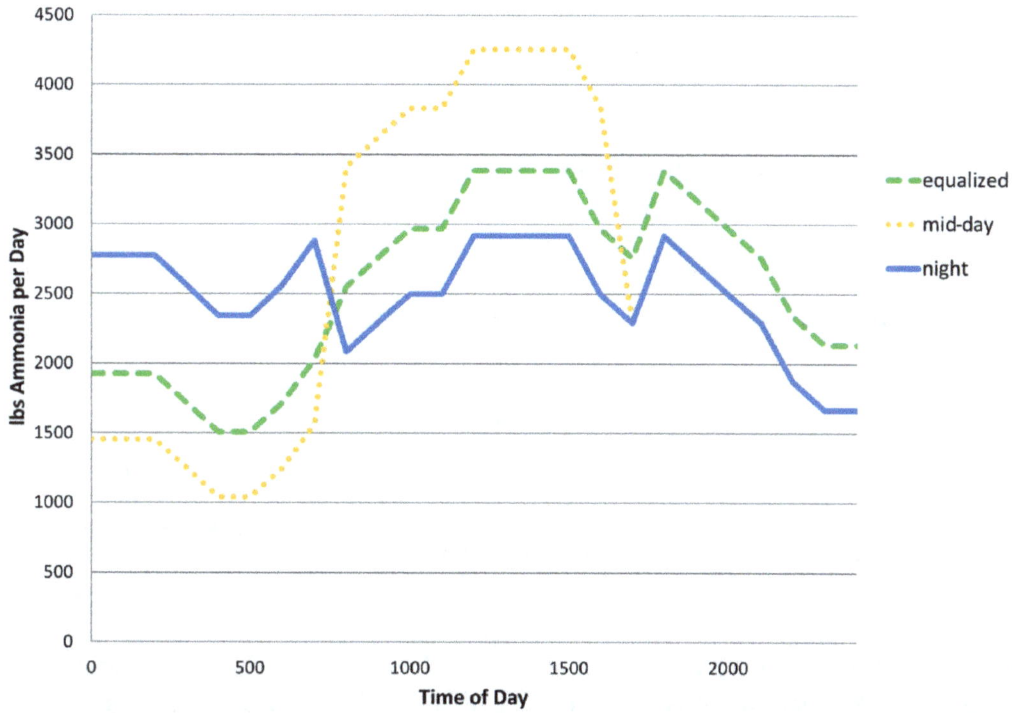


FIGURE 12.1 Centrate Equalization

the early morning recycle loads. For example, in a facility with primary clarifiers, nutrient concentrations in the primary clarifiers can increase significantly and be "washed out" during the early morning hours. This can cause a significant increase in ammonia loading to the BNR process.

This example illustrates that careful examination of each facility's operation and the potential effect of different dewatering schedules and operating schemes can be beneficial to the BNR treatment process. Monitoring of recycle loads and daily diurnal loadings of facility influent can provide valuable information on how to optimize recycle stream management. Similarly, facilities that accept significant septic tank loads and other high-nutrient wastes should consider the effect of this operation on recycle loads to the biological treatment process.

## 4.0 SIDESTREAM TREATMENT OPTIONS

This section provides a general overview of sidestream treatment processes and the significant benefits and tradeoffs of each. There are numerous management and treatment options available to control the effect of recycle loading on the BNR treatment process. Figure 12.2 shows various options available for managing nitrogen and phosphorus in dewatering recycle streams at the time of publication of these standards. The process configuration that works best for a particular facility depends on multiple factors, including influent loading, effluent quality requirements, existing infrastructure and facilities, and the economics of treatment. Recycle streams discussed in this section are those for stabilization processes that generate significant nutrients (whether ammonia, phosphorus, or both) and are typically associated with anaerobic digestion processes. The key factor in sidestream treatment processes is taking advantage of the high loading and low volume of the recycle stream. The type and degree of sidestream treatment is facility specific and must always be weighed against alternatives. The remainder of this section describes principles behind the various sidestream technologies.

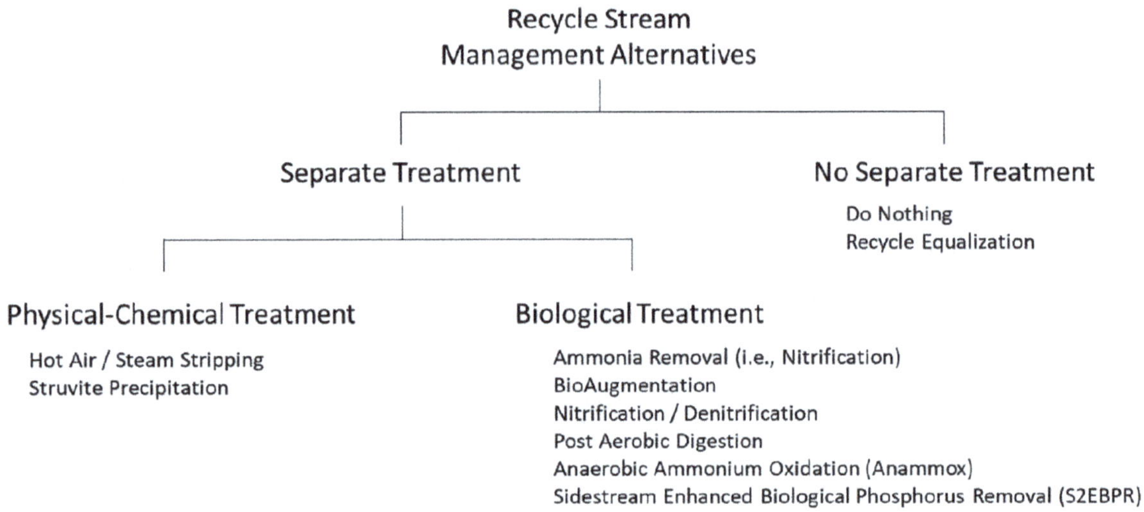


FIGURE 12.2 Overview of Centrate Processes (Adapted from Constantine et al., 2005)

Although a detailed discussion of each process cannot be provided in these standards, a general description of these treatment options is included. Further details can be obtained from *Standards for Design of Water Resource Recovery Facilities* (WEF, 2023), *Nutrient Removal* (WEF, 2010), and vendor and technology-owner websites. These technologies (some of which are proprietary) can be as simple as equalization tanks and as complex as physicochemical/biological treatment.

## 4.1 Alternatives

### 4.1.1 *Recycle Equalization*

Recycle equalization is one of the first tools for mitigating the effect of recycle streams on the BNR treatment process. As shown in Figure 12.1, equalizing the dewatering recycle loading on the BNR treatment process may be achieved by modifying dewatering operations; however, this may not be possible because of labor availability. The incorporation of an equalization tank to hold the dewatering recycle stream volume is one of the simplest operating strategies that permits a controlled discharge back to the BNR treatment process at the most opportune time. A well-mixed tank (mechanical or hydraulic) will not only ensure a homogeneous mixture but can also reduce the potential for stripping odors.

### 4.1.2 *Recycle Equalization and Semi-treatment*

#### 4.1.2.1 *Solids Removal*

The equalization tank should be designed to allow solids removal because solids in the dewatering recycle stream typically settle very rapidly because of the polymer present. This equalization tank can be mixed, but incorporating an operational strategy to operate unmixed for a portion of the day will allow the solids to settle in the equalization tank, thereby providing for solids removal. This mode of equalization tank operation is designed to reduce the solids recycled to the BNR treatment process.

#### 4.1.2.2 *Partial Nitrification*

Some facilities have incorporated aeration of equalization tanks to allow for some nitrification to occur. Partial nitrification can occur in the equalization tank, with the maximum amount of nitrification being approximately 50% of the recycle stream ammonia load because of alkalinity limitations. However, aeration of the equalization tank can promote struvite formation because of the stripping of carbon dioxide from the liquid, thus causing an increase in pH and precipitation of struvite. Without close monitoring for struvite, extensive coating of concrete wall, aeration diffusers, and other equipment has been observed.

## 4.2 Ammonia Removal (Ammonia Oxidation Only)

For facilities with ammonia and/or total nitrogen limits, the additional ammonia in the recycle stream will reduce overall facility capacity (vs. no recycle loading). Having a higher ammonia concentration in the BNR influent will require additional oxygen and time to achieve complete nitrification and, for total nitrogen removal processes, additional nitrate that must be removed. For those facilities that may be borderline in meeting effluent discharge limits, use of a sidestream nitrification process, particularly warm anaerobic digester dewatering liquor, represents a cost-effective approach to managing the higher ammonia loading.

The concept for a separate nitrification process is to nitrify the warmer recycle stream and discharge the oxidized nitrogen to the main facility. One example is the centrate treatment process at the 91st Avenue Water Pollution Control Plant in Phoenix, Arizona (Coughenhour II et al., 2010). The main BNR treatment process is required to reduce a total nitrogen influent (with centrate) of 35 mg/L down to less than 8 mg/L. Because of the high concentration of ammonia within the facility's centrate, pilot studies were conducted that demonstrated that a separate, complete-mix activated sludge-style reactor with a detention time of approximately 1.8 days could oxidize more than 95% of the ammonia within the centrate, with the addition of alkalinity. Actual operation has demonstrated that nitrification without supplemental alkalinity has been possible, even with a reactor pH of less than 6. In addition to reducing the ammonia loading on the biological treatment process, the discharge of oxidized nitrogen to the headworks also reduced the cost of odor control chemicals (from $460 per day to $150 per day) because of the oxidation of influent hydrogen sulfide.

Another benefit of separate, centrate nitrification is the use of the WAS from the centrate nitrification process as a seed source for nitrifiers for the main BNR treatment process; this process is referred to as *bioaugmentation*. Nitrification bioaugmentation with WAS from sidestream treatment has been widely demonstrated in full-scale facilities using various process configurations (Bowden et al., 2016; Parker & Wanner, 2007). For example, a patented process by M2T Technologies, Lotepro Environmental Systems & Services, called In-Nitri® (Kos et al., 2000), uses WAS from the separate, centrate nitrification reactor to supplement the activated sludge system of the main BNR treatment process with nitrifiers, which allows facilities to operate at lower SRTs while maintaining nitrification. Conversely, there are some bioaugmentation processes that send a portion of the return activated sludge (RAS) from the main BNR treatment process to the sidestream process (e.g., BABE®). This offers the advantage of using nitrifiers already acclimated to conditions within the biological treatment process to continue to grow using ammonia in the recycle stream.

Another process that has gained widespread use (but may not be strictly considered a sidestream process) is performed using RAS from the main BNR treatment process within the existing infrastructure of the main facility. Essentially, a portion (or all) of the RAS from the main facility's BNR treatment process is mixed with the recycle stream and aerated in the first pass of a step-feed and contact stabilization activated sludge process. This process of RAS reaeration and centrate treatment, known as BioAugmentation Reaeration (BAR), uses the first zone of the biological treatment tank to treat the high-strength ammonia recycle stream with recycled RAS sludge, while main process influent is introduced to the downstream zone (Parker & Wanner, 2007). Other variations include the common centrate and RAS reaeration basin process (CaRBB) (Luna et al., 2010) developed at the Robert W. Hite Treatment Facility in Denver, Colorado. The centrate generated in the combined solids handling facility is distributed equally to the head of each sludge reaeration tank (CaRRB unit) along with a portion of RAS. The CaRRB effluent (partially nitrified centrate plus RAS) is discharged into the upstream end of the main influent channel to be mixed with primary effluent. The portion of RAS not routed through CaRRB is sent to the head of the main influent channel. It should be noted that some facilities have installed a new, separate RAS reaeration tank before the main activated sludge tank to take advantage of this process.

The advantages of BAR are the relatively high concentrations of MLSS (RAS concentration) and ammonia, which increase the nitrification population (bioaugmentation) in comparison to similar environmental conditions treating raw wastewater. In addition, the detention time in this reactor can be significantly longer than the remainder of the biological reactor, further increasing the nitrification population and oxidation of ammonia.

## 4.3 Nitrogen Removal

There are numerous sidestream biological treatment processes available to remove nitrogen. Typically, these processes take advantage of the warmer temperatures and higher ammonia concentrations of the recycle stream to more efficiently remove nitrogen in a sidestream reactor, rather than performing the same function in the main BNR treatment process. Using a conventional approach, sidestream nitrification and denitrification can achieve nitrogen removal in a smaller volume than returning the recycle stream to the main biological treatment process. To minimize costs associated with sidestream nitrification and denitrification, facilities may consider using mainstream carbon to supplement denitrification, rather than an external carbon source (Lacroix et al., 2020). The following subsections describe treatment processes that further improve on this "traditional" sidestream nitrogen removal process.

### 4.3.1 *Post-aerobic Digestion*

Post-aerobic digestion (PAD) of anaerobically digested sludge can be applied to BNR treatment facilities to reduce the concentration of nitrogen recycled back to the main BNR treatment process without the need for supplemental carbon or alkalinity. Soluble, biodegradable carbon leaving a conventional anaerobic digester is typically low, but in PAD, aerobic endogenous decay occurs, resulting in additional volatile solids destruction. The solids content of PAD is high compared to a mainstream BNR treatment process, resulting in higher oxygen uptake rates and low or non-detectable oxygen concentrations within the reactor. This creates a combination of oxic and anoxic conditions that allow for simultaneous nitrification and denitrification within the reactor (McNamara et al., 2022). Technologies such as Digestivore-PAD® can remove up to 90% of nitrogen in the sidestream. Other benefits of PAD include volatile solids reduction, odor reduction, and struvite stabilization.

### 4.3.2 *Anammox*

The anaerobic ammonium oxidation (anammox) reaction harnesses a specific species of autotrophic bacteria to achieve partial nitritation-deammonification (i.e., the conversion of nitrite and ammonia to nitrogen gas). The anammox reaction is described as follows (Strous et al., 1998, 1999):

$$NH_3 + 1.32\ NO_2^- + 0.066\ HCO_3^- + 0.13\ H^+ \Rightarrow 1.02\ N^+\ 0.26\ NO_3^- + 0.066\ CH_2O_{0.5}N_{0.15}\ 2.03\ H_2O$$

In a partial nitritation-deammonification process, approximately 50% of the available ammonia is converted to nitrite under aerobic conditions by ammonia oxidizing bacteria (AOB). Next, the remaining ammonia and nitrite are converted to nitrogen gas and a small amount of nitrate under anoxic conditions by the (eponymous) anammox bacteria. This process offers the following benefits over conventional nitrification and denitrification and nitritation and denitritation processes:

- Only approximately one-half of the ammonia needs to be converted to nitrite, thus reducing the energy required for aeration.
- The resulting nitrite and remaining ammonia are converted to nitrogen gas, thus eliminating the need for an external carbon source for denitrification.
- There are lower sludge yields, in comparison to nitrification-denitrification.

Challenges associated with this process include the following:

- Anammox bacteria have a slow growth rate, with a doubling of population in the range of 6 to 12 days. Startup of an anammox process without seed sludge requires from 100 to 180 days. Accordingly, these microorganisms must be handled with care, and operational configurations must be optimized to prevent the loss of anammox bacteria.
- The operation of nitritation over nitrification can be challenging. The primary operational strategy to stop nitrification at nitrite is to maintain a minimal SRT and a high temperature (25 to 35 °C). At this temperature, AOB, which have a faster growth rate than nitrite oxidizing bacteria (NOB), will survive in the system while NOB will be washed out of the system.

Partial nitritation-anammox has been effectively implemented in sidestreams with ammonia concentrations greater than 100 mg/L, and temperatures above 26 °C. The effectiveness of the process depends on the successful pairing of AOB and anammox, as well as an effective NOB out-selection strategy. To this end, several process configurations that combine nitritation and anammox exist and can be categorized as having either "separate" or "integrated" nitritation and anammox steps (Wett, 2005).

In the first step of a "separate" process (such as SHARON), a nitritation reactor is typically operated without addition of supplemental alkalinity, which results in conversion of approximately half of the ammonia in the dewatering centrate to nitrite. The effluent from the nitritation reactor, which typically is composed of approximately equal parts of ammonia and nitrite, is fed into the anammox reactor, which is operated at relatively high temperatures. The anammox bacteria present in this second reactor convert ammonia and nitrite to nitrogen gas, typically without carbon addition.

In an "integrated" nitritation and anammox process scheme, nitritation and anammox occur in the same reactor. There are several "integrated" nitritation and anammox technologies available. DEMON is a sequencing batch reactor process that uses intermittent aeration to allow for cycling between nitritation and anammox. A pH-based control system determines the length of aeration intervals depending on the production of hydrogen ions (pH depression) or nitrite (Wett, 2005). Another successful technology, ANITAMox is a single stage, continuously aerated moving bed process that uses carrier media to retain anammox within the reactor. Other anammox-based technologies include ELAN, Cleargreen, ANAMMOX, and TerraMOX.

At the time of publication, combined partial denitrification and anammox as opposed to partial nitritation-anammox, is beginning to take hold. This process can be accomplished without relying on extensive nitrite oxidizing bacteria outselection.

### 4.3.3 Physicochemical Nitrogen Removal

Ammonia stripping can be achieved by raising the pH of the recycle stream so that the aqueous ammonia exists in equilibrium with its gaseous counterpart in accordance with Henry's law. When the pH exceeds 9.5, un-ionized ammonia prevails and can be stripped from the recycle stream. This requires a large volume of air to achieve high ammonia removal. In addition, the exhaust stream should be captured to eliminate odor problems. This requires condensing the exhaust gas and capturing the ammonia. Stripping via pH can cause scaling problems and is subject to freezing during cold weather.

A second option is to steam strip the ammonia from the recycle stream. Studies conducted by the New York City Department of Environmental Protection (Gopalakrishnan et al., 2000) indicated that steam stripping is more cost effective than hot air stripping and can achieve 70% to 90% removal of ammonia from the recycle stream. The adjustment of pH (typically 7.0 to 7.5) is not required in a stream stripper, thereby reducing scaling issues. The stripped ammonia associated with exhaust gas is sent to a distillation column, where the exhaust gas is cooled and a pure ammonia solution is formed. Clogging did occur in heat exchangers as a result of organic material, including hair, calcium, and magnesium in the recycle centrate stream. Pretreatment requirements for steam stripping include fine screening and suspended solids settling to prevent clogging issues.

A third option is to use a vacuum distillation process to remove most of the ammonia in the recycle stream. A proprietary vacuum distillation system (RCAST®) that uses a combination of physical separation methods has reportedly demonstrated the ability to remove ammonia from municipal centrate and filtrate (Orentlicher et al., 2009). This process includes pretreatment of the centrate (i.e., removal of hair and other material interfering with the mechanical process) and conversion of aqueous soluble ammonium ions to ammonia gas by elevation of pH before introduction to the RCAST® system. The RCAST® system liberates soluble ammonia gas from the recycle stream by increasing the temperature and lowering the operating pressure of the recycle stream. The liberated gaseous ammonia is entrained in a sulfuric acid solution and converted to highly soluble and stable ammonia sulfate.

## 4.4 Phosphorus Removal

Recycle streams from anaerobic digestion processes and dewatering can contain a significant amount of soluble phosphorus because of the hydrolysis of microorganisms and solids. Typically, the phosphorus concentration in microorganisms is approximately 2% on a dry-weight basis to satisfy metabolic requirements and is organically bound in the microorganisms. Elevated phosphorus concentrations between 4% and 10% on a dry-weight basis can result from EBPR treatment processes.

### 4.4.1 Chemical Precipitation

Highly concentrated orthophosphate recycle streams can be managed using metal salts to precipitate the phosphorus. The advantage of adding metal salts to the more concentrated recycle stream is that the ratio of metal to phosphorus is closer to theoretical demand because of the higher concentration of orthophosphate, lower pH (which must be managed for subsequent treatment), and lower competing reactions with other constituents. Whereas use of metal salts is a method to precipitate and capture the recycle stream phosphorus, this practice requires a significant amount of chemicals and increased dewatering volume because of the increased chemical solids generated.

### 4.4.2 Sidestream Enhanced Biological Phosphorus Removal

By diverting a portion of biomass from activated sludge to a sidestream reactor, fermentative conditions can be created, which can produce a more diverse EBPR microbiome and increase phosphorus removal performance and stability.

### 4.5 Nutrient Recovery

Sustainable nutrient recovery and reuse is gaining traction as wastewater utilities look for ways to decrease energy costs and reduce greenhouse gas emissions, while simultaneously meeting more stringent regulatory requirements. One of the most promising areas is chemical recovery from recycle streams.

Struvite (magnesium ammonium phosphate [$MgNH_4PO_4$]) is typically a nuisance for WRRFs that operate anaerobic digestion because it precipitates on piping and pumps from the anaerobic digester effluent through the dewatering facilities. Once formed, it is an extremely hard substance that can cause piping to clog without an easy removal method. Often, facilities will need to replace piping (vs. the expense of cleaning the pipe). However, struvite is also an excellent fertilizer with the ability to release nutrients over an extended period of time, reducing costs to the agricultural user while also reducing nutrient runoff from farmland. Extensive work on physicochemical treatment processes was carried out in the late 1980s and early 1990s with processes to induce struvite precipitation.

Struvite precipitation had limited applicability because orthophosphate and magnesium often are present in relatively low levels in anaerobically digested dewatering recycle streams. Typically, these concentrations are approximately 10% of the mass loading of the ammonia concentration. The stoichiometric requirements for the precipitation of 1 g of $NH_4$-N as struvite are 1.7 g/L of $Mg^{2+}$ and 2.2 g/L of $PO_4$-P. For facilities that have an EBPR biological treatment process, the orthophosphate concentration of the dewatering recycle stream is typically much higher. Magnesium addition is also typically required because magnesium is the limiting element in the formation of struvite. Magnesium is typically added in the form of magnesium chloride or magnesium hydroxide to maximize the capture of phosphorus. In addition, the ammonia concentration remaining after struvite formation will remain high, thereby requiring additional management.

There are a number of struvite precipitation processes that have been commercially developed. In North America, a process using a fluidized bed reactor for struvite precipitation (marketed by Ostara under the commercial name Pearl®) has been demonstrated to be technically and commercially viable at multiple full-scale facilities (Baur, 2011). A recirculation flow suspends the struvite particles in a reactor. The particles are segregated by size by the upflow velocity. The heaviest, large particles are in the lowest, smaller section, where the velocity is the greatest, the magnesium, ammonia, and phosphorus concentrations are the highest, and precipitation of struvite on existing particles occurs. The desired particle size is taken from the reactor, dried, and can be sold as a fertilizer supplement.

## 5.0 OPERATIONS SUMMARY

The following are key operating takeaways for recycle stream nutrient management systems:

- Recycle streams represent a small fraction of a WRRF's flow but can represent a significant fraction of a WRRF's nutrient load. It is vital to establish a sampling program that quantifies nutrient loads in each of the recycle streams.
- Intermittent recycle schedules can have a significant effect on nitrogen loading in the mainstream process. Place careful consideration on the time of day that recycle streams are returned to the mainstream process, and/or equalization of the recycle stream.
- Many options are available for nutrient load reductions in recycle steams. Consider effluent goals, energy, and space demands when selecting the best alternative for a WRRF.

## 6.0 REFERENCES

Baur, R. (2011, January 9–12). Results of the first year of operation of North America's first full scale nutrient removal facility. *Proceedings of the WEF/IWA Specialty Conference—Nutrient Management and Recovery,* Miami, FL, United States. Water Environment Federation.

Bowden, G., Tsuchihashi, R. H., & Stensel, D. (2016). *Technologies for sidestream nitrogen removal.* IWA Publishing.

Constantine, T., Murthy, S., Bailey, W., Benson, L., Sadick, T., & Daigger, G. T. (2005, October 30–November 2). Alternatives for treating high nitrogen liquor from advanced anaerobic digestion at the Blue Plains AWTP. *Proceedings of the 78th Annual Water Environment Federation Technical Exposition and Conference* [CD-ROM], Washington, DC, United States. Water Environment Federation.

Coughenour II, J. R., Walz, T., Blatchford, G., Phillips, J., & Husband, J. (2010, October 2–6). Innovative approach to centrate nitrification accomplishes multiple goals—Nitrogen removal, odor control, and potential for $40 million savings. *Proceedings of the 83rd Annual Water Environment Federation Technical Exhibition and Conference* [CD-ROM], New Orleans, LA, United States. Water Environment Federation.

Gopalakrishnan, K., Anderson, J., Carrio, L., Abraham, K., & Stinson, B. (2000, October 14–18). Design and operational considerations for ammonia removal from centrate by steam stripping. *Proceedings of the 73rd Annual Water Environment Federation Technical Exposition and Conference* [CD-ROM], Anaheim, CA, United States. Water Environment Federation.

Kos, P., Head, M. A., Oleszkiewicz, J., & Warakomski, A. (2000, October 14–18). Demonstration of low temperature nitrification with a short SRT. *Proceedings of the 73rd Annual Water Environment Federation Technical Exposition and Conference* [CD-ROM], Anaheim, CA, United States. Water Environment Federation.

Lacroix, A., Mentzer, C., & Pagilla, K. R. (2020). Full-scale N removal from centrate using a sidestream process with a mainstream carbon source. *Water Environment Research, 92,* 1922–1934.

Luna, B., Narayanan, B., Rogowski, S., & Walker, S. (2010, October 2–6). Metro's CaRRB diet—Centrate treatment process tackles big challenges in small package. *Proceedings of the 83rd Annual Water Environment Federation Technical Exhibition and Conference* [CD-ROM], New Orleans, LA, United States. Water Environment Federation.

McNamara, P., Sabba, F., Redmond, E., Dunlap, P., Worley-Morse, T., Marks, C., & Downing, L (2022). Post aerobic digestion (PAD) is a solids sidestream nutrient removal process that utilizes native carbon: Performance and key operational parameters from two full-scale PAD reactors. *Environmental Science: Advances, 1,* 216–228.

Orentlicher, M., Ginzburg, H., & Grey, G. (2009, October 10–14). Greenhouse gas and energy savings from integration of centrate ammonia reduction with BNR operation: Simulation of a New York City WPCP. *Proceedings of the 82nd Annual Water Environment Federation Technical Exhibition and Conference* [CD-ROM], Orlando, FL, United States. Water Environment Federation.

Parker, D., & Wanner, J. (2007, October 13–17). Review of methods for improving nitrification through bioaugmentation. *Proceedings of the 80th Annual Water Environment Federation Technical Exhibition and Conference* [CD-ROM], San Diego, CA, United States. Water Environment Federation.

Strous, M., Fuerst, J. A., Kramer, E. H. M., Logemann, S., Muyzer, G., van de Pas-Schoonen, K. T., Webb, R., Kuenen, J. G., & Jetten, M. S. M. (1999). Missing lithotroph identified as new planctomycete. *Nature, 400,* 446–449.

Strous, M., Heijnen, J. J., Kuenen, J. G., & Jetten, M. S. M. (1998). The sequencing batch reactor as a powerful tool for the study of slowly growing anaerobic ammonium-oxidizing microorganisms. *Applied Microbiology and Biotechnology, 50,* 589–596.

U.S. Environmental Protection Agency. (1987). Sidestreams in wastewater treatment plants (U.S. EPA Design Information Report). *Journal Water Pollution Control Federation, 59,* 54–59.

Water Environment Federation. (2010). *Nutrient removal* (Manual of Practice No. 34).

Water Environment Federation. (2023). *Standards for design of water resource recovery facilities* (WEF 8-23).

Wett, B. (2005, September 19–21). Solved scaling problems for implementing deammonification of rejection water. *Proceedings of the IWA Specialized Conference on Nutrient Management,* Krakow, Poland. International Water Association.

## 7.0 SUGGESTED READINGS

U.S. Environmental Protection Agency. (2007). Side stream nutrient removal. *Wastewater technology fact sheet.*

# Chapter 13

# Process Control, Instrumentation, and Automation

# 1.0 INTRODUCTION

Biological nutrient removal (BNR) processes often require more sophisticated operations than conventional processes that do not involve nutrient removal; in addition, capital and operational costs of a BNR facility may be higher than a conventional activated sludge facility. Therefore, BNR system operation can significantly benefit from using online analyzers for both monitoring and automatic control. The BNR facility should have adequately trained staff, or contracts in place, to maintain the instrumentation and control systems. This chapter provides an overview of online analyzers and sensors, process parameter optimization, automatic control, and requirements for control systems.

# 2.0 PROCESS CONTROL PARAMETERS

## 2.1 General Considerations

External conditions, such as flowrate and ammonia load variations, vary significantly at wastewater facilities. For example, the ratio of maximum to minimum process oxygen demand, consisting of demand for both biochemical oxygen demand (BOD) and ammonia removal, typically varies from 1.5 to 1 to 3 to 1 between peak and off-peak hours. For small water resource recovery facilities (WRRFs), this ratio is sometimes as high as 10 to 1, depending on the amount of infiltration, inflow, and other extraneous sources to the collection system. To maintain optimum process performance, it is necessary to compensate for changing ammonia and carbonaceous oxygen demand (COD) load by changing airflow supply, waste flow, internal recycling flow, and other control parameters. This can be accomplished through the implementation of control schemes discussed in the previous chapters.

## 2.2 Selecting Optimum Setpoints

Selecting the optimum setpoint for each of the control approaches discussed in the previous chapters is a complicated task. Table 13.1 presents considerations that must be considered when selecting the setpoint for each parameter.

## 2.3 Basic Automatic Control

Optimum performance of BNR systems can be achieved only when the solids retention time is maintained at an optimum value. Below this value, a system can experience low dissolved oxygen filamentous bulking, poor ammonia removal and nitrite breakthrough, dispersed growth of biomass, and an overloaded thickening facility. At a solids retention time (SRT) above the optimum value, a system may experience a low F/M ratio, filamentous bulking and foaming, increased oxygen demand, and increased clarifier loading.

Wasting mixed liquor is the simplest method to implement SRT because it does not require measurement of the waste activated sludge solids concentration. In addition, such wasting can help control foam. However, this simplicity comes at a price. As a result of the low solids concentration, the volume of excess sludge may be several times larger than the volume of sludge to be wasted from the return activated sludge (RAS) line. As a result, the pumping cost of the wasted sludge and recycled water from the thickening facility is also higher. Pipes, pumps, and thickening facilities may need to be upsized to accommodate an increased flow, which is why wastage from the RAS line alone, or in combination with wasting foam from mixed liquor, is more widespread. At the same time, SRT control for this

TABLE 13.1 Considerations for Selecting Setpoints

| Process | Parameter | Potential problem is that value is: Lower than optimum | Higher than optimum |
|---|---|---|---|
| Primary treatment | Sludge depth | Reduction of volatile fatty acid production and, as a result, reduction of influent denitrification potential; reduction of efficiency of subsequent sludge treatment processes as a result of excess water in thin sludge discharge from primary clarifiers | Increased effluent total suspended solids (TSS), soluble and particulate BOD, and hydrogen sulfide ($H_2S$) generation; phosphate release in case of co-thickening of primary and BNR wasted sludge; increased floating solids resulting from gasification; increased mechanical stress on sludge and scum collection mechanism |
| Biological nutrient removal process | Solids retention time | Chlorination problems resulting from the presence of nitrite in the BNR effluent; inadequate removal of phosphate, ammonia, and nitrate; deterioration of waste sludge thickening as a result of thin sludge discharged from secondary clarifiers | Low food-to-microorganism (F/M) ratio of foaming and bulking; increased clarifier solids loading; phosphate release; increased energy demand to sustain endogenous respiration |
| Biological nutrient removal process | Dissolved oxygen concentration | Chlorination problems resulting from the presence of nitrite in the BNR effluent; inadequate removal of ammonia; foaming problems caused by *Microthrix*; gasification in the clarifiers resulting from the presence of nitrogen ($N_2$) | Inhibition of denitrification; high energy cost resulting from excessive airflow supply; breakup of floc and, as a result, increased effluent TSS |
| Biological nutrient removal process | Internal recycle | Reduction of the amount of nitrate to be denitrified | Decreased denitrification rate resulting from increased oxygen concentration in the anoxic compartment; increased pumping cost |

popular wasting method is somewhat more challenging than wasting methods previously mentioned in these standards.

## 2.4 Nitrification

Nitrification performance is affected by a complex relationship between several key parameters, including SRT, F/M ratio, dissolved oxygen concentration in the aeration basin, mixed liquor volatile suspended solids (MLVSS), hydraulic retention time, temperature, and available alkalinity, as described in detail in Chapter 3. Each parameter has an optimum value for achieving the desired treatment level. When one parameter is outside of its optimum range, other parameters can often be adjusted to compensate. Some parameters, such as available alkalinity, cannot be compensated through manipulation of other parameters and would need outside correction such as adding a chemical to meet the process requirement.

### 2.5 Denitrification

Denitrification is affected by the F/M ratio, dissolved oxygen concentration, temperature, MLVSS, and hydraulic detention time, as described in detail in Chapter 5. In some nitrified wastewaters, there is not an adequate source of carbon for denitrification to occur. In these situations, a supplemental source of food (carbon) is required. Most commonly, the carbon source is methanol; however, other chemicals that are less hazardous to handle have been introduced to the marketplace. Some facilities that do not use a supplemental carbon source purposely reduce their F/M ratio to allow more naturally occurring carbon to be available for denitrification. This approach is particularly well suited to facilities using a chemically enhanced primary process and those using two stages of treatment, such as a trickling filter followed by a conventional aeration basin.

The denitrification process requires good mixing to ensure that any supplemental chemical is well dispersed, biomass remains in suspension, and anoxic conditions are maintained (the minimum dissolved oxygen available can quickly be depleted in a settled sludge environment, leading to anaerobic conditions). The mixing process allows surface agitation to provide for the minimum dissolved oxygen required for an anoxic environment.

### 2.6 Enhanced Biological Phosphorus Removal

Available phosphorous is typically only partially removed in conventional biological treatment processes. The mass fraction of phosphorous in MLVSS is typically between 1.5% and 2.5%. Generally, this represents less than half of the available phosphorous. However, implementing an enhanced biological phosphorus removal (EBPR) process, as described in Chapters 6 and 8, can increase this mass fraction to between 6% and 10%.

### 2.7 Chemical Phosphorus Removal

Chemical phosphorous removal, as described in detail in Chapter 10, is often used when the required effluent level is below 1 mg/L or to supplement or back up EBPR where lower regulatory limits are required. The two most commonly used chemicals are ferric chloride and aluminum salts, which are both well-known and effective coagulants. When added to advance treated wastewater upstream of polishing filters, effluent levels of 0.1 mg/L are typically achievable.

## 3.0 ONLINE AND IN-LINE INSTRUMENTATION

Instruments to measure flow, temperature, pH, TSS, COD, dissolved oxygen, oxidation-reduction potential (ORP), ammonia-ammonium, nitrate-nitrite, and phosphorus-orthophosphate are commonly used at BNR facilities for monitoring, control, and automation. The principles of operation, accuracy and repeatability, installation, maintenance requirements, and specific applications for each of these instruments are discussed in this chapter. Installation and maintenance of flow meters are discussed in Section 3.1. Installation and maintenance of all other instruments are discussed in Sections 3.2 through 3.8. For more detailed information on these topics, refer to *Standards for Automation of Water Resource Recovery Facilities* (Water Environment Federation [WEF], 2024).

### 3.1 Flow

There are many types of flow meters used at WRRFs, including flumes; weirs; and electromagnetic, ultrasonic, differential-pressure, mechanical, variable-area, and mass flow meters.

### 3.1.1 Principle of Operation

Electromagnetic flow meters operate based on the principle that the voltage induced by a conductive fluid moving through a magnetic field is proportional to the velocity of the fluid. The flowrate is then calculated based on the velocity of the fluid and the size of the conduit carrying the fluid.

There are two types of ultrasonic flow meters: transit time and Doppler. Transit-time ultrasonic flow meters measure fluid velocity by calculating the difference in travel time for one sonic pulse to go a specific distance downstream and another sonic pulse to go the same distance upstream. In a Doppler ultrasonic flow meter, a signal of a specific frequency is sent into the fluid. The suspended solids of gas bubbles in the fluid then reflect the signal back to a transducer. If the fluid is moving, the reflected signal will be of a different frequency than the original signal. The magnitude in shift in frequency between the original signal and the reflected signal is proportional to flowrate.

Weirs and flumes can be used to measure flow when combined with a level-sensing device. Fluid flows through the weir or flume, and the water depth is measured at a specific point. The flow is then determined using characteristic head-to-flow relationships that have been developed for specific weir and flume geometries.

Differential-pressure flow meters operate based on the principle that when a fluid passes through a pipe, it will cause a pressure drop that is proportional to the square of the flowrate. Pitot tubes and Venturi meters are two types of differential-pressure flow meters.

Mechanical flow meters operate based on the principle that the movement of an object that has been moved by a fluid can be correlated to the flowrate of the fluid. Turbines and propellers are examples of mechanical flow meters.

Variable-area flow meters typically consist of a glass flow tube and a stainless-steel float. Fluid passes vertically through the tube. The level of the stainless-steel float rises in proportion to the flowrate through the tube.

Mass flow meters directly measure the mass flow rate of a material. Coriolis meters and thermal-dispersion flow meters are examples of mass flow meters. A Coriolis meter measures mass flow by inducing a vibration and comparing it to the measured vibration of the material flowing through the meter. Thermal-dispersion mass meters use heat transfer and the known characteristics of the fluid being measured to determine the mass of material flowing through the meter per unit of time.

### 3.1.2 Accuracy and Repeatability

Each style of meter provides a different level of accuracy. Most applications do not require an extreme amount of precision in flow measurement. Typically, 0.5% is sufficient for most wastewater applications. If the meter is being used for control, repeatability is typically more important than accuracy.

### 3.1.3 Installation

To obtain the most accurate readings, electromagnetic flow meters should be installed in a portion of a pipe that has a low amount of turbulence. A manufacturer typically specifies the minimum distance that the flow meter should be installed from any fittings or valves. The distance varies based on the diameter of the pipe and whether the fittings or valves are located upstream or downstream of the instrument. Flow meters should be installed on the discharge side of pumps and on the upstream side of throttling valves. Pipe fittings should be kept at the minimum specified distances on the upstream and downstream side of the meter. Flow meters that require a full-flowing pipe should be installed in pipes where the flow is going up or upward. Horizontally mounted flow meters need air-bleed valves in many applications. All flow meters should be installed in areas where there is enough space to access the meter or to remove the meter body for calibration or maintenance. Providing bypass around the flow meter is critical in applications where the process cannot be shut down.

### 3.1.4 Maintenance Requirements

Maintenance requirements vary based on the type of flow meter and the manufacturer. The manufacturer should be consulted for the required maintenance of any flow meter under consideration. Maintenance requirements can include periodic inspection, periodic cleaning of devices such as orifice plates, cleaning of pipes where grease has accumulated, altering the pipe diameter, replacing moving parts that wear, and calibration checks to maintain accuracy.

## 3.2 Temperature

### 3.2.1 Principle of Operation

Thermal couples and thermal bulbs, resistance temperature detectors (RTDs), and thermistors are the three main types of temperature sensors that are commonly used. Thermocouples operate based on the principle that heating a metallic junction composed of two dissimilar metals will cause a flow of current, also known as the *Seebeck effect*. Resistance temperature detectors consist of a wire made of a pure metal, typically platinum, coiled around a ceramic or glass core. The RTD measures the temperature by correlating the resistance in the metal coil to temperature. Thermistors are semiconductors whose resistivity changes in response to temperature change. The resistance of most thermistors decreases with increasing temperature.

### 3.2.2 Accuracy and Repeatability

Temperature sensors typically have a resolution of 0.1 °C and an accuracy of about 0.5 °C.

## 3.3 pH Measurement

### 3.3.1 Principle of Operation

The heart of a pH sensor is the glass membrane. An electrical potential, varying with pH, is generated across the membrane. The difference between this potential and a reference electrode is measured and amplified by an electronic signal conditioner. The complete electric circuit includes the glass electrode wire, glass membrane, process fluid, reference electrode fill solution, and reference electrode wire.

A pH electrode assembly, or *sensor*, as it is sometimes called, consists of the following two primary parts:

- Measuring electrode: The measuring electrode is sometimes called the *glass electrode* and is also referred to as a *membrane* or *active electrode*.
- Reference electrode: The reference electrode is also referred to as a *standard electrode*.

The measuring and reference electrodes can be in one of the following two forms: two physically separate electrodes, known as an *electrode pair*, or joined together in a single glass body assembly, known as a *combination electrode*. The combination pH electrode is the most widely applied.

### 3.3.2 Accuracy and Repeatability

Manufacturer claims for pH meter accuracy range from ±0.02 to ±0.2 pH units. This represents the combined accuracy of the electrodes and the signal conditioner or transmitter. Without temperature compensation, an additional error of 0.002 pH units per 1 °C difference from the calibration temperature can be expected. The repeatability of pH meter measurements varies by manufacturer from 0.02 to 0.04 pH units. Stability (drift) is an important performance parameter that indicates how often meters must be recalibrated. Manufacturer claims for stability vary from 0.002 to 0.2 pH units of drift per week. With flow-through probe mounts, the velocity of the sample can cause a shift (0.2 to 0.3 pH) in measured values. High velocity creates non-uniformity around the glass, displacing some of the charges and affecting the pH measurement.

Methods of reporting performance specifications vary among manufacturers. Adjustment of the method of reporting performance specifications to equal units of measure shows that there is large variance in the accuracy and stability claimed by different manufacturers. Typically, good pH meters achieve the following performance standards in WRRFs: accuracy, ±0.1 pH units; repeatability, ±0.03 pH units; and stability, ±0.02 pH units per week.

### 3.3.3 Applications

In wastewater treatment, pH sensors are used to monitor facility conditions and biological treatment process conditions and to control acid or base additions for pH adjustment. Regulatory agencies require measurement of facility influent and effluent pH to extrapolate overall facility conditions.

It may also be necessary to monitor the pH of specific industrial discharges to give advance warning of possible toxic conditions. Whereas the activated sludge and most other biological processes can tolerate a pH variance of 5 to 9, some, such as anaerobic digestion, are pH sensitive. Normal monitoring of facility influent and primary effluent or mixed liquor suspended solids (MLSS) (if applicable) is typically sufficient to detect impending toxic conditions.

An online pH monitor can provide feedback for control of other processes requiring pH adjustment. For example, pH adjustment may be required to neutralize low-pH industrial wastes, enhance phosphorus removal by alum addition, or adjust pH to optimum ranges for nitrification and denitrification.

## 3.4 Total Suspended Solids

### 3.4.1 Principle of Operation

Currently, most TSS instruments use near-infrared technology. However, instruments that use microwave, nuclear, or ultrasonic technologies are also available.

There are two main types of analyzers using near-infrared technology: transmittance and reflectance. Transmittance analyzers are available in both two-beam and four-beam configurations. The two main types of reflectance analyzers are those that measure beams backscattered at 90° from the source and those that measure forward-scattered beams. There are some reflectance analyzers that have multiple detectors. For example, one analyzer has detectors positioned 90 and 140° from the light source, and another analyzer uses as many as six channels of multi-angle measurement. Depending on the application, one method might perform better than another; therefore, facility operators may need to be diligent to identify the proper sensor selection.

The most important feature for any technology is color compensation or color-independent analysis. Blackish, highly concentrated sludge is more difficult to measure than light-colored sludge with a lower suspended solids concentration. Multichannel analyzers may provide some advantages for these applications.

Another important feature is a method for sensor cleaning. The following automatic cleaning methods are available: water or air purging, ultrasonic, and wiper cleaning. Some manufacturers claim that their products are not susceptible to active biofouling and, as a result, do not require a self-cleaning system. According to information from these manufacturers, infrequent manual cleaning is still required. However, it is always advisable to consider self-cleaning systems, even for those analyzers for which cleaning is not specified by the manufacturer.

### 3.4.2 Accuracy and Repeatability

The accuracy of a suspended solids analyzer is typically 5% of reading, and several ranges of operation are available. The repeatability of solids analyzers is typically ±1% of reading or ±0.1 g/L, whichever is greater. It is important to note that different sensors and different manufacturers have different accuracy, precision, and error specifications, and investigation of these parameters is necessary to find the best TSS sensor for the application.

### 3.4.3 Applications

Suspended solids analyzers are used to control sludge wasting to maintain a desired SRT and MLSS.

## 3.5 Carbonaceous Oxygen Demand

The laboratory method used for direct measurement of COD, which involves wet chemistry, can be completed in a few hours. The method uses hazardous chemicals (potassium dichromate, silver sulfate catalyst, and mercuric sulfate). Therefore, the laboratory COD test has limited use in online analysis. Online wet chemical COD analyzers are available, but these are not commonly used. In addition, online methods for measuring COD are available, but these methods are surrogate methods in which COD is estimated using a correlation with the laboratory method. The three main methods of online measurement of COD are ultraviolet (UV) absorbance, thermal combustion, and electrochemical oxidation. These methods are discussed in the following subsections.

### 3.5.1 Principle of Operation

In the UV absorbance method of measuring COD, a spectrophotometer measures the intensity and wavelength of UV light that passes through a sample. The intensity and wavelength of light passing through the sample is compared to the intensity and wavelength of light being shined on the sample to determine the intensity and wavelength of light absorbed by the sample. The intensity and wavelength of UV light absorbed by the sample is then correlated to the COD concentration in the sample.

In thermal combustion, a sample is pumped from a process area and injected to a furnace that is operated at temperatures up to 1200 °C. Once the sample enters the furnace, it is completely oxidized. The gas from the furnace goes through a cooler and an absorption column, then into a gas analyzer. The gas analyzer determines the amount of COD in the sample.

In the electrochemical oxidation method, an electric current is passed through an electrode producing hydroxyl radicals. The sample is brought into contact with the hydroxyl radicals, where it is oxidized, and the hydroxyl radicals are converted into hydroxyl ions. This process causes a change in the electrode current. The COD concentration is correlated to the change in current.

### 3.5.2 Accuracy and Repeatability

Although the accuracy of COD analyzers varies with the type of instrument used, most instruments can achieve accuracies of ±3%.

## 3.6 Dissolved Oxygen

The laboratory method used for direct measurement of dissolved oxygen, the Winkler test, involves wet chemistry and can be completed in a few minutes. The method uses hazardous chemicals (manganese II sulfate, potassium iodide in a hydroxide solution, acid, and thiosulfate). Therefore, the laboratory dissolved oxygen test has limited, if any, use in online analysis. There are two main types of dissolved oxygen sensors: optical and electrochemical.

### *3.6.1 Principle of Operation*

Luminescent dissolved oxygen analyzers use optical technology using luminescent material on a sensor, light-emitting diodes, and a photo diode detector to measure dissolved oxygen. The sensors, typically factory calibrated, require no sensor calibration and minimal maintenance. Membrane dissolved oxygen analyzers use an electrochemical measurement from one electrode to a second electrode across a galvanic cell and using a dissolved oxygen selective membrane. They typically require a minimum of weekly calibration, and the membranes require routine cleaning/replacement.

### *3.6.2 Accuracy and Repeatability*

Although the accuracy of dissolved oxygen analyzers varies with the type of instrument used, most dissolved oxygen concentration instruments can achieve accuracies of ±0.05% to 0.1%. depending on range and manufacturer. Handheld dissolved oxygen analyzers should be used for routine checks to determine if cleaning or calibration is needed by the field instruments, in addition to profiling dissolved oxygen throughout the BNR zones. Inaccurate dissolved oxygen readings can cause energy loss due to inaccurate aeration control and potential for excess dissolved oxygen to be carried over to anoxic zones.

## 3.7 Oxidation-Reduction Potential

The laboratory and field method used for indirect measurement of ORP is dependent on the reference electrode (silver vs. platinum) in the ORP sensor (probe). Different manufacturers use different reference electrodes with specific reference potentials. To compare ORP values for two instruments, manufacturers should have the same reference electrode, or one measurement should be compensated for the difference in reference potential.

### *3.7.1 Principle of Operation*

Oxidation-reduction potential is determined by measuring the potential of the sensing electrode relative to a reference electrode. The rate of electron transfer is measured in millivolts (mV).

### *3.7.2 Accuracy and Repeatability*

Because of certain wastewater characteristics, the types of ORP sensors used and reference electrodes affect ORP readings. Exact values cannot be assigned for aerobic, anoxic, and anaerobic conditions. The accuracy of ORP analyzers varies with the type of instrument and reference electrode used. The ORP sensor should be calibrated regularly. Generally, ORP sensors should be calibrated every few months or as necessary. The proper standard solution for calibration will depend on the manufacturer's recommendation for specific probes.

### *3.7.3 Applications*

In wastewater treatment, ORP sensors are used to monitor biological treatment process conditions and to provide feedback for control of aerobic, anoxic, and/or anaerobic environments, for example, in swing zones for BNR of nitrogen and phosphorous. Oxidation-reduction potential can be used as a quick indicator of aeration basin conditions; a high ORP value is an indicator of conditions that favor nitrification, and low ORP value is a symptom that other process indicators should be looked at, such as dissolved oxygen, ratios of COD to total Kjeldahl nitrogen (TKN) or BOD to TKN, or toxicity. Autotrophic nitrification requires aerobic ORP greater than +50 mV (Zehnder & Stumm, 1988). During nitrification, the oxidation of ionized ammonia ($NH_4$) to nitrate ($NO_3$) is performed by nitrifying bacteria when the ORP of the wastewater is +100 to +350 mV (Grissop, 2010). Lower ORP could indicate an inactive sludge in the aerobic zone even though dissolved oxygen levels could be around 2.0 mg/L or higher. Occasional ORP measurement by a portable ORP meter is sufficient to ensure a positive millivolt potential.

Denitrification occurs at ORPs in the range of −100 mV (Zehnder & Stumm, 1988) to +25 mV (Zhao et al., 1999). Dabkowski (2008) identified ORP ranges in the anoxic tanks for five of the most common combined BNR processes (i.e., five-stage Bardenpho, anaerobic, anoxic, oxic [A2/O], University of Cape Town, Virginia Initiative Process, and modified Johannesburg; see Chapter 9 for detailed process descriptions). The ORP in the anoxic reactors was in the range of −100 to +100 mV for all five BNR processes. It should be noted that the ORP values suggested for denitrification are general recommendations that may differ with changing operating conditions and facility design. The suggested ranges are starting points; therefore, the optimal range of ORP values for a specific location should be based on additional criteria such as sludge settleability and effluent total nitrogen concentrations. One of the most valuable uses for an online ORP sensor is trending the process to see any sudden changes in ORP value that indicate the environment is no longer anoxic.

In enhanced phosphorus removal processes, an ORP probe is considered a reliable indicator to determine aerobic and anaerobic conditions. Sudden changes in the ORP value, for example, could help indicate whether the process is no longer aerobic. During the biological phosphorus release stage (the anaerobic zone), an ORP range of −100 to −225 mV is typical.

## 3.8 Nutrient Analyzers

Ammonia-ammonium, nitrate-nitrite, and phosphorus-orthophosphate analyzers are discussed in detail in Sections 3.6.1 through 3.6.4. Regarding the accuracy of nutrient analyzers, it is important to note that consistency is often more important than accuracy. Analyzer results are often discredited because they do not match laboratory tests. However, full-scale experience has shown that the information they provide is just as valuable if they are consistent because operators will learn how to interpret the readings.

### *3.8.1 Ammonia and Ammonium*

Ammonia ($NH_3$) is a gas that, when dissolved in wastewater, exists in equilibrium with ammonium ($NH_4$). The equilibrium between the two molecules is dependent on pH and temperature. When applicable, most WRRFs have a permit limit expressed in terms of ammonia. However, depending on the measuring principle used, instrument manufacturers may express results as either ammonia or ammonium. To avoid confusion, most results are reported in terms of nitrogen (ammonia-nitrogen [$NH_3$-N] or ammonium-nitrogen [$NH_4$-N]). Instruments for measuring ammonia and ammonium are discussed in the following subsections.

#### *3.8.1.1 Principle of Operation*

The three principal methods for measuring ammonia-ammonium are ion-selective electrode (ISE), gas-sensitive electrode (GSE), and colorimetry. In principle, ammonium ISEs are similar to a pH electrode. However, instead of a glass bulb, the sensors have a specialized membrane that allows ammonium to pass through the membrane into the cell. Ammonium is destroyed in the cell through a chemical reaction. The voltage created by the reaction is proportional to the ammonium concentration in the sample. Ammonium ISEs are the most common method used for online measurement of ammonium.

An ammonia GSE is, effectively, a pH electrode behind a gas-permeable membrane. Sodium hydroxide is added to the sample to raise the pH above 11 and to drive all the ammonium to ammonia. The ammonia gas is allowed to pass across the GSE membrane and then redissolves in an electrolyte surrounding a pH electrode. The ammonia gas increases the pH of the electrolyte. The analyzer then converts this change in pH to an ammonia reading. A GSE is not suitable for online measurement because the sample has to be withdrawn from the tank and taken to the laboratory.

Ammonia can be measured using the colorimetric method. In the colorimetric method, a chemical reaction causes a color change in the sample. The color change is proportional to the ammonia concentration in the sample. There are three main colorimetric methods: Indophenol blue method; Monochloramine-F method; and a third colorimetric method in which ammonium in the sample is converted to ammonia gas, stripped from the sample, and then redissolved in a liquid pH indicator solution. The change in color of the indicator solution is converted to an ammonia reading. Colorimetric methods are not commonly used at wastewater facilities.

#### 3.8.1.2 Accuracy and Repeatability

Ammonia ISEs measure between 0.2 and 1000 mg/L $NH_4$-N, with an error of 5% of the measured value, or 0.2 mg/L $NH_4$-N, whichever is greater. Gas-sensitive electrodes are more precise, measuring between 0.02 and 1000 mg/L $NH_4$-N with an error of 2% of the measured value or 0.02 mg/L $NH_4$-N, whichever is greater.

#### 3.8.1.3 Applications

Applications of ammonia and ammonium analyzers are discussed in Section 3.8.4.

### 3.8.2 Nitrate and Nitrite

#### 3.8.2.1 Principle of Operation

Online measurement of nitrate ($NO_3$) in wastewater is most commonly achieved using UV absorbance. Nitrate and nitrite in water absorb UV light between 215 and 240 nm. Online instruments direct UV light through the sample and measure the amount of light that passes through to a detector. The absorbance is calculated and converted to a nitrite-nitrate value. The result is provided in milligrams per liter of $NO_x$-N (where $NO_x = NO_3 + NO_2$) because both nitrate and nitrite absorb light in that spectrum. In wastewater facilities, the concentration of $NO_2$ is generally insignificant and considered part of the error of the measurement. In some instruments, light of a different wavelength (a reference wavelength) is also passed through the sample to allow for compensation for interference resulting from solids and organic material that also absorb UV light.

Another method of measuring nitrate uses ISE. The ISE works in the same manner as was described for ammonium. This technology is not frequently applied because of the convenience of the direct UV absorbance method.

Instruments that measure nitrite using UV absorbance have recently been developed. These instruments measure the UV absorbance of a sample and then separate the absorbance of nitrite from that of other compounds that absorb UV light by either using a proprietary algorithm or comparing the spectra of the sample to a library of known spectra. Nitrite can also be measured using online colorimetric methods.

#### 3.8.2.2 Accuracy and Repeatability

Ultraviolet nitrate probes typically provide a lower detection limit of 0.05 to 0.1 mg/L $NO_x$-N. The measurement range may extend as high as 50 mg/L $NO_x$-N, although this may require a longer path length. Claimed accuracy varies from 2% to 5% of the reading, and repeatability is in the same range.

#### 3.8.2.3 Applications

Applications of nitrate and nitrite analyzers are discussed in Section 3.8.4.

### 3.8.3 Phosphorus and Orthophosphate

Total phosphorus consists of orthophosphate, organic phosphate, and polyphosphate. There are instruments available that measure total phosphorus. Most total phosphorus in wastewater is typically present as orthophosphate ($PO_4$). Therefore, most instruments measure orthophosphate. Results are typically reported in terms of orthophosphate as phosphorus ($PO_4$-P).

#### 3.8.3.1 Principle of Operation

All online orthophosphate instruments use colorimetry. The most commonly used methods are the following: the molybdovanadate method, also referred to as the *yellow method*, and the ascorbic acid method, also referred to as the *blue method* (or *molybdenum blue method*). In practice, the yellow method is most widely applied because of the relative simplicity of the instrument and related reagents. In instances in which low detection levels are required (<0.1 mg/L $PO_4$-P), the blue method is used.

Online total phosphorus instruments first convert polyphosphate and organic phosphorus compounds to orthophosphate. Orthophosphate is then measured colorimetrically. The conversion step requires high pressures and temperatures; it is this step that makes online total phosphorus instruments relatively expensive and maintenance intensive.

#### 3.8.3.2 Accuracy and Repeatability

Most orthophosphate analyzers that use the yellow method are capable of measuring over a range from 0.05 mg/L to as high as 50 mg/L $PO_4$-P. The measurement error is approximately 2% or 0.05 mg/L, whichever is higher. Analyzers that use the blue method can measure down to 2 mg/L $PO_4$-P. The measurement error is typically ±4% of measured value or ±0.02 mg/L $PO_4$-P, whichever is greater. Total phosphorus analyzers are available with measuring ranges from as low as 0.3 mg/L to as high as 100 000 mg/L. The measurement error is typically 3% of the measured value.

#### 3.8.3.3 Applications

Most discharge permits for phosphorus in the United States are written in terms of total phosphorus. Although online total phosphorus instruments are available, they are relatively expensive and require a high level of maintenance. As a result, orthophosphate instruments are more widely accepted, particularly for process monitoring and control, and, in many instances, for effluent monitoring. Applications of phosphorus and orthophosphate analyzers are discussed further in Section 3.8.4.

### 3.8.4 Nutrient Analyzer Applications

Online nutrient analyzers discussed in Sections 3.8.1 through 3.8.3 are used in the following applications:

- Influent monitoring for feed-forward control of chemical dosing and mass-load balancing;
- Mass-load equalization;
- Aeration control;
- Recirculation control;
- At the exit of an anaerobic zone to ensure adequate release of phosphorus from phosphate-accumulating organisms; it can also be used to limit excessive phosphate release by controlling recirculation (containing nitrate) to the anaerobic zone;
- At the exit of the aeration basin to ensure adequate uptake of phosphorus; and
- Effluent monitoring (facility effluent, clarifier effluent, etc.) for feedback control of a chemical dose.

## 3.9 Instrument Installation

There are two methods to install the instruments discussed in Sections 3.1 through 3.8: in-tank or open-channel installation and free-standing installation. With the in-tank or open-channel approach, the analyzer is installed directly to the process tank or channel. This provides a faster response time and alleviates the need to transport samples to the analyzer with a pump. The analyzer should be installed in a well-mixed zone to provide a representative sample of the process. If the probe is installed in an open channel, it should be located in a free-flowing zone.

The main disadvantage of the in-tank or open-channel approach is the inconvenience of having to remove the analyzer for reagent replacement and maintenance. This task may become overwhelming if analyzers require frequent maintenance.

Under the free-standing approach, the instrument is installed in a sampling station located away from the process tank or channel. Samples must be delivered to the analyzer by a pump and are often required to be filtered. The design of sample pumping, piping, and filtration has critical implications on the quality of data being produced by the overall system. An important factor to consider is the detention time in the piping, which should be kept to a minimum. One approach to minimize detention time is to deliver a high-volume sample stream to the analyzers. Only a small portion of the sample stream will be transferred to the analyzer, and the rest of the stream will be wasted. Another important consideration is the potential buildup of biomass in the piping. Biomass can react with wastewater and alter the characteristics of the sample stream before it reaches the analyzer. Therefore, it is necessary to provide some means of periodically flushing the line (often with a hypochlorite solution) to strip biomass from the pipe walls. Installation site lists, which are typically provided by an analyzer manufacturer, should be checked to learn more details about successful design of a sample delivery and preparation system.

Free-standing analyzers are generally wall mounted and sometimes require a shelter. When this approach is implemented, all analyzers and other high-maintenance instruments are often located in the same area to provide easier service. Buffer solutions needed for standardization and other analytical instrument reagents can also be conveniently stored in the analyzer area.

## 3.10 Instrument Maintenance

Most instrument manufacturers provide a variety of features to reduce maintenance, such as automatic cleaning and calibration and the use of peristaltic pumps and pinch valves (so the sample stays inside the tubing and does not contact any moving parts). Nevertheless, some of the more sophisticated instruments may require extensive routine maintenance.

The following maintenance steps and frequencies are common to most online analyzers:

- Routine replacement of reagents typically occurs monthly or quarterly, but can vary according to the measurement, calibration, and cleaning intervals chosen.
- Cleaning or replacement of electrodes (for pH probes and ISE and GSE instruments) on a monthly basis. Automatic sensor cleaning systems are available on some instruments.
- Replacement of sample tubing every 3 to 6 months.
- Regular inspection (and cleaning, if necessary) of the sample filtration system.
- Regular cleaning of optical windows (for spectrometer instruments). Although this procedure is sometimes performed manually, there are instruments that have automatic cleaning systems.
- Regular calibration (weekly calibration is a reasonable frequency); calibration should be based on laboratory analysis of the same grab sample processed in triplicate.

With the increased complexity and sophistication of control system hardware, it has become increasingly important to train and retrain control system staff to ensure that systems are operated correctly and that instruments and network systems are functioning properly. Many facilities that are upgrading from conventional treatment to BNR systems are not staffed to handle the new instrumentation that goes along with the process conversion. Facilities that are upgrading to BNR should familiarize instrumentation and control staff with the new process by involving them in construction and startup of the process.

## 4.0 OPTIMIZATION AND ADVANCED CONTROLS

### 4.1 Diurnal Load Variations

The effects of diurnal variations in flow and loadings do not allow facilities to achieve true steady-state operating characteristics. For certain process technologies, this can be particularly challenging. A batch process like sequencing batch reactors (SBRs) (see Chapter 7), repeating several times a day, may experience highly variable loads over the fixed process cycle time. In these instances, without sophisticated control systems to manage the aeration blowers for dissolved oxygen control and allowing for adequate oxygen depletion time in transitioning to anoxic conditions, it is likely that either incomplete nitrification will occur because of oxygen limitations or incomplete denitrification will result from a lack of available carbon or excessive residual dissolved oxygen levels.

Conventional aeration basins provide a naturally occurring buffering of flow and load variations. It is also easier to maintain targeted dissolved oxygen levels at certain points in the treatment process. However, separate clarifiers, waste and return MLSS pumping systems, and other factors generally result in more costly construction to achieve high-quality effluent characteristics.

### 4.2 Respirometry

Respirometry is the continuous measurement of the specific oxygen uptake rate of microorganisms in the activated sludge process as they consume organic matter. Respirometry has been widely used to characterize waste and as a method of determining kinetics. Development of respirometric control has been made possible by the use of an online respirometer. This instrument has been used to conduct surveys of WRRFs, to develop diurnal load profiles, and to optimize the operation of many facilities. Several control approaches have been suggested, and they are dependent on the type of activated sludge facility that is to be controlled. Oxidation ditches, completely mixed tanks, SBRs, and plug flow reactors are all suitable candidates for respirometric control.

Respirometric control was shown to be the most effective approach for improving nitrogen removal in an oxidation ditch in Beemster, Holland (Draaiger et al., 1997). In this application, an in situ respirometer was used to measure the oxygen load present in the ditch and to simply control the number of aerators in operation to match demand.

Respirometric control is an obvious choice for SBRs, and some promising investigations have been carried out in this field by Cohen et al. (2003) and Yoong et al. (2000). The simplest method of respirometric control, proposed by Shaw and Watts (2002), is to discontinue aeration once the respirometric measurement indicates that the biomass is endogenous and allow an idle phase that makes use of endogenous denitrification to improve overall nitrogen removal (Shaw & Watts, 2002). Some SBR facilities have implemented a dual-process cycle approach in which two aerated and anoxic cycles are used per batch to allow for initial nitrogen removal, followed by denitrification, and then repeated a second time within the overall treatment cycle. This allows for more readily available carbon during the initial denitrification step.

Plug flow reactors provide the greatest challenge to implementation of respirometric control, although they were one of the first activated sludge processes considered for advanced control by the foremost proponent of respirometric control, Dr. John Watts (Watts & Garber, 1993). Watts and Garber proposed that anoxic and aerobic swing zones could be used with respirometric measurements to adjust the overall aerobic volume to match influent load. Rapid respirogram generation is required to facilitate dynamic control; alternatively, respirometry can be used to build a picture of typical load variations, and this can be used to adjust anoxic and aerobic volumes.

Regardless of which BNR process is used, respirometric control can be used to improve the performance of the facility. There is great potential to provide better nitrogen removal, to increase nitrifier viability by reducing the amount of time that the biomass is endogenous, and to reduce energy costs by matching aeration to the nitrogenous oxygen demand in the facility.

## 4.3 Intermittent Aeration

Intermittent aeration in a continuous-flow system (as opposed to SBRs) is an effective way to achieve nitrification and denitrification in a single reactor. In this process, as described in Chapter 7, the biomass in the main reactor (or reactors) is subjected to a cycling of aerobic and then anoxic conditions as the aeration supply is switched on and off to that particular basin. Because the reactor repeatedly changes from an aerobic state to an anoxic state and back again, one measurement option for control is ORP. Several examples of ORP control systems exist, including those described by Caulet et al. (1997), Charpentier et al. (1998), and Mori et al. (1997). In these systems, ORP is used to determine when denitrification is complete under anoxic conditions and when nitrification is complete under aerobic conditions. Carucci et al. (1997) also recognized the potential for using ORP to control intermittent aeration but contended that pH was a more suitable control parameter based on their experiments at the Cisterna di Latina Wastewater Treatment Plant (Cisterna di Latina, Italy).

Another approach to controlling intermittent aeration is to use ammonia and nitrate analyzers. A control scheme runs the aeration system to drive ammonia to a low target (e.g., 0.2 mg/L) while nitrate increases, and then it stops aeration to result in anoxic conditions to reduce the nitrate back to its low target.

A third approach to controlling intermittent aeration is to simply use dissolved oxygen and timers to achieve the required treatment. In their publication, *Introduction of Japanese Advanced Environmental Equipment*, the Japan Society of Industrial Machinery Manufacturers (2001) described a simple control scheme that runs the aeration system to low dissolved oxygen (0.2 mg/L) for a set period to provide mixing, but no nitrification, and then increases the dissolved oxygen to 2.5 mg/L to provide fully aerobic conditions for nitrification.

## 4.4 Cooperation in the Field of Scientific and Technical Research Model for Control Strategy Development

Much of the current development of control algorithms uses process models. To standardize the way in which models are used to develop control approaches, in 2001, the benchmark group of the European Cooperation in the Field of Scientific and Technical Research (COST) Action 624 published guidelines for several benchmark models that can be used to test control strategies (Copp, 2001). This work is now continued under the International Water Association Task Group on Benchmarking of Control Strategies for WRRFs.

## 4.5 Model-Based Process Control

Some facilities are working to combine the use of modern process simulation models to be a predictive tool for active process control. This can be approached in two ways. The first is to use historical trend data and to input that information as a range of conditions that the WRRF sees on a daily, weekly, or seasonal basis and to project the best combination of process control parameters to use based on the details of the information and the goal of the effort.

The second approach is to take online information such as MLSS, dissolved oxygen, ammonia, and nitrate from a field instrument and feed that information in real time to a simulation model. This allows for fine-tuning both the model and process performance simultaneously. Although this approach is more labor and capital intensive, it may make sense for large facilities to meet stringent regulatory compliance standards.

## 4.6 Supplemental Carbon and Chemical Phosphorus Removal

Additional optimization opportunities include control approaches for supplemental carbon addition, chemical phosphorus removal, and ammonia-based aeration control. Carbon supplementation using a source such as methanol, acetate, glycerol, or molasses to support denitrification in anoxic reactors or denitrification filters may be controlled based on flow and effluent nitrate analyzer trim. Overfeeding chemicals is a waste and will increase the BOD of the process effluent stream. Underfeeding reduces the amount of nitrate removed and may cause an elevation of nitrite levels. The initial dosage may be calculated based on the carbon source, but the actual dosage will be determined in the field. Once an optimal dosage is determined, the chemical feed systems are typically flow-paced to increase or decrease with changes in flow. More advanced control could be achieved with nitrate analyzers, although this is not typical because flow control is generally sufficient. Additional information on this topic can be found in Chapters 5 and 7.

Chemical phosphorus removal is typically controlled based on flow-pacing of chemical addition. The optimal ratio of chemical to phosphorus is determined in the field and can be fine-tuned periodically as conditions change. This assumes that the influent phosphorus and the phosphorus needed for biological growth are consistent over a period of time, and therefore, flow pacing is appropriate. A more detailed discussion on chemical phosphorus removal is provided in Chapter 10.

## 4.7 Ammonia-Based Aeration Control

Ammonia-based aeration control (ABAC) is the control of aeration based on online ammonia concentration, rather than dissolved oxygen. These systems have the potential to reduce aeration costs significantly, as they should result in substantial aeration system turndown under most process loading conditions. There are two main types of ammonia-based control: feedback control and feed-forward control. In feedback control, aeration is adjusted to maintain an ammonia setpoint in the bioreactor. In feed-forward control, aeration is adjusted based on monitoring the influent ammonia concentration. Feed-forward control may provide a faster system response time but also requires more installed analyzers. Care must be taken regarding optimal placement of the ammonia analyzer used for process control.

Ammonia-based aeration control systems can be inserted into an existing dissolved oxygen–control system, and the measured ammonia cascades to set a (now variable) dissolved oxygen setpoint. If effluent ammonia is measured higher than the effluent ammonia setpoint, the ABAC logic will increase the dissolved oxygen setpoint, with the aeration system responding to that dissolved oxygen setpoint. If effluent ammonia is below the setpoint, then the dissolved oxygen setpoint is decreased. Ammonia-based aeration control systems can also "skip" dissolved oxygen control, with effluent ammonia used to directly control blower output (WEF, 2021).

### 4.8 Ammonia-vs NOx-N Control

Ammonia-vs NOx-N (AvN™) Control is a patented approach that targets minimizing effluent total inorganic nitrogen (the sum of ammonia-N and NOx-N) by controlling aeration based on both ammonia and NOx concentrations. This control scheme seeks to promote ammonia-oxidizing bacteria or nitrite-oxidizing bacteria depending on which parameter is at a higher concentration (WEF, 2021).

## 5.0 CONCLUSION

It is important to remember that each WRRF has a specific performance objective to meet and that they seek to exceed, and that an approach that is appropriate for one facility may not be suitable for another. All the factors discussed in these standards, whether related to facility size, pollutant loadings, process technology, variations in flows and loadings, or availability of capital and operating funds, must be considered in terms of what is the most effective and efficient way to achieve treatment objectives of the facility.

## 6.0 REFERENCES

Carucci, A., Rolle, E., & Smurra, P. (1997, July). Experiences of on-line control at a wastewater treatment plant for nitrogen removal. *IAWQ Specialist Group on Instrumentation, Control and Automation 7th International Workshop,* Brighton, UK. International Association on Water Quality.

Caulet, P., Lefevre, F., Bujon, B., Reau, P., Philippe, J. P., & Audic, J. M. (1997, July). Automated aeration management in waste water treatment: Interest of the application to serial basins configuration. *IAWQ Specialist Group on Instrumentation, Control and Automation 7th International Workshop,* Brighton, UK. International Association on Water Quality.

Charpentier, J., Martin, G., Wacheux, H., & Gilles, P. (1998). ORP regulation and activated sludge: 15 years of experience. *Water Science & Technology, 38*(3), 197–208.

Cohen, A., Hegg, D., de Michele, M., Song, Q., & Kasbov, N. (2003). An intelligent controller for automated operation of sequencing batch reactors. *Water Science & Technology, 47*(12), 57–63.

Copp, J. (2001). *The COST simulation benchmark-description and simulator manual.* COST (European Cooperation in the Field of Scientific and Technical Research).

Dabkowski, B. (2008, October 18–22). Applying oxidation–reduction potential sensors in biological nutrient removal systems. *Proceedings of the 81st Annual Water Environment Federation Technical Exhibition and Conference* [CD-ROM], Chicago, IL, United States. Water Environment Federation.

Draaiger, H., Bunnen, A. H. M., & van Dijk, J. W. (1997, July). Full-scale respirometric control of an oxidation ditch. *IAWQ Specialist Group on Instrumentation, Control and Automation 7th International Workshop,* Brighton, UK. International Association on Water Quality.

Grissop, G. (2010, December 1). *Biological nutrient removal operation* [Paper presented at the Shenandoah Valley Pure H2O Forum]. Wastewater Treatment Plant Network. http://www.acsawater.com/sites/default/files/websitefiles/SVWWTPN/BNR%20Operations.pdf

Japan Society of Industrial Machinery Manufacturers. (2001). *Introduction of Japanese advanced environmental equipment.*

Mori, Y., Sasaki, K., Yamamoto, Y., Tsumura, K., & Ouchi, S. (1997, July). Countermeasures for hydraulic load variation in intermittently aerated 2-tank activated sludge process for simultaneous removal of nitrogen and phosphorus. *IAWQ Specialist Group on Instrumentation, Control and Automation 7th International Workshop,* Brighton, UK. International Association on Water Quality.

Shaw, A., & Watts, J. (2002, September 28–October 2). The use of respirometry for the control of sequencing batch reactors—Principles and practical application. *Proceedings of the 75th Annual Water Environment Federation Technical Exhibition and Conference* [CD-ROM], Chicago, IL, United States. Water Environment Federation.

Water Environment Federation. (2024). *Standards for automation of water resource recovery facilities* (WEF 21-24).

Water Environment Federation. (2021). *Energy in water resource recovery facilities* (2nd ed., Manual of Practice No. 32).

Watts, J., & Garber, W. (1993, June 17–23). On-line respirometry: A powerful tool for ASP operation and design. *6th IAWQ Specialized Conference, Sensors in Waste Water Technology,* Hamilton, Canada. International Water Association.

Yoong, E. T., Lant, P. A., & Greenfield, P. F. (2000). In situ respirometry in an SBR treating wastewater with high phenol concentrations. *Water Research, 34*(1), 239–245.

Zehnder, A. J. B., & Stumm, W. (1988). Geochemistry and biochemistry of anaerobic habitats. In *Anaerobic microbiology*. Wiley & Sons.

Zhao, H. W., Mavinic, D. S., Oldhan, W. K., & Koch, I. A. (1999). Controlling factors for simultaneous nitrification and denitrification in a two stage intermittent aeration process treating domestic sewage. *Water Research, 23*(4), 961–970.

## 7.0 SUGGESTED READINGS

Geary, S. (2004, October 2–6). Selection of automation equipment. *Proceedings of the 77th Annual Water Environment Federation Technical Exhibition and Conference* [CD-ROM], New Orleans, LA, United States. Water Environment Federation.

Guisasola, A., Marcelino, M., Lemaire, R., Baeza, J. A., & Yuan, Z. (2010). Modeling and simulation revealing mechanisms likely responsible for achieving the nitrite pathway through aeration control. *Water Science & Technology, 61*(6), 1459–1465.

Ingildsen, P., Rosen, C., Gernaey, K. V., Nielsen, M. K., Guildal, T., & Jacobsen, B. N. (2006). Modelling and control strategy testing of biological and chemical phosphorus removal at Avedore WWTP. *Water Science & Technology, 53*(4/5), 105–113.

Jahan, K., Hoque, S., Ahmed, T., Türkdoan, L., Arslankaya, E., & Patarkine, V. (2011). Activated sludge and other suspended culture processes. *Water Environment Research, 83,* 1092–1149.

Joh Kang, S., Olmstead, K. P., Takacs, K. M., Collins, J., Wheeler, J., & Zharaddine, P. (2009, October 10–14). Energy sustainability and nutrient removal from municipal wastewater. *Proceedings of the 82nd Annual Water Environment Federation Technical Exhibition and Conference* [CD-ROM], Orlando, FL, United States. Water Environment Federation.

Kestel, S., Duffy, G. J., Gregory, J., Gray, M., Stahl, T., & Lee, G. (2009, October 10–14). DO control based on on-line ammonia measurement. *Proceedings of the 82nd Annual Water Environment Federation Technical Exhibition and Conference* [CD-ROM], Orlando, FL, United States. Water Environment Federation.

Kim, Y. M., Cho, H. U., Lee, D. S., Park, D., & Park, J. M. (2011). Influence of operational parameters on nitrogen removal efficiency and microbial communities in a full-scale activated sludge process. *Water Research, 45*(17), 5785–5795.

Kiser, P., DeSantis, G., & Rousey, M. (2005). Monitoring treatment plants automatically: Automated monitoring makes wastewater treatment more efficient. *Public Works, 136*(5), 84–86.

Kobylinski, E., Shaw, A., Steichen, M., & Analla, H. (2004, October 2–6). The basics of automated activated sludge process control: Keep it simple. *Proceedings of the 77th Annual Water Environment Federation Technical Exhibition and Conference* [CD-ROM], New Orleans, LA, United States. Water Environment Federation.

Littleton, H. X., Daigger, G. T., Amad, S., & Strom, P. F. (2009, October 10–14). Develop control strategy to maximize nitrogen removal and minimize operation cost in wastewater treatment by online analyzer. *Proceedings of the 82nd Annual Water Environment Federation Technical Exhibition and Conference* [CD-ROM], Orlando, FL, United States. Water Environment Federation.

Liu, W., Lee, G. J. F., & Goodley, J. J. (2003, October 11–15). Using online ammonia and nitrate instruments to control modified Ludzack–Ettinger (MLE) process. *Proceedings of the 76th Annual Water Environment Federation Technical Exhibition and Conference* [CD-ROM], Los Angeles, CA, United States. Water Environment Federation.

Ma, T., Zhao, C., Peng, Y., Liu, X., & Zhou, L. (2009). Applying real-time control for realization and stabilization of shortcut nitrification-denitrification in domestic water treatment. *Water Science & Technology, 59*(4), 787–796.

Olsson, G. (2006). Instrumentation, control and automation in the water industry—State-of-the-art and new challenges. *Water Science & Technology, 53*(4/5), 1–16.

Olsson, G. (2011). Automation development in water and wastewater systems. *Environmental Engineering Research, 16*(4), 197–200.

Powell, R., & Thompson, R. (2004, October 2–6). South west water (Cornborough WWTP) automatic sludge age control benefits. *Proceedings of the 77th Annual Water Environment Federation Technical Exhibition and Conference* [CD-ROM], New Orleans, LA, United States. Water Environment Federation.

Rosen, C., Jeppsson, U., & Vanrolleghem, P. A. (2004). Towards a common benchmark for long-term process control and monitoring performance evaluation. *Water Science & Technology, 50*(11), 41–49.

Schmit, C. G., Jahan, K., Debik, E., & Pattarkine, V. (2010). Activated sludge and other aerobic suspended culture processes. *Water Environment Research, 82,* 1073–1123.

Turney, H. G., & Wells, C. D. (2004, October 2–6). The use of automated oxygen uptake rate measurements to improve the reliability of a nitrification/denitrification activated sludge treatment plant. *Proceedings of the 77th Annual Water Environment Federation Technical Exhibition and Conference* [CD-ROM], New Orleans, LA, United States. Water Environment Federation.

Vrecko, D., Hvala, N., Stare, A., & Burica, O. (2006). Improvement of ammonia removal in activated sludge process with feedforward–feedback aeration controllers. *Water Science & Technology, 53*(4–5), 125–132.

# Chapter 14

# Laboratory Analyses

## 1.0 INTRODUCTION

Facility operators with supervisory experience who are responsible for overall facility performance and compliance need valid and legally defensible data to be developed by the facility's internal analytical laboratory. Therefore, all staff at a water resource recovery facility (WRRF)—from managers who are trained to have extensive knowledge of federal, state, and local wastewater treatment and discharge regulations to laboratory technicians who have the specific education, experience, and training to competently generate valid and legally defensible analytical data—should understand the fundamental concepts underlying the analytical procedures performed to meet these objectives. This chapter provides background information on the parameters that are encountered in wastewater treatment for process control and regulatory compliance. In addition, a brief introduction to the importance of wastewater microbiology is provided, including a description of recent advances in this field, as nutrient removal is becoming a higher priority for many facilities. More detailed information on this topic can be found in *Basic Laboratory Procedures for the Operator-Analyst* (Water Environment Federation [WEF], 2022).

### 1.1 Regulatory Compliance

Satisfying the requirements of the National Pollution Discharge Elimination System (NPDES) permit or the requirements of a specific reuse application for treated effluent, which will vary from state to state (U.S. Environmental Protection Agency [U.S. EPA], 2004), is possible only with accurate and defendable laboratory results. The analyses necessary for compliance require that quality assurance, which is critical to producing sound and defensible data, is necessary to provide the empirical evidence on which compliance is determined. Therefore, in addition to the description of the parameters that are necessary for regulatory compliance, a discussion is provided in this chapter on considerations for sample collection, the appropriate use of sample statistics, reporting, and common pitfalls in sample collection.

### 1.2 Process Control

A WRRF must operate properly to consistently achieve regulatory compliance. However, with increasingly stringent limits on nutrient discharges, operation of facilities is becoming more complex. Gathering information about the performance of a nutrient removal facility is the key to optimizing its operations.

In many instances, the analyses that are required for permit compliance may not be comprehensive enough for process control. Therefore, many facilities will collect analytical data that are not associated with compliance monitoring or will implement real-time online monitoring to aid in developing the necessary information for process control. Permit language must be carefully checked for sampling and reporting requirements because some permits may require sample data to be reported to a regulatory agency, even if the sampling was not specified in the permit. For example, biological phosphorus removal may require the addition of supplemental carbon source if adequate concentrations of volatile fatty acids (VFAs) are not available. Thus, some facilities monitor this parameter to help manage this biological process. This chapter provides a description of a range of analytical parameters and associated laboratory analyses that, although not critical to permit compliance, may be helpful or necessary for biological and chemical process control that is important to effective nutrient removal.

### 1.3 Calibration, Validation, and Comparison to Online Instruments

During any given workday, utility operators are busy tending to multiple tasks and customer needs. For this reason, online instrumentation is increasingly being used for process control monitoring. While online instrumentation can provide continuous information, there are several factors to consider, including calibration and validation of the instruments and a comparison of the readings with grab samples that are analyzed using a laboratory method.

Although the required steps to calibrate and maintain specific online analyzers are covered here, a pragmatic approach to the procedural requirements for operation and calibration of the online analyzer is necessary for proper operation. To maximize the value of information obtained from online instrumentation, a procedures manual should be developed to guide the operator through the following quality-assurance steps:

- Conduct initial verification of the online analyzer.
- Demonstrate capability of the online analyzer.
- Conduct initial grab sample comparison when the online analyzer is used.
- Conduct routine analyzer checks with grab samples at an appropriate frequency.

### 1.4 Modeling

Wastewater process modeling currently represents the industry's best understanding of the complex relationships between chemical, physical, and biological processes that provide successful wastewater treatment. Nearly 50 years of research have gone into characterizing the behavior of approximately 60 of the most critical, intricately related wastewater treatment processes. However, the numerical models that have been developed to describe the observed chemical, physical, and biological reactions related to these processes continue to evolve as understanding of these processes improves. Thus, the ability of any model to accurately represent reality depends on the quality of the inputs. Measuring actual conditions at a WRRF is, arguably, the component of modeling that is most susceptible to error. There is uncertainty inherent to sampling and analysis, which suggests that models might not always reflect that which is measured at a facility. Therefore, reliable agreement between model results and facility data is only possible and meaningful if the data are verified to be of the best quality. A description of best practices for developing information for wastewater modeling is included in this chapter.

## 2.0 BIAS AND VARIATION IN OPERATION, SAMPLING, AND ANALYSIS

Optimizing nutrient removal requires characterization of the various streams into, within, and from a WRRF to make informed and effective process control decisions. Although online instrumentation is being used more often, the traditional wet chemistry approach is still commonly implemented for characterization and instrument calibration. Mass can be calculated by measuring both flow and concentration. Samples for constituent quantity and quality can be collected from influent, primary effluent, final effluent, return activated sludge (RAS), waste activated sludge (WAS), mixed liquor suspended solids (MLSS), sidestreams, and a host of other locations. Samples can be measured for flow, solids, dissolved oxygen, organics, nutrients, metals, pesticides, and compounds of emerging concern. The sampling location, constituent list, sampling mode (grab or composite), and sampling frequency are dependent on the ultimate purpose of the information (i.e., reporting, process control, optimization, design, or modeling) and how critical and sensitive the result is for each use (e.g., a costly fine for an effluent violation or calibration of a process model).

Because sample collection and analyses are time consuming and costly, only the minimum number of samples and replicates that can provide a meaningful result for informed decision making should be collected. The question then becomes, "What is a meaningful result?" What was meaningful a generation ago when NPDES limits were focused only on total suspended solids (TSS) and biochemical oxygen demand (BOD) is far from what may be considered a meaningful result with today's nutrient removal requirements.

Arriving at a reasonably accurate (i.e., minimum difference between the measured and the "true" value) and precise (i.e., minimum difference between replicates of measured values) quantification of constituents is often more complicated and challenging than is widely acknowledged. Accuracy and precision are a function of bias (systemic error) and variation (random error). Bias should be identified so minimization efforts can be taken. Variation is inherent in all material systems because they are heterogeneous, especially wastewater. The concept of heterogeneity is fundamental to sampling theory (Gy, 1992). Variation should be identified to determine how many samples are needed to provide a mean value. This issue is complicated by the fact that both wastewater flow and composition vary disproportionately over the course of a day, which makes variation especially difficult to characterize.

Bias and variation are collectively known as *uncertainty*. The following are three areas where uncertainty is found at a WRRF:

- Analysis, which occurs with analytical methods or equipment in a laboratory; efforts to minimize and quantify laboratory analytical error have been thorough and have become standardized; this is called *analytical uncertainty*.
- Sample collection, either in mode, equipment, method, or location; this is called *sampling uncertainty*.
- Operation, such as parallel trains or unit processes (bioreactors, clarifiers, or dewatering units), which inevitably lead to unintended differences in hydraulic and/or mass distributions; this is called *operational uncertainty*.

Whereas thorough effort has gone into quantifying and minimizing analytical uncertainty in facility laboratories, comparatively little is typically done to quantify and minimize operational and sampling uncertainty at the facility itself. Some of the same concepts for quantifying and minimizing uncertainty at the laboratory scale can also be applied to sampling and operational scales.

## 2.1 Analytical Uncertainty

Good laboratory practices that characterize and minimize analytical uncertainty are presented elsewhere, namely Section 1000 of *Standard Methods for the Examination of Water and Wastewater* (American Public Health Association [APHA] et al., 2023). Good laboratory practices involve establishing and following standard operating procedures when performing analytical methods; calibrating measurement equipment such as scales, pipettes, and spectrophotometers on a regular basis; and including reagent blank samples, laboratory-fortified blank samples, laboratory-fortified matrix samples and duplicates, duplicate analyses of the unknown sample, and splitting samples between laboratories, all at prescribed regularity. Good bench-scale practices involve such things as shaking samples brought into the laboratory (pseudo-homogenization) before subsampling an aliquot to be used for actual analysis and preventing cross contamination of samples. Good data-processing practices include plotting control charts of results for accuracy and precision; reporting to the correct number of significant figures; calculating standard deviations for both repeatability (i.e., running multiple analysis on the same sample without changing the instrument, analyst, laboratory, or day) and reproducibility (i.e., running the analysis on the same sample, but changing either the instrument, analyst, laboratory, or day); and calculating and reporting method detection limits. All of these practices, if done deliberately and methodically, will allow the analyst to quantify and minimize analytical uncertainty of a sample once it crosses the threshold of the laboratory.

## 2.2 Sampling Uncertainty

*Standard Methods* (APHA et al., 2023) states, "Sampling error due to population variability . . . is usually much larger than analytical error components. Unfortunately, sampling error is usually not available, and the analyst is usually left with only the published error of the measurement system." This concept is typically forgotten, and the result of a single analysis determined under the auspices of good laboratory practices is then assumed to be the average (mean) concentration of whatever is being sampled or it is assumed that the average measurement will be revealed over time through continued sampling (e.g., assuming the average measurement can be determined by averaging daily values over the course of a month). This is not the case, however, and fundamental statistics must not be overlooked. A single analysis can only be representative of a mean of the population if it is shown to be so—that is, many samples must be taken from a single population (i.e., replicates) to determine the distribution of the population, the standard deviation, and the confidence level of the actual mean. Figure 14.1 quantifies how many samples are typically required to determine the mean value with a given level of confidence, where

$N$ = number of samples required,
$s$ = overall standard deviation, and
$U$ = acceptable level of uncertainty.

For example, for influent BOD concentration, if

$s = \pm 30$ mg/L and
$U = 20$ mg/L, then
$N = 20$ samples are required for a 95% level of confidence.

This means that 20 random samples are needed to determine the mean for each day with 95% confidence if the standard deviation of the 20 influent BOD analysis results was ±30 mg/L. Unfortunately, characterizing and minimizing the influencing factors of sampling uncertainty and determining a mean value in this way is rarely done. Important sources of uncertainty include the following:

- Unrepresentative sampling location—compositional differences within an influent channel, bioreactor, distribution box, or launder;
- Unrepresentative composite sampling mode—compositional versus flow differences with time and the effect of the selected sampling period, duration, and mode (continuous or discrete, time proportional, flow proportional, or volume proportional);

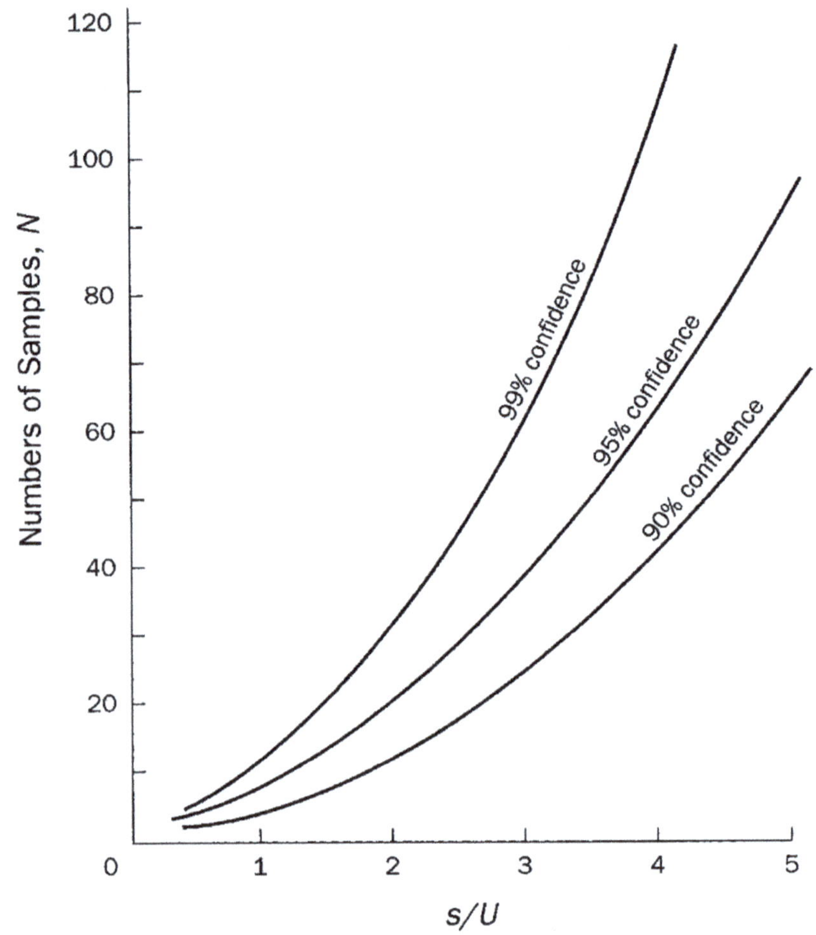


FIGURE 14.1 Approximate Number of Samples Required in Estimating a Mean Concentration (Her Majesty's Stationery Office, 1980)

- Inherent bias in sampling equipment—exclusion of larger particles through sample tubes and preferential settling in composite sample containers;
- Inherent bias in sampling method—grab samples from the surfaces of tanks and the beginning, middle, or end of the solids handling operation; and
- Random error because of the heterogeneity of wastewaters, sludges, and effluents.

## 2.3 Sampling Mode for Compositing Samples

Whereas wastewater flow and composition are highly variable, both will generally follow repeatable patterns for a given facility under base- or dry-weather conditions (i.e., when storm flows are not contributing to the influent). Time-paced (discrete, uniform aliquot per unit time), discrete flow-paced (discrete, uniform aliquot per given volume), and continuous flow-paced (continuous sample proportional to flow rate) samples will likely provide different results. For many facilities, discrete flow-paced samples are taken with a sampler collecting a volume at given time intervals (e.g., every 15 minutes), and that volume is based on a flow meter. Continuous flow-paced sampling provides a better representation of the wastewater but is typically impractical because a separate continuous flow line, sampling pump, and adequate sample storage are required. This distinction is important to identify constituents that are present in low volumes or for micropollutants. For example, in a study by Ort and Gujer (2006), continuous flow-paced sampling was needed to identify pharmaceuticals in an influent stream.

Commercially available composite samplers typically operate in either time-paced or discrete flow-paced sampling modes. Unless flowrates are constant, discrete flow-paced composites are most accurate between these two options. For example, take a generic domestic wastewater influent that represents the typical diurnal variability of flow, concentration, and load (refer to Figure 2.2). Hourly uniform volume aliquots result in an average concentration that is 10% less than that of hourly flow-proportional volume aliquots. The actual difference between time-paced versus discrete flow-paced sampling will depend on the variability of the flow and load and the offset between the two at each WRRF. It will also depend on what is being sampled.

### 2.4 Operational Uncertainty

Operational uncertainty is simply the uncertainty associated with differences in operation between different process trains. Although the intent might be to operate two parallel trains at a selected split hydraulic or solids loading (50 to 50, 60 to 40, etc.), actual distributions may be different than the intent, hydraulic and solids distributions may be disproportionate, distributions might be variable, or all these factors. For example, an attentive operator may have the presence of mind to sample MLSS from different trains to determine an average, but it may be difficult to determine if the flow distribution between clarifiers is equal.

## 3.0 PROCESS PARAMETERS

This section describes many parameters and how they affect the operation of nutrient removal processes. Refer to Section 4.0 for a description of the procedures necessary to analyze these parameters.

### 3.1 Flow

Flow is an important wastewater parameter that quantifies the bulk of fluid movement in facility streams. Flow of streams must be measured, along with concentrations of various parameters, to determine mass loading rates. Essential flow measurements include process influent, effluent, RAS, WAS, and internal recycles. Flow measurements for many other streams, such as sidestreams, are also necessary depending on the facility configuration and unit treatment processes included.

The two main types of flow measurement are open channel and closed pipe. One of these types may be preferred for flow measurements for a certain location and purpose. Open-channel flow measurement consists of a weir or flume measuring flow in open channels or partially full pipes. Parshall flumes are examples of this type of flow-measurement device. Closed-pipe flow measurement devices include magnetic, mechanical, ultrasonic, and mass flow meters.

### 3.2 Mixed Liquor Suspended Solids, Mixed Liquor Volatile Suspended Solids, Return Activated Sludge, and Waste Activated Sludge

The TSS concentration in mixed liquor, MLSS, RAS, and WAS is measured for process monitoring and control. Waste activated sludge concentration, along with MLSS concentration, is typically measured daily to maintain the desired solids retention time (SRT), which is an important process parameter. The volatile fraction of the mixed liquor is measured with mixed liquor volatile suspended solids (MLVSS) measurements. All of the samples should be collected from a well-mixed, representative location. Ideally, samples to determine MLSS, MLVSS, RAS, and WAS should be collected several times a day, as they may vary depending on influent flow variations, RAS pumping rates, and WAS wasting conditions. These samples should be tested using grab samples (or online instrumentation), and results should be correlated with influent and process flows.

The WAS test is run on a sample taken from the WAS pipe (or channel or box) in a location that is well mixed and representative. When the WAS is pumped from a single hopper, composite sampling may be necessary over the duration of the WAS pumping cycle to accurately determine the mass of solids wasted out of the process. The selection of sample type and timing for WAS should consider the facility's wasting strategy (i.e., continuous or intermittent).

The MLSS concentration is also used to calculate the sludge volume index (SVI) in conjunction with settleability test results (refer to Section 3.3). The MLVSS result divided by the MLSS result provides the volatile fraction. The volatile fraction is a rough indicator of active biomass in suspended growth activated sludge processes. In typical biological nutrient removal (BNR) facilities, the volatile fraction is approximately 70% to 80%, depending on the relative concentration of inert material in the process influent and the SRT. Longer SRTs will result in a lower volatile fraction because volatile material is reduced in the mixed liquor through oxidation of organic material in the biomass (endogenous respiration). Changes to the volatile fraction of MLSS can be an indicator of process problems and should be considered.

The MLSS and WAS solids testing results (with WAS flow) are typically used in conjunction with the volumes of anaerobic, anoxic, and aerobic zones in the treatment process to calculate each phase and total process SRT. Aerobic SRT is important for nitrification because the nitrifiers are obligate aerobic organisms that grow slower than other organisms used in BNR (refer to Chapter 3). The MLSS and RAS results can also be used to evaluate whether the RAS rate is appropriate in relation to process influent flow and clarifier limitations. The required RAS flowrate is typically selected based on maintaining a target ratio with the influent flowrate and is the result of controlling sludge blanket depth (0.6 to 0.9 m) and settled MLSS detention time in the secondary clarifiers. Sludge volume index conditions have a major influence on RAS rates and resulting clarifier performance.

## 3.3 Settleability and Sludge Volume Index

A settleability measurement is used to indicate the settling and compaction characteristics of mixed liquor in suspended growth treatment processes at the instantaneous MLSS concentration. Settleability is reported as milliliters per liter, which represents the volume corresponding to the top of the settled solids interface from a 1-L sample of mixed liquor after 30 minutes of settling. The SVI "normalizes" the settleability using the MLSS concentration and is reported as milliliters of settled volume per gram of MLSS. The SVI is generally interpreted as follows:

- An SVI less than 80 indicates good settling and compacting characteristics.
- An SVI from 80 to 150 indicates moderate settling and compacting characteristics.
- An SVI greater than 150 indicates impacted settling and compacting characteristics.

A "good" settling sludge is not necessarily the same as a "fast" settling sludge; nonetheless, the settleability and SVI tests verify that the BNR facility has stable operation and may not experience problems with solids carryover at the clarifiers under high-flow conditions. It should be noted that clarifier capacity and performance depend on many factors. The SVI test serves as an indicator of how the settleability of the mixed liquor may change over time. The SVI should then be trended against other treatment indices (SRT, temperature, aeration, denitrification) to evaluate how these parameters affect settleability and its trends.

Solids settling and compacting characteristics can vary during the day as a result of diurnal variations in organic or nutrient loadings. Samples can be pulled manually once (or more, if necessary) per 8-hour shift and tested for settleability and MLSS concentration, and the SVI can be calculated.

To determine SVI, a sample of MLSS is pulled from a well-mixed representative location of the aeration basin upstream of the clarifiers. This sample should be tested quickly. The sample is gently mixed and poured into a 1-L settleometer, calibrated beaker, or a graduated cylinder, the latter of which is less desirable. One type of container should be selected and consistently used. The settled volume should be recorded approximately every 5 minutes up to 30 minutes as milliliters per liter. The 30-minute reading is generally used for settleability and SVI. One of the reasons for recording the settled volume every 5 minutes is that the mixed liquor in some nitrifying facilities denitrifies, and solids float up before the 30-minute period is complete. Another reason to pay attention during the settling test is to detect if excessive settling rates are occurring, which causes higher levels of solids to be left behind and suspended in the supernatant. Sludge volume index results should routinely be compared to conditions occurring in the secondary clarifiers. Trending of SVI values over time and relating the results to process conditions is important for understanding cause and effect relationships in the WRRF process. This information can provide early indication of when conditions are developing to enable operators to take proactive steps to prevent impacts to downstream treatment units and effluent quality.

## 3.4 pH

pH is a measurement of the acid, neutral, or basic condition of wastewater. A pH of 7 is neutral, while a pH below 7 is acidic and a pH above 7 is basic. The pH of incoming wastewater is a function of the drinking water supply; acids or bases added through household, commercial, or industrial activity; residence time in the collection system; and alkalinity/buffering capacity (refer to Section 3.5). The pH can be changed through chemical addition, byproducts of biological activity, and interaction of the wastewater with air. The pH should be measured in wastewater on at least a daily basis; in addition, many facilities may find it advantageous to install a permanent in situ pH analyzer to provide continuous indication of potential upset conditions or unusual wastewater discharges.

The pH in the BNR process should be within a narrow range for stable operation. If the pH is too low or varies significantly, the nutrient removal activity can be slowed to where nutrient limits are not met. pH can be used with alkalinity to determine if wastewater characteristics and/or the treatment process are responsible for a pH drop and whether further pretreatment by industrial contributor(s), process modification, or alkalinity addition is needed.

The optimum pH range for nitrifying bacteria is pH 7.5 to 8.6, while an acceptable pH range is 7.0 to 9.0. Nitrifying bacteria can acclimate to a pH in the 6.5 to 7.0 range, with little decrease in activity, if the pH does not vary significantly. The optimum pH for denitrification is between 6.5 and 8.5 (WEF, 2023).

For enhanced biological phosphorus removal (EBPR), a pH of 6.5 to 7.5 is common. Although the optimum pH for chemical removal of phosphorus is 5.3 for ferric chloride or 6.3 for aluminum sulfate, chemical removal of phosphorus will occur at higher pH values, with increased chemical dosage. The pH of the wastewater is rarely adjusted to accomplish chemical removal of phosphorus, unless lime is used. All that said, maintaining a consistent pH is important for the biological system. Operations should normally target a pH in the range of 6.8 to 7.2.

The level of pH monitoring at a particular WRRF is relative to site-specific conditions and process sensitivities to pH variability. Although pH is generally not used for control purposes, it can be a quick indicator of a significant industrial constituent in the wastewater entering a BNR facility. Typically, pH in BNR facility effluent represents a compliance parameter. In larger facilities, pH is monitored continuously in the facility effluent; some facilities also monitor pH in the BNR process.

## 3.5 Alkalinity

Alkalinity is a measurement of the buffering ability of wastewater and is a rough indicator of carbonate and bicarbonate concentration in the wastewater. Alkalinity can result from hydroxides, carbonates, bicarbonates, and, less commonly, phosphates, silicates, borates, and similar compounds in the wastewater. With sufficient alkalinity, water can take in acid (or base) with no change or only a small change in pH. An alkalinity over 50 mg/L is necessary to prevent pH from dropping sharply when small amounts of acid are introduced to the water by either chemical addition or biological activity. Wastewater alkalinity results from alkalinity in the water supply plus alkalinity added during domestic use (e.g., cleaners and detergents). A typical alkalinity value for raw wastewater to a WRRF is 150 mg/L, although it may range from 60 to 250 mg/L (or more widely), depending on regional alkalinity in the drinking water and the amount of infiltration and inflow entering the collection system. Samples for alkalinity testing are typically taken from the raw wastewater entering a BNR facility.

Alkalinity is most important during the nitrification process, which reduces alkalinity through the metabolism of inorganic carbon by nitrifying bacteria and produces acid when ammonia is oxidized. The nitrification reaction predicts that, for each gram of ammonia-nitrogen oxidized to nitrate-nitrogen, 7.08 g of alkalinity as calcium carbonate ($CaCO_3$) is destroyed. When too much alkalinity is destroyed, the pH can decrease to a level outside the targeted pH range for good activity of the nitrifying bacteria.

The denitrification process replenishes alkalinity by destroying acid and producing carbon dioxide. The denitrification reaction predicts that, for each gram of nitrate-nitrogen reduced to nitrogen gas, 3.5 g of alkalinity (as $CaCO_3$) is produced. Where alkalinity in the raw wastewater is low, the denitrification step can be performed before, or simultaneously with, nitrification to reduce net alkalinity destruction.

Although alkalinity is not used for process control in most BNR facilities, alkalinity addition is often necessary in situations with low background alkalinity concentrations and/or with significant nitrification needs. It is tested when pH problems are experienced, or initially when a BNR process is being designed, to identify whether alkalinity feed systems are needed.

## 3.6 Temperature

The relevant temperature of wastewater in suspended growth systems is that at which the microorganisms function and oxygen transfer is occurring. In some biofilm systems, temperature is a function of both wastewater temperature and air temperature. Wastewater temperature typically ranges from 3 to 27 °C in the United States and as high as 30 to 35 °C in areas of Africa and the Middle East (Metcalf & Eddy, Inc./AECOM, 2013). Wastewater temperature rises from the temperature of the drinking water supply; warm water from households, commercial, and industrial water uses; and soil temperature. In addition, biological activity and aeration may affect water temperatures, and heat is lost from surfaces and turbulent areas (hydraulic drops and surface mechanical aerators) open to the atmosphere.

Temperature affects the growth rate and metabolism of microorganisms. Nitrification and denitrification rates are affected by temperature, and nitrifying bacteria are particularly sensitive to temperature variations. Low temperatures reduce the biochemical reaction kinetic rates and longer SRT is needed to maintain adequate population of slow-growing microorganisms (e.g., nitrifying bacteria) in BNR facilities (refer to Chapter 3). Temperature also affects oxygen transfer into the wastewater because warmer temperatures reduce the solubility of oxygen in water and, consequently, the oxygen-transfer rate. At the same time, when the temperature is high, biological activity is higher, thereby increasing oxygen demand for treatment.

## 3.7 Dissolved Oxygen

In water, the dissolved oxygen concentration is limited by the maximum equilibrium concentration, or saturation concentration, which depends on the gas used (air or pure oxygen), water temperature, dissolved solids in the water, and elevation (atmospheric pressure). The dissolved oxygen saturation concentration decreases with increasing temperature, increasing dissolved solids concentration, and decreasing atmospheric pressure. Oxygen-uptake rates are higher with higher temperature due to increased microbiological activity. The critical time for maintaining adequate dissolved oxygen is during high-temperature periods at peak loading. It should be noted that dissolved oxygen, as measured in an activated sludge system, provides the dissolved oxygen concentration in the bulk solution and may not be directly representative of dissolved oxygen concentrations within the floc. The dissolved oxygen within the floc will generally be lower than that measured in the bulk solution. The relationship of the bulk versus floc dissolved oxygen relationship is unique due to microbiological differences at each WRRF.

Dissolved oxygen is essential for aerobic microorganisms that are responsible for $BOD_5$ removal, nitrification, and EBPR. In aeration zones, dissolved oxygen concentrations must be sufficient to meet oxygen requirements of the reactions taking place and should be high enough to achieve the necessary removal rates. Removal of $BOD_5$ can occur with dissolved oxygen concentrations of 0.5 mg/L or less; however, low dissolved oxygen conditions can lead to bulking conditions due to excessive growth of filamentous bacteria. In nitrification processes, a minimum dissolved oxygen concentration of approximately 2.0 mg/L is typically adequate, with the nitrification rate decreasing with lower dissolved oxygen concentrations. Operation at lower dissolved oxygen concentrations is increasingly being considered as an energy-savings measure because oxygen transfer is more efficient at lower dissolved oxygen concentrations. Dissolved oxygen should not be present in any significant quantity in anoxic zones because denitrification will be inhibited. Similarly, dissolved oxygen should not be present in the anaerobic (fermentation) zone because VFA uptake/phosphate release and subsequent phosphate uptake in the aerobic zone will be inhibited.

Dissolved oxygen in aeration zones should be monitored a minimum of several times per day and, preferably continuously, if online instrumentation is used. Dissolved oxygen is commonly used to control aeration to enhance BNR and reduce energy costs. Although dissolved oxygen concentrations in anoxic zones, anaerobic zones, and internal recycle or RAS streams are typically not monitored, they should be checked if an upset condition is experienced or as part of a process optimization effort. Chapter 13 contains a detailed discussion of process control using dissolved oxygen.

## 3.8 Oxidation-Reduction Potential

Oxidation-reduction potential (ORP) is a measurement of the oxidative state of a liquid. Dissolved oxygen, as described in the preceding section, is simply the concentration of dissolved oxygen in the bulk solution of the mixed liquor. Oxidation-reduction potential, on the other hand, indicates if aerobic, anoxic, or anaerobic biochemical reactions are occurring in the mixed liquor. The highest positive ORP values indicate the most oxidative conditions (i.e., aerobic), while the highest negative ORP values indicate the most reductive conditions (i.e., anaerobic) in mixed liquor. Chapter 13 contains a detailed description of the meaning, interpretation, and process control using ORP.

## 3.9 Ammonia and Total Kjeldahl Nitrogen

Total Kjeldahl nitrogen (TKN) is the sum of organic nitrogen and ammonia-nitrogen. The TKN is measured in the influent to determine the nitrogen load to the BNR facility, as municipal wastewater typically does not contain nitrate-nitrogen unless there is an industrial source. A large fraction (60% to

75%) of the TKN is ammonia-nitrogen. The rest of the influent nitrogen is organic nitrogen that is associated with soluble and particulate organic matter in municipal influent.

Some of the organic nitrogen in the influent is hydrolyzed by microorganisms to ammonia-nitrogen at the WRRF. Nitrifying organisms oxidize ammonia to nitrate under aerobic conditions. Nitrogen from the TKN serves as a critical nutrient required for new cell growth by the biomass. The oxidation of each milligram of ammonia-nitrogen requires 4.6 mg of oxygen. The nitrogen uptake in cell growth results in approximately 0.08 to 0.12 mg N/mg volatile suspended solids (VSS) in MLVSS. Considering that BNR facilities have an MLVSS-to-MLSS ratio of 0.70 to 0.80, the nitrogen content of WAS is estimated to be in the range of 5% to 10%.

Ammonia-nitrogen and TKN should be measured in the influent wastewater before the return point of sidestreams, influent to the BNR process, and BNR effluent. The TKN in facility influent is typically run at the same frequency as compliance monitoring in facilities that have a total nitrogen limit; ammonia-nitrogen in WRRF influent can also be run at the same time. The TKN and ammonia-nitrogen should also be periodically checked in sidestreams from solids handling processes to evaluate their effect on BNR process performance. If the BNR facility uses anaerobic digestion, the sidestreams may contain substantial ammonia-nitrogen.

Ammonia-nitrogen is monitored at the end of aerobic zones with nitrite-nitrogen and/or nitrate-nitrogen to evaluate and tune nitrification and denitrification performance. These measurements can also be used to help control the aeration system. Aeration can be reduced to levels such that the ammonia-nitrogen setpoint at the end of the aeration basins is not exceeded. This type of control can be implemented with automation and real-time measurements using online ammonia-nitrogen analyzers and probes.

## 3.10 Nitrate-Nitrogen

Influent wastewater typically has little or no nitrate-nitrogen, although, in some instances, nitrate-nitrogen can enter the collection system with industrial discharges or by chemical addition for hydrogen sulfide control. Nitrate-nitrogen can be generated through complete nitrification in the aeration basins. In addition, nitrate-nitrogen can be measured in facility influent, aeration zones, clarifier effluent, and final effluent after disinfection.

Nitrate-nitrogen in WRRF effluent can be limited by total nitrogen effluent limits or nitrate-nitrogen limits for groundwater discharge. The nitrate-nitrogen profile through the BNR process indicates performance of nitrification and denitrification and can be used to control aeration or internal recycle rates and to make decisions regarding SRT.

## 3.11 Nitrite-Nitrogen

In wastewater, nitrite-nitrogen can enter the collection system with industrial discharges and can be generated through incomplete nitrification at the WRRF. An increase in nitrite-nitrogen is more commonly due to incomplete nitrification caused by insufficient aerobic SRT at low temperatures, low dissolved oxygen, low pH, and possible substances inhibiting nitrite oxidating bacteria. Incomplete nitrification or nitritation should be avoided unless the WRRF is specifically designed for this process. Nitrite-nitrogen can be measured in facility influent, aeration zones, clarifier effluent, and final effluent after disinfection. Increasing levels of nitrite-nitrogen indicate the need for addressing SRT, dissolved oxygen conditions, pH levels, and potential inhibitory wastewater components.

Nitrite-nitrogen in wastewater effluent can have a toxic effect on microorganisms, macroinvertebrates, and fish in surface waters. In addition, nitrite-nitrogen will significantly increase chlorine demand for disinfection.

## 3.12 Refractory Dissolved Organic Nitrogen

Refractory dissolved organic nitrogen (rDON) is the fraction of influent nitrogen load that is resistant to chemical or biological treatment (Stensel et al., 2008). Compounds that contribute to rDON are largely unknown. Organic nitrogen levels in facility effluent depend on industrial wastewater contributions, organic nitrogen present in the drinking water, and microbial decay that occurs during biological treatment. Organic nitrogen resistant to further biological degradation can represent a significant fraction of the effluent total nitrogen when low effluent total nitrogen limits are required (Bott & Parker, 2011).

Refractory dissolved organic nitrogen becomes particularly important when the facility is required to achieve low levels of nitrogen. Levels of rDON in municipal WRRF effluents can vary between 0.5 to 2.8 mg/L, as shown by a survey of 30 municipal WRRF effluents conducted by Pagilla (2007). Dissolved organic nitrogen levels in facility effluent are often used as indicators of rDON. Dissolved organic nitrogen is estimated by subtracting effluent ammonia from effluent soluble TKN.

## 3.13 Total Phosphorus

Like nitrogen, phosphorus is a macronutrient needed by all living things to reproduce and grow cell mass. Phosphorus is generally regarded as the limiting factor for algae growth in freshwater streams and lakes. For this reason, many WRRFs have phosphorus effluent limits that attempt to reduce the total phosphorus load to receiving water. When algae are produced in streams and lakes, they eventually die and are decomposed by organisms, resulting in a reduction of dissolved oxygen (eutrophication). In addition to the potential immediate effects of low dissolved oxygen, decomposed algae settle to the bottom of the body of water and contribute to the process of filling the stream or lake.

Total phosphorus should be monitored as required by the facility permit and as often as is practical from industry at the influent, from any sidestream, and at the effluent for process control. It is important that enough data are collected when the facility is running well to form a reliable baseline of information.

Because phosphorus is needed for reproduction and cell growth, there is a need for influent phosphorus to biological wastewater treatment systems. Activated sludge phosphorus content is generally 0.025 to 0.035 mg P/mg VSS without EBPR. Phosphorus content of activated sludge can be substantially larger when EBPR is achieved. Most municipal WRRFs will have more than enough phosphorus as a nutrient to satisfy cell growth, and the remainder needs to be removed through treatment enhancements.

Phosphorus can occur in many forms and typically changes form through biological or chemical wastewater treatment. Among the forms of phosphorus, some will be dissolved and some will be part of the suspended solids. The reactive soluble fraction is generally in the orthophosphorus state and measured as orthophosphate (refer to Section 3.14); although these two forms of phosphorus (reactive soluble phosphorus and orthophosphorus) are not the same, they are often used interchangeably.

## 3.14 Orthophosphorus

Orthophosphorus is a soluble form of inorganic phosphorus that is common in treated BNR process effluent. Most of the particulate phosphorus will be contained within the activated sludge system and, therefore, should not be discharged from the facility in the effluent, with the exception of phosphorus included in effluent suspended solids. For this reason, the majority of effluent phosphorus in a facility with low effluent TSS will be in the orthophosphorus form.

Orthophosphorus can also be used within EBPR processes to quickly monitor various stages of the system for orthophosphate release in the anaerobic zone and subsequent luxury uptake in the aerobic zone. Solids should be removed by allowing the samples to settle or by filtering them; in addition, they should be analyzed as quickly as possible.

## 3.15 Recalcitrant Dissolved Organic Phosphorus

Total phosphorus consists of particulate phosphorus and soluble phosphorus. Particulate phosphorus must be hydrolyzed and converted to reactive orthophosphorus form to be removed by EBPR or by chemical precipitation with coagulants (metal salts). Similar to rDON, there is also a refractory fraction of phosphorus, which is known as rDOP. Compounds that contribute to rDOP are largely unknown.

Recalcitrant dissolved organic phosphorus (rDOP) becomes a greater concern when WRRFs are required to reduce effluent total phosphorus to low levels (i.e., below 0.10 mg/L). Advanced treatment technologies may be required to achieve low levels of effluent total phosphorus because biological and chemical treatment will not be effective for rDOP. The rDOP levels in WRRF effluent may be in the range of 0.05 mg/L or lower. Analysis and quantification of rDOP is also an issue because of such low levels of effluent total phosphorus.

## 3.16 Chemical Oxygen Demand

Chemical oxygen demand (COD) is a test that measures the amount of oxygen required to chemically oxidize the organic material and inorganic nutrients in water. At approximately 3 hours, COD analysis is much faster than the $BOD_5$ test. Variations of COD analysis have been developed in an attempt to better characterize the effect of COD on biological systems by isolating certain fractions of COD that are directly used by polyphosphate-accumulating organisms (PAOs) that are biologically selected for in the EBPR processes. Figure 14.2 shows these influent COD subdivisions, as presented by Ekama et al. (1984).

The influent total COD is first subdivided into biodegradable and un-biodegradable fractions. The un-biodegradable soluble fraction will typically pass through the treatment system in the facility effluent while the un-biodegradable particulate COD fraction becomes part of the MLSS and is eventually wasted from the system in WAS. The biodegradable fraction is broken into readily biodegradable and slowly biodegradable. The readily biodegradable fraction is the key component when dealing with any BNR system evaluation.

Samples for total COD should be collected from facility influent, just before the influent reaches the BNR process, and effluent from the BNR process. Other locations may be sampled to quickly evaluate organic contributions to loading on the BNR process. Samples for soluble COD should be collected at

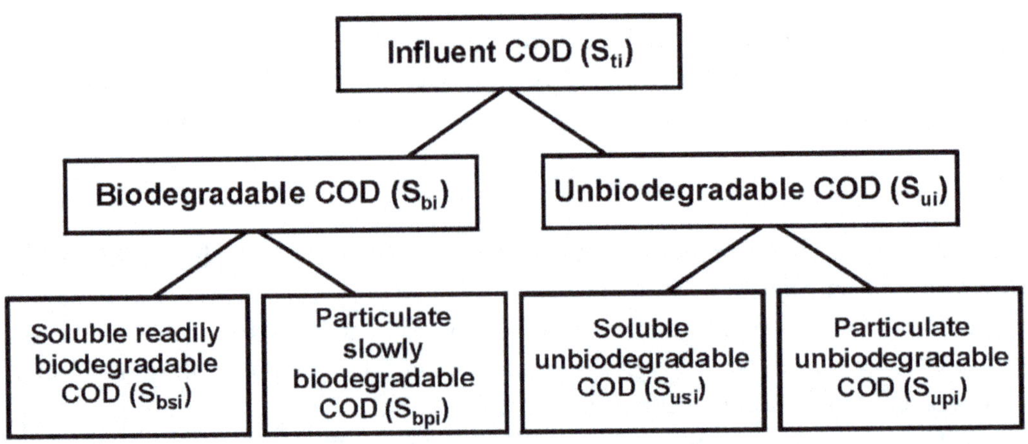


FIGURE 14.2 Division of Total Influent COD in Municipal Wastewater

locations just before the influent reaches the BNR process (readily biodegradable) and the effluent from the BNR process (soluble un-biodegradable).

In the anaerobic zone, typically only the readily biodegradable COD component is susceptible to fermentation to form VFAs within the short retention time (1 to 2 hours). Ekama et al. (1984) found that phosphorus release in the anaerobic zone increased as the influent readily biodegradable COD increased. They concluded that a readily biodegradable COD concentration of at least 25 mg/L was necessary surrounding the PAO in the anaerobic zone for phosphorus release to occur.

Other research indicates that a readily biodegradable COD influent concentration of 60 mg/L is needed for reliable EBPR to occur. Every facility has its own unique wastewater characteristics and readily biodegradable COD concentrations that correspond to good EBPR performance. Operations staff can develop a database to determine if changes in treatment efficiencies are related to changes in influent wastewater characteristics. These data can be trended against effluent quality and other operating parameters to better understand conditions that affect treatment.

A facility can elect to use total COD in their sampling protocol; however, variations in the amount of inorganic oxygen demand may reduce the value of this data. The readily biodegradable COD analysis may be the preferred parameter for tracking feed to the anaerobic zone for promoting phosphorus release and anoxic zones for denitrification.

The total COD or soluble readily biodegradable COD data are most useful when they are collected regularly so that results obtained during good treatment can be compared to results obtained during poor treatment. Although sample frequency will likely depend on the resources of each facility, it is recommended to occur at least weekly to monthly. Additional sampling should be conducted when treatment is poor and during seasonal changes.

## 3.17 Volatile Fatty Acids

The performance of an EBPR process will vary with the specific VFAs available in the anaerobic zone. Based on past research, COD consumption during EBPR has been estimated to be 50 to 60 mg/L COD per milligram per liter of phosphorus removed from municipal wastewater. According to Abu-ghararah and Randall (1991), acetic acid (HAc) is the most efficient VFA for EBPR. Fortunately, HAc is typically the primary VFA formed from wastewater fermentation, with propionic acid as the secondary VFA. Sufficient VFA production can occur in wastewater collection systems, especially in long force mains or gravity sewers with long detention times in warmer climates. Several facilities observed improved reliability of EBPR with the addition of a VFA-laden waste stream, such as a long force main serving an area several miles from the WRRF. Fermentation of wastewater or primary sludge involves the reduction of proteins and carbohydrates into, most commonly, acetates, propionates, and butyrates.

Samples should be collected from the raw wastewater influent. For facilities with primary clarifiers, the sample should be primary effluent; for facilities without primary clarifiers, the sample should be the effluent from preliminary treatment. Samples should also be collected on the elutriate from prefermentation facilities to determine the performance of these units in generating additional VFAs that are directed to the EBPR anaerobic zone(s).

Enhanced biological phosphorus removal requires the presence of VFAs in the anaerobic zone of any BNR wastewater treatment system. When the influent wastewater is weak and not septic, VFA production can potentially be accomplished in the anaerobic zone of the BNR process or outside the BNR system in a separate process called prefermentation. A facility can elect to continue monitoring VFAs as a process control parameter, which may prove especially useful at BPR facilities that do not have an abundance of VFAs in the influent to the anaerobic zones. These facilities can potentially be optimized by tracking the abundance of VFAs and adjusting their processes accordingly.

Samples should be taken during times of adequate EBPR to obtain a baseline to compare to periods of poor performance. Typically, approximately 40 to 50 mg/L of HAc as COD is desired in the influent to the anaerobic zone; however, this can vary by facility. If the anaerobic zone is larger (greater than 1.5 hours of detention time), then VFA production can potentially be generated within the zone.

An alternative method to determine sufficient fermentation of influent wastewater is to compare the ratio of COD to total phosphorus to the ratio of $BOD_5$ to total phosphorus. If the ratio of COD to total phosphorus is considerably higher than 40 to 1 and the ratio of $BOD_5$ to total phosphorus is considerably lower than 20 to 1, then the wastewater has likely not undergone sufficient fermentation (Randall et al., 1992).

Whereas ample VFA is critical for EBPR performance, excessive VFAs can promote excessive filamentous and zoogloeal bacteria that can affect aeration basins and secondary clarifiers with excessive foaming and scum accumulation and/or promote poor settling MLSS.

## 3.18 Soluble Biochemical Oxygen Demand

Soluble $BOD_5$ ($SBOD_5$) is the $BOD_5$ of the filtered wastewater sample and is analogous to the soluble readily biodegradeable COD. The soluble fraction of $BOD_5$ that is loaded to the anaerobic zone of a biological phosphorus removal facility may be more accurate in predicting the response of PAOs than total $BOD_5$. This is because PAOs are able to use only the soluble fraction of $BOD_5$ in gaining their competitive advantage while in the anaerobic zone. Although $SBOD_5$ should not replace $BOD_5$ for calculating oxygen demand exerted on the biological system, it may be useful in predicting EBPR success.

The EBPR process is dependent on having a readily degradable food source available to the PAOs to remove phosphorus. The greater the ratio of $BOD_5$ to total phosphorus or $SBOD_5$ to soluble phosphorus, the better the chance that the system will be capable of adequately removing enough phosphorus to achieve a high-quality effluent. Review of this data over time will allow a facility to determine if variations in effluent quality are the result of changes in the amount of food available to the system. It should be noted that the ratio of $SBOD_5$ to soluble phosphorus offers the advantage of being in the correct form to predict how the system will react and is likely a better control parameter than the ratio of $BOD_5$ to total phosphorus; however, it may not be as good an indicator as readily biodegradable COD or short-chain VFAs. The ratio of $SBOD_5$ to soluble phosphorus may be the best indicator for a facility that does not have the equipment and staff needed to run COD or VFA analysis. The key disadvantage of using $BOD_5$ or $SBOD_5$ is that it takes 5 days to get the results; therefore, it is not a good predictor of the immediate future but instead could be used to identify long-term trends. Designers may use ratios of 15 to 1 ($SBOD_5$ to soluble phosphorus) or greater and, possibly, a ratio of 20 to 1 ($BOD_5$ to total phosphorus) as a threshold to determine if a facility is a good candidate for EBPR. If the ratios being measured at a given facility are at or greater than those values, it is unlikely that the facility is $BOD_5$ limited.

## 3.19 Nitrification Rate

The purpose of the nitrification test is to initially determine a baseline indicator value of nitrification rate for a specific facility and to determine the performance of the facility relative to that baseline under the various operating conditions encountered. Samples for the nitrification test should be collected from the influent to the BNR process before the introduction of RAS and from RAS.

The performance of the nitrification process depends on the concentration of nitrifiers in the system, the ammonia concentration, and environmental factors such as wastewater temperature, dissolved oxygen concentration, pH and alkalinity, and the presence or absence of inhibitory compounds in the influent. The nitrification test provides adequate supply of ammonia, dissolved oxygen, and alkalinity. Influent pH, influent wastewater temperature, and biomass concentration (MLSS and MLVSS) are recorded for the test, and the biomass is presumably acclimated to the influent and contains a sufficient proportion of nitrifiers to achieve nitrification.

Establishing a baseline indication of nitrification performance is useful at the various temperatures encountered and concentrations of MLSS and MLVSS controlled by facility operations. Periodic use of the test reinforces the baseline indicator values of nitrification rate and will indicate if there is some factor, such as wastewater temperature, inhibitory compounds, or nitrifier population, that is adversely affecting the nitrification process if poor process performance occurs. If the nitrification test shows that adequate nitrification can occur but that facility performance is lacking, then facility operators should look for conditions related to insufficient dissolved oxygen, inadequate detention time or inadequate mixing, insufficient alkalinity and lowered pH in the process, or intermittent slugs of inhibitory compounds. If the nitrification test shows poor nitrification compared to the baseline, then facility operators should look for lowered SRT, insufficient dissolved oxygen, high organic loading, low wastewater temperature, and the presence of inhibitory compounds.

The test should be performed regularly to maintain baseline information and to assist in judging the effectiveness of operational adjustments necessary to maintain nitrification during seasonal changes in wastewater temperature. The test also can be performed in multiple reactors several times a day if slugs of inhibitory compounds are suspected based on observations of full-scale nitrification performance.

There are three basic tests that determine nitrification maximum growth rate. These are the low food-to-microorganism (F/M) sequencing batch reactor method, the washout method, and the high F/M method, as described in the Water Environment Research Foundation's (2003) *Methods for Wastewater Characterization in Activated Sludge Modeling.*

### 3.20 Microbiological Activity and Examination

Monitoring microbiological activity and population trends is an important activity for understanding the health and performance conditions of a WRRF. Microscopic examination of MLSS in concert with process data presents operators with a more comprehensive view of the outcome of their process decisions and the conditions they are promoting (selecting for) and maintaining. Trending process control data in parallel with observations of relative abundance of key indicator microorganisms in the MLSS—protozoans, metazoans, and filamentous and zoogloeal (floc-forming) bacteria—provide operators a more comprehensive view and an important tool for making process control decisions and troubleshooting for optimizing performance in WRRFs.

Water resource recovery facilities are unique, so the optimum operating conditions and microbiological profile must be determined for each. Routine microscopic observations compared to process data and facility performance provides trending relationships that help operators understand operational boundary conditions due to seasonal changes and/or special waste loading conditions.

Excessive growth of filamentous bacteria that affect settleability and/or cause excessive foaming are indicative of operating parameters falling outside of optimum ranges. Key operating parameters are SRT, pH, dissolved oxygen, F/M, and excessive VFA/septic conditions. Filaments characteristic to a particular WRRF are generally always present in low-moderate levels that help promote a well-flocculating MLSS. However, filaments adapt quickly to adverse conditions and can grow to over-abundant levels that can overwhelm facility process units and impact treatment performance. Identification of the type of filament that has grown to significant abundance presents clear insight to the causative conditions thereby presenting the needed corrective actions. These relationships and causes are discussed in depth in the *Manual on the Causes and Control of Activated Sludge Bulking, Foaming, and Other Solids Separation Problems* (Jenkins et al., 2004). Laboratories experienced in microorganism/filament identification should be used to routinely profile the filaments in a WRRF to understand the conditions of the microbiological populations during optimum and affected operational conditions.

## 4.0 LABORATORY PRACTICES

This section reviews laboratory practices for many of the parameters discussed in Section 3.0. The information presented here is mainly based on *Standard Methods* (APHA et al., 2023). Detailed laboratory protocols, sample preservation, and hold times can be found in *Standard Methods* and are listed in this section, as appropriate.

### 4.1 Temperature

Depending on the location, the mean annual temperature of wastewater influent varies between 3 and 27 °C and is kept between 10 and 30 °C throughout biological treatment processes to support optimal microbial activities. General kinetic models for BNR processes are temperature dependent. Biological reaction rates approximately double with every 10 °C increase until inhibitory temperature values are reached. Nitrification reaches a maximum rate at temperatures between 30 and 35 °C, and denitrification occurs efficiently between 5 and 30 °C. Typical temperature ranges can also be found in Chapter 11. Besides the effects on microbial activity, temperature of wastewater also influences the solubility of oxygen and, consequently, affects the oxygen-transfer capacity of the aeration equipment in nitrification.

A calibrated, mercury-filled Celsius thermometer is typically used for temperature measurements (Method 2550; APHA et al., 2023). At a minimum, the thermometer should have a scale marked for every 0.1 °C, with markings etched on the capillary glass, and should be periodically checked against a precision thermometer.

### 4.2 pH

pH is one of the most important and frequently tested parameters in water chemistry. It represents the concentration of hydrogen ions in solution and is defined by the following equation:

$$pH = -\log_{10}[H^+] \tag{14.1}$$

where $[H^+]$ is the concentration of hydrogen ions in moles per liter.

The pH of typical wastewater influent varies between 6 and 9. It is important to keep pH within the range of 7 to 8.5 during biological nutrient removal processes for microorganisms to remain sufficiently active. pH is typically reported as unitless or with standard units. Most WRRFs can effectively nitrify with a pH of 6.5 to 7.0, but not below 6.0. Optimum pH values for denitrification are between 7.0 and 8.5. In addition, both nitrification and denitrification will cause a change of pH, and acid-base neutralization is commonly required. Typical pH ranges can also be found in Chapter 11.

The basic principle of the electrometeric pH measurement (Method 4500-H+; APHA et al., 2023) is based on the potentiometric measurement of hydrogen ion activity within the solution using a standard hydrogen electrode and a reference electrode or a standard pH meter. A circuit is completed through the potentiometer when the electrodes are immersed in the test solution. Many pH meters are capable of reading pH and millivolts with a precision of 0.01 pH (Figure 14.3). pH electrodes require periodic cleaning and calibration to ensure accurate readings.

A pH electrode is relatively free from interference except for a sodium error at pH above 10, which can be reduced by using a special "low-sodium error" electrode. pH measurements are affected by temperature in the following two ways: mechanical effects that are caused by changes in properties of the electrodes and chemical effects caused by equilibrium changes. Standard pH buffers have a specified pH at indicated temperatures.

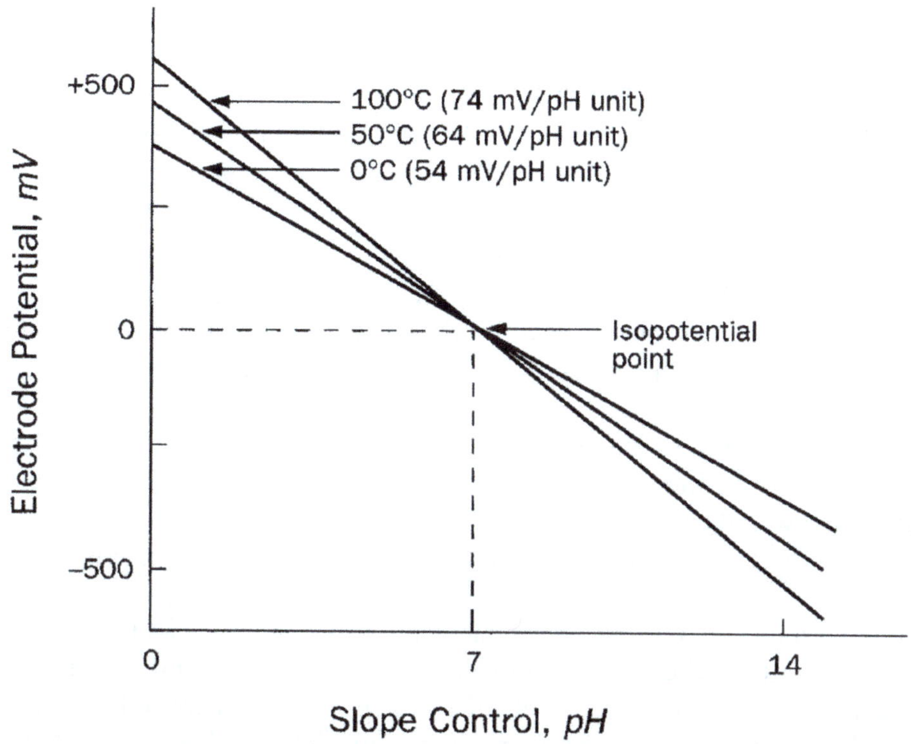


FIGURE 14.3 Typical pH Electrode Response as a Function of Temperature

## 4.3 Dissolved Oxygen

Dissolved oxygen measures the amount of molecular oxygen present in water or wastewater, typically in milligrams of oxygen per liter. Dissolved oxygen is particularly important to wastewater nutrient removal processes because it fully oxidizes ammonia to nitrate at a typical concentration greater than 1.0 mg/L. Dissolved oxygen also affects the growth of other bacteria in the system, including filamentous organisms that may cause sludge bulking. Therefore, dissolved oxygen concentrations must be maintained in a range that favors desired organisms in different configurations. Dissolved oxygen is a function of temperature, the degree of hardness, and the demand for oxygen in the body of wastewater. The two methods for measuring dissolved oxygen are an electrometric method using a dissolved oxygen electrode and an iodometric method (also called *Winkler titration*). The most common method uses the membrane dissolved oxygen electrode, which compensates for pressure and temperature variations in water, and is the preferred method for measuring dissolved oxygen in polluted and highly colored waters.

The membrane electrode method (Method 4500-O G; APHA et al., 2023) offers the advantage of analyzing dissolved oxygen in situ and, therefore, is also preferred for in-stream analysis in the field. Oxygen-sensitive electrodes of the polarographic type are composed of two solid metal electrodes in contact with supporting electrolyte separated from the test solution by an oxygen-permeable plastic membrane, which serves as a diffusion barrier against interfering impurities. The diffusion current is linearly proportional to the concentration of molecular oxygen and can be easily converted to concentration units. Because of the effects of temperature changes on membrane permeability, membrane electrodes require temperature compensation by using thermistors in the electrode circuit. There are also other dissolved oxygen measuring technologies, such as the luminescent (optical) technology, that are popular and are approved for reporting purposes in all states. Additional information on this technology is provided in Chapter 13.

Plastic films in membrane electrode systems are permeable to a variety of gases besides oxygen, such as hydrogen sulfide ($H_2S$). The interference can be eliminated by frequently changing the membrane and calibrating the electrode.

## 4.4 Oxidation-Reduction Potential

Certain substances release or lose electrons when dissolved in solution. The intensity or ease of electron loss (oxidation) or electron gain (reduction) is ORP. Raw wastewater influent has a typical ORP of –200 mV, and ranges from –400 (strong) to –50 mV (weak). Common ORP values found in nitrification and denitrification systems are –50~ +50 mV and +100~ +325 mV, respectively.

Electrometric measurements (Method 2580; APHA et al., 2023) are made by potentiometric determination of electron activity using an inert indicator electrode and a suitable reference electrode. The potential difference between the two electrodes equals the ORP of the test solution. Before measuring any samples, ORP electrodes need to be calibrated with a standard solution at the temperature of the sample.

Temperature determines the reference potential of a particular solution and an electrode pair and affects the reversibility of the redox (reduction-oxidation) reaction, the magnitude of the exchange current, and the stability of the apparent redox potential reading. Oxidation-reduction potential is also sensitive to pH (if hydrogen or hydroxide ions are involved in the redox half cells), and the value increases as pH decreases.

## 4.5 Alkalinity

Alkalinity is a measure of the ability of the wastewater to neutralize acid. Alkalinity is reported as milligrams of calcium carbonate ($CaCO_3$) per liter. The main sources of alkalinity in wastewater streams are carbonate, bicarbonate, and hydroxide compounds (borates, silicates, and phosphates may also contribute). The full nitrification reaction (conversion of ammonia to nitrate) consumes 7.1 mg/L of alkalinity as $CaCO_3$ for each milligram per liter of ammonia-nitrogen oxidized. Conversely, denitrification produces 3.57 mg of alkalinity as $CaCO_3$/mg of nitrate-nitrogen reduced to nitrogen gas. Therefore, a treatment system that supports both nitrification and denitrification can regain some of the lost alkalinity. In general, a minimum alkalinity of 50 to 100 mg/L is required to ensure adequate buffering during both nitrification and denitrification.

When alkalinity is caused by carbonate or bicarbonate, the pH at the equivalence point of the titration is determined by the concentration of $CO_2$ at that stage. The end-point pH value varies according to alkalinity concentrations, which are, in turn, determined by the potentiometric titration to the end-point pH, as follows:

$$\text{Alkalinity (mg CaCO}_3\text{/L)} = \frac{A \times N \times 500\,000}{\text{mL sample}}$$

Where

$A$ = mL standard acid used, and
$N$ = normality of standard acid.

Soaps, oily matter, suspended solids, or precipitates may coat the glass electrode and cause a sluggish response of the pH electrode.

## 4.6 Solids

Solids are matters suspended or dissolved in water or wastewater. The suspended and dissolved fractions are further subdivided into volatile (organic) and nonvolatile (fixed or inert) fractions. In raw domestic wastewater, solids are about 50% organic.

### 4.6.1 *Total Solids*

Total solids in raw wastewater consist of all solids that remain after a sample has been evaporated and subsequently dried in an oven at a defined temperature. Coarse solids, such as rags and debris, are removed before analysis. Total solids affect water clarity. Higher solids decrease the passage of light through water, thereby slowing photosynthesis by aquatic plants. Water will heat up more rapidly and hold more heat; this, in turn, might adversely affect aquatic life that has adapted to a lower-temperature environment.

A well-mixed sample in a weighed dish should be evaporated and dried to a constant weight at 103 to 105 °C in an oven. The increase in weight over that of the empty dish represents the total solids. Highly mineralized water (high concentrations of calcium, magnesium, chloride, and sulfate) may require prolonged drying, proper desiccation, and rapid weighing.

### 4.6.2 *Total Dissolved Solids*

Total dissolved solids (TDS) are material in the water that will pass through a filter with a 2.0-μm nominal average pore size. Smaller pore sizes have also been used to characterize TDS, such as the Whatman glass fiber filter, with a nominal pore size of 1.58 μm. The TDS is comprised of both colloidal and dissolved solids. The concentration of TDS affects the water balance in the cells of aquatic organisms.

According to Method 2540 C (APHA et al., 2023), the well-mixed sample should be filtered through a standard glass fiber filter and the filtrate should be evaporated to a dryness in a weighed dish and dried to a constant weight at 180 °C. Highly mineralized water containing high concentrations of calcium, magnesium, chloride, and sulfate may require prolonged drying, proper desiccation, and rapid weighing.

### 4.6.3 *Total Suspended Solids*

The raw wastewater suspended solids concentration can range from 100 to 350 mg/L. Suspended solids are approximately 50% organic solids and 30% inorganic solids. Higher concentrations of suspended solids can serve as carriers of toxics, which readily cling to suspended particles.

According to Method 2540 D (APHA et al., 2023), a well-mixed sample should be filtered through a weighed standard glass fiber filter, and the residue retained on the filter should be dried to a constant weight at 103 to 105 °C (the sample size should be limited to yield no more than 200 mg of residue). The increase in weight of the filter represents TSS. The micron rating of the filter should be provided with the result.

Excessive residue on the filter may form a water-entrapping crust; therefore, it may be necessary to exclude large floating particles or submerged agglomerates of nonhomogeneous materials from the sample and to limit the sample size to that yielding no more than 200 mg of residue.

### 4.6.4 *Fixed and Volatile Solids*

Fixed solids are the residue of total, suspended, or dissolved solids after being heated to dryness for a specified time at a specified temperature (typically, 550 °C). Volatile solids are calculated by subtracting fixed solids from the total solids value. Determinations of fixed and volatile solids do not distinguish between inorganic and organic matter because the loss upon ignition is not confined to organic matter but also includes losses because of decomposition or volatilization of some mineral salts. However, the volatile solids fraction is still useful in the controlling WRRF operation because it offers a rough approximation of the amount of organic matter present in the wastewater and activated sludge. Figure 14.4 shows the various solids fractions graphically.

According to Method 2540 E (APHA et al., 2023), the residue should be ignited from total solids, TDS, and TSS tests to a constant weight at 550 °C; the remaining solids represent the fixed fraction of total solids, TDS, and TSS, respectively. Low concentrations of volatile solids in the presence of high fixed solids may result in considerable error. Highly alkaline residues may react with silica in the sample or silica-containing crucibles.

## 4.7 Carbonaceous Oxygen Demand

*Carbonaceous oxygen demand* (COD) is defined as the amount of oxygen required for the chemical oxidation of both biodegradable and non-biodegradable organic substances. The COD test uses dichromate to oxidize the organic material in wastewater.

The principal fractions are particulate and soluble (or filtered) COD. In biological processes, the two are further fractionated to readily biodegradable soluble COD (rbCOD), slowly biodegradable colloidal and particulate COD, non-biodegradable soluble COD, and non-biodegradable colloidal and particulate COD. The rbCOD is often fractionated further into complex COD that can be fermented to VFAs and short-chain VFAs. Flocculated and filtered COD (ffCOD) is a better measure of truly soluble COD. During ffCOD tests, coagulants (e.g., zinc) are added to enhance precipitation of colloidal solids. The sample is subsequently filtered, and the filtrate is subjected to COD analysis, as follows:

$$\text{rbCOD} = \text{ffCOD—Non-biodegradable COD} \tag{14.2}$$

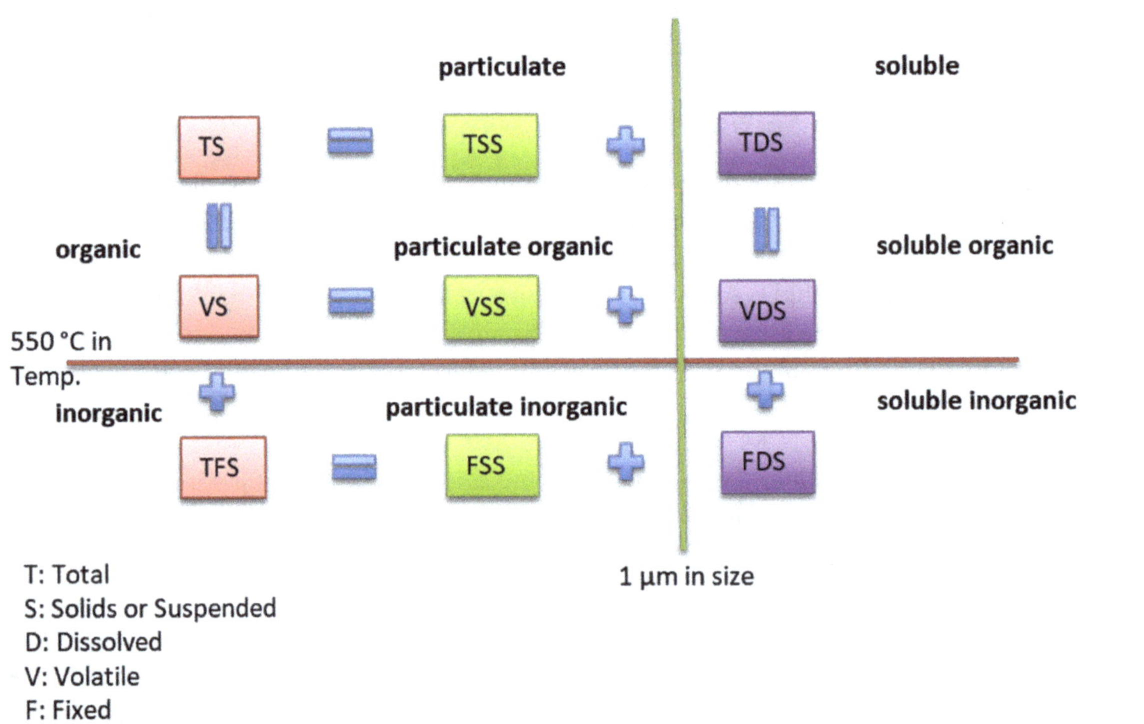


FIGURE 14.4 "Solids Matrix"

According to Method 5220 (APHA et al., 2023), when a sample is digested (at 150 °C for 2 hours), the dichromate ion oxidizes COD and changes from the hexavalent state (IV) to the trivalent (III) state. The $Cr_2O_7^{2-}$ absorbs strongly in the 400-nm region, whereas $Cr^{3+}$ absorbs strongly in the 600-nm region. Higher COD (100 to 900 mg/L) is determined based on the increase in $Cr^{3+}$, and low COD values (<90 mg/L) are determined by following the decrease in $Cr_2O_7^{2-}$ at 420 nm.

Chloride, bromide, or iodide can react with dichromate to produce the elemental form of the halogen and the chromic ion. The interference of chloride can be overcome by complexing with mercuric sulfate before the refluxing procedure. Nitrite exerts a COD of 1.1 mg $O_2$/mg $NO_2$-N, and 10 mg of sulfamic acid for each milligram of $NO_2$-N can be added to eliminate this interference.

## 4.8 Biochemical Oxygen Demand

Biochemical oxygen demand of wastewater is defined as the amount of oxygen required for the biological decomposition of biodegradable organic matter under aerobic conditions. Theoretically, at 20 °C, an infinite time is required for complete biological oxidation of organic matter; however, for all practical purposes, the reaction may be considered complete in 20 days. Typically, a 20-day period is too long to wait for results in most instances. Experience has shown that a reasonably large percentage (70% to 80%) of total BOD is exerted or oxidized in 5 days.

The rate of biological consumption of organic materials is governed by the following first-order function:

$$\frac{d\text{BOD}r}{dt} = -k_1\text{BOD}r \tag{14.3}$$

$$\text{BOD}r = \text{BOD}u(e^{-k1t}) \tag{14.4}$$

Where

BOD$r$ = amount of waste remaining at time $t$, in oxygen equivalents, mg/L;
$k_1$ = first-order reaction rate constant, $d^{-1}$;
BOD$u$ = ultimate CBOD, mg/L; and
$t$ = time, d.

Therefore, the BOD exerted up to time $t$ (BOD$t$) is given by

$$\text{BOD}t = \text{BOD}u - \text{BOD}r = \text{BOD}u - \text{BOD}u(e^{-k1t}) = \text{BOD}u(1 - e^{-k1t}) \tag{14.5}$$

Carbonaceous biochemical oxygen demand (CBOD) is a method-defined test measured by the depletion of dissolved oxygen by the biological organisms in a body of water in which the contribution from nitrogenous bacteria has been suppressed by a nitrification inhibitor (the remaining steps are the same as a regular BOD test). It is important to note that the use of a nitrification inhibitor has also been associated with inhibiting CBOD demand. It is important to indicate the use of a nitrification inhibitor and to investigate any potential effects on CBOD removal. On the other hand, the oxygen demand associated with the oxidation of ammonia to nitrite is called *nitrogenous biochemical oxygen demand* (BODn or NBOD). Figure 14.5 shows the difference between various BOD determinations.

According to Method 5210 (APHA et al., 2023), a well-mixed sample is first suitably diluted with prepared dilution water and then placed in a BOD bottle. This bottle is then filled with dilution water that is saturated with oxygen and contains required nutrients for biological growth. If the testing

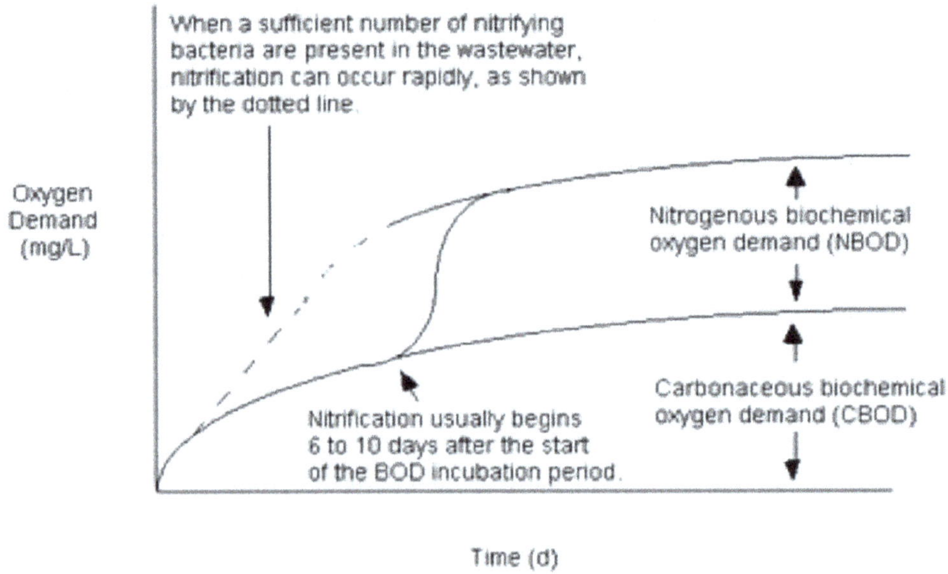


FIGURE 14.5 Components of Total BOD

samples contain low concentrations of microorganisms, seed is needed for the test. After incubation for 5 days at 20 °C, the dissolved oxygen concentration is measured again. The BOD is calculated by

$$\text{BOD, mg/L} = \frac{(D_1 - D_2) - (B_1 - B_2)f}{P} \tag{14.6}$$

Where

$D_1$ = dissolved oxygen of diluted sample immediately after preparation, mg/L;
$D_2$ = dissolved oxygen of diluted sample after 5-day incubation at 20 °C, mg/L;
$B_1$ = dissolved oxygen of seed control before incubation, mg/L;
$B_2$ = dissolved oxygen of seed control after incubation, mg/L;
$F$ = fraction of seeded dilution water volume in the sample to that in the seed control; and
$P$ = fraction of wastewater in the combined volume.

In raw influent wastewater, 5-day CBOD is approximately 0.85 × $\text{BOD}_{t5}$. However, there is typically variability in BOD tests that can make this ratio not apply to individual samples.

## 4.9 Volatile Fatty Acids

Low-molecular mass carboxylic acids (C2-C7 monocarboxylic aliphatic acids) are important intermediates and metabolites in biological processes. The most common form of short-chain volatile fatty acids (SCVFAs) in domestic wastewater is acetic acid; however, propionic, butyric, valeric, caproic, and heptanoic acids can also be present in wastewater. The concentration and type of SCVFAs in raw wastewater are particularly important because they have a direct effect on biological phosphorus removal and denitrification performance. In addition, VFAs constitute one of the chemical classes responsible for unpleasant odor generated in wastewater, together with amines and sulfur compounds.

Traditionally, the VFA content in wastewaters and foods has been analyzed by titrimetric or gas chromatographic methods preceded by solvent extraction. The rapid colorimetric determination of organic acids is also in use. Volatile fatty acids contain negative charges when they are ionized (solubilized) in water. Therefore, more recently, some SCVFAs, including acetate, may be measured through ion chromatography without further derivatization and purification. High-pressure liquid chromatography can also be used to determine VFAs in wastewater.

According to Method 5560 (APHA et al., 2023), the gas chromatographic procedure can be used to determine the individual concentrations of acetic, propionic, butyric, isobutyric, valeric, and isovaleric acids. Samples are prepared by acidification, centrifugation, and filtration before injection into the gas chromatogram equipped with a flame ionization detector. Figure 14.6 shows an image of a gas chromatogram. The blank amount of certain fatty acids might be higher than the sample analyte (caused by the buildup of contaminants in the injector and guard column) and needs to be corrected if it exceeds the reporting limits.

## 5.0 NUTRIENT ANALYSIS

### 5.1 Nitrogen

The common types of species of nitrogen in wastewater are ammonium ion, un-ionized ammonia, nitrite, nitrate, and organic nitrogen. Fresh samples are always ideal for laboratory analysis. If analysis cannot be done within a short period of time, sample preservation is required to ensure the accuracy of laboratory analysis. This can be done by adding sulfuric acid ($H_2SO_4$) to drop pH of the sample to 1.5 to 2 and stop any biological reaction (APHA et al., 2023). Neutralization of preserved samples before laboratory analysis by adding sodium and potassium hydroxide (NaOH/KOH) is required. During preservation, the sample should be kept refrigerated at a temperature of 4 °C.

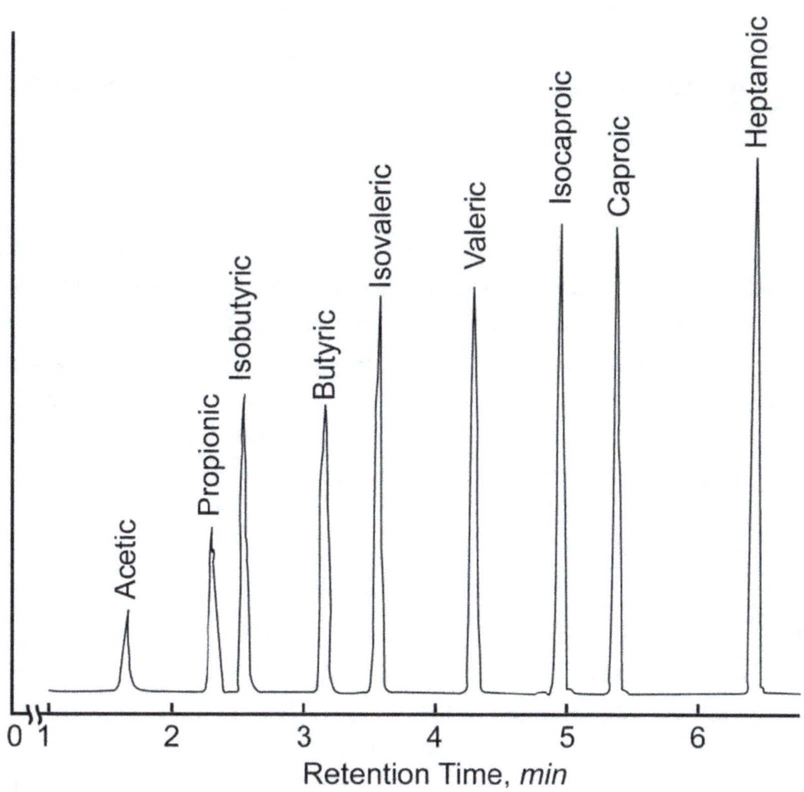


FIGURE 14.6 Gas Chromatogram of a Fatty Acid Standard. Column DB-FFAP, 0.53-mm-ID, 30-m, 0.50-µm film thickness, temperature-programmed as described in 5560D.2d (APHA et al., 2023)

### 5.1.1 Ammonia

Ammonia-nitrogen is a significant nutrient in raw wastewater and is typically present in the following two forms: ammonium ion and un-ionized ammonia. Measurement of ammonia-nitrogen in the wastewater stream is essential to determine the performance of the BNR facility. Techniques such as colorimetric method, titrimetric method, ion-selective method, and ion chromatography method can be used to determine ammonia-nitrogen. Measuring the range, difficulty, and presence of interference should be taken into consideration when choosing an appropriate method.

#### 5.1.1.1 Colorimetric Method

The Nessler and Phenate methods are the two most common methods for colorimetric determination of ammonia. In each method, chemical reagents with ammonia produce a distinct color and are measured in a spectrophotometer. The Nessler method (Method 4500-NH3 F; APHA et al., 2023) typically is used to determine ammonia concentration in treated wastewater effluents with an ammonia-nitrogen concentration of less than 5 mg/L. Turbidity and color interfere with the method and need to be removed by distillation or filtration. If the filtrate is turbid, zinc sulfate should be used to eliminate turbidity (Method 4500-NH3 F; APHA et al. 2023). For the detailed procedure and calculation, refer to Method 4500-NH3 F (APHA et al., 2023).

#### 5.1.1.2 Titrimetric Method

The titrimetric method for ammonia is used only on samples carried through the primary distillation step described in Method 4500-NH3 C (APHA et al., 2023). Mixed indicators containing methyl red and methylene blue are added to the distillated samples and blanks. The samples are titrated with $H_2SO_4$ until a pale lavender color is observed (APHA et al., 2023). The ammonia concentration is calculated by taking the difference between the volume of $H_2SO_4$ titrated for samples and the volume of $H_2SO_4$ titrated for blanks. For the detailed procedure and calculation, refer to Method 4500-NH3 C (APHA et al., 2023).

#### 5.1.1.3 Ion-Selective Method

The ammonia-selective electrode is equipped with a gas-permeable membrane that only allows un-ionized ammonia to diffuse and changes the internal solution pH that is measured by a pH meter with an expanded millivolt scale. Strong base (i.e., NaOH or KOH) needs to be added to the sample to raise the pH to above 11 to convert all ammonium to free ammonia to be detected by the electrode.

The ion-selective method is applicable to analyses of ammonia-nitrogen in wastewaters within the range of 0.03 to 1400 mg/L. High concentrations of other dissolved ions may interfere with the measurements. The method is not affected by high turbidity and color. For the detailed procedure and calculation, refer to Method 4500-NH3 D (APHA et al., 2023).

#### 5.1.1.4 Ion Chromatography Method

Ion chromatography is a rapid and simple method to analyze many ions in wastewater. It is accurate, and it eliminates preparation of hazardous reagents. The wastewater sample is injected to a stream of carbonate-bicarbonate and passed through a series of ion exchangers. Each ion has a unique affinity to a separator column and exhibits a different separation time.

If the pH of the wastewater sample is 7.5 or less, pH correction is not necessary. However, pH values of 8.0 to 8.5 require a correction factor of 1.04 and 1.12, respectively. The minimum detection level for ammonia-nitrogen is approximately 0.05 mg/L. Dilution is recommended if the ammonia concentration is higher than 40 mg/L. This method can be directly applied to measure ammonia concentration in raw wastewater, activated sludge, and industrial wastewaters. The following steps must be performed:

- Prepare at least seven standard solutions in the range of low to high (0.2, 1.0, 5.0, 10, 20, 30, and 40 mg/L).
- Use ion-free deionized water during standard solution preparation.
- Take 2 to 3 mL of volume from each standard solution and inject it to the ion chromatograph.
- Prepare a standard curve for ammonia calculations.
- Filter all samples with a 0.45-µm membrane filter.
- Take a 2- to 3-mL sample with a disposable syringe.
- Inject the samples to the ion chromatograph.
- Calculate the ammonia concentrations based on the standard curve.

### 5.1.2 Nitrite and Nitrate

Nitrite is an intermediate product during nitrification and the final product of nitritation. The nitrite ion is rarely found in a well-operated WRRF. Nitrite formation occurs during the startup period of facility operation and can occur intermittently but is typically quickly converted to a more stable form of nitrogen.

#### 5.1.2.1 Colorimetric Method

The colorimetric determination of nitrite relies on the formation of a reddish-purple azo dye at pH 2.0 to 2.5 and its photometric measurement at 540 nm. The method is suitable to measure low nitrite concentration in microgram levels of 10 to 1000 µg $NO_2^-$-N/L.

Suspended solids, free chlorine, and nitrogen trichloride interfere with the measurements. The presence of ions may also interfere with color development. Before analyses, all samples must be filtered through a 0.45-µm membrane filter. Analysis of fresh samples is recommended for accurate results. If sample storage is necessary, filtering and freezing of samples is sufficient to preserve the samples for a couple of days. Acid preservation should never be used for nitrite analysis (APHA et al., 2023). For the complete procedure and calculation, refer to Method 4500-$NO_2^-$ B (APHA et al., 2023).

There is also a proprietary spectrophotometer that is commercially available for nitrite measurements. However, the details of any proprietary system should be reviewed for U.S. EPA approval.

#### 5.1.2.2 Ion Chromatography Method

Ion chromatography is a rapid, easy, and accurate method to determine nitrite-nitrogen and nitrate-nitrogen of wastewaters. It eliminates the use of hazardous chemicals. The anions are separated based on their relative affinities for a strongly basic anion exchanger (separator column). The separated anions are directed onto a strongly acidic cation exchanger (suppressor column), where they are converted to their acidic form. The anions, in their acidic form, are measured by conductivity. The quantification is performed based on the peak area for each ion (APHA et al., 2023). The common anions in wastewater treatment are chloride ($Cl^-$), nitrite ($NO_2^-$), phosphate ($PO_4^{3-}$), nitrate ($NO_3^-$), and sulfate ($SO_4^{2-}$). Some organic acids have been reported to interfere with measurements (APHA et al., 2023). The following procedure is recommended for nitrite and nitrate determination by ion chromatography:

- Prepare a series of nitrite and nitrate standards by weighing sodium nitrite and sodium nitrate (deionized water should be used during standard preparation).
- Take 2 to 3 mL from the combined standard solutions and inject them to the ion chromatographer.
- Repeat standard injection at least twice; produce a standard curve based on area and concentration relationship.
- Filter samples with a 0.45-µm pore diameter filter (if prompt determination is not possible, freeze the filtered samples at –10 °C or below).
- Make dilution, if necessary (use deionized water during dilution).
- Inject samples into the ion chromatographer.
- Calculate nitrite and nitrate calculations from the standard curve.

Nitrate is the highest oxidation state of nitrogen. Nitrate nitrogen is part of total nitrogen and total inorganic nitrogen.

#### 5.1.2.3 Colorimetric Method (Cadmium Reduction Method)

A proprietary spectrophotometer is commercially available for nitrate measurements. More than 16 substances can cause interference effects at certain concentration levels. The test can only be used if COD is less than 500 mg/L. The details of any proprietary system should be reviewed for U.S. EPA approval.

#### 5.1.2.4 Nitrate Electrodes Screening Method

This method uses a selective sensor for the nitrate ion that develops a potential across a thin, inert membrane (Method 4500-$NO_3^-$ D; APHA et al., 2023). The electrode responds only to the nitrate ion in the range of 0.14 to 1400 mg/L (APHA et al., 2023). The electrode responds to nitrate activity rather than nitrate concentration. Interference includes chloride, bicarbonate, and nitrite ions. Chloride and bicarbonate concentrations exceeding the nitrate concentration by 5 and 10 times, respectively, may interfere with the nitrate. Using a buffer solution containing silver sulfate ($Ag_2SO_4$) may minimize such interferences. Reducing the pH to 3 eliminates bicarbonate ions. Adding sulfamic acid will minimize the nitrite interference. For the detailed procedure and calculation, refer to Method 4500-$NO_3^-$ D (APHA et al., 2023).

### 5.1.3 Organic Nitrogen

Organic nitrogen in wastewater does not represent all the nitrogen found in organic form. Rather, it is the portion related to protein, urea, and amino compounds. Organic nitrogen decomposition releases organic nitrogen into the medium. In a well-operated BNR facility, effluent organic nitrogen concentration is typically 0.5 to 2.0 mg/L.

To determine the organic nitrogen concentration, the Kjeldahl method is performed after removing ammonia-nitrogen in the medium. Otherwise, the organic nitrogen concentration is estimated by taking the difference between TKN and ammonia-nitrogen.

### 5.1.4 Total Kjeldahl Nitrogen

In wastewater, TKN is the sum of unionized ammonia, ammonium ion, and organic nitrogen. For total nitrogen, the nitrite and nitrate concentration is added to TKN. The macro-Kjeldahl method and semi-micro-Kjeldahl method are available to determine Kjeldahl nitrogen. The most reliable results are obtained from fresh samples. If prompt analysis is not possible, acid preservation should be used followed by 4 °C storage. Mercury chloride addition is not recommended because it reduces the ammonia in the sample.

#### 5.1.4.1 *Macro-Kjeldahl Method*

The macro-Kjeldahl method is applicable for samples containing a broad range of organic nitrogen. It requires a relatively large volume for analysis and is recommended if enough sample volume can be obtained (APHA et al., 2023). The three main steps of the process include digestion, distillation, and ammonia determination. For the complete procedure and calculation, refer to Method 4500-$N_{org}$ B (APHA et al., 2023).

#### 5.1.4.2 *Semi-Micro-Kjeldahl Method*

The semi-micro-Kjeldahl method is applicable to samples with high concentrations of organic nitrogen. For the detailed procedure and calculation, refer to Method 4500-$N_{org}$ C (APHA et al., 2023). Proprietary test kits are also available for quick TKN analysis; however, it is important to note that the kits are not based on *Standard Methods*. Carbonaceous oxygen demand and chloride will interfere at certain concentration levels.

## 5.2 Phosphorus

Phosphorus is essential to the growth of biological organisms. However, excessive phosphorus content in surface water causes issues such as accelerated plant growth, algal bloom, low dissolved oxygen, and the death of certain invertebrates and fish. Typical forms of phosphorous in wastewater include orthophosphates, condensed phosphates, and organically bound phosphate. Orthophosphates are reactive phosphates in wastewater that respond to the colorimetric test or ion chromatography without preliminary hydrolysis or digestion, such as $PO_4^{3-}$, $HPO_3^-$, $H_2PO_4^-$, and $H_3PO_4$. Organically bound phosphates are phosphate fractions that are converted to orthophosphates by digestion of organic matter (APHA et al., 2023).

Orthophosphate and total phosphorus are of particular interest when it comes to phosphorus analysis. If the soluble form of phosphate is of interest, the sample should be filtered promptly after collection and stored at –10°C. If longer storage is desired, 40 mg/L of mercury chloride should be added. If total phosphorus is of interest, then 1 mL of concentrated HCl should be added to the unfiltered samples. If the analysis is performed within 48 hours, then freezing without acid addition will be sufficient to preserve the samples (APHA et al., 2023). Use of phosphate-containing detergents to clean phosphorus glassware should be avoided (APHA et al., 2023).

### 5.2.1 *Total Phosphorus*

Total phosphorus includes all orthophosphates and condensed phosphates, both soluble and particulate, and organic and inorganic fractions. It is determined by using digestion methods and orthophosphate measurement. There are three digestion methods: persulfate digestion, which is the most common, perchloric acid digestion, and sulfuric-nitric acid digestion (Method 4500-P B; APHA et al., 2023). Unfiltered activated sludge samples require 1 to 50 to 1 to 200 times dilution before the aforementioned digestion step. Following digestion, all complex phosphorus is converted to orthophosphate form and is determined by colorimetric, spectrophotometric, or chromatographic methods.

Although proprietary test kits are available for a quick total phosphorus analysis and continuous process analyzers are available for directly measuring phosphorus in wastewater. These kits and analyzers are not based on *Standard Methods* but are helpful for providing a quick phorphorus measurement for controlling chemical feed points based on measured phorphorus levels. The details of any proprietary system should be reviewed for U.S. EPA approval.

### 5.2.2 Dissolved Phosphorus

The dissolved phosphorus test measures that fraction of total phosphorus that is in solution in the water (as opposed to being attached to suspended particles). It is determined by first filtering the sample through a 0.45-μm pore diameter filter, then analyzing the filtered sample for total phosphorus.

### 5.2.3 Orthophosphate

Orthophosphate can be measured by directly adding a complex agent such as ammonium molybdate, which will form a colored complex with phosphate, followed by performing a colorimetric test using a spectrophotometer. Although proprietary test kits are also available for quick orthophosphate analysis, it is important to note that the kits are not based on *Standard Methods.* The details of any proprietary system should be reviewed for U.S. EPA approval.

#### 5.2.3.1 Vanadomolydophosphoric Acid Colorimetric Method

Orthophosphate reacts with ammonium molybdate under acidic conditions to form molybdophosphoric acid. Vanadium is added to promote vanadomolybdophosphate, which produces a yellow color. The color intensity is proportional to phosphorus in the sample. The color is spectrometrically detected at a wavelength of 470 nm (recommended) for measurements (APHA et al., 2023). A standard curve is prepared in a phosphorus range of 0 (blank) to 20 mg/L. The detection limit of the method is 200 μg P/L. If the digestion step is included, the measured phosphorus values will reflect the total phosphorus. For the complete procedure and calculation, refer to Method 450-P C (APHA et al., 2023).

#### 5.2.3.2 Ascorbic Acid Method

Like the method described in Section 5.2.3.1, in the ascorbic acid method (Method 450-P E; APHA et al., 2023), orthophosphates react with ammonium molybdate and potassium antimonyl tartrate in a highly acidic medium to form intensely colored molybdenum blue by ascorbic acid. The developed color should be measured within 30 minutes following the color development at 880 nm. An arsenate concentration as low as 0.1 mg As/L interferes with the color development. Hexavalent chromium and nitrite may also interfere with the results. The minimum detectable concentration of phosphorus is approximately 10 μg/L. For the complete procedure and calculation, refer to Method 450-P E (APHA et al., 2023).

### 5.2.4 Refractory Phosphorus

When faced with low total phosphorus limits ($<$0.2 mg/L total phosphorus), it may be helpful to identify the soluble component that is not responsive to biological treatment (conventional or enhanced biological phosphorus removal), advanced phosphorus removal methods (tertiary processes that include adsorption onto metal hydroxide solids, flocs, or coated silica sand), and high solids capture. The soluble phosphorus that passes through both biological and advanced treatment is considered refractory or recalcitrant phosphorus. For most domestic wastewater, the background concentration of recalcitrant phosphorus is between 0.005 and 0.02 mg/L. However, with certain industrial contributions, concentrations may be higher. The magnitude of recalcitrant phosphorus is only significant when it might be a dominant portion of the effluent total phosphorus because it makes meeting the effluent objective difficult.

If a facility has a low total phosphorus limit and is able or planning to achieve high solids capture (<5 mg/L TSS) and near-complete adsorption of orthophosphate (nondetect, which can be as low as 0.005 mg/L phosphorus, depending on the method) through some tertiary process, recalcitrant phosphorus can be determined by a test. This is done by taking a secondary effluent sample and treating it in a way that simulates a tertiary phosphorus removal process by adding sufficient metal salt, mixing it to adsorb the orthophosphate, passing it through a 0.45-µm filter, and analyzing the filtrate for total phosphorus and orthophosphate. The difference will be the sum of the dissolved acid hydrolyzable and dissolved organic phosphorus, which will be what remains following biological treatment and advanced phosphorus removal. This is considered recalcitrant phosphorus. The analytical method that provides the lowest detection limit should be used, which is best done with long cell path cuvettes and a spectrophotometer. The test kits do not provide a sufficiently low detection limit.

Excessive metal dose can cause a matrix interference with the low detection method. In addition, if the metal salt used is a strong acid, or even a weak acid, but too much is used, the pH becomes so low that solubility limitations are exceeded. A metal salt that is a weaker acid, such as sodium aluminate, is recommended for use, or a buffer should be added to ensure that the pH is not too low. The total phosphorus analysis goes through persulfate digestion followed by colorimetry. The orthophosphate analysis, which is sometimes considered "reactive," is simply direct colorimetry.

## 6.0 MICROBIOLOGY

Just as microorganisms play a significant role in biological processes, the same is true for BNR in WRRFs. Microorganisms enable the removal of most nitrogenous and phosphoric compounds found in wastewater. However, like many other treatment processes, microorganisms have long been viewed as the "black boxes" of a facility (Oehmen et al., 2007; Seviour et al., 2003). The knowledge of engineering and operational features of BNR processes predates an understanding of its microbiology.

Today, after much advancement in the field of wastewater microbiology, engineers and operators agree that a thorough knowledge of the structure and function of these complex communities is essential to successful operation and process optimization of BNR processes. Understanding how these microorganisms function, monitoring their growth and abundance in activated sludge, and proactively predicting their behavior can help minimize operational upsets. Thus, microbiological examinations of activated sludge samples are important and provide an opportunity for engineers and operators to decide on ways to maintain healthy activated sludge conditions and to prevent and correct unhealthy conditions (Gerardi, 2008). Recent advancement in molecular biology, like next-generation sequencing, has enabled operators to gain insight of various microbial populations like ammonia-oxidizing bacteria, nitrite-oxidizing bacteria, PAOs, and glycogen-accumulating organisms (Roy et al., 2021) (see Section 6.5).

Bacteria are the group of microbes most directly involved in BNR processes. They dominate the microbial composition of activated sludge, both in numbers and biomass. However, their presence also allows for the growth and existence of several eukaryotic predator organisms such as protozoa and metazoa (Curds, 1982). Although protozoa and metazoa are not directly responsible for the primary degradation of pollutants in wastewater, they contribute to the elimination of particulate matter and act as indicators of environmental conditions in the treatment process. For example, the presence of stalked ciliates, rotifers, and nematodes indicate long SRTs or low organic loading. A predominance of metazoa is often described as "old sludge" and a predominance of flagellates and amoeboids is described as "young sludge." Because ammonia removal is associated with sensitive nitrifying bacteria, microscopic examination can help determine SRT, which can be modified, if needed, to provide nitrification. This section provides laboratory methods for microbiological examinations of activated sludge processes focusing on BNR.

## 6.1 Microscope Use

All microbiological examinations require the use of microscopes. The two basic types of microscopes used for analyzing activated sludge samples are light, or optical, microscopes and electron microscopes.

Light, or optical, microscopes require light waves to provide illumination and are the most commonly used devices for all general wastewater laboratory work. Light microscopes can be bright field, dark field, phase contrast, or fluorescence, with lenses that can accommodate magnifications ranging from 10 to 1000×, depending on specific laboratory analytical needs. The most commonly used laboratory microscopes in WRRFs are the bright-field and phase-contrast microscopes. The fluorescence microscope uses ultraviolet (UV) light as its light source and is used mainly in advanced molecular analytical methods such as fluorescent in situ hybridization (FISH) and fluorescence staining techniques.

Electron microscopes use electrons to provide the illumination. Because the beam has an exceptionally short wavelength, it strikes most objects in its path and increases the resolution of the microscope, enabling extremely small objects such as subcellular components or viruses to be viewed. Electron microscopes can be either transmission or scanning and are much more expensive than light microscopes. These microscopes are not commonly used in WRRF laboratories; rather, they are often used for research purposes. For more details on the microscopes, types of lenses, filters, microscopic measurements, and how to use them, refer to literature by Csuros and Csuros (1999), Gerardi (2008), and Jenkins et al. (2004).

## 6.2 Protozoa and Metazoa

Biological wastewater treatment relies on the activity of a mixed community of bacterial culture for the removal of organic substances and nutrients. However, their presence also allows for the growth of predator organisms such as protozoa and metazoa. Whereas these organisms are of little value to the process of BNR directly, microscopic examinations of protozoa and metazoa are a common and widespread practice as indicator organisms. In addition, they help with the removal of nonflocculated and loosely flocculated bacteria from wastewater producing more clarified effluent and may contribute to biomass flocculation through the production of fecal pellets and mucus (Jenkins et al., 2004). Protozoa and metazoa constitute approximately 5%, by weight, of the activated sludge biomass and are represented by about 200 species (Curds, 1975).

Protozoa are a large collection of organisms with considerable morphological and physiological diversity. They are all eukaryotic and are unicellular. Common protozoa found in activated sludge include amoeba, flagellates, and ciliates. A useful guide to identifying protozoa in wastewater treatment systems can be found in literature by Patterson (1998).

Metazoa are multicellular eukaryotic organisms that include worms such as rotifers and nematodes and arthropods such as crustacea and insects. Most of them are aerobic, but some organisms can survive under low dissolved oxygen conditions. The main metazoa present in activated sludge are rotifers and nematodes (Richard, 1989).

To observe these organisms under a microscope, Jenkins et al. (2004) recommend placing one drop of activated sludge on a microscopic slide, adding a cover slip on it, and examining it at 100× magnification using phase-contrast illumination. All protozoa and other higher life forms should be counted by scanning the entire cover slip area using the mechanical stage of the microscope; an average of the results for four to five separate preparations should be used for purposes of enumeration. For further details, refer to literature by Jenkins et al. (2004).

### 6.2.1 Flagellates

Flagellates are typically small, oval, or elongated forms that possess a whip-like structure or flagella that provide locomotion for the organism. Their presence can indicate significant soluble BOD levels. Many of these occur at low dissolved oxygen concentrations accompanied by high organic loads. Jenkins et al. (2004) and Patterson (1998) provided illustrations of common flagellates found in activated sludge.

### 6.2.2 Amoebae

Amoebae, which represent the simplest life forms, vary in shapes and sizes. They are slow-moving organisms and have a cytoplasm or gut content that flows against a flexible cell membrane, permitting locomotion. Two types of amoebae are known to exist in activated sludge: naked and testate. Amoebae grow well on particulate organic matters and can tolerate low dissolved oxygen conditions. Jenkins et al. (2004) and Patterson (1998) provided illustrations of amoeba found in activated sludge.

### 6.2.3 Ciliates

Ciliates are free-swimming, crawling, or stalked organisms. Ciliates are typically found under good floc-forming conditions and are generally indicators of satisfactory activated sludge operations. They are sensitive organisms, and their presence or absence may indicate toxicity. Jenkins et al. (2004) and Patterson (1998) provided illustrations of the different kinds of ciliates found in activated sludge.

### 6.2.4 Rotifers

These metazoa have a variety of shapes and are much larger and more complex in structure than protozoa. Most of them are motile and can attach themselves to activated sludge flocs with contractile feet. They occur over a wide range of SRTs, and some species are indicative of high SRT. Jenkins et al. (2004) provided illustrations of common rotifers found in activated sludge.

### 6.2.5 Higher Invertebrates

Higher forms of organisms found in activated sludge include nematodes, tardigrades, gasterotrichs, and annelids. They are generally observed at higher SRTs. Tardigrades, gasterotrichs, and annelids appear to occur only in nitrifying systems. Jenkins et al. (2004) provided illustrations of common higher invertebrates found in activated sludge.

## 6.3 Bacteria

Biological nutrient removal processes involve the metabolic activities of various bacteria to remove nitrogenous and phosphoric compounds from wastewater. Common, favorable ones include PAOs, ammonia oxidizers (ammonia-oxidizing bacteria), nitrite, and nitrate oxidizers (nitrate-oxidizing bacteria), denitrifiers, denitrifying PAOs, and anaerobic ammonia oxidizers. Common, unfavorable ones include filamentous bacteria. Several species of bacteria have been identified to fall into each of these categories. Moreover, the contributions of interdisciplinary research in wastewater microbiology have led to the development of numerous advanced molecular methods, such as FISH, cloning of 16S rDNA, denaturant gradient gel electrophoresis, and real-time polymerase chain reactions, to characterize and understand the structure and function of these bacteria. However, only a handful of WRRF laboratories are equipped to run these advanced methods. They are more commonly used for research and not for monitoring the health of the treatment process, although they are highly capable of accomplishing that purpose.

### 6.3.1 Filamentous Bacteria

The presence of filamentous bacteria in activated sludge is known to cause sludge bulking and foaming. These problems are commonly reported in BNR systems, and several publications have provided evidence that BNR favors the growth of undesirable and excessive filamentous bacteria (Eikelboom, 2000; Lee et al., 1996; Metcalf & Eddy, Inc./AECOM, 2013). The causes for the growth of filamentous bacteria include several operational conditions, such as low oxygen concentration, low F/M, septicity, nutrient deficiency, low pH, and high grease and oil. Different types of filamentous bacteria are known to thrive under different operating conditions. Microscopic examination of filamentous bacteria can provide a wealth of information on the nature and causes of sludge foaming and bulking and the ability to correct them.

Determining the number and type of filamentous microorganisms is the first step in controlling this problem. Such examination requires a phase contrast microscope with the capability of 100 to 1000× magnification. Details on sample collection, transport, storage, and sampling frequency are provided in Jenkins et al. (2004).

Common microscopic methods of enumerating and identifying filamentous bacteria involve counting and staining methods. Counting methods include

- simplified filament counting technique;
- total extended filament length measurement method; and
- nocardioform organism filament counting technique.

Staining methods include

- gram stain;
- Neisser stain;
- sulfur stain;
- India ink reverse stain;
- polyhydroxyalkanoate (PHA) stain; and
- crystal violet sheath stain.

For more details on the counting methods, refer to Jenkins et al. (2004).

### 6.3.2 Polyphosphate-Accumulating Organisms

The fluorescent stain, 4', 6-diamidino-2-phenylindol (DAPI), binds strongly to DNA in cells and fluoresces blue when excited with light at 340 nm using a microscope. However, DAPI also stains polyphosphate granules accumulated within any bacteria to emit yellow fluorescence. This staining technique is commonly used to differentiate between PAOs and non-PAOs. For direct polyphosphate staining using DAPI, mixed liquor samples collected from the aeration tank should be washed twice with 1X phosphate-buffered saline (PBS) (a mixture of 8 g NaCl, 0.2 g KCl, 1.44 g $Na_2HPO_4$, and 0.24 g $KH_2PO_4$ in 1 L of water; pH 7.4 and filter sterilized) and purged at least 30 times using a 26-gauge needle to disrupt flocs. The samples should then be fixed with 4% paraformaldehyde solution in 1X PBS for 45 minutes at room temperature. The fixed cells should be filtered through a 0.22-µm polycarbonate filter paper, and the filter paper containing the fixed cells should be transferred onto gelatin-coated sterile glass slides. The cells on the slide should then be stained with 5 µg/mL of DAPI in the dark for 30 minutes, and the polyphosphate positive cells can be viewed using an epifluorescence microscope.

## 6.4 Microbiological Activity

Monitoring microbiological activity can be useful in optimizing the operation of a bioreactor, particularly for systems operating in nutrient removal mode. It can aid in detecting influent toxicity or inhibition or a potential slug loading into the facility. In addition, having the ability to quickly and easily measure active biomass concentration creates many opportunities for optimizing bioreactors, including management of biomass inventory, which, consequently, aids in controlling the F/M to improve settleability and the optimization of aeration. This allows operators to adjust oxygen delivery to the active biomass, leading to more efficient operation. There are available methods for monitoring at least two compounds that are indicators of microbiological activity. These include nicotinamide-adenine dinucleotide (NADH) and adenosine triphosphate (ATP).

### 6.4.1 *Nicotinamide-Adenine Dinucleotide*

Microbiological activity can be monitored using the concentration of a coenzyme NADH, which is found in all living cells. In cellular metabolism, nicotinamide-adenine dinucleotide (NAD+) is involved in redox reactions, carrying electrons from one reaction to another. The coenzyme is, therefore, found in the following two forms: NAD+ (oxidizing agent that accepts electrons) and NADH (reducing agent to donate electrons). These electron transfer reactions are the main function of NAD+, although it is involved in other cellular processes. This metabolic intermediate can be characterized by fluorescence at particular wavelengths. For a given biomass, the NADH fluorescence signal reflects the ratio between NADH and NAD+, allowing determination of the redox state of the biomass.

Monitoring NADH involves the use of a sensor (probe) to monitor the strength of a fluorescence signal reflecting from the biomass; this, in theory, allows for more reliable monitoring and control of the biological state or activity in the aeration tanks at low dissolved oxygen concentrations than with a dissolved oxygen probe. The fluorescence signal is dependent on the concentration of NADH in the biomass in the aeration tanks. The fluorescence signal reflected from the biomass originates from the probe, which emits a UV light with a wavelength of 340 nm; NADH has a unique property in that it fluoresces at 460 nm when excited with UV irradiance at a wavelength of 340 nm. There is a proprietary control system available on the market that has been tested at full-scale WRRFs: The SymBio® process by Eimco Water Technologies uses online monitoring of NADH to determine changes in biological demands. Using the online NADH signal, airflow to the basin is controlled to promote simultaneous nitrification-denitrification of wastewater. The NADH monitor uses a fluorescence sensor to detect changes in NADH, which, in turn, provides information on the status of biological wastewater treatment processes. Weerapperuma and de Silva (2004) reported that the NADH sensor requires minimal maintenance and can provide real-time information for process control. The manufacturer claims a 25% to 30% energy savings compared to nitrifying facilities without this control technology; however, no independent data from full-scale facilities has been published to verify these claims (U.S. EPA, 2010).

### 6.4.2 *Adenosine Triphosphate*

Adenosine triphosphate is the keystone of metabolic activity in living cells. Most of the energy within microorganisms is stored and transmitted via ATP. Adenosine triphosphate is produced as microbial food and is consumed and subsequently used for cell maintenance and the synthesis of new cells and biochemical reactions. As such, ATP can be used for monitoring biological processes. The value of monitoring ATP in biological wastewater treatment was recognized 50 years ago (Patterson et al., 1970). More recently, in a study using a suite of respirometric tests on mixed liquor from paper mill activated sludge processes, Archibald et al. (2001) concluded that ATP measurements provided a useful monitor of the proportion of viable cells and a toxicity indicator in an activated sludge process. These and other studies have provided direct measurement of ATP as a monitoring tool in biological treatment processes; however, most studies have not used methods that distinguish between extracellular ATP and ATP contained only within microorganisms. In efforts to distinguish between these forms, a related stress index that is based on the ratio of dissolved ATP to total ATP, referred to as the *biomass stress index*, has been developed as part of a proprietary monitoring tool. This commercially available tool has been demonstrated in laboratory and full-scale reactors; as stresses such as suboptimal pH, anoxia, toxicity, and nutritional deficiencies were applied to the microbial populations, the stress index increased (Cairns et al., 2005).

## 6.5 Next-Generation Sequencing

Next-generation sequencing (NGS) has revolutionized the study of microbial communities in various environmental contexts, including wastewater treatment systems. When it comes to understanding how microbial communities contribute to nutrient removal from wastewater, NGS techniques provide powerful tools for characterizing the diversity, composition, and functional potential of these microbial populations.

Next-generation sequencing can be applied in the context of studying microbial communities involved in nutrient removal from wastewater:

- Sample collection and DNA extraction:
  - Collect representative samples from different points in the wastewater treatment process, such as influent, primary treatment, secondary treatment, and effluent.
  - Extract DNA from these samples to obtain genetic material from the microbial communities.
- 16S rRNA gene sequencing:
  - One common approach is to use 16S rRNA gene sequencing to analyze the bacterial and archaeal diversity in the samples.
  - This technique amplifies and sequences a specific region of the 16S rRNA gene, allowing for taxonomic classification of the microbial species present.
  - It provides insights into the composition of the microbial community and helps identify key microbial players involved in nutrient removal.
- Metagenomic sequencing:
  - Metagenomic sequencing involves sequencing all the genetic material (DNA) present in a sample, providing a comprehensive view of the microbial community's functional potential.
  - It can be used to identify specific genes and pathways related to nutrient removal processes, such as denitrification, nitrification, and phosphorus removal.
  - Metagenomics also allows for the reconstruction of individual microbial genomes, enabling the study of specific metabolic capabilities of different microbial species.
- Metatranscriptomics and metaproteomics:
  - To gain insights into the active metabolic processes within the microbial community, metatranscriptomics (sequencing of RNA) and metaproteomics (analysis of proteins) can be used.
  - These techniques can help identify which genes and proteins are actively involved in nutrient removal at a given point in time.
- Bioinformatics analysis:
  - After sequencing, bioinformatics tools and pipelines are used to process and analyze the large datasets generated by NGS.

  - Taxonomic classification, functional annotation, and pathway analysis are performed to interpret the data and identify key microbial players and functional genes involved in nutrient removal.
- Longitudinal studies:
  - NGS can be used in longitudinal studies to track changes in microbial community composition and functional potential over time, helping to optimize and improve wastewater treatment processes.
- Integration with process data:
  - NGS data can be integrated with operational data from the WRRF to correlate microbial community dynamics with process performance and identify opportunities for process optimization.

In summary, NGS techniques provide a comprehensive and high-resolution approach to studying microbial communities involved in nutrient removal from wastewater. By characterizing the composition and functional potential of these communities, researchers and wastewater treatment operators can gain valuable insights for optimizing treatment processes and improving nutrient removal efficiency.

## 7.0 REFERENCES

Abu-ghararah, Z. H., & Randall, C. W. (1991). The effect of organic compounds on biological phosphorus removal. *Water Science & Technology, 23*, 585–594.

American Public Health Association, American Water Works Association, & Water Environment Federation. (2023). *Standard methods for the examination of water and wastewater* (24th ed.). American Public Health Association.

Archibald, F., Me´thot, M., Young, F., & Paice, M. G. (2001). A simple system to rapidly monitor activated sludge health and performance. *Water Research, 35*(10), 2543–2553.

Bott, C., & Parker, D. S. (2011). *Nutrient management volume II: Removal technology performance and reliability.* Water Environment Research Foundation.

Cairns, J. E., Whalen, P. A., Whalen P. J., Tracey, D. R., & Palo, R. E. (2005, October 29–November 2). Dissolved ATP—A new process control parameter for biological wastewater treatment. *Proceedings of the 78th Annual Water Environment Federation Technical Exhibition and Conference* [CD-ROM], Washington, DC, United States. Water Environment Federation.

Csuros, C., & Csuros, M. (1999). *Microbiological examination of water and wastewater.* CRC Press.

Curds, C. R. (1975). Protozoa. In C. R. Curds and H. A. Hawkes (Eds.), *Ecological aspects of used water treatment: The organisms and their ecology*. Academic Press.

Curds, C. R. (1982). The ecology and role of protozoa in aerobic sewage treatment processes. *Annual Review of Microbiology, 36*, 27–46.

Eikelboom, D. H. (2000). *Process control of activated sludge plants by microscopic investigation*. IWA Publishing.

Ekama, G. A., Marais, G. v. R., Siebritz, I. P., Pitman, A. R., Keay, G. F. P., Buchan, L., Gerber, A., & Smollen, M. (1984). *Theory, design and operation of nutrient removal activated sludge process.* Water Research Commission: Pretoria, South Africa.

Gerardi, M. H. (2008). *Microscopic examination of the activated sludge process.* Wiley & Sons.

Gy, P. M. (1992). *Sampling of heterogeneous and dynamic material systems—Theories of heterogeneity, sampling, and homogenizing.* Elsevier Science.

Jenkins, D., Richard, M. C., & Daigger, G. T. (2004). *Manual on the causes and control of activated sludge bulking and foaming, and other separation problems* (3rd ed.). Lewis Publishers.

Lee, N. M., Carlsson, H., Aspegren, H., Welander, T., & Andersson, B. (1996). Stability and variation in sludge properties in two parallel systems for enhanced biological phosphorus removal operated with and without nitrogen removal. *Water Science & Technology, 34*(1–2), 101–109.

Metcalf & Eddy, Inc./AECOM. (2013). *Wastewater engineering, treatment and resource recovery* (5th ed.). McGraw-Hill.

Her Majesty's Stationery Office. (1980). *Methods for the examination of waters and associated materials: General principles of sampling and accuracy of results.*

Oehmen, A., Lemos, P. C., Carvalho, G., Yuan, Z., Keller, J., Blackall, L. L., & Reis, M. A. M. (2007). Advances in enhanced biological phosphorus removal: From micro to macro scale. *Water Research, 41,* 2271–2300.

Ort, C., & Gujer, W. (2006). Sampling for representative micropollutant loads in sewer systems. *Water Science & Technology, 54*(6–7), 169–176.

Pagilla, K. (2007, September 27–28). Organic nitrogen in wastewater treatment plant effluents. *Presented at the Water Environment Research Foundation and Chesapeake Bay Science and Technology Advisory Committee Workshop*, Baltimore, MD, United States (p 97). Chesapeake Research Consortium, Inc.

Patterson, D. J. (1998). *Free-living freshwater protozoa: A colour guide.* Manson Publishing.

Patterson, J. W., Brezonik, P. L., & Putnam, H. D. (1970). Measurement and significance of adenosine triphosphate in activated sludge. *Environmental Science & Technology, 4*(7), 569–575.

Randall, C. W., Barnard, J. L., & Stensel, H. D. (1992). *Design and retrofit of wastewater treatment plants for biological nutrient removal.* Technomic Publishing.

Richard, M. (1989). *Activated sludge microbiology.* The Water Pollution Control Federation.

Roy, S., Guanglei, Q., Zuniga-Montanez, R., Williams, R. B., & Wuertz, S. (2021). Recent advances in understanding the ecophysiology of enhanced biological phosphorus removal. *Current Opinion in Biotechnology, 67,* 166–174.

Seviour, R. J., Mino, T., & Onuki, M. (2003). The microbiology of biological phosphorus removal in activated sludge systems. *FEMS Microbiology Review, 27,* 99–127.

Stensel, H. D., Bronk, D., Khan, E., Love, N., Makinia, J., Neethling, J. B., Pelligrin, M., Pitt, P., Sedlak, D., & Sharp, R. (2008). *Dissolved organic nitrogen (DON) in biological nutrient removal wastewater treatment processes.* Water Environment Research Foundation.

U.S. Environmental Protection Agency. (2004). *EPA guidelines for water reuse.*

U.S. Environmental Protection Agency. (2010). *Evaluation of energy conservation measures for wastewater treatment facilities.*

Water Environment Federation. (2022). *Basic laboratory procedures for the operator-analyst* (6th ed.).

Water Environment Federation. (2023). *Standards for design of water resource recovery facilities* (WEF 8-23).

Water Environment Research Foundation. (2003). *Methods for wastewater characterization in activated sludge modeling.*

Weerapperuma, D. & de Silva, V. (2004, October 2–6). *On-line analyzer applications for BNR control. Proceedings of the 77th Annual Water Environment Federation Technical Exhibition and Conference* [CD-ROM], New Orleans, LA, United States. Water Environment Federation.

## 8.0 SUGGESTED READINGS

California State University. (2003). *Operation of wastewater treatment plants* (Vol. 2, 6th ed.).

Curds, C. R. (1982). The ecology and role of protozoa in aerobic sewage treatment processes. *Annual Review of Microbiology, 36,* 27–46.

Dold, P., Bye, C., Chapman, K., Brischke, K., White, C., Shaw, A., Barnard, J., Latimer, R., Pitt, P., Vale, P., & Brian, K. (2010, March 28–30). What do we model, how should we model? *Proceedings of the Wastewater Treatment Modeling 2010 Seminar (WWTmod 2010)* (pp. 133–150), Monte St. Anne, Quebec, Canada.

Grady, Jr., C. P. L., Daigger, G. T., Love, N. G., & Filipe, C. D. M. (2011). *Biological wastewater treatment* (3rd ed.). CRC Press.

# Chapter 15

# Case Studies—Nitrification and Denitrification

# 1.0 INTRODUCTION

This chapter documents eight case studies of facilities that operate nitrification and denitrification processes. These case studies cover full-scale municipal water resource recovery facilities (WRRFs) that are specifically designed and operated for nitrogen removal. All facilities used in the case studies in this chapter have been in operation for a minimum of 2 years, so the reader can gain insight to the challenges encountered by facility operators during the startup and commissioning phases of the project and the design and operational changes that were required following facility startup.

# 2.0 CASE STUDY 1 (MODIFIED LUDZACK–ETTINGER)—GRASS ISLAND WASTEWATER WATER TREATMENT PLANT, GREENWICH, CONNECTICUT

## 2.1 Facility History

The Grass Island Wastewater Treatment Plant (GIWWTP) in Greenwich, Connecticut, provides nitrogen removal for an average annual design flow of 47.3 megaliters per day (ML/d) (12.5 million gallons per day [mgd]). The facility is configured as a modified Ludzack–Ettinger (MLE) process to target the facility's annual effluent nitrogen allocation as issued by the Connecticut Department of Energy and Environmental Protection (CTDEEP). The MLE process consists of five bioreactors, each equipped with a five-compartment pre-anoxic zone followed by three aerobic zones.

Before the upgrade to its current MLE configuration, the GIWWTP had been providing treatment for at least a portion of the town's wastewater since the 1920s. In the 1960s, the facility was upgraded to full secondary treatment and was further upgraded to increase treatment capacity in the 1990s, and additional capital improvements projects have occurred over the life of the facility to improve solids handling, disinfection, and other facilities. The facility was upgraded to the MLE process in 2000. A 2015 upgrade renovated and modified the five bioreactors and the aeration system (blowers and diffusers). Figure 15.1 is an aerial of the GIWWTP with the bioreactors in the foreground right.

## 2.2 Process Description

The GIWWTP consists of preliminary, primary and secondary treatment, disinfection, and solids-handling (thickening and dewatering) processes. The facility influent flow is largely domestic, and the facility also receives a significant amount of hauled septage. Table 15.1 summarizes typical influent characteristics. After treatment the effluent is discharged into Greenwich Harbor, a part of the larger Long Island Sound. Along with its design average daily flow of 47.3 ML/d (12.5 mgd), the GIWWTP is designed to handle a wet-weather peak flow of 117 ML/d (31 mgd).

Figure 15.2 presents a process flow diagram of the GIWWTP. Following primary treatment, biological treatment is provided by five MLE bioreactors, each with a total volume of 11.7 ML (3.1 mil. gal). Table 15.2 summarizes operating criteria for the MLE process. The bioreactor volume is 25% anoxic and 75% aerobic. Each 2.9 ML (0.78 mil. gal) anoxic zone is further divided into five subzones, and these are mixed by intermittent, alternating coarse-bubble aeration per a timed sequence at a maximum aeration rate of 2.1 $m^3$/min (75 scfm) per subzone. The aerobic volume is divided into three zones, each equipped with fine bubble diffusers and automatically operated based on maintaining operator-set dissolved oxygen concentrations of 3.0, 1.5, and 1.0 mg/L, sequentially. Dissolved oxygen is monitored continuously in each zone of bioreactor nos. 3 and 5.

FIGURE 15.1 Grass Island Wastewater Treatment Plant Aerial

TABLE 15.1 Grass Island Wastewater Treatment Plant Influent Wastewater Characteristics (2018–2022)

| | |
|---|---|
| Influent flow | 30.3 ML/d (8 mgd) |
| Total suspended solids | 121 mg/L |
| Biochemical oxygen demand | 107 mg/L |
| Total Kjeldahl nitrogen | 25 mg/L |
| Total phosphorus | 3.1 mg/L |

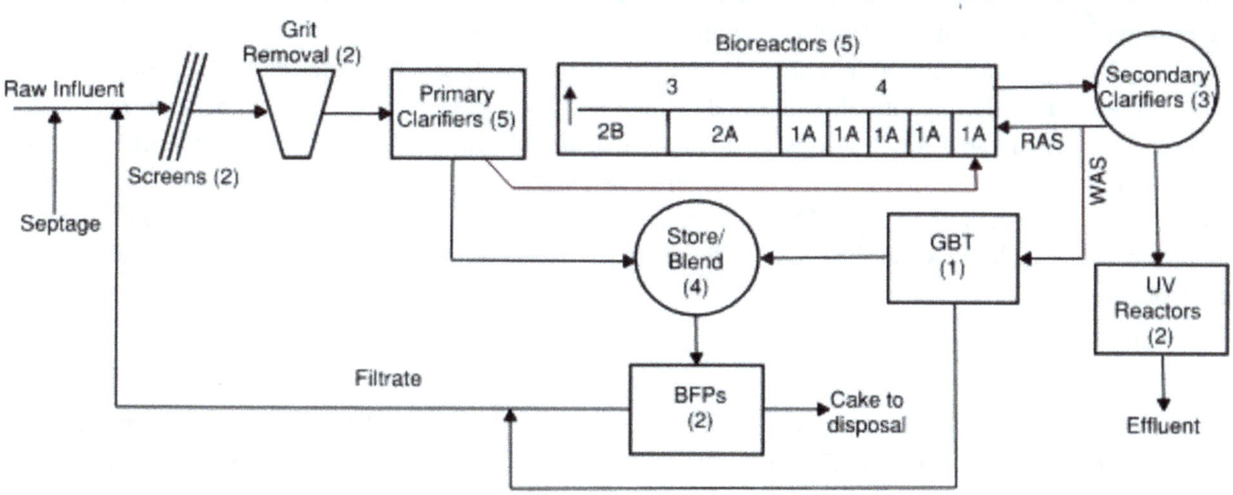


FIGURE 15.2 Grass Island Wastewater Treatment Plant Flow Schematic

TABLE 15.2 Grass Island Wastewater Treatment Plant MLE Process Operating Criteria

| | |
|---|---|
| Effluent total nitrogen goal | 8 mg/L |
| Aerobic SRT | 10–11 days |
| Minimum monthly water temperature | 15 °C |
| MLE process basins | 5 |
| Total anoxic volume | 14.5 ML (3.8 mil. gal) |
| Total aerobic volume | 43.5 ML (11.5 mil. gal) |
| Final clarifiers (35-m [115-ft] diameter) | 3 |

Aeration is provided by three single-stage, gear-driven centrifugal blowers, each with a range of 128 $m^3$/min (4500 acfm) to 280 $m^3$/min (10 000 acfm). These blowers also provide air to the facility's aerated channels and grit system.

An in situ ammonia analyzer is installed at the effluent end of bioreactor no. 3, normally used for monitoring process performance.

The GIWWTP does not have an effluent total nitrogen limit in its National Pollutant Discharge Elimination System (NPDES) permit. Rather, the facility is assigned an effluent total nitrogen waste-load allocation via a statewide general permit as part of CTDEEP's Long-Island Sound nitrogen credit trading program. The GIWWTP's nitrogen allocation equates to 218 kg/d (479 lb/d). If the GIWWTP's annual nitrogen discharge is less than their allocation, the town of Greenwich receives a payment from the program. If the facility exceeds its annual allocation, the town must pay into the program.

## 2.3 Design and Operation Modifications

From June through December 2022, the GIWWTP piloted aeration-based ammonia control (ABAC) in one of its five bioreactors. An in situ ammonium analyzer was installed at the effluent end of the bioreactor. Due to predicted turndown limitations of the existing main aeration blowers, a 100 $m^3$/min (3500 acfm) "jockey" blower was installed to better match predicted oxygen requirements during low-load periods. The goal of the pilot was to build upon the facility's existing automatic dissolved oxygen control system by modulating the bioreactor's dissolved oxygen setpoints depending on the measured effluent ammonia concentration (rather than maintaining a constant dissolved oxygen setpoint). Energy savings, with perhaps improved effluent total nitrogen, were anticipated.

The pilot operation did not produce the benefits anticipated, largely because: (1) the process oxygen demand fluctuated significantly during the day, such that the jockey blower was only able to "keep up" with demand during low-load periods; and (2) the existing blower control system was not amenable to an easily modified control approach, so the operating "mode" had to switch between ABAC and standard dissolved oxygen control daily. This constraint prevented long-term stable ABAC operation. Implementation of reliable ABAC at the GIWWTP will require a more holistic upgrade to the overall process control scheme.

## 2.4 Performance

The GIWWTP effluent total nitrogen concentrations from 2020 and 2022 are shown in Figure 15.3. Effluent total nitrogen averaged 6.75 mg/L over the 3-year period. The facility sees its highest effluent total nitrogen concentrations seasonally in the autumn, perhaps because this is a dry time of year and the facility's influent nitrogen concentration tends to be higher.

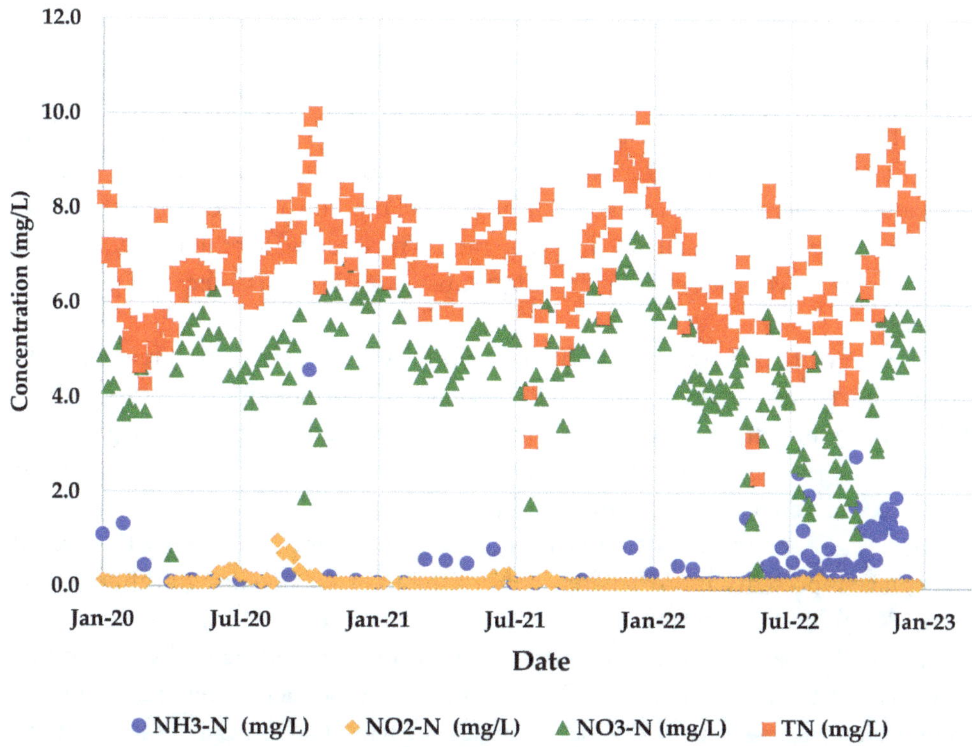


FIGURE 15.3 Grass Island Wastewater Treatment Plant Performance (Effluent Total Nitrogen)

Of note is the period from June 2022 through December 2022, which is the period during which ABAC operation was piloted. During this period, effluent ammonia was higher than historic levels, though it stayed less than 2 mg/L. Effluent nitrate decreased at the same time but not to the extent that the facility's effluent total nitrogen was greatly impacted from previous performance.

## 3.0 CASE STUDY 2 (SEQUENCING BATCH REACTOR)—KINGSTON, MASSACHUSETTS, WASTEWATER TREATMENT PLANT

### 3.1 Facility History

The Kingston Wastewater Treatment Plant (WWTP) is located in the town of Kingston, Massachusetts. Construction of the original facility was completed in 2002 as a biological nutrient removal (BNR) treatment process to provide treatment for an average daily flow of 1.42 ML/d (0.375 mgd) and a peak-hour flow of 5.67 ML/d (1.50 mgd). An expansion project was recently completed, increasing the biological process capacity to an average daily flow of 2.63 ML/d (0.694 mgd) and a peak-hour flow of 8.52 ML/d (2.25 mgd). The facility discharges its effluent in accordance with a permitted groundwater discharge permit, or the effluent may be pumped to a holding pond for possible use as supplemental irrigation water at a nearby golf course.

### 3.2 Process Description

The preliminary treatment process provides for measurement of the facility influent flow and grit and screenings removal. The equipment used in the preliminary treatment process includes an aerated inlet tank, a mechanically cleaned rotary bar screen, and an aerated grit tank.

Raw wastewater enters the influent box at the head end of two inlet tanks through two 200-mm (8-in.) force mains. Septage and facility recycles are pumped from a nearby holding tank and also discharged at this point. The raw wastewater flows through two sections of the inlet tank, each of which is aerated to keep solids in suspension and to strip off odors. Coarse-bubble aeration is provided.

Flow is then directed to a 6-mm (0.25-in.) spaced, mechanically cleaned influent screen. If the mechanically cleaned influent screen is taken out of service, the flow can be directed to a bypass channel, which is equipped with a manually cleaned bar rack. The wastewater then enters an aerated grit tank. An airlift pump removes the settled grit and pumps it to a grit classifier. The grit tank can be bypassed, if necessary.

Effluent from the grit tank flows by gravity to one of four sequencing batch reactors (SBRs), two from the original construction and two built in the recent expansion. Before entering an SBR, magnesium hydroxide may be introduced to the flow stream to increase alkalinity. The magnesium hydroxide system replaced the facility's original soda ash system, which operators found required excessive maintenance efforts due to scaling/clogging of piping and machinery.

The facility is designed to receive a significant quantity of septage considering the overall capacity of the facility. Septage represents approximately 35% of the total average design influent biochemical oxygen demand (BOD) load and approximately 4% of the total average design influent total Kjeldahl nitrogen (TKN) load. The septage-receiving facility consists of two holding tanks with approximately 91 000 L (24 000 gal) of total storage with diffused aeration. The septage-receiving facility consists of two offloading stations with 44-mm (1.75-in.) manual bar racks. The tanks are used to store septage, filtrate from the gravity belt thickener (GBT), and other facility drainage.

The Kingston facility now can use a four-SBR tank system to provide continuous treatment, though given current influent flows and loads, only two are typically in service. One SBR is filled at a time, and while the one tank is in midcycle, the second online tank is in filling, and vice versa. Each SBR is 18.3 m (60 ft) by 18.3 m (60 ft), with a low water level (LWL) of 5.4 m (17.6 ft) and a high-water level of 5.6 m (21 ft), providing a LWL-treatment volume of 1.79 ML (0.474 mil. gal) per SBR.

Each SBR typically runs through five cycles per day, although, occasionally, a six-cycle-per-day operation is used. When running at five cycles per day, each cycle is set to take 4.8 hours and is broken into five basic stages during normal operation to achieve the desired degree of nitrification and denitrification. These stages are fill (including an anoxic mixing period and an aerobic mixing period), react (aerobic treatment), settle, decant, and idle. Each treatment cycle operates at the stage times shown in Table 15.3. These cycle times are similar to the original design intent, except for slightly reduced react and settle time, to allow for a longer decant period. Each SBR is equipped with a fixed fine-bubble diffuser system, and three 56-kW (75-hp) rotary-displacement blowers are installed to supply the required process air.

The Kingston facility has a two-disk effluent cloth filter. This system is used to remove suspended solids and to provide polishing of the SBR effluent before ultraviolet (UV) disinfection and discharge. The filter unit consists of two disks; each disk is comprised of six segments. The disk filter has an

**TABLE 15.3** Stage Times for Treatment Cycles at the Kingston, Massachusetts, Facility

| Stage | Time (min.) |
|---|---|
| Fill static and mix (anoxic) | 60 |
| Fill (aerobic) | 84 |
| React (aerobic) | 50 |
| Settle | 50 |
| Decant | 44 |
| Total | 288 |

average-daily design hydraulic capacity of 1.77 ML/d (0.467 mgd) and a design peak hydraulic capacity of 3.53 ML/d (0.933 mgd).

The UV disinfection system uses two UV reactors. Each reactor has two banks that are contained within a single flow channel; each bank contains four modules, and each module accommodates two lamps. Disinfection is typically not required, because the facility's effluent is currently discharged below ground. Disinfection is only required if the effluent is sent to the golf course.

The facility uses two GBTs (including one standby) to thicken waste activated sludge (WAS) from the SBRs. Polymer is injected to the sludge feed through a four-port tangential injection ring by a polymer metering pump and blended by a static mixer. A process flow diagram is shown in Figure 15.4.

The Kingston facility has NPDES maximum monthly average limits of 30 mg/L on 5-day BOD ($BOD_5$) and total suspended solids (TSS). A maximum monthly nutrient limit on total nitrogen requires an effluent total nitrogen of 10 mg/L. Additionally, more stringent discharge limits must be met when discharging to the holding point for effluent reuse as golf course irrigation water. Effluent daily average $BOD_5$ and total nitrogen must not exceed 10 mg/L, and effluent daily TSS must not exceed 5 mg/L. Table 15.4 summarizes the Kingston WWTP's influent design loading for the maximum monthly design condition, and Table 15.5 summarizes the Kingston WWTP's significant unit process design criteria.

### 3.3 Performance

The Kingston WWTP has been in compliance with its NPDES permit since startup, and the BNR process has performed well (Figure 15.5). Monthly average total nitrogen, ammonia-nitrogen, nitrite-nitrogen, and nitrate-nitrogen are plotted. The facility's average monthly effluent total nitrogen concentration has ranged from 3 to 7 mg/L.

## 4.0 CASE STUDY 3 (FOUR-STAGE BARDENPHO)—BROCKTON, MASSACHUSETTS, ADVANCED WATER RECLAMATION FACILITY

### 4.1 Facility History

The Brockton Advanced Water Reclamation Facility (AWRF) is owned by the City of Brockton, Massachusetts, and serves Brockton and portions of the neighboring cities of Abington and Whitman. The original facility was commissioned in the early 1960s. The facility underwent a major upgrade

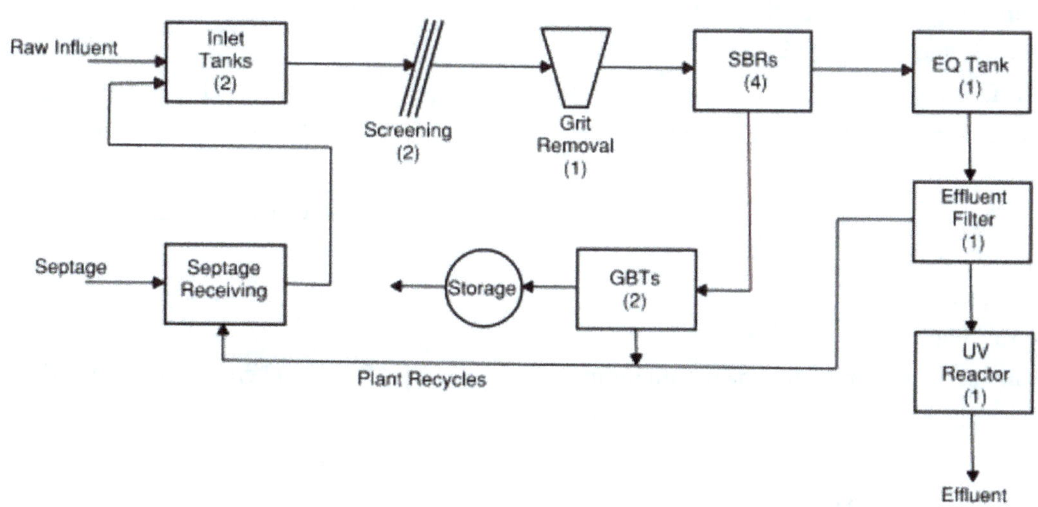


FIGURE 15.4 Kingston WWTP Facility Process Flow Diagram

**TABLE 15.4** Kingston WWTP Influent Design Maximum Monthly Loads

| Contaminant | Load (kg/d)[(lb/d)] |
|---|---|
| Biochemical oxygen demand | 1270 (2800) |
| Total suspended solids | 1440 (3160) |
| Total Kjeldahl nitrogen | 227 (449) |

in the 1980s to achieve ammonia and phosphorus removal and to expand to an annual average treatment capacity of 68 ML/d (18 mgd). An early 2000s upgrade project converted the process to the MLE configuration, targeting year-round nitrification and an average effluent total nitrogen concentration of 5.5 mg/L The AWRF recently completed a series of upgrades to achieve compliance with a new seasonal 205 kg/d (450 lb/d) mass limit for effluent total nitrogen, equating to 3 mg/L total nitrogen at the average design for of 68 ML/d (18 mgd), and an average effluent total phosphorus limit of 0.1 mg/L. These

**TABLE 15.5** Kingston WWTP Design Criteria

| Description | Value |
|---|---|
| Preliminary | |
| • Number of inlet tanks | 1 |
| ○ Dimensions | 8.1 × 2.1 × 2.4 m (26.5 × 7 × 7.75 ft) sidewater depth |
| • Influent screen | |
| ○ Number of screens | 1 |
| ○ Bar spacing | 6 mm (0.25 in.) |
| • Number of grit chambers | 1 |
| ○ Dimensions | 2.4 × 2.4 × 3.7 m (8 × 8 × 12 ft) sidewater depth |
| SBRs | |
| • Number of units | 4 |
| ○ Dimensions (each) | 18.3 × 18.3 × 5.6 m high water level (HWL) (60 × 60 × 21 ft) |
| • Aerobic SRT at maximum month load | 12 days |
| • Mixed liquor suspended solids (MLSS) (at low water level) | 4700 mg/L |
| Effluent filtration | |
| • Number of filters | 1 |
| • Filter surface area | 10.0 $m^2$ (108 sq ft) |
| • Hydraulic loading rate (average and maximum) | 7.3/14.6 m/h (3/6 gpm/sq ft) |
| Disinfection | |
| • Type | Ultraviolet medium pressure |
| • Number of reactors | 2 |
| • Total number of lamps | 16 |
| Sludge processing | |
| • Waste sludge load (average and maximum monthly) | 1160/1750 kg/d (2550/3860 lb/d) |
| • Type of thickening | GBT |
| • Number of GBTs | 2 |
| • Thickened sludge concentration | 4% |

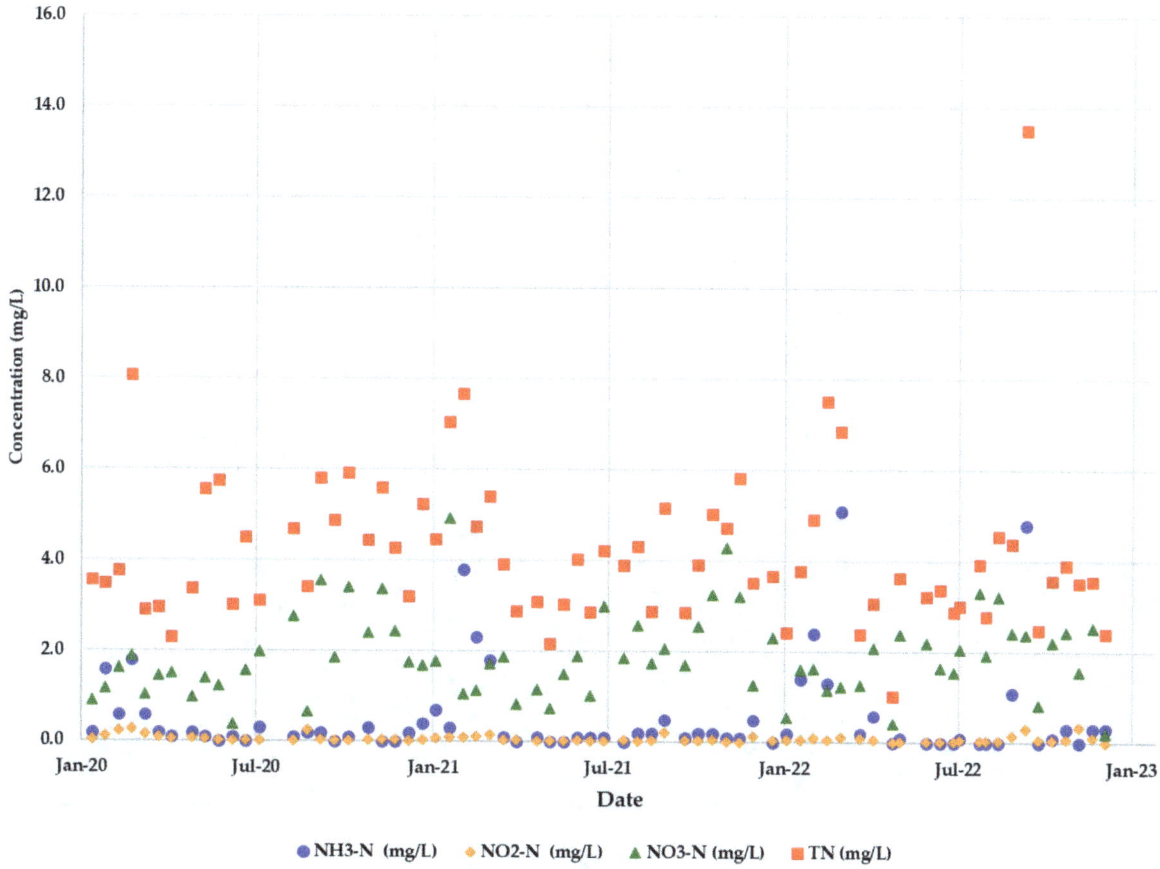


FIGURE 15.5 Kingston Facility BNR Process Performance Results

upgrades included converting the MLE process to a four-stage Bardenpho configuration, provision of supplemental carbon facilities, and aeration control improvements.

## 4.2 Process Description

The Brockton AWRF provides preliminary and chemically enhanced primary, secondary, and tertiary treatment, achieving compliance with its stringent limits for total nitrogen and total phosphorus and disinfection. Of particular interest is the bioreactor configuration used to achieve compliance with the total nitrogen limit. The AWRF has three separate biological processes comprising a total of seven process trains, with three different bioreactor geometries. Recently all seven trains were upgraded and are now configured as plug flow reactors with the flexibility to operate as either an MLE system for nitrification and denitrification or a four-stage Bardenpho system (refer to Figure 15.6a and 15.6b for the facility's liquid process flow schematic and the bioreactor configuration for train 1 showing anoxic, aerobic, and swing zone layout).

The inlet end of each basin is configured as a three-stage anoxic reactor, with the first two stages sized to promote selection of preferred microorganisms and to "deselect" problematic microorganisms. The third anoxic selector is sized to provide increased volume for nitrate removal. Following the anoxic selector is a swing zone that can run in either anoxic or aerobic mode.

A main aerobic reactor volume provides nitrification year-round with sufficient aerobic SRT. Downstream of the aerobic volume are either one or two swing-zone stages (depending on that train's tank geometry) that can be operated either anoxically or aerobically. During periods when the wastewater

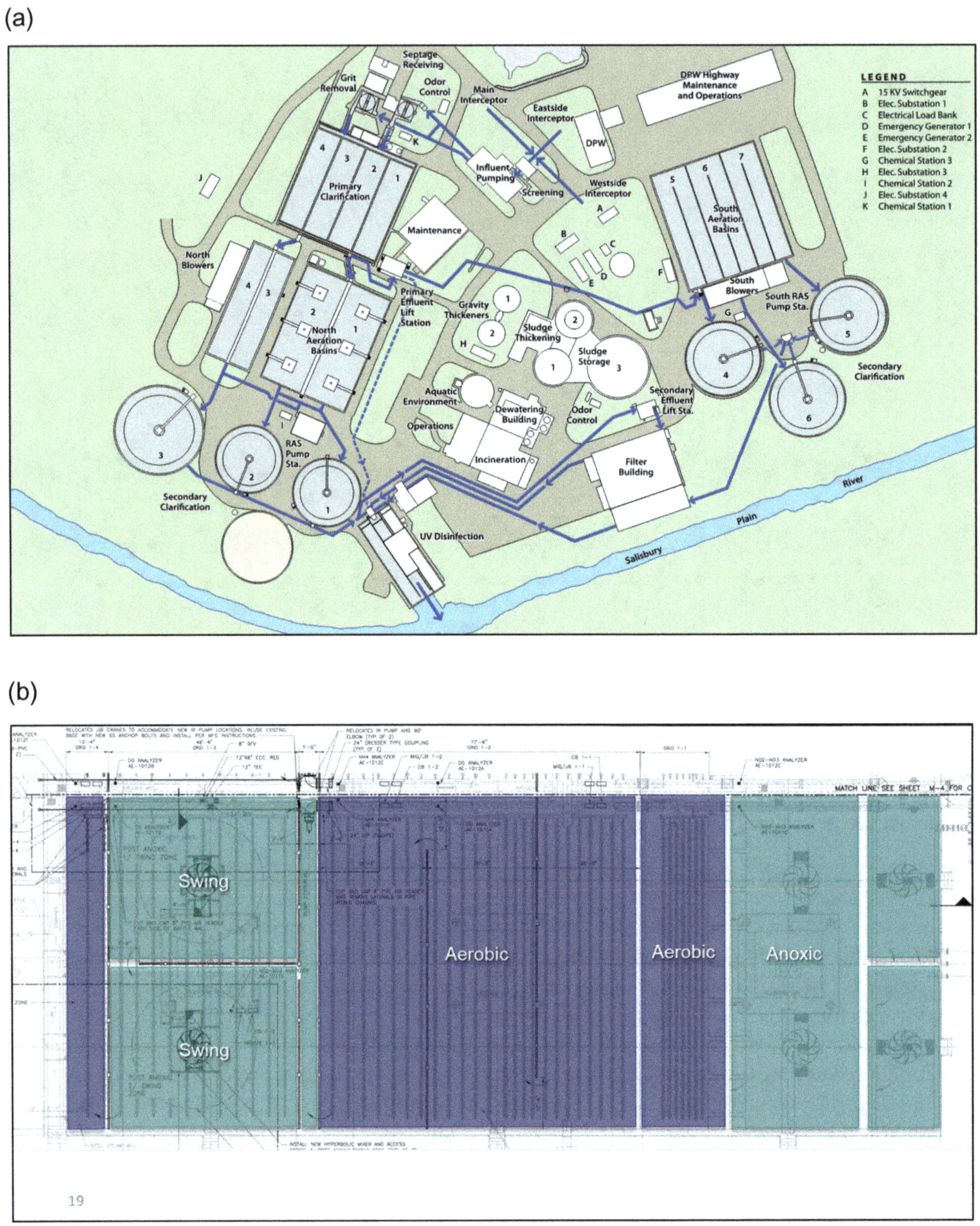


FIGURE 15.6 Brockton AWRF (a) Liquid Process Flow Schematic; (b) Bioreactor Layout (Train 1)

is coldest, operators have the flexibility to run these zones aerobically to maintain nitrification through worst-case conditions. The swing zones are operated as anoxic zones during warmer conditions, enabling the process to provide very low effluent total nitrogen. The facility can add MicroC2000™ as a supplemental carbon source to the post-anoxic zones, but to date supplemental carbon addition has not been required to meet the effluent total nitrogen limit.

Downstream of the final clarifiers that serve the suspended growth system, tertiary 5-micron cloth filtration polishes the effluent to very low effluent TSS. These filters enable the AWRF to comply with its monthly average effluent total phosphorus limit. A UV-light disinfection facility is used to disinfect the treated effluent. Before discharge to Salisbury Plain River, a tributary of the Taunton River, the effluent is reaerated through a cascade. Solids processing includes gravity thickening for primary sludge, gravity-belt thickening for WAS, and dewatering with centrifuges. The AWRF is equipped with a sludge incinerator, but this has been removed from service, and dewatered cake is hauled off-site.

The Brockton AWRF employs chemical addition with ferric chloride, combined with its effluent filters, to achieve compliance with its total phosphorus limit. Ferric chloride can be added upstream of the primary clarifiers to the mixed liquor channels between the bioreactors and the secondary clarifiers and, rarely if needed, upstream of the effluent filters.

Tables 15.6 and 15.7 summarize the Brockton AWRF's influent and primary effluent design loading, and Table 15.8 summarizes the facility's significant liquid unit process design criteria.

## 4.3 Performance

The Brockton AWRF's BNR process has performed very well since commissioning and has maintained compliance with the facility's effluent nutrient limits. Monthly average total nitrogen, ammonia-nitrogen, nitrite-nitrogen, and nitrate-nitrogen ($NO_x$-N) are shown in Figure 15.7. Despite reduced efficiency in the winter months, these results show effluent total nitrogen concentration of about 3 mg/L during the permit season (May–October), which results in compliance with the AWRF's mass limit for effluent total nitrogen, even with no supplemental carbon addition. Effluent nitrate typically ranges from 1.5 to 2.0 mg/L, so lower total nitrogen concentrations should be achievable if/when supplemental carbon is added.

As part of the recent BNR upgrade, ABAC was initiated for four of the seven bioreactors, as a test case to determine how well the control scheme would work before implementing for the whole AWRF. Bioreactor nos. 1 through 4 (which comprise process trains 1 and 2) were equipped with the online analyzers and aeration control valves needed, and the AERprocess™ABAC control scheme was integrated into the facility's aeration controls. The system uses a proprietary algorithm to control blower output and air distribution based on maintaining a target effluent ammonia concentration. Though this system has only been in operation a short time, it is showing the ability to reduce the dissolved oxygen setpoint (and aeration rate) while showing a slight reduction in effluent total nitrogen.

**TABLE 15.6** Brockton AWRF Influent Design Loads

| Parameter | Average Annual Load (kg/d) [(lb/d)] | Maximum Day Load (kg/d) [(lb/d)] |
|---|---|---|
| $BOD_5$ | 14 790 (32 540) | 22 930 (50 440) |
| Total suspended solids | 11 030 (24 260) | 21 390 (47 060) |
| Total Kjeldahl nitrogen | 1870 (4120) | 2910 (6400) |
| Total phosphorus | 370 (824) | 580 (1280) |

TABLE 15.7 Brockton AWRF NPDES Primary Effluent Design Loads (Includes Sidestreams)

| Parameter | Average Annual Load (kg/d) [(lb/d)] | Maximum Day Load (kg/d) [(lb/d)] |
|---|---|---|
| $BOD_5$ | 14 850 (32 660) | 23 010 (50 630) |
| Total suspended solids | 8330 (18 320) | 16 060 (35 340) |
| Total Kjeldahl nitrogen | 2200 (4850) | 3410 (7510) |
| Total phosphorus | 390 (860) | 600 (1330) |

## 5.0 CASE STUDY 4 (OXIDATION DITCH)—THREE OAKS WATER RECLAMATION FACILITY

### 5.1 Facility History

The Lee County Utilities (LCU) owns and operates the Three Oaks Water Reclamation Facility (WRF), which has a permitted treatment capacity of 22.7 ML/d (6 mgd) annual average daily flow. The current process train consists of screening and grit removal, odor control, and biological treatment with nitrogen removal using simultaneous nitrification/denitrification in four oxidation ditches, five clarifiers, four deep bed filters, and two chlorine contact tanks. Biosolids are held in aerobic storage tanks and dewatered with belt filter presses for composting at the county landfill. The facility is permitted to discharge treated effluent to an unrestricted public access reuse system and a deep injection well.

### 5.2 Process Description

Raw influent enters the facility and return activated sludge (RAS) is returned at the headworks, where it is screened with mechanical bar screens and degritted using a stacked tray grit removal system. After preliminary treatment, the following unit process provide secondary and tertiary treatment:

- Oxidation ditches (4);
- Secondary clarifiers (5);
- Deep bed filters (4); and
- Disinfection used sodium hypochlorite.

Effluent is stored and can be sent either to the county's reclaimed water distribution system or to deep well injection. Solids handling facilities and processes include a six WAS storage tanks and two belt filter presses for sludge dewatering. Figure 15.8 shows a process flow diagram of the facility.

There are four oxidation ditches that operate in parallel at the current permitted design capacity of 22.7 ML/d (6 mgd). Flow to each oxidation ditch is split at the downstream side of the headworks through weir gates. Oxidation ditch nos. 1 and 2 were each rated at design at 3.8 ML/d (1 mgd) capacity, while oxidation ditch nos. 3 and 4 were each rated at design at 7.6 ML/d (2 mgd). Operational experience has demonstrated that the ditches can treat at least 10% more flow than the original design. Oxidation ditch nos. 1 through 3 are equipped with brush-type aerators that are not equipped with variable frequency drives (VFDs) and are instead controlled by timers. Oxidation ditch no. 4's aerators are equipped with VFDs to provide oxygen transfer flexibility.

Permit requirements for the Three Oaks WRF include average monthly effluent limits on BOD and TSS of 20 mg/L and 30 mg/L, respectively, when discharging underground, and a maximum TSS of

TABLE 15.8 Brockton AWRF Liquid Process Design Criteria

| Description | Value | | |
|---|---|---|---|
| Preliminary treatment | | | |
| • Number of screens | 2 | | |
| ○ Type | Multi-rake | | |
| ○ Bar spacing | 12 mm (0.50 in.) | | |
| • Number of grit chambers | 2 | | |
| ○ Type | Vortex | | |
| ○ Diameter | 3.7 m (12 ft) | | |
| Primary clarifiers | | | |
| • Number of units | 4 | | |
| • Dimensions (L × W) | 55 m × 13.5 m (180 ft × 44 ft) | | |
| • Surface overflow rate (average and peak) | 1.3/3.2 m/h (740/1890 gpd/sq ft) | | |
| Bioreactors | Trains 1–2 | Trains 3–4 | Trains 5–7 |
| • Number of basins | 2 | 2 | 3 |
| • Volume per basin (total) | | | |
| ○ Pre-anoxic volume per basin | 6.6 ML (1.73 MG) | 3.9 ML (1.03 MG) | 5.8 ML (1.54 MG) |
| ○ Aerobic volume per basin | 1.5 ML (0.39 MG) | 0.87 ML (0.23 MG) | 1.3 ML (0.35 MG) |
| ○ Post-anoxic (swing) volume | 3.5 ML (0.92 MG) | 2.1 ML (0.55 MG) | 3.2 ML (0.85 MG) |
| per basin | 1.3 ML (0.34 MG) | 0.76 ML (0.20 MG) | 1.1 ML (0.28 MG) |
| ○ Reaeration volume per basin | 0.31 ML (0.082 MG) | 0.19 ML (0.049 MG) | 0.24 ML (0.064 MG) |
| • Aerobic SRT at maximum monthly load | 9.8 days | | |
| • Design MLSS (average) | 3740 mg/L | 4510 mg/L | 4500 mg/L |
| • Internal recycle pumping capacity | 6 × average daily flow (ADF) | | |
| Secondary clarifiers | Trains 1–2 | Trains 3–4 | Trains 5–7 |
| • Number of units | 2 | 1 | 3 |
| • Diameter | 29 m (110 ft) | 29 m (150 ft) | 29 m (130 ft) |
| • Surface overflow rate (average and peak) | 0.65/1.1 m/h (380/630 gpd/sq ft) | 0.5/0.8 m/h (290/480 gpd/sq ft) | 0.5/0.8 m/h (290/490 gpd/sq ft) |
| Effluent filtration | | | |
| • Number of filters | 4 | | |
| • Filter rating | 5 micron | | |
| • Filter surface area | 800 $m^2$ (8580 sq ft) | | |
| • Hydraulic loading rate (average and max) | 8.3/15.9 m/h (2.6/6.5 gpm/sq ft) | | |
| Disinfection | | | |
| • Type | Ultraviolet low-pressure, high-output | | |
| • Number of reactors | 2 | | |
| • Modules per reactor | 9 | | |
| • Total number of lamps | 864 | | |

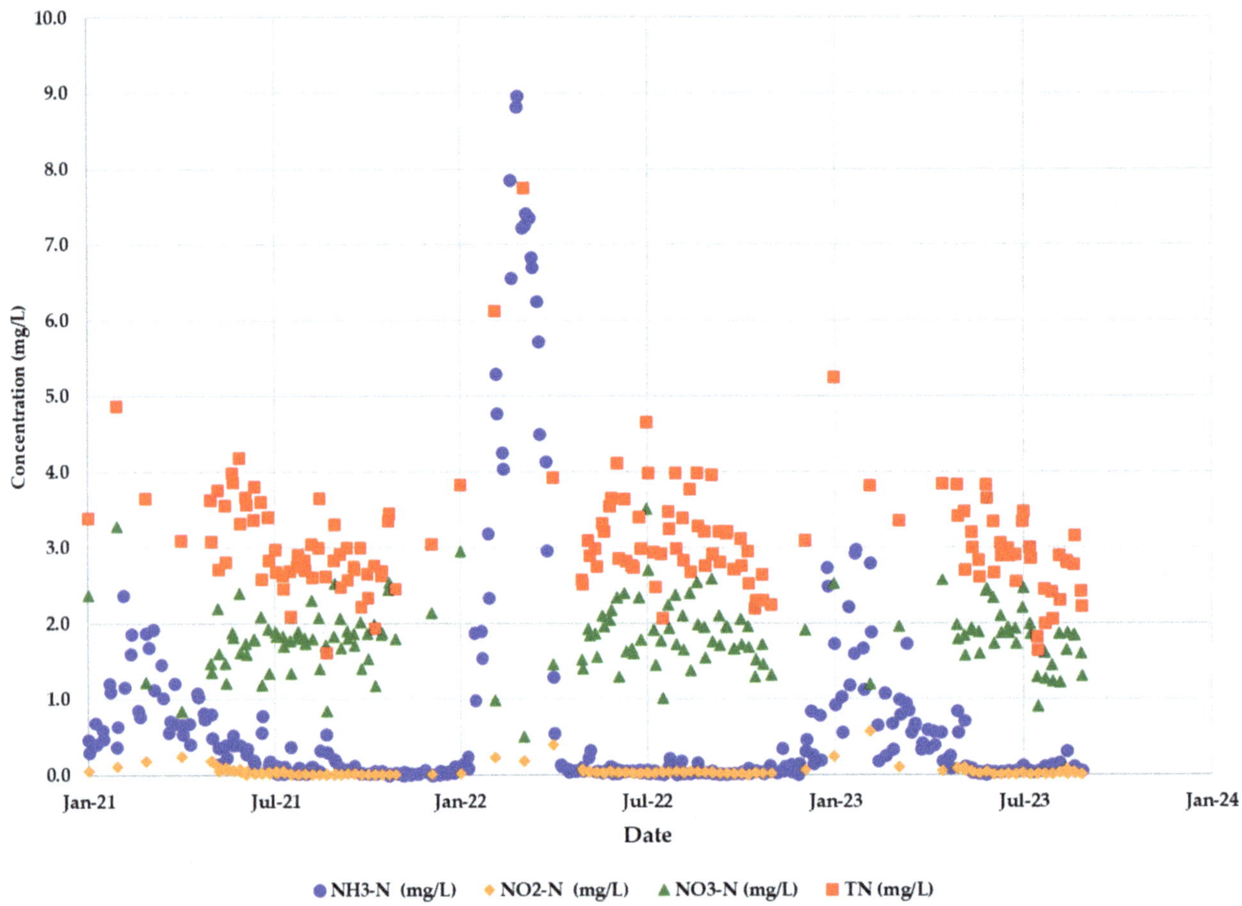


FIGURE 15.7 Brockton AWRF Effluent Performance

5 mg/L for reclaimed water. Though the permit does not contain nutrient limits, the county's staff operates the facility to minimize total nitrogen and total phosphorus in the effluent.

The Three Oaks WRF is treating the influent wastewater characteristics given in Table 15.9. Process design criteria for the secondary treatment process are provided in Table 15.10.

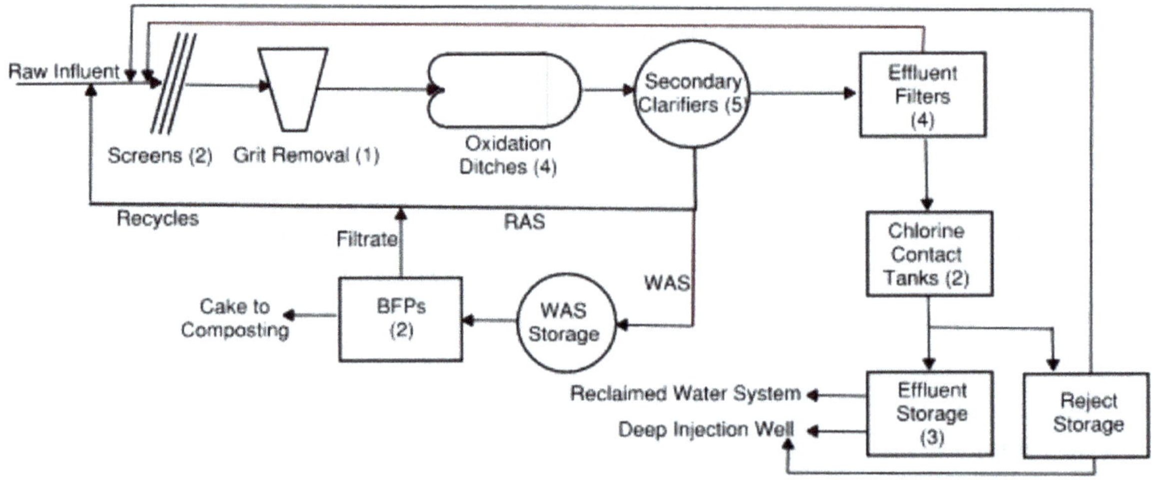


FIGURE 15.8 Three Oaks WRF Process Flow Diagram

TABLE 15.9 Three Oaks WRF Influent Parameters

| Parameter | Design Value |
|---|---|
| Average daily flow | 12.75 ML/d (3.4 mgd) |
| Average 5-day BOD, mg/L | 190 |
| Average TSS, mg/L | 193 |
| Average total Kjeldahl nitrogen, mg/L | 45 |
| Average total phosphorus, mg/L | 5.1 |

## 5.3 Design and Operation Modifications

The Three Oaks WRF is presently in construction of a capital improvement program that will result in an increase in rated capacity of the oxidation ditches, from a total of 27.7 ML/d (6 mgd) to 34 ML/d (9 mgd). The existing ditches do not require modifications, as these processes have demonstrated the ability to treat more flow than originally designed—such that the total capacity of the four ditches will be 25.5 ML/d (6.75 mgd). A new oxidation ditch no. 5 will be constructed to reach a total capacity of 34 ML/d (9 mgd). Based on operating experience with the current oxidation ditch no. 4, the new ditch will be controlled by the same control system. Oxidation ditch no. 4 was provided with three 75 HP platform mounted surface aerators with VFDs. The VFDs will provide turn-down capability and more flexibility with maintaining the target residual dissolved oxygen profile in the oxidation ditch. The surface aerators will be designated as primary, secondary, and tertiary aerators for the automated control loop. The operation of the surface aerators will be controlled automatically based on either dissolved oxygen or oxidation-reduction potential (ORP) measurements taken in the oxidation ditch. The proposed automated aeration system control will maximize the simultaneous nitrification and denitrification (SND) achieved in the oxidation ditch while providing power cost savings.

## 5.4 Performance

The Three Oaks WRF has typically has non-detect effluent BOD and TSS. The facility has averaged 0.06, 7.0, and 3.7 mg/L of effluent ammonia-nitrogen, total nitrogen, and total phosphorus, respectively, over

TABLE 15.10 Three Oaks WRF Process Design Criteria

| Parameter | Design Value |
|---|---|
| **Oxidation Ditch design** | |
| Oxidation ditch nos. 1–2 volume per basin, ML (mil. gal) | 5.37 (0.75) |
| Oxidation ditch nos. 3–4 volume per basin, ML (mil. gal) | 5.37 (1.5) |
| Number of brush aerators/basin | Nos. 1–3: 4 |
| | No. 4: 3 |
| Horsepower per aerator | Nos. 1–2: 22 KW (30 HP) |
| | No. 3: 37 KW (50 HP) |
| | No. 4: 56 KW (75 HP) |
| **Process design** | |
| Solids retention time (SRT) for nitrification, days | 12 |
| MLSS, mg/L | 4500 |

the last 3 years. The historical effluent nitrogen and total phosphorus performance for the facility is shown in Figures 15.9 and 15.10.

Of note is the summer 2022 period, when effluent nitrogen and phosphorus were noticeably and more consistently lower than typical for the Three Oaks WRF. During this period, operators were regularly adjusting aerator timers in oxidation ditch nos. 1 through 3 to target optimization and turned off the downstream aerator in oxidation ditch no. 4. The data show that the facility's effluent total nitrogen reached 3 mg/L and total phosphorus reached less than 0.5 mg/L during this period. This operational practice was suspended in late September 2022 with the arrival of Hurricane Ian.

## 6.0 CASE STUDY 5 (INTEGRATED FIXED-FILM ACTIVATED SLUDGE)—FIELD'S POINT WASTEWATER TREATMENT FACILITY

### 6.1 Facility History

The Field's Point Wastewater Treatment Facility (FPWWTF) is one of the oldest activated sludge WRRFs in the United States and today provides preliminary and primary treatment for up to 760 ML/d (200 mgd) and secondary treatment for up to 291 ML/d (77 mgd) and has recently treated an average of about 159 ML/d (42 mgd). The facility is owned and operated by the Narragansett Bay Commission (NBC) and serves about 250,000 people in Providence, Rhode Island, and surrounding communities

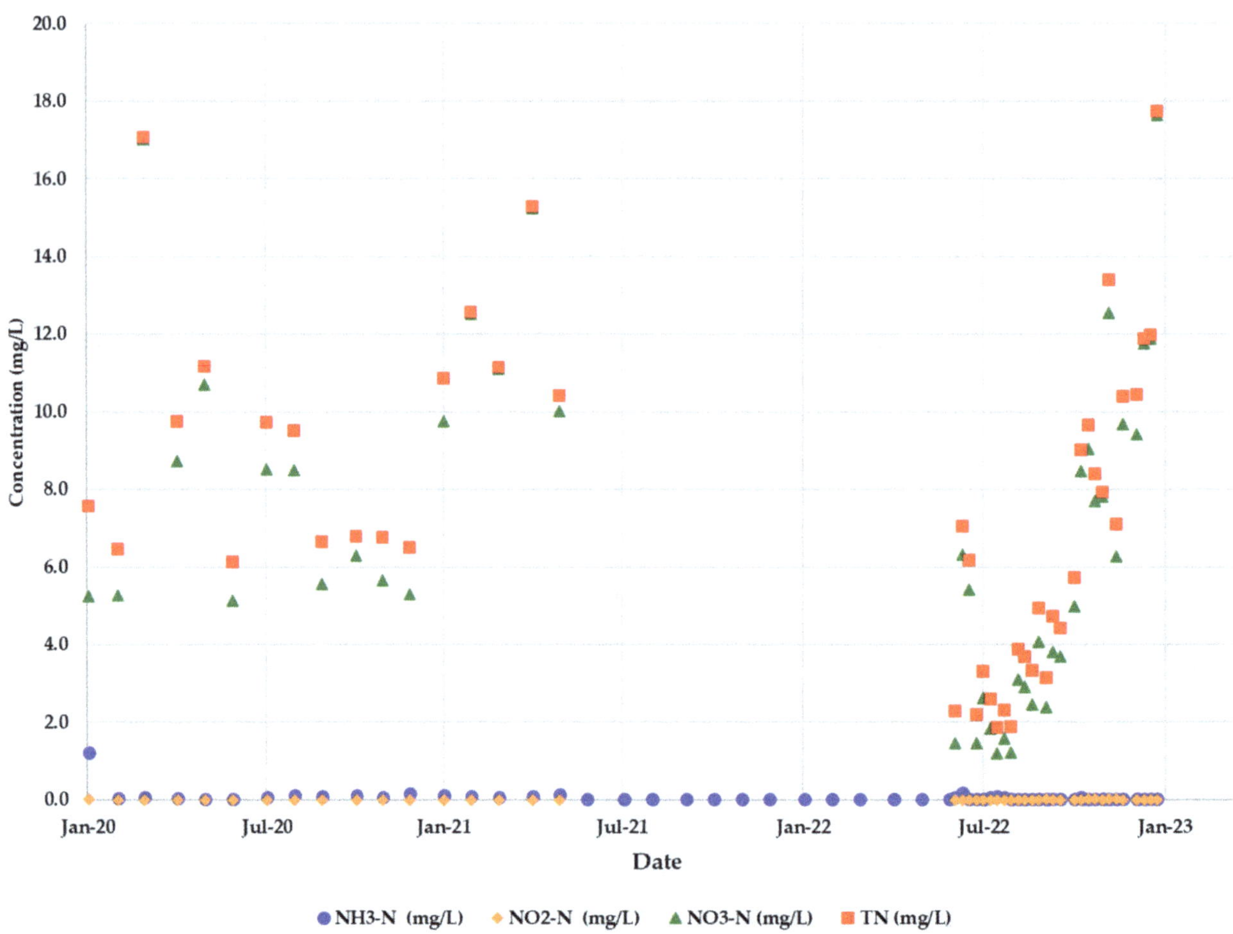


FIGURE 15.9 Three Oaks WRF Effluent Nitrogen Performance

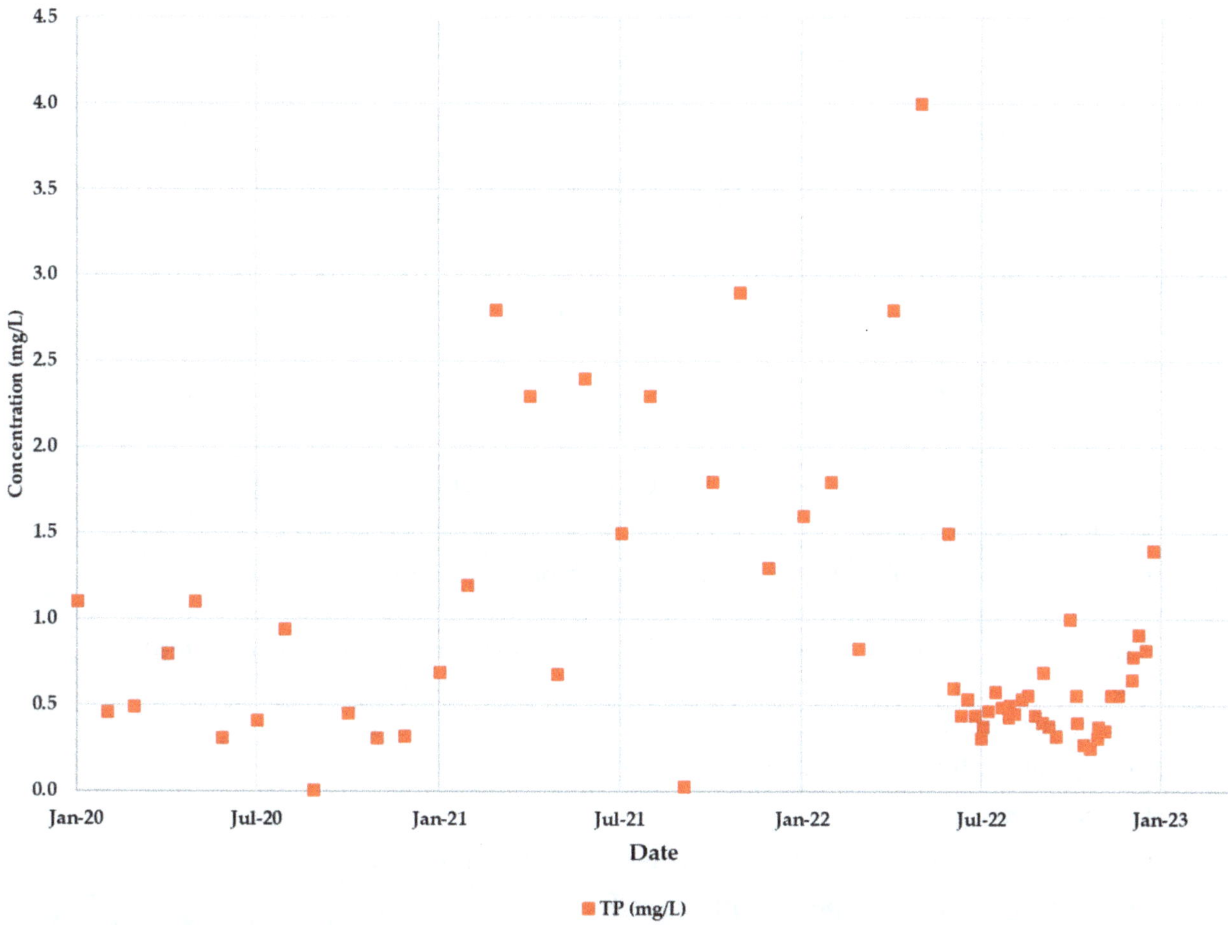


FIGURE 15.10 Three Oaks WRF Effluent Total Phosphorus Performance

and discharges into the northern end of Narragansett Bay. In 2013, the NBC completed an upgrade of the FPWWTF to achieve compliance with a new average monthly permit requirement of 5 mg/L total nitrogen, in place seasonally from May through October. The upgrade included conversion of 10 conventional activated sludge aeration tanks into an integrated fixed film activated sludge (IFAS)-based, four-stage Bardenpho configuration.

## 6.2 Process Description

The FPWWTF operates as a four-stage Bardenpho nitrification/denitrification process, using IFAS media in the main aerobic zone. Primary effluent and RAS combine as mixed liquor in the first anoxic zone. Mixed liquor then enters the main aerobic (IFAS) zone to provide BOD removal and nitrification. An internal recycle pump system returns nitrified mixed liquor from a space aerobic zone downstream of the IFAS media retention screens back to the first anoxic zone. Supplemental carbon can be added to the following two-cell post-anoxic zone, followed then by an aerobic re-aeration zone (Figure 15.11). The FPWWTF provides the following unit processes:

- Preliminary treatment, including screening and grit removal;
- Primary clarifiers;
- Fine screening;

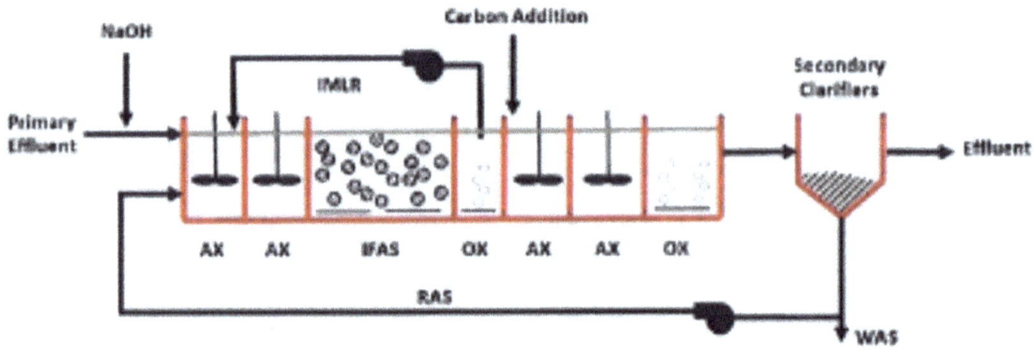


FIGURE 15.11 Field's Point WWTF Bioreactor Process Flow Diagram (Courtesy of Veolia Water Technologies)

- Ten trains of biological treatment configured as described above;
- Secondary clarifiers; and
- Disinfection using sodium hypochlorite with dechlorination with sodium bisulfite.

The IFAS biological process upgrade included the following in addition to the IFAS media:

- Media retention screens;
- Coarse-bubble diffusion in the aerobic zones;
- Fine screens;
- Concrete baffle walls to separate the various anoxic and aerobic zones;
- Aeration blower system improvements;
- Submersible mixers for anoxic zone mixing;
- Supplemental carbon (glycerin-based) addition;
- Alkalinity (sodium hydroxide) addition; and
- System controls.

The FPWWTF has average monthly effluent permit limits on carbonaceous biochemical oxygen demand (CBOD), TSS, and total nitrogen of 25, 45, and 5 mg/L (seasonal), respectively. The design primary effluent (biological process influent) wastewater characteristics are provided in Table 15.11.

## 6.3 Process Design Criteria

The design for the biological process upgrade included the addition of IFAS media at a 50% fill fraction in the aerobic (IFAS) zone. Design criteria of the treatment process are provided in Table 15.12.

TABLE 15.11 Field's Point WWTF Biological Process Influent Design Data

| Parameter | Design Average | Design Maximum Month |
|---|---|---|
| Flow | 189 ML/d (50 mgd) | 291 ML/d (77 mgd) (max) |
| Carbonaceous $BOD_5$ | 19 860 kg/d (43 700 lb/d) | 21 450 kg/d (47 200 lb/d) |
| TSS | 11 950 kg/d (26 300 lb/d) | 14 550 kg/d (32 000 lb/d) |
| TKN | 3440 kg/d (7560 lb/d) | 4440 kg/d (9770 lb/d) |

TABLE 15.12 Field's Point WWTF Biological Process Design Criteria

| Parameter | Value |
|---|---|
| IFAS fill fraction in aerobic zones | 50% |
| Suspended growth SRT (not including attached biomass) | 4–6 days |
| MLSS (design maximum) | 2500 mg/L |
| Mixed liquor recycle rate | 300% of influent flow |
| Design wastewater temperature (May–October) | 14 °C |

### 6.4 Design and Operation Modifications

Since the installation of IFAS media in 2013, operators have refined their operation and maintenance of the treatment process. At times the operator notes that thicker biomass on the media doesn't correlate with better treatment. Operators periodically take samples of media from each IFAS train, and when there is too much biomass on the media such that the media sinks (it is otherwise bouyant), the tank is removed from service and drained to allow the biofilm to dry out and slough off. The effects of this on the process are not established. Taking tanks off-line on a 3-year rotation optimizes the excess biofilm on the media.

### 6.5 Performance

The FPWWTF has been successfully operating in an IFAS configuration since 2013. From 2020 to 2022, the facility effluent total nitrogen averaged 2.5 mg/L during the May to October permit season, at an average flow of 151 ML/d (39.9 mgd). The IFAS system has been able to achieve treatment objectives while maintaining a mixed liquor concentration of 2900 mg/L. Effluent performance of the facility is shown in Figure 15.12. Note that the facility's effluent ammonia concentration increases during the winter, when the total nitrogen limit is not in effect; however, the nitrifying biomass is not "lost" as it might be with a suspended-growth-only system—the biofilm maintains a population of nitrifying bacteria. Nitrification efficiency is reduced, but the population remains, such that the process is quick to recover full nitrification in the spring, prior to the seasonal total nitrogen removal requirement that commences in May.

## 7.0 CASE STUDY 6 (CYCLIC AERATION)—TOWN OF CARY, NORTH CAROLINA, WATER RECLAMATION FACILITY

### 7.1 Facility History

The wastewater collection system and water reclamation facilities of Cary, North Carolina, have evolved over the years. As a small town, until 1970 the primary means of wastewater disposal was via septic tank. As the town grew, wastewater pumping stations began collecting sewer influent in several areas and then conveying the wastewater to a neighboring city for treatment. During the 1980s, small sections of the town were served by small WRRFs, until 1984, when Cary completed construction and put into operation the North Cary Water Reclamation Facility (NCWRF). The original facility had a maximum permitted flow of 15.1 ML/d (4 mgd). In the late 1980s, the rapid growth of Cary and the implementation of new regulations limiting the volume of nutrients, specifically nitrogen and phosphorous, into the receiving river basin prompted Cary to undertake a significant expansion of the NCWRF. In 1994, a completely

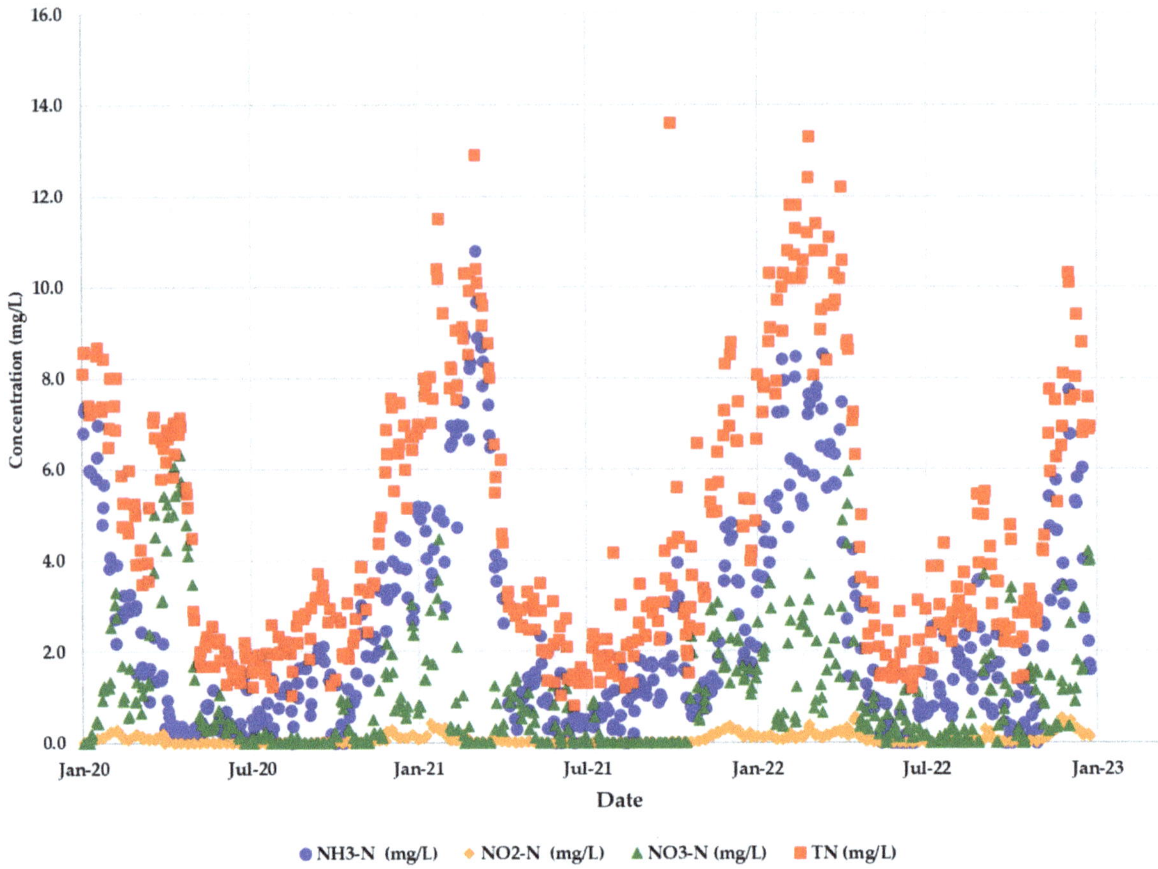


FIGURE 15.12 Field Point's WWTP Effluent Performance

new 5-stage BioDenipho BNR facility with a permitted capacity of 30.3 ML/d (8 mgd) was put into service. With continued growth came the need for expansion of the BNR facility, and in 2005 a third treatment process was added that allowed for an increase in permitted capacity to 45.4 ML/d (12 mgd). In 2021, the facility aeration system was upgraded from surface aeration to fine-bubble diffusers.

## 7.2 Process Description

The 45-ML/d (12.0-mgd) North Cary facility uses the Biodenipho BNR process, which consists of three trains of paired oxidation ditches where the active process volume is continuously alternated between aerobic and anoxic conditions to achieve BNR. The current influent flow rate is approximately 25 ML/d (6.5 mgd), or 55% of design flow. Therefore, one train remains offline. Each Biodenipho process train is followed by multistage post-denitrification reactors and a post-aeration zone.

Each biological process train at the North Cary facility consists of an anaerobic selector for biological phosphorous removal, a pair of oxidation ditches, and a secondary anoxic zone for further nitrate reduction. Each of the oxidation ditches is equipped with three grids of fine-bubble diffusers and two submersible mixers. Air flow to the air grids is controlled by automatic valves on the main header and on each diffuser header. Air for the entire system is provided by four VFD-controlled centrifugal blowers: two 224 KW (300 hp) and two 336 KW (450 hp). In addition, each oxidation ditch is equipped with analyzers for dissolved oxygen, ammonia, and nitrate for automatic process control and treatment optimization. Supplemental carbon addition to the secondary anoxic zone is currently not required to meet total nitrogen removal goals.

The town of Cary facility has an effluent permit limit on total nitrogen of 64 975 kg/yr (143 246 lb/yr), which is equivalent to 3.92 mg/L at the facility's permitted capacity. The facility also has an effluent quarterly average total phosphorus limit of 2 mg/L.

## 7.3 Process Design

Current treatment processes at the North Cary facility consist of the following: coarse bar screens; two influent pump stations; fine bar screens; grit removal facilities; three separate biological treatment trains, each consisting of a four-stage anaerobic selector, two oxidation ditches, a three-stage secondary anoxic zone, and a reaeration zone; two secondary clarifiers; effluent filters; UV disinfection facilities; and a cascade aerator for reaeration before discharge. Figure 15.13 is a process flow diagram for the facility.

Waste activated sludge at the North Cary facility is pumped to two GBTs for thickening, where it is stored in two aerated biosolid holding tanks until it is hauled to the South Cary Water Reclamation Facility (SCWRF) and blended with their WAS. The blended product is then dewatered and processed through the SCWRF regional thermal biosolids dryer into a class A biosolids pellet that is sold as a soil amendment product. Table 15.13 includes a summary of process design parameters.

## 7.4 Design and Operation Modifications

The BNR trains use the Veolia-Kruger BioDenipho Phased Isolation Ditch technology for treatment. Each process train has 10 phases (A through J) controlled by a programmable logic controller (PLC) receiving inputs from the nutrient and dissolved oxygen analyzers. The PLC will cycle through the phases based on the nutrient loading in the ditch alternating aerobic and anoxic conditions by cycling the air control valves to the diffusers, providing excellent nutrient removal at reduced energy costs. This change in dissolved oxygen conditions will take the microorganisms through nitrification and denitrification cycles to remove ammonia and reduce NOx within the process. Concurrently, phosphorous released in the anaerobic selectors is taken back up in the aerated environment of the oxidation ditches.

The low total nitrogen concentrations are achieved with a final anoxic tank similar to a five-stage Bardenpho process. The wastewater is then reaerated before the secondary clarifiers to prevent denitrification, which could cause rising sludge. The RAS rate is 55% of the average flow. The RAS enters a holding tank with a 30-minute detention time to minimize the concentration of nitrates that enter the anaerobic selector. The sludge age is approximately 16 to 18 days. The secondary effluent passes through a deep-bed sand filter.

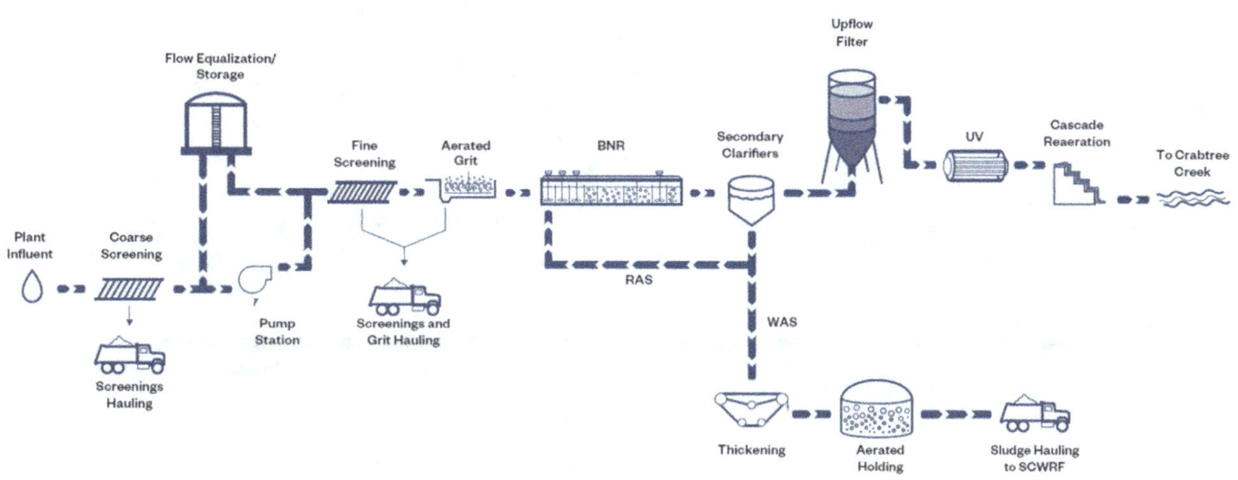


FIGURE 15.13 Town of North Cary, North Carolina, Facility Process Flow Diagram

TABLE 15.13 Town of North Cary, North Carolina, Facility Process Design Parameter Summary

| Tank/Process | Number | Size (each) | Size (total) |
|---|---|---|---|
| Anaerobic selectors | 4 each train | 0.35 ML (0.093 mil. gal) | 1.41 ML (0.372 mil. gal) |
| Oxidation ditch | 2 each train | 5.7 ML (1.5 mil. gal) | 11.4 ML (3 mil. gal) |
| Oxidation ditch aeration | 3 diffuser grids per ditch | 500 per grid<br>1500 per ditch<br>3000 per train | 9000 diffusers total across all process trains |
| Secondary anoxic zone | 3 each train | 0.42 ML (0.111 mil. gal) | 1.26 ML (0.333 mil. gal) |
| Reaeration zone | 1 each train | 0.42 ML (0.111 mil. gal) | 0.42 ML (0.111 mil. gal) |
| Clarifiers | 2 each | 39.6-m (130-ft) diameter | n/a |

Figure 15.14 shows the North Cary facility nitrogen removal performance. In the period from 2020 through 2023, the facility averaged 2 mg/L effluent total nitrogen. The facility has noted the following concerning operation of the process:

- The process operation focuses on nitrogen removal. If dissolved oxygen setpoints are too low, there may not be enough aerobic treatment to remove orthophosphate adequately.

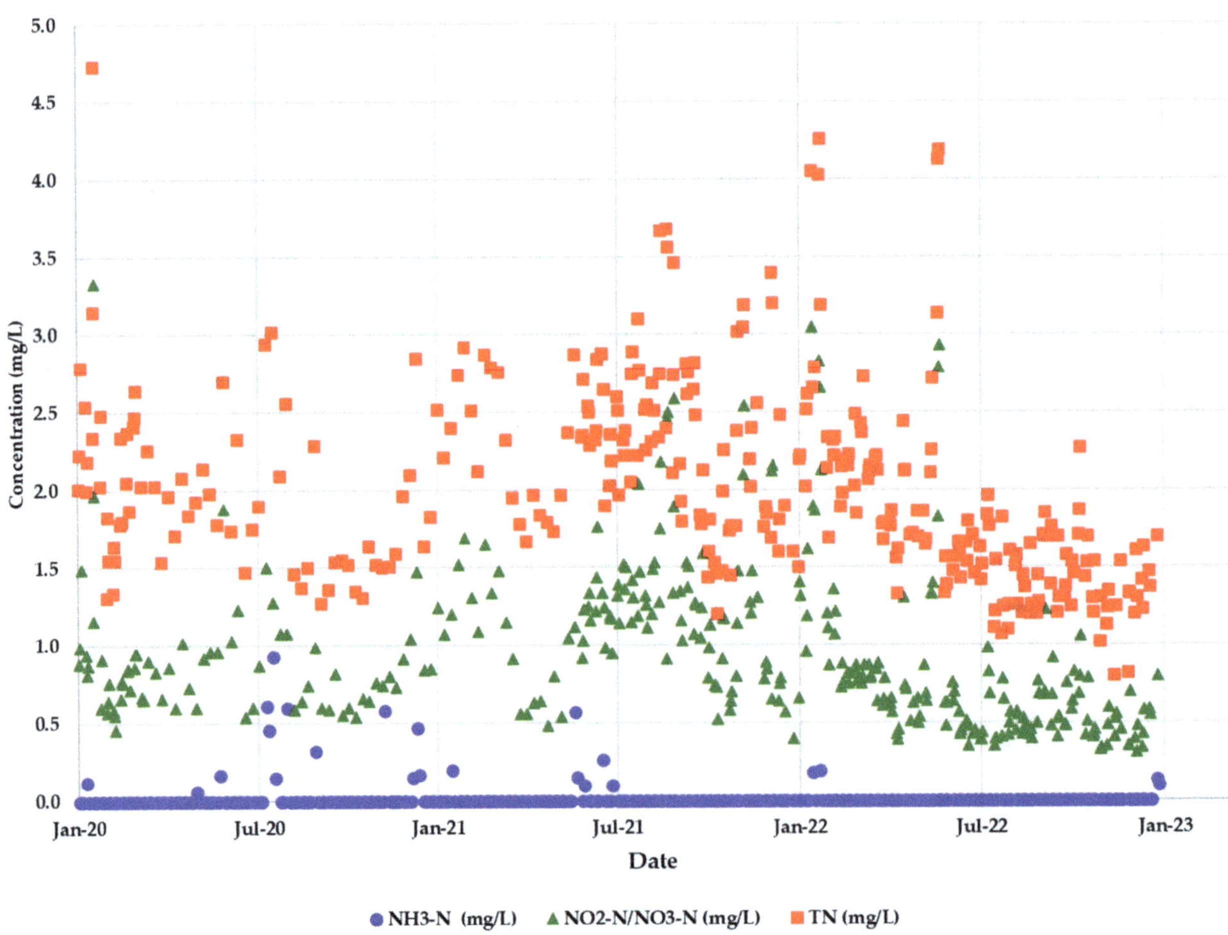


FIGURE 15.14 Town of North Cary, North Carolina, Facility Effluent Nitrogen Data

- The second anoxic zones can be very low in NOx concentrations, so nitrate and phosphate analyzers were added that can be used to automatically actuate a diffuser grid to prevent phosphorus re-release.
- If the installed analyzers that control the treatment fail, the system can be set to a timed phased isolation treatment approach with manually selected phases.
- The BioDenipho aeration process can be set to three different modes:
  - Dynamic mode— dissolved oxygen setpoints automatically increase if a large concentration of ammonia is detected.
  - Dissolved oxygen mode— dissolved oxygen setpoints are preselected to control blowers.
  - SCFM mode—SCFM setpoints are preselected to control blowers.
- Air headers must be periodically drained of condensate.

## 8.0 CASE STUDY 7 (MEMBRANE BIOREACTOR)—CITY OF NORTH LAS VEGAS, NEVADA, WATER RECLAMATION FACILITY

### 8.1 Facility History

The City of North Las Vegas Water Reclamation Facility is an advanced tertiary level WRRF designed to remove conventional pollutants, such as nitrogen and phosphorous, through either enhanced BNR or chemical addition. On a 40-ac site, the facility has average daily and peak-hour flow capacities of 95 ML/d (25 mgd) and 190 ML/d (50 mgd), respectively. The facility uses a membrane bioreactor (MBR) and provides reclaimed water that is disinfected before being discharged. The MBR process removes all substances from the wastewater that are larger than 0.1 μm. Coupled with disinfection, the treatment process provides a pathogen-safe reclaimed water product that can be used for irrigating areas such as public golf courses or that can be discharged back to Lake Mead.

In the past, all of the wastewater generated within the city limits of North Las Vegas was conveyed to the City of Las Vegas and the Clark County Water Reclamation District for treatment. Once conveyed to these agencies, the potential for water reuse by the City of North Las Vegas was gone, and the water commodity was owned by those agencies. In addition, almost 68% of the wastewater fund's operating budget paid for treatment costs to these third-party agencies.

In January 2004, the City of North Las Vegas city council authorized an in-depth analysis of wastewater treatment options for the city. In October 2004, the city council directed staff to pursue construction of a wastewater reclamation facility and authorized the city manager to pursue a site for the facility. After obtaining all necessary approvals and before construction, the proposed wastewater reclamation facility and all associated infrastructure was estimated to cost approximately $321.3 million and was to be funded with bond proceeds and capital reserves.

Construction of this facility began in November 2008, with the facility being substantially completed in August 2011. The City of North Las Vegas facility has been in operation since June 2011, and influent flows currently average 76 ML/d (20 mgd). This allows independence from the city of Las Vegas wastewater treatment rates and provides North Las Vegas with the ability to maximize efficiency of operations and water resources.

When it was placed online, the City of North Las Vegas facility was the largest MBR facility treating wastewater in North America and one of the largest MBR facilities in the world. Advanced nutrient removal is used to remove phosphorous and nitrogen before discharging the highly treated wastewater to the Sloan Channel, where it is conveyed to Lake Mead. The City of North Las Vegas facility uses enhanced biological phosphorus removal (EBPR) and chemical phosphorus precipitation for phosphorus removal. A modified $A^2/O$ process is used for simultaneous nitrification and denitrification. The City of North Las Vegas facility has a staff of 22 highly trained personnel who operate and maintain the facility 24 hours a day, 365 days a year, through an extensive computerized supervisory control and data-acquisition operational system.

## 8.2 Process Description

The incoming wastewater first enters the facility through a 10.65-m-deep influent pump station and then through coarse screens. After grit removal, the flow proceeds through fine screens and then 12 MBRs for treatment. The treated water is disinfected with sodium hypochlorite at the chlorine contact building and then dechlorinated. The City of North Las Vegas facility uses EBPR and chemical phosphorus precipitation (Figure 15.15).

The liquid treatment processes include preliminary treatment, secondary treatment, and disinfection. The following subsections discuss each process in detail.

The headworks facility (HWF) contains all the preliminary treatment for the facility. Preliminary treatment includes coarse screening, influent pumping, grit removal, and fine screening. Processing units for the collected screenings and grit are also located in the HWF.

Biological treatment is performed in six parallel bioreactor basins. Each basin is divided into a series of zones that are isolated from each other by submerged baffle weir walls. The configuration and sizing of these zones is based on the basic principles of BNR using a configuration adapted for use in the MBRs that optimizes the enhanced biological removal of nitrogen and phosphorus while recognizing the unique differences in the quality and flow of RAS in MBR systems. Membrane bioreactor air scour is controlled by an airflow setpoint.

Facility effluent can be discharged as a surface water discharge to the Las Vegas Wash, pumped to reuse applications and uses, or pumped to the facility water system. Sodium hypochlorite is injected to the 1.22-m (48-in.) chlorine contact basin (CCB) influent pipe upstream of a jet mixing system to ensure even distribution of chlorine throughout the CCB influent. Membrane permeate enters the chlorine contact basin via one of two 0.91-m (36-in.) pipes from the MBR system. Chlorine residual is monitored at both the upstream end of the CCB (for the influent) and the downstream end (for the

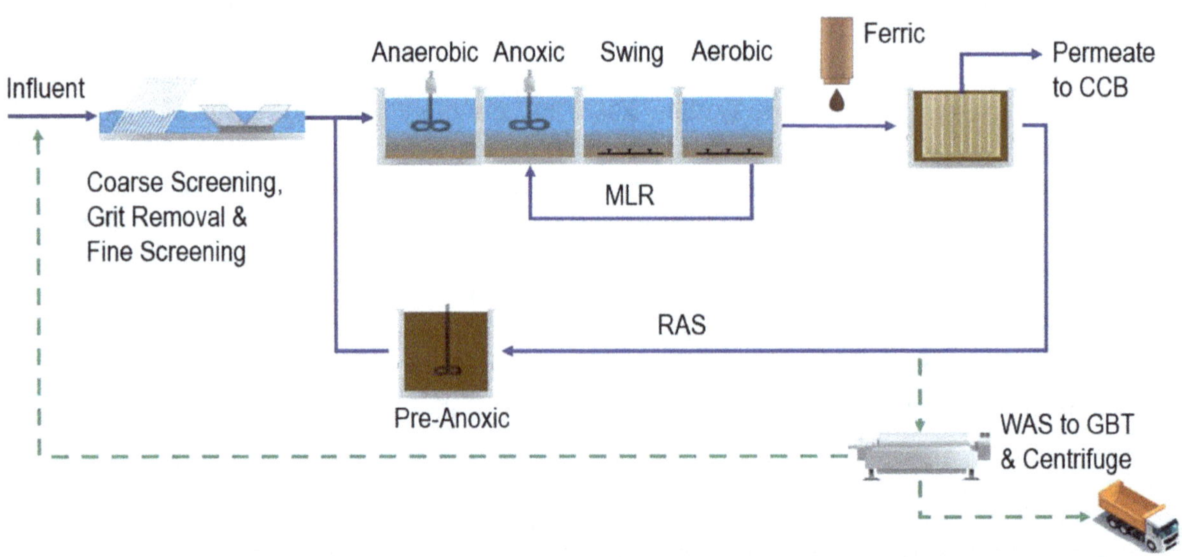


FIGURE 15.15 City of North Las Vegas Facility Process Flow Diagram

effluent). Any residual chlorine in the effluent discharged to the wash is dechlorinated using sodium bisulfite. Effluent for facility water is withdrawn after dechlorination.

Thickening and dewatering equipment is designed to remove excess water from the solids to reduce the volume and weight of waste transported to a nearby landfill. The solids thickening equipment consists of GBTs. Waste activated sludge is pumped from the membrane bioreactor to the solids handling facility. The WAS is injected with diluted polymer before being discharged onto the GBT's rotating belt. The thickened WAS (TWAS) is then pumped to horizontal bowl centrifuges. The TWAS is injected with diluted polymer before entering the centrifuge. The separated water is collected and transported back to the facility's headworks for treatment. The cake is stored in the fully enclosed, odor-controlled hoppers before being loaded and hauled to a landfill for disposal.

The City of North Las Vegas facility has effluent permit limits for BOD and TSS of 30 mg/L. Additionally, the facility must meet a 30-day average effluent ammonia-nitrogen concentration of 200 kg/d (91 lb/d), which is equivalent to 0.43 mg/L at the facility's design capacity of 94.6 ML/d (25 mgd). The facility has a total inorganic nitrogen limit of 20 mg/L and a total phosphorus limit of 14.5 kg/d (32 lb/d), which is equivalent to 0.15 mg/L at the design capacity.

## 8.3 Process Design Criteria

The following are process design criteria for the City of North Las Vegas facility: GE/Zenon Model ZeeWeed (ZW 500D):

- Design flows:
  - Design average daily flow of 94.6 ML/d (25.0 mgd), expanding to flow of 132.5 ML/d (35 mgd);
  - Design peak hour flow of 189.3 ML/d (50.0 mgd), expanding to peak hour flow of 227.1 ML/d (60 mgd); and
  - Membrane bioreactor sludge recycle flow (4 times the average day maximum month [ADMM] of 454.3 ML/d [120 mgd]);
- Process tankage:
  - Twelve trains;
  - Basin side water depth of 6.35 m (20.83 ft) W × 17.4 m (57 ft) L × 3.7 m (12 ft);
  - A working volume for each train of 404 000 L (106 700 gal); and
  - Total working volume of 4 841 070 L (1 278 875 gal);
- Membrane data:
  - Sixteen cassette slots per train, 192 total cassettes slots (plus 4 extra future cassette spaces per train);
  - Forty-eight modules per cassette, 9216 total modules (10 cassettes each train with 48 modules, 2 cassettes each train with 32 modules, and 6528 total modules);
  - Fifty-two module cassette replacements beginning fall 2022, 3 trains per year, 604 modules per train (11 cassettes each train with 52 modules, 1 cassette each train with 32 modules);
  - Membrane area per module of 31.6 $m^2$ (340 sq ft);
  - Replacement membrane area per module of 40 $m^2$ (430 sq ft);
  - Total membrane area of 206 200 $m^2$ (2 219 520 sq ft);
  - Total membrane area of 289 545 $m^2$ (3 116 640 sq ft) by fall 2025;
  - Design maximum total membrane pressure (TMP) of 82.7 kPa (12 psig);

- Operating TMP of 6.9 to 69.0 kPa (1 to 10 psig);
- Two 1305-kW (1750-hp) process blowers and two 932-kW (1250-hp) process blowers; and
- Membrane scour (two) at 634 kW (850 hp).

Table 15.14 summarizes the influent wastewater characteristics.

## 8.4 Design and Operation Modifications

The facility operating SRT has been reduced from approximately 20 days at startup to a target SRT of 8 to 11 days. This was done to increase the amount of phosphorus being removed from the facility by WAS.

The facility was started up with a RAS rate of 61 ML/d (16 mgd) per bioreactor (four bioreactors were used at startup and a RAS flow of 4 times the influent flow, totaling 242 ML/d [64 mgd]) and an internal mixed liquor rate (MLR) of 17 ML/d (4.5 mgd) per bioreactor. Because of the low MLR rate, both the anoxic and swing zones were anaerobic, which caused additional phosphorus release in the anoxic zone and continued into the swing zone, downstream of the anoxic zone. The relatively high RAS rate maximized the amount of nitrate being recycled into the anaerobic zone, reducing the quantity of volatile fatty acids (VFAs) available to drive EBPR. Return activated sludge flow rates have since been reduced to facilitate dissolved oxygen quenching in the pre-anoxic zone. Current RAS rates are 2 times the influent flow, with a minimum of 161 ML/d (42.5 mgd) and a maximum of 238 ML/d (63 mgd), to give tighter control of bioreactor hydraulic retention time (HRT) in the absence of flow equalization.

The facility diurnal flow variability has been more severe than anticipated. The facility can experience influent swings between 132 ML/d (35 mgd) during late morning to early afternoon to flow of less than 23 ML/d (6 mgd) during the early morning hours. This variation causes the organic-to-nutrient ratios in the facility influent to show their highest levels in the early afternoon through the evening and then rapidly drop overnight. These wide variations cause wide swings in bioreactor dissolved oxygen levels, anaerobic zone ORP, and overall process HRT. An 8.3 ML (2.2 mg) flow equalization basin is scheduled for completion in 2024, which should mitigate issues associated with hydraulic loading rates and broadly attenuate those associated with variable waste strength.

## 8.5 Performance

The City of North Las Vegas facility has had excellent performance since startup and has not exceeded any conventional effluent quality parameters except total phosphorus. Biochemical oxygen demand,

**TABLE 15.14** City of North Las Vegas Facility Wastewater Characteristics

| Parameter | Influent Average | Influent Maximum |
|---|---|---|
| 5-day BOD | 300 mg/L | 600 mg/L |
| TSS | 300 mg/L | 850 mg/L |
| TKN | 45 mg/L nitrogen | 60 mg/L nitrogen |
| Ammonia–nitrogen | 34 mg/L nitrogen | 40 mg/L nitrogen |
| Total phosphorus | 6.5 mg/L phosphorus | 11.5 mg/L phosphorus |
| Minimum temperature | 18 °C | 18 °C |
| Average temperature | 24 °C | 24 °C |
| Maximum temperature | 30 °C | 30 °C |

TSS, and fecal coliforms results have been at or below method-detection limits since startup, with and ammonia, total inorganic nitrogen (TIN), and total phosphorus results averaging below 0.5, 7, and 0.11 mg/L, respectively (see Figure 15.16 and 15.17).

The only operations issue that has affected the City of North Las Vegas facility has been the facility's ability to remove total phosphorus using EBPR. Since the facility came online in 2011, EBPR efficiency has not consistently met the facility's permit requirement, particularly during high flow periods when process temperatures exceed 30 °C. Optimization strategies focusing on SRT, membrane air scour rates, and VFA production did not provide effluent total phosphorus concentrations below the 0.195 mg/L average that they currently require. Beginning in March 2023, the facility adopted a strategy focused on reducing pre-anoxic dissolved oxygen levels and increasing anaerobic HRT at high flows by manipulating RAS pumping rates. Historically the facility used a RAS rate of 3 times the influent flow, resulting in an average pre-anoxic dissolved oxygen concentration of 0.5 mg/L and an average anaerobic HRT of 0.3 hours. Through reduction of the RAS rate to 2 times the influent flow, bounded by a 161 ML/d (42.5 mgd) minimum and 238 ML/d (63 mgd) maximum, average pre-anoxic dissolved oxygen concentration has fallen to 0.2 mg/L, and average anerobic HRT increased to 0.4 hr. Since March 2023, the average composite total phosphorus concentration has been less than 0.11 mg/L, and the maximum monthly composite average has been less than 0.13 mg/L. This shows an improvement over the recent historical value of between 33% and 60%. After the planned commission of the flow equalization basin in 2024, RAS rates will be further reduced to 1.5 times the influent flow. Modeling suggests that this configuration can achieve pre-anoxic dissolved oxygen concentration <0.1 mg/L, anaerobic HRT of 0.5 hrs, and average daily total phosphorus concentration of 0.08 mg/L.

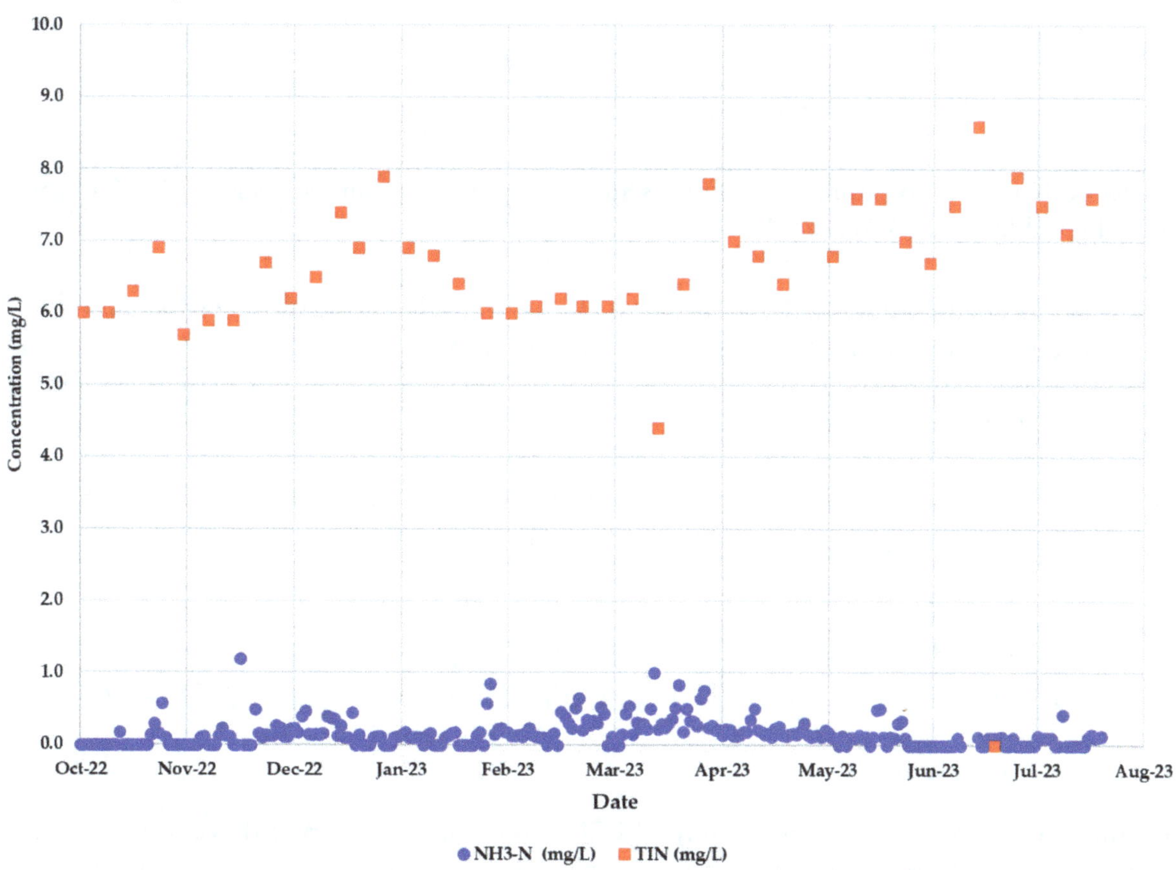


FIGURE 15.16 City of North Las Vegas Facility Effluent Ammonia-Nitrogen and Total Inorganic Nitrogen

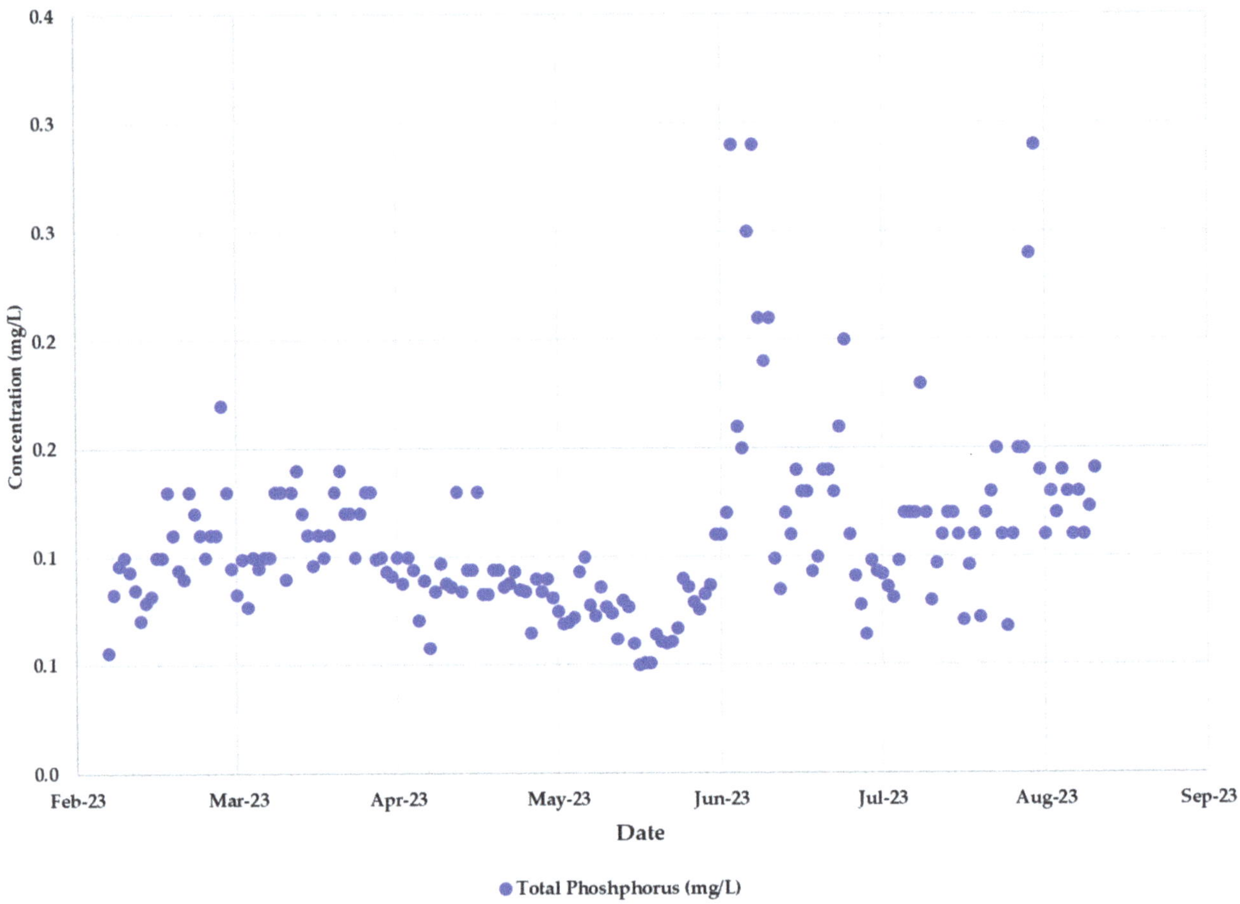


FIGURE 15.17 City of North Las Vegas Facility Effluent Phosphorus

Chemicals are used to clean the membranes and to keep them within the operating TMP range. Sodium hypochlorite (10.3%) is used for

- maintenance cleans twice per week per train (12 trains)—132 000 L/yr (35 000 gal/yr); and
- recovery cleans semiannually per train—52 000 L/yr (13 600) gal/yr.

Citric acid (50%) is used for

- one maintenance clean weekly per train (12 trains)—33 000 L/yr (8750 gal/yr); and
- recovery cleans semiannually per train—10 000 L/yr (2650 gal/yr).

## 9.0 CASE STUDY 8 (FOUR-STAGE BARDENPHO)—PUTNAM, CONNECTICUT, WATER POLLUTION CONTROL FACILITY

### 9.1 Facility History

The Putnam Water Pollution Control Facility (WPCF) was originally constructed as a primary treatment facility in the 1930s and was upgraded to a conventional activated sludge facility using mechanical aeration that was placed into operation in the early 1970s. The WPCF was upgraded in 2011 to include BNR and chemically enhanced phosphorous removal (depicted in Figure 15.18). The Putnam WPCF treats domestic wastewater, industrial wastewater, and septage (currently averaging 64 000 L/d [17 000 gal/d]). Effluent is pumped to the Quinebaug River, a tributary to the Long Island Sound. The Putnam WPCF participates in Connecticut's watershed-based trading program for discharges to the Long Island Sound. Additionally, three sections of the Quinebaug River have been designated National Recreation Trails by the National Park Service, some of the first water trails to receive this designation.

## 9.2 Process Description

The 2011 upgraded facility processes and unit operations include preliminary treatment, primary settling, and carbon, nitrogen, and phosphorus removal via a four-stage activated sludge process (primary anoxic, primary aerobic, secondary anoxic, and reaeration) followed by secondary clarification and then UV disinfection. Solids handling systems include gravity thickening of primary and WAS and dewatering using a belt filter press, as depicted in Figure 15.19.

The monthly average effluent criteria for the Putnam WPCF are as follows:

- Design flow rate 11 ML/d (2.91 mgd);
- 5-day biochemical oxygen demand ($BOD_5$), 30 mg/L;
- Total suspended solids, 30 mg/L;
- Total phosphorous [seasonal April 1 through October 31], 1.09 mg/L with an average seasonal daily load cap of 3.8 kg/d (8.41 lb/d);
- Total nitrogen and nitrogen speciation are not limited—however, the Putnam WPCF participates in Connecticut's watershed-based trading program for discharges to the Long Island Sound, and the targeted effluent TN concentration for the biological process is 3 to 5 mg/L;

FIGURE 15.18 Putnam WPCF After the 2011 BNR Upgrade, With One BNR Train Operating (Courtesy of the town of Putnam)

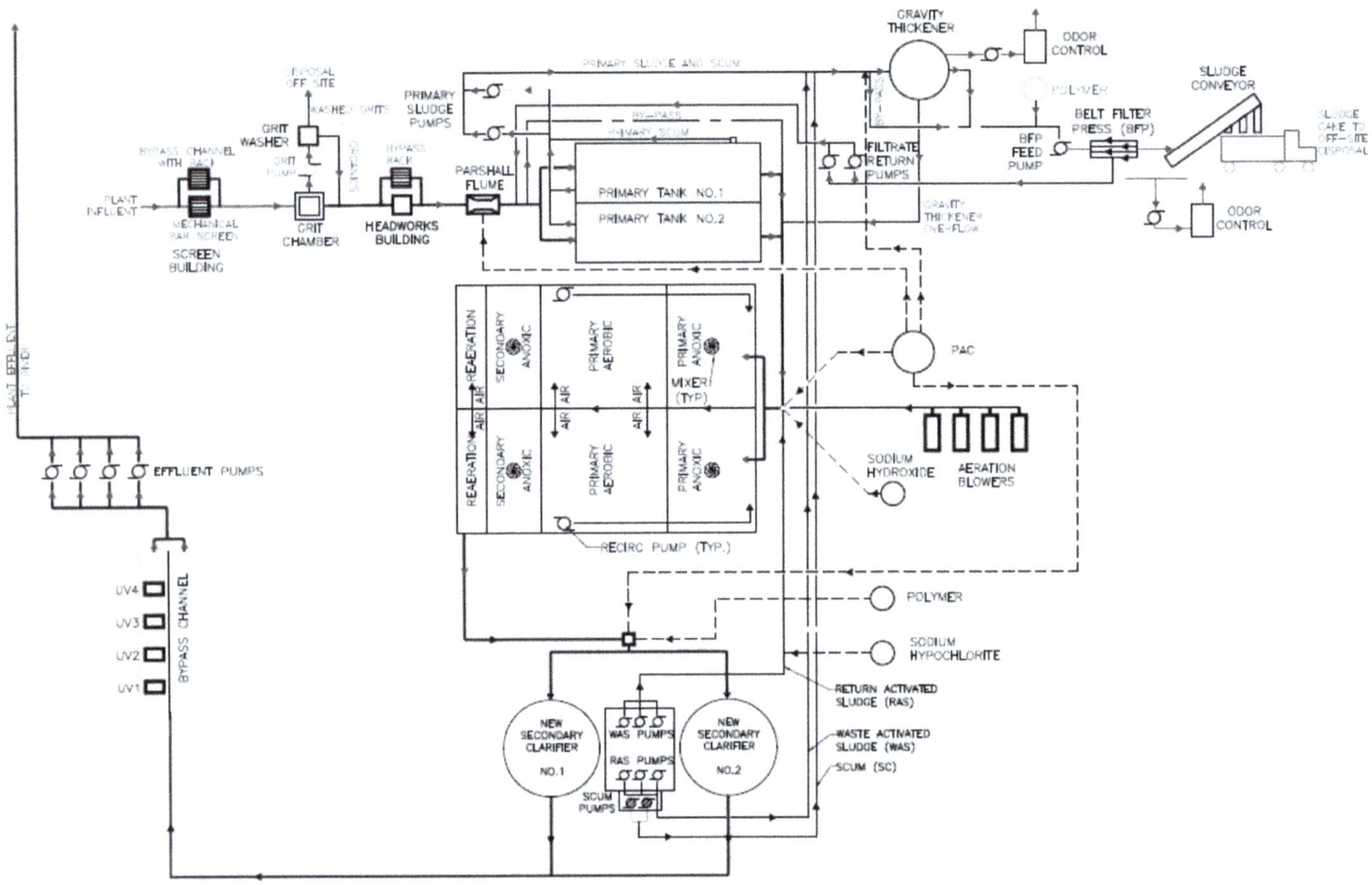


FIGURE 15.19 Putnam WPCF Process Flow Diagram (Courtesy of the town of Putnam)

- The UV disinfection system must maintain an instantaneous UV lamp intensity ≥ 6.10 mW/cm$^2$ on a seasonal basis from May 1 through September 30.

At the current average daily flow of 4.1 ML/d (1.09 mgd), the total phosphorus concentration must be below 0.9 mg/L, which is less than the permitted average monthly concentration limit. If the daily load limit is exceeded for any 2 consecutive years or any 2 of 3 consecutive calendar years, Putnam will need to submit plans for additional phosphorus controls.

The liquid treatment stream consists of primary clarification, followed by primary anoxic, primary aerobic, secondary anoxic, and secondary aerobic zones. The anoxic zones are fitted with hyperbolic mixers to keep the MLSS in suspension and in continuous contact with the wastewater. Air is provided to the two aerobic zones through fine-bubble diffusers. Air to the diffusers is provided by positive displacement blowers located in the blower building. Dissolved oxygen probes located in the primary and secondary aerobic zones provide feedback to control blower speed and to optimize air flow supplied to aerobic zones. An internal recycle pump recirculates nitrate-rich wastewater from the end of the primary aerobic zone to the beginning of the primary anoxic zone for denitrification. The RAS and internal recycle rates average 60% and 70% of the facility's average daily flow, respectively. The system is comprised of the following major components:

- Two BNR reactors to provide year-round BOD removal and nitrification;
- Two internal recycle pumps, each located in the primary aerobic zones to recycle nitrate-rich mixed liquor to the primary anoxic zones for denitrification;

- Two 230-m (75-ft) diameter secondary clarifiers with 4.3-m (14-ft) side water depth;
- Four positive displacement blowers located in the blower building;
- Return activated sludge/waste activated sludge pump station located between the two new clarifiers and housing three RAS pumps, three WAS pumps, and two secondary scum pumps; and
- Chemical feed systems including poly-aluminum chloride (PAC) dosing pumps, feeding BNR effluent/secondary clarifier influent, and the gravity thickener.

### 9.3 Process Monitoring and Control

Key parameters that are monitored and used to control the biological process include the following:

- Primary effluent flow and pollutant characteristics ($PO_4$-P, total phosphorus, $NH_3$-N, $NO_3$-N, $NO_2$-N, TKN, total nitrogen, TSS, and $BOD_5$).
- Anoxic zone ORP;
- Aerobic zone dissolved oxygen and MLSS concentrations;
- Blower speed and airflow using dissolved oxygen setpoints and most-open modulating valve control;
- Secondary effluent characteristics ($PO_4$-P, total phosphorus, $NH_3$-N, $NO_3$-N, $NO_2$-N, TKN, total nitrogen, TSS, and $BOD_5$);
- Biological process SRT and sludge volume index (SVI) based on RAS, WAS, and internal recycle pumping rates; and
- Flow-paced PAC dose control for the liquid process with a post-BNR and pre-secondary clarifiers dosing point.

### 9.4 Operating Data

Figure 15.20 shows the monthly average effluent total phosphorus concentrations for 2020, 2021, and 2022 through May 31, 2022. The facility's average effluent total phosphorus concentration was 0.10 mg/L. Figure 15.21 compares the facility's seasonal effluent total phosphorus total phosphorus load limit of 3.82 kg/d (8.41 lb/d) to actual data spanning April 1 through October 31 in 2020 and 2021. Final effluent total phosphorus load averaged 0.36 kg/d (0.79 lb/d) in 2020 and 0.35 kg/d (0.77 lb/d) in 2021. The average PAC dose to the facility's main liquid stream is 0.23 $m^3$/d (60 gpd) with an average facility flow of 4129 $m^3$/d (1.09 mgd) for an approximate dose of 17 ppm aluminum, presuming a specific gravity of 1.34 and 23% weight fraction of aluminum (70 kg Al/d).

### 9.5 Operating Experience

The facility's control systems have allowed operating staff to continually meet permit requirements and tune biological nutrient removal and the chemically enhanced phosphorous removal systems to outperform those requirements, resulting in the sale of nitrogen credits in the Long Island Sound trading program and limiting effluent phosphorus load to and potential eutrophication in the Quinebaug River.

## 10.0 SUMMARY

Biological nitrogen removal occurs in a variety of configurations and has evolved over the last couple of decades. The configuration of the treatment basins, internal recycle locations, and flows, swing zones, tertiary filtration, chemical addition, and solids handling processes discussed herein are unique to each WRRF and have various advantages and disadvantages. However, the core reactions of BNR, autotrophic

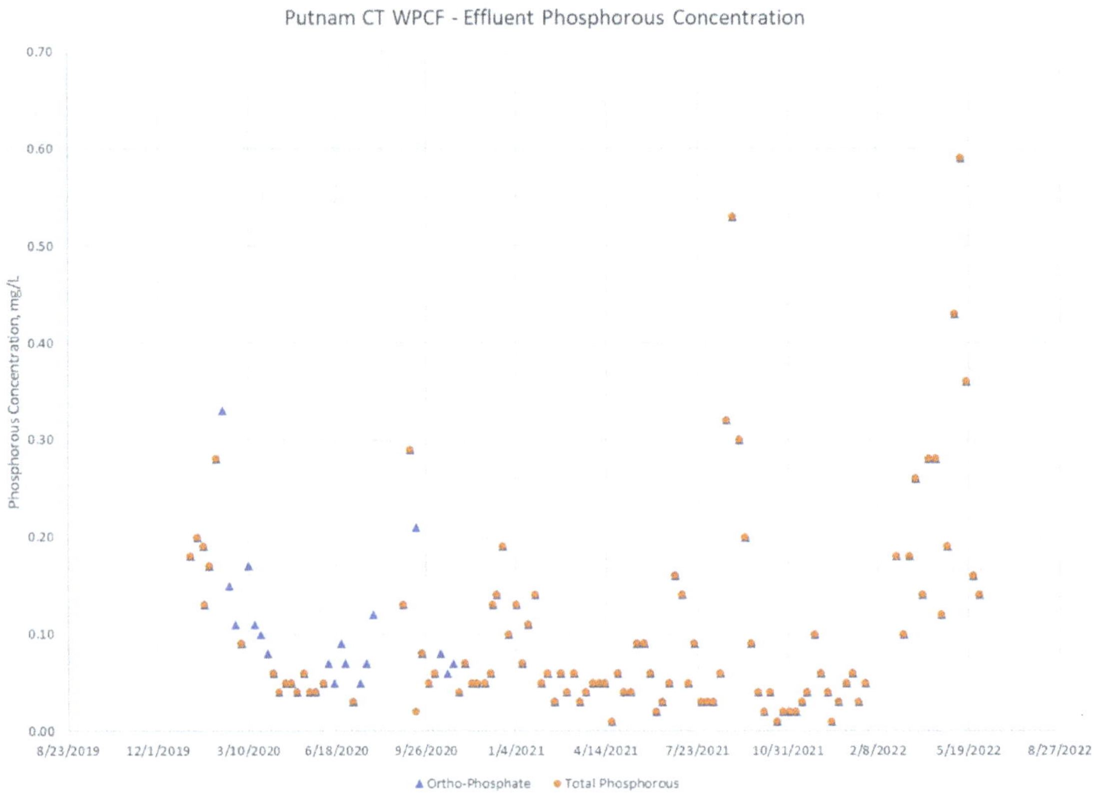


FIGURE 15.20 Putnam WPCF Operating Data (Effluent Total Phosphorus) (Courtesy of the town of Putnam)

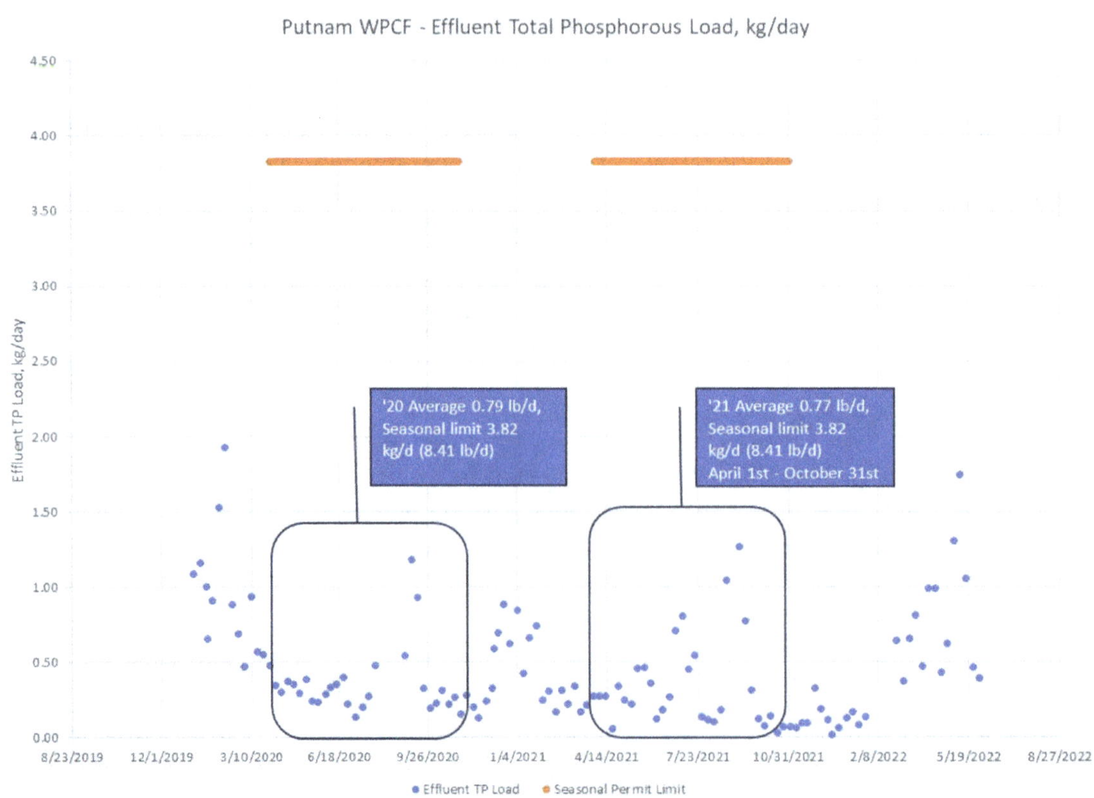


FIGURE 15.21 Putnam WPCF Operating Data (Effluent Total Phosphorus Load) (Courtesy of the town of Putnam)

aerobic nitrification, and heterotrophic anoxic denitrification fueled by a carbon source have remained virtually unchanged since its first use in water resource recovery facilities. The level of total nitrogen removal is highly dependent on site-specific conditions (e.g., temperature) and nitrogen characteristics in the raw wastewater. The level of treatment that is appropriate for each facility is dependent on several cost and noncost factors. Current and anticipated future permit limits on nitrogen species, physical constraints such as land availability and topography, local energy and operating costs, and ease of operation must all be considered to select the most efficient and cost-effective treatment process for BNR.

# Chapter 16

# Case Studies—Enhanced Biological Phosphorus Removal



## 1.0 INTRODUCTION

This chapter presents six case studies of full-scale facilities that are specifically designed and operated for optimal enhanced biological phosphorus removal (EBPR) from municipal wastewaters to meet effluent total phosphorus limits between 0.05 and 1.5 mg/L. It should be noted that most EBPR facilities in North America are also designed to simultaneously achieve some degree of nitrogen removal in a single sludge system, that is, to meet either an effluent ammonia-nitrogen ($NH_3$-N) standard though nitrification or an effluent total nitrogen standard through nitrification and denitrification. All facilities used as case studies in this chapter have been in operation for a minimum of 2 years to provide the reader with insight to the challenges encountered by facility operators during the startup and commissioning phases of the project and the design and operational changes that were required following commissioning to ensure reliable and effective EBPR.

## 2.0 CASE STUDIES

### 2.1 Case Study 1—Flamingo Water Resource Center, Clark County Water Reclamation District, Las Vegas, Nevada

#### *2.1.1 Facility History*

The Clark County Water Reclamation District is a regional wastewater collection system and treatment service provider. Formed under Nevada judicial decree in 1954, the agency was originally named the Clark County Sanitation District (operated as a general improvement district under Nevada Revised Statutes) and was renamed in 2003. The district is responsible for wastewater collection and treatment for all unincorporated areas of Clark County, which includes the Las Vegas Valley and outlying communities of Blue Diamond, Indian Springs, Laughlin, Moapa Valley, and Searchlight.

The district owns and operates the Flamingo Water Resource Center (FWRC) located in Las Vegas, Nevada. The FWRC currently treats approximately 400 ML/d (106 mgd) annual average flow (AAF) of wastewater by using a combination of physical, chemical, and biological treatment. Approximately 97% of reclaimed water is discharged to the Las Vegas Wash, which flows to Lake Mead. The remaining 3% of highly treated effluent is reused onsite at the FWRC and distributed for other reuse purposes (e.g., irrigation of parks, golf courses, and playing fields). The Flamingo Water Resource Center is currently going through the expansion to a design annual average treatment capacity of 570 ML/d (150 mgd) AAF.

Final effluent from FWRC is discharged through two outfalls (Outfall 001 and Outfall 002) to the environmentally sensitive Las Vegas Wash ("Wash"), which flows into Lake Mead. The lake is an important source of potable and irrigation water and recreational lake water for several communities in Arizona, California, Nevada, and Mexico.

#### *2.1.2 Process Description*

Effluent total phosphorus limits are applied from March 1 to October 31, while effluent ammonia-nitrogen limits are applied from April 1 to September 30. The FWRC limits are part of a total limit set for the Wash, which receives effluent from North Las Vegas, Las Vegas, City of Henderson, and the Clark County Water Reclamation District (CCWRD), along with couple of others smaller entities. In 2001, there was an algae bloom in Lake Mead and the dischargers volunteered to meet effluent total phosphorus and ammonia-nitrogen limits during each month. The Nevada Department of Environmental Protection accepted this offer and kept the seasonal limits in the National Pollutant Discharge Elimination System (NPDES) permits.

The effluent criteria for the FWRC are as follows:

- Five-day biochemical oxygen demand ($BOD_5$), 30 mg/L;
- Total suspended solids (TSS), 30 mg/L;
- Total phosphorus, <79 kg/d (174 lb/d);

- Ammonia-nitrogen, <229 kg/d (505 lb/d); and
- Fecal coliforms, <200 MPN per 100 mL.

The Flamingo Water Resource Center implements a comprehensive wastewater treatment process to treat receiving wastewater (Figure 16.1). The process begins with influent screening, removing large debris. Primary treatment involves primary clarifiers where solids settle, followed by biological treatment in activated sludge reactors to break down organic pollutants. Secondary clarifiers separate remaining solids from treated water, leading to tertiary treatment involving filtration and disinfection.

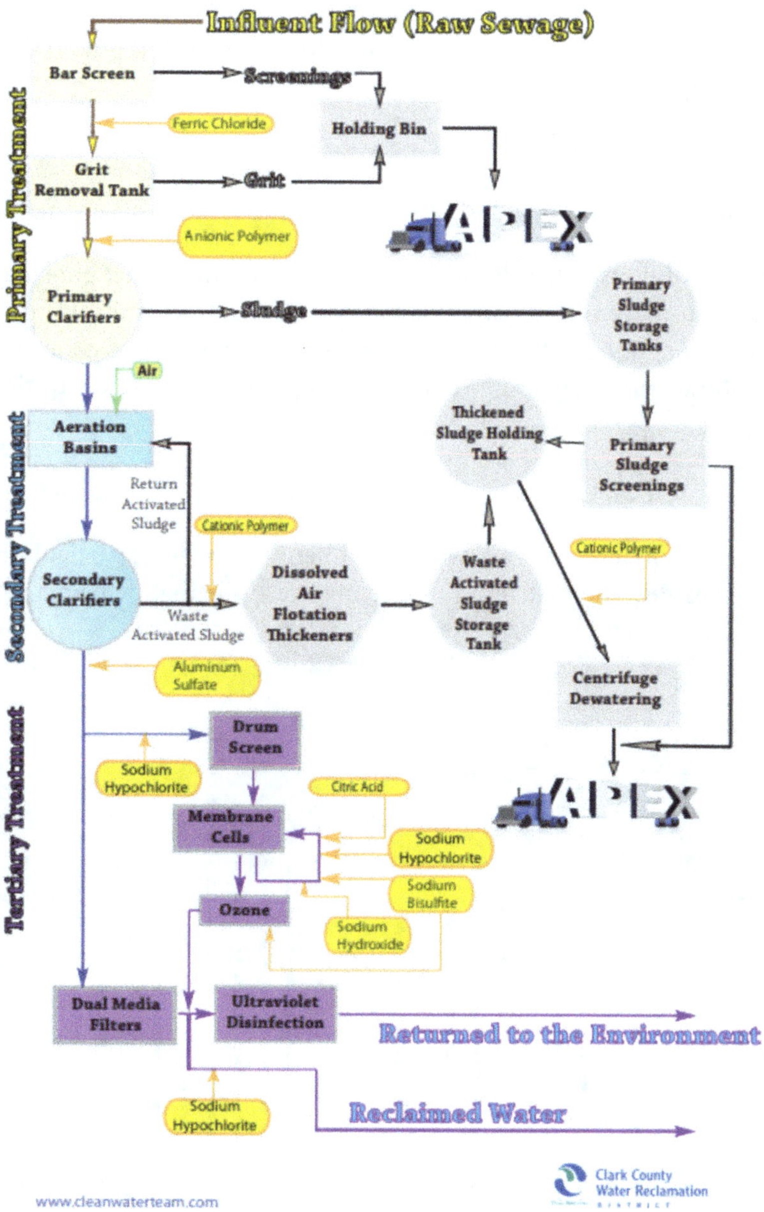


FIGURE 16.1 Flamingo Water Resource Center Process Flow Schematic (Courtesy of Clark County Water Reclamation District)

### 2.1.3 Process Monitoring and Control

Key parameters that are monitored and used to control the biological nutrient removal (BNR) process include the following:

- Wastewater flow and characteristics (BOD, TSS, volatile fatty acids [VFAs], $NH_3$, total Kjeldahl nitrogen [TKN], total phosphorus, orthophosphate);
- Anaerobic, anoxic, aerobic zone, and supernatant characteristics ($PO_4$, $NH_3$, and $NO_x$);
- Secondary effluent and filter effluent characteristics (total phosphorus, $PO_4$, $NH_3$, $NO_x$, TSS, and pH);
- Aerobic zone dissolved oxygen, mixed liquor suspended solids (MLSS), and settleability;
- Primary sludge solids retention time (SRT) and hydraulic retention time (HRT);
- Activated sludge SRT; and
- Chemical dose (alum and polymer).

### 2.1.4 Operating Data

Since the implementation of phosphorus limits in early 2000s, effluent total phosphorus and ammonia-nitrogen concentrations have consistently been well below effluent limits. In the last several years, the FWRC has consistently maintained a monthly average total phosphorus concentration below 0.1 mg/L. Figure 16.2 shows monthly average effluent total phosphorus concentrations for 2020, 2021, 2022, and 2023.

### 2.1.5 Operating Experience

The FWRC has been at the forefront of wastewater treatment innovation, particularly in the realm of substrate generation for EBPR. Their current practice involves utilizing the primary clarifiers for the

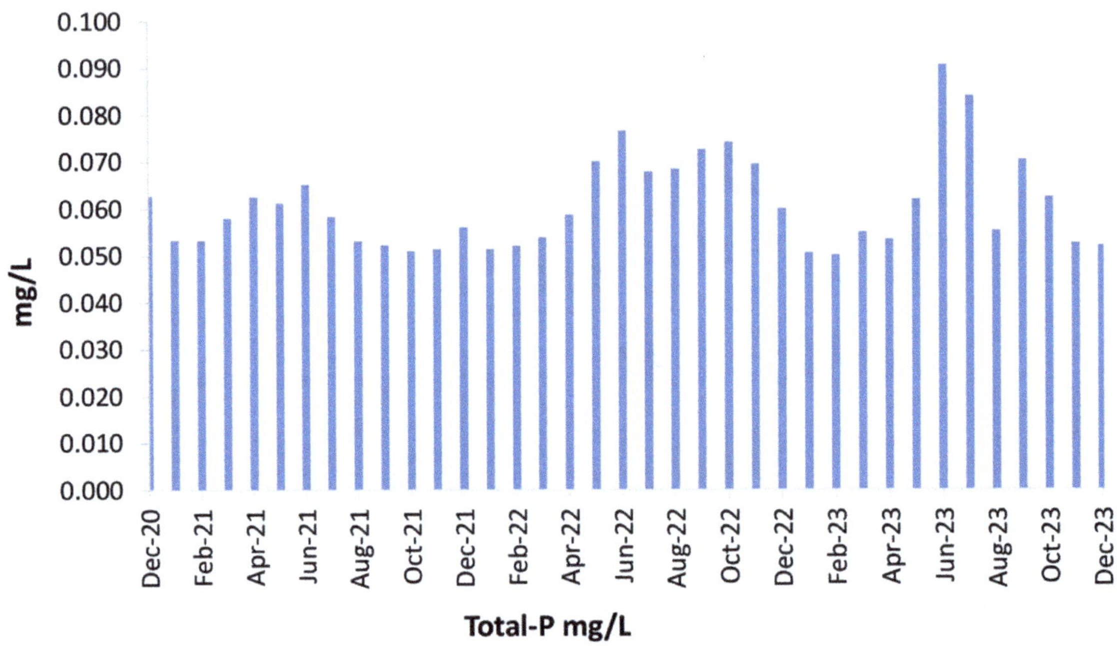


FIGURE 16.2 Flamingo Water Resource Center Effluent Total Phosphorus Concentrations From 2020 to 2023 (Courtesy of the Clark County Water Reclamation District)

generation of VFAs, essential for enhancing phosphorus removal efficiency. However, a strategic shift is on the horizon, as the facility is gearing up for a future in which VFAs will be generated in the gravity thickener, ushering in a more efficient and streamlined process.

At the heart of this innovation is the fermentation of primary sludge, a process integral to the production of VFAs. The primary sludge, composed of organic matter from the initial treatment stages, serves as the raw material for the fermentation process. This approach not only addresses the need for sustainable resource utilization but also aligns with the facility's commitment to environmentally conscious practices.

Operators experimented with primary clarifier operation and found that carrying a sludge blanket of approximately 1 m (3 ft) increased the VFA concentration (measured as acetate) from 10 mg/L to between 60 and 80 mg/L. This, in turn, increased the ratio of VFA to total phosphorus in the BNR influent from less than 2 to approximately 10—a condition quite favorable for EBPR dynamics.

In the current operational approach, the primary clarifiers play a pivotal role in VFAs generation. The facility uses a carefully calculated approach by maintaining a certain sludge blanket depth (approximately depth ranging between 0.5 m [1.5 ft] to 2 m [6.5 ft]), providing an ideal environment for the microbial community to thrive and contribute to the fermentation process. This equates to an SRT ranging between 1 and 1.5 days. This optimal timeframe ensures the biological activity within the clarifiers is conducive to the production of VFAs.

The FWRC is proactively planning for a future transition in VFAs generation. The forthcoming shift to the gravity thickener as the primary site for VFA production represents a strategic move toward process optimization. This transition is anticipated to offer several advantages, including improved control over the fermentation process, enhanced resource recovery, and increased phosphorus removal efficiency.

Several other strategies have also been implemented through different processes by the FWRC to increase the efficiency of phosphorus removal. In the secondary treatment process, the initial activated sludge process was retrofitted to include a return activated sludge conditioning zone to minimize the effect of nitrate recycled from the secondary clarifiers to the anaerobic zone. The secondary clarifiers were upgraded with rapid sludge removal systems, Stamford baffles, and optimized energy dissipation inlets (EDIs). An automated SRT control was implemented in all basins and a solids transfer system was designed. These improvements provided operators the ability to rapidly shut down and start up basins. The aeration system also was upgraded to allow for zone control. These strategies were key to reducing orthophosphate concentrations in the secondary effluent from a typical maximum month of 0.7 mg/L in the 1990s to less than 0.1 mg/L in the 2010s.

Regarding tertiary treatment, the CCWRD has invested in infrastructure replacement, decommissioning the 1979 advanced wastewater treatment facility and replacing it with new media filtration and a 110-ML/d (30-mgd) microfiltration facility to treat secondary effluent. Operators also developed an alum addition strategy—called the "Alum Kick"—to provide effluent polishing, thus increasing the process's reliability and robustness.

Finally, the CCWRD implemented centrate recycle stream management. By adding ferric chloride to the centrate return, the operators reduced soluble phosphate from being recirculated from solids handling processes back to primary treatment.

## 2.2 Case Study 2—Nine Springs Wastewater Treatment Plant, Madison, Wisconsin

### *2.2.1 Facility History*

The 160-ML/d (42-mgd) Nine Springs Wastewater Treatment Plant serves the City of Madison, Wisconsin, and several surrounding communities in the Dane County area. The Madison metropolitan area surrounds a chain of four ancient glacial lakes, and in 1959, the district began pumping treated effluent 5 miles to Badfish Creek to avoid discharge to the chain of lakes. The facility was initially constructed on the current site in 1929. The first facility was a trickling filter facility with Imhoff tanks for sludge treatment; within 5 years, an activated sludge train and anaerobic digestion were added. Several facility upgrades have been constructed in response to population growth and changes in effluent discharge limits. There have been 11 significant upgrades to the facility since 1929, and several additional major upgrades are currently in planning and under construction.

### 2.2.2 Process Description

In the early 1980s, the activated sludge system was modified for single-stage nitrification, and the trickling filter was removed. The last significant change to address revised discharge limits occurred in 1997, when the facility was modified for EBPR without the construction of additional tankage. Because the facility did not have an effluent total nitrogen standard, the selected process configuration was the University of Cape Town process without nitrified mixed liquor recycle. Figure 16.3 presents a process flow diagram of the facility operation from 1997 to 2013. A summary of some of the effluent permit criteria for the Nine Springs facility until 2019 were as follows:

- Five-day BOD, 7 and 16 mg/L (during summer and winter, respectively);
- Total suspended solids, 10 and 16 mg/L (during summer and winter, respectively);
- Total phosphorus, 1.5 mg/L;
- Ammonia-nitrogen, 1.3 and 3.8 mg/L (during summer and winter, respectively); and
- Fecal coliforms, 400 per 100 mL (during summer).

The Wisconsin Pollutant Discharge Elimination System permit for effluent discharge was modified in 2019. The permit addressed more stringent phosphorus regulations in Wisconsin, which would have required an effluent total phosphorus concentration for the District to be 0.075 mg/L. The alternative

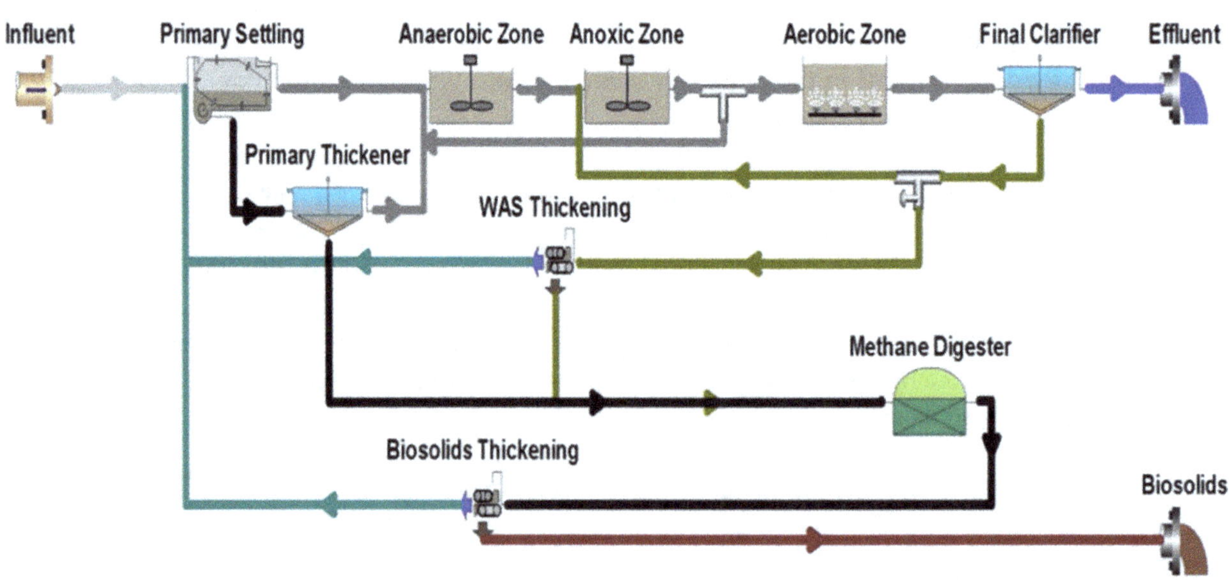


FIGURE 16.3 Nine Springs Process Configuration 1997–2013 (Courtesy of the Madison Metropolitan Sewerage District)

used by the District and incorporated in the permit was an adaptive management strategy to reduce phosphorus loadings in the watershed that would avoid the much more stringent effluent limits. The following is a partial summary of some of the new criteria relative to the EBPR process:

- Five-day $CBOD_5$, 19 mg/L (monthly average);
- Total suspended solids, 20 mg/L (monthly average);
- Total phosphorus, 1.0 mg/L (monthly average); 0.6 mg/L (6-month average);
- Ammonia-nitrogen, 1.1 mg/L (monthly average May to September); 3.8 mg/L (monthly average October to April); and
- Fecal coliforms, 400 per 100 mL (monthly geometric mean March 1 to November 30).

The Nine Springs Treatment Plant had experienced struvite precipitation problems in the anaerobic digestion system pipes and digesters since the 1970s. This was in part due to the area using 100% groundwater for drinking water with a hardness of 18 to 21 grains/gal and a magnesium concentration of approximately 45 mg/L. The water source provided plenty of magnesium to combine with the ammonium and phosphorus released in anaerobic digestion to produce struvite (magnesium ammonium phosphate). This problem was exacerbated with the process modification to biological phosphorus removal in 1997. In response to the struvite issue, as well as other anaerobic digestion and biosolids issues, the facility was modified to incorporate a struvite harvesting process beginning in about 2013. The process revision schematic is shown in Figure 16.4.

### 2.2.3 Process Monitoring and Control

The facility uses dissolved oxygen control for biological process control and energy savings. There are 10 EBPR trains with two dissolved oxygen probes in each train. This is the main operational control used for the process. The laboratory performs single daily colorimetric tests for effluent ortho-phosphorus Monday through Friday for a quick check on EBPR process performance. The 24-hour composite

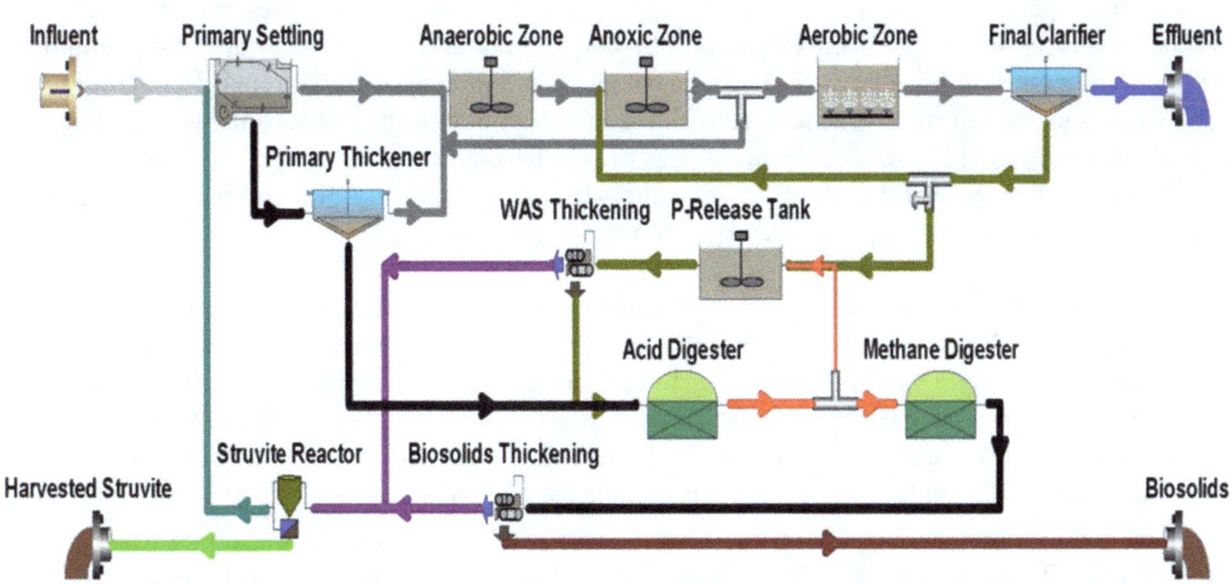


FIGURE 16.4 Nine Springs Process Configuration 2013–2023 (Courtesy of the Madison Metropolitan Sewerage District)

samples are held and batched for auto-analyzer analyses for daily regulatory reporting requirements. Phosphorus removal has been consistent, and this testing regime has proved adequate to ensure stable EBPR operation.

Key operational parameters that are monitored and used to control the EBPR process include the following:

- Primary effluent characteristics (total phosphorus, $NH_3$, $BOD_5$, and TSS) (daily monitoring);
- Anaerobic, anoxic, and aerobic zone supernatant characteristics ($PO_4$, $NH_3$, $NO_x$, etc.) (checked occasionally in conjunction with research projects or other studies);
- Aerobic zone dissolved oxygen and MLSS concentrations; and
- Bioreactor SRT.

A portion of return activated sludge (RAS) is wasted using hydraulic controls to maintain bioreactor SRT. Wasting rates are based on influent flows and RAS flows. The 24-hour composite samples of mixed liquor and RAS are analyzed for TSS concentration three times a week to verify that the EBPR process is being maintained at the target SRT. The target SRT varies somewhat seasonally but has averaged approximately 10 days.

### 2.2.4 Operating Data

The facility has produced an effluent exceeding discharge permit requirements including those for phosphorus and ammonia nitrogen. The facility averaged 0.39 mg/L from 1997 to 2015 without chemical addition or tertiary filtration. The effluent ammonia-nitrogen and nitrate-nitrogen ($NO_3$-N) concentrations averaged 0.16 and 14.8 mg N/L, respectively, while the effluent TSS concentration averaged 4.8 mg/L. Effluent concentrations from 2015 to 2023 have remained similar to those shown from 1997 to 2015.

The successful application of EBPR beginning in 1997 was attributable, in part, to a conservative secondary clarifier design surface overflow rate of 12.2 $m^3/m^2 \cdot d$ (300 gpd/sq ft) at an AAF of 190 ML/d (50 mgd). Before upgrading to EBPR, the facility was frequently subject to filamentous bulking problems, with sludge volume indexes (SVIs) in the 200- to 300-mL/g range. The highest monthly average SVI value after the facility's conversion to EBPR was 140 mL/g, with average values being around 100 mL/g. Average monthly effluent and settling data is shown in Figures 16.5 through 16.8. The figures show the significant improvements realized with the incorporation of the EBPR configuration. The main changes since 2013 have been for mitigating struvite problems.

Operating experience at the Nine Springs facility after the upgrade to EBPR and before incorporation of the struvite harvesting process can be summarized as follows:

- Waste activated sludge (WAS) production and handling: After the upgrade to EBPR, WAS production increased by approximately 9% at the same SRT. This was attributable to an increased uptake of phosphorus and other ions into the activated sludge cell mass and improved TSS removal. The increase in WAS production was offset, to some degree, because the sludge thickened better at lower SVIs. The dissolved air flotation (DAF) thickened WAS concentration increased from an average of 3.7% solids to 4.2% solids without the use of polymer.
- Ultraviolet disinfection: Maintenance requirements for the ultraviolet disinfection system decreased significantly with EBPR. Before EBPR, quartz tube scaling required cleaning with citric or phosphoric acid approximately every 3 weeks. After the upgrade, the quartz tubes no

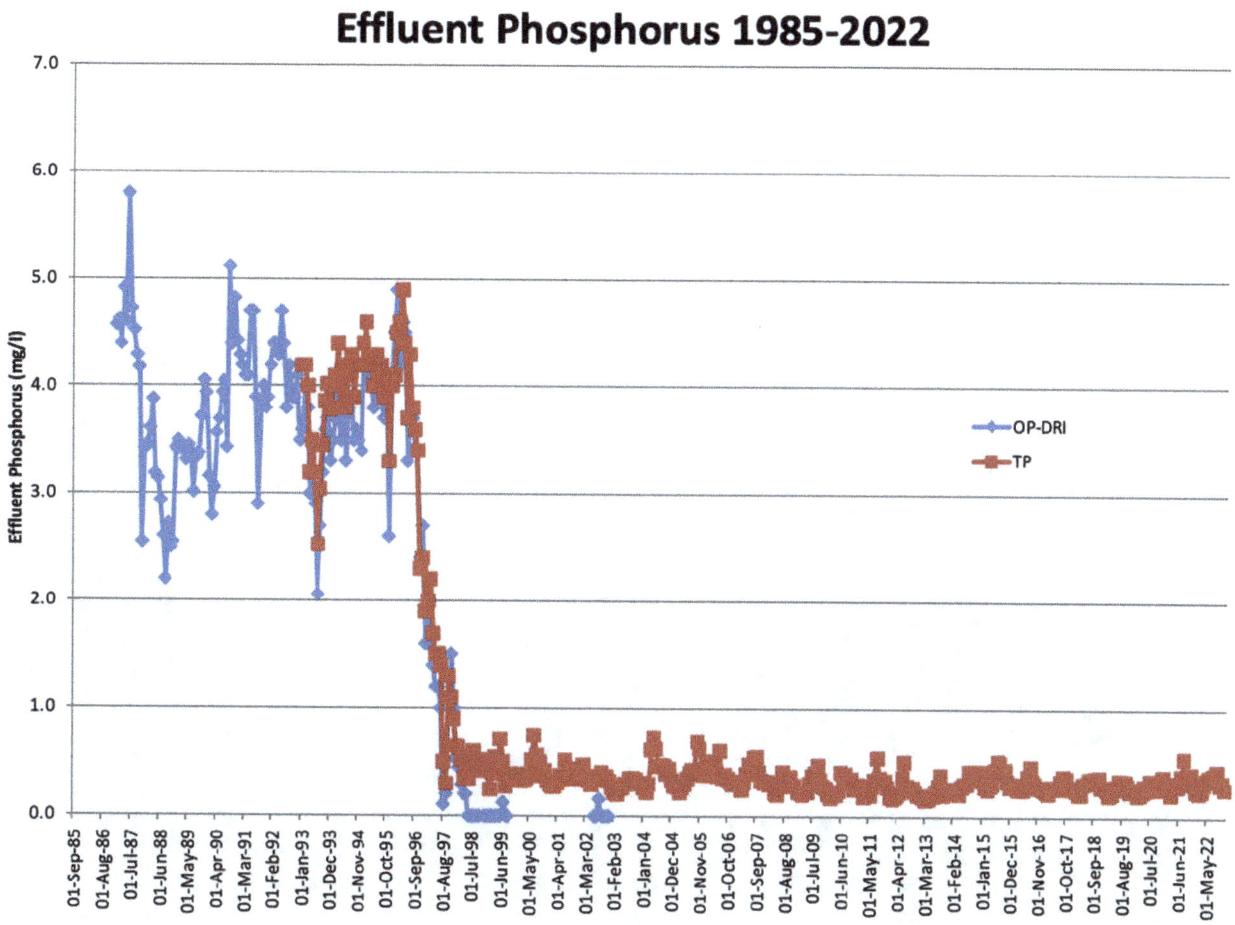


FIGURE 16.5 Nine Springs Facility Effluent Orthophosphorus and Total Phosphorus Concentrations From 1985 to 2022 (Courtesy of the Madison Metropolitan Sewerage District)

longer required cleaning during the 6-month disinfection season. The reason was thought to be lower effluent TSS and the luxury uptake of phosphorus, magnesium, and potassium in the EBPR process.

- Aeration system: Off gas testing of the aeration system demonstrated that, in spite of the uptake of soluble organics in the anaerobic and anoxic zones, the alpha values of the fine-bubble aeration system were not significantly greater than at the head end of the process before incorporation of the anaerobic and anoxic zones. Alpha values were 0.3 to 0.4 at the head end of the aerobic zone and increased to ~0.7 as aeration proceeded. This was similar to the off gas testing results obtained before the EBPR upgrade. Low dissolved oxygen operation (target aeration tank dissolved oxygen = 0.3 mg/L) was successfully tried in one of the four facilities in 2003. Maintaining low dissolved oxygen concentrations throughout the aerobic zone did not affect EBPR, had a small effect on ammonia-nitrogen removal, and increased the total nitrogen removal. Aeration savings were significant compared to a control facility.
- Recycle rate pacing: The mixed liquor recycle from the anoxic zone to the anaerobic zone is paced with the influent flowrate. The recycle rate typically does not affect process performance significantly except under wet weather flow conditions. It was discovered early during operation of the EBPR process that unless the recycle flow was increased proportionately to influent flow during

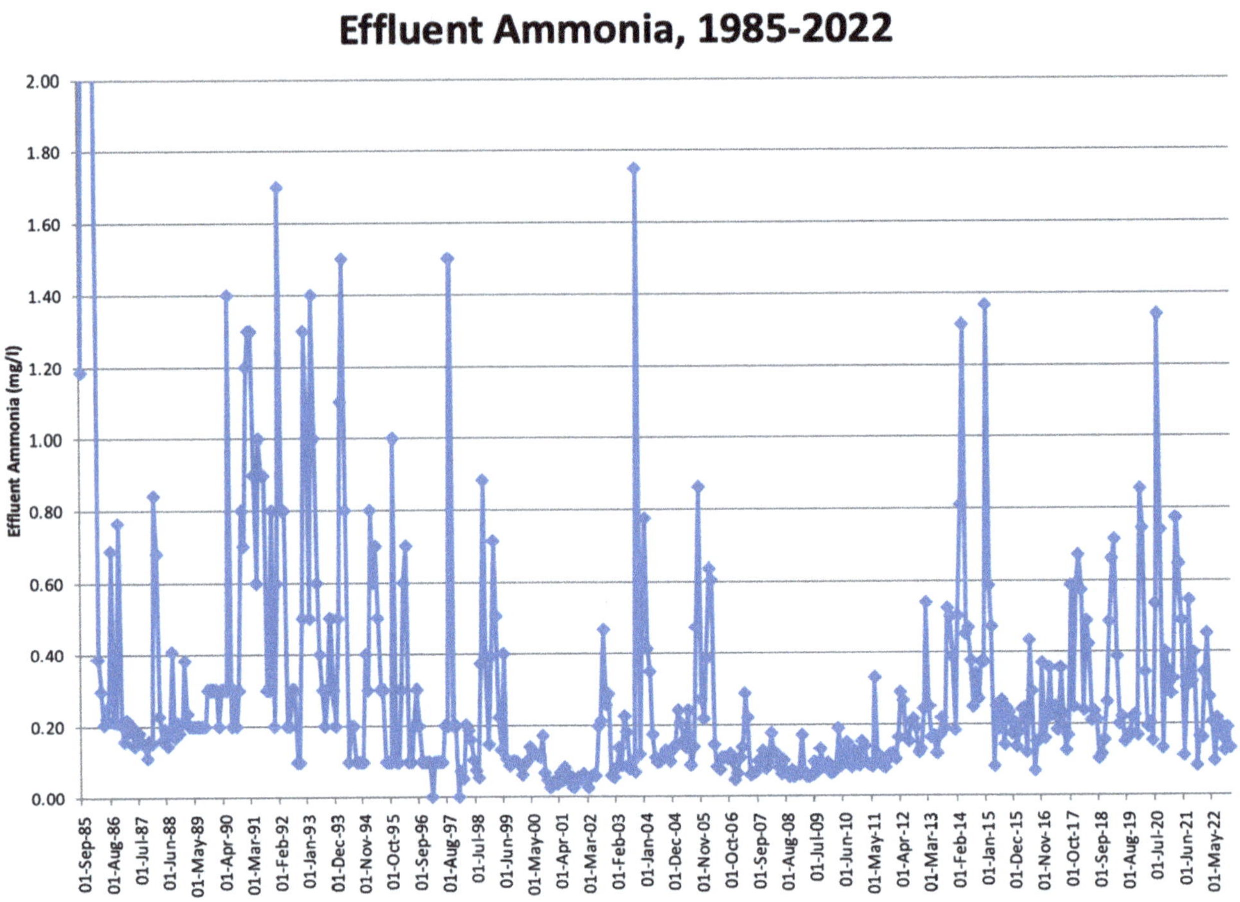


FIGURE 16.6 Nine Springs Facility Effluent Ammonia-Nitrogen Concentrations From 1985 to 2022 (Courtesy of the Madison Metropolitan Sewerage District)

high-flow events, the solids mass fraction in the anaerobic zone can drop significantly and EBPR can suffer. However, pacing the recycle rate with the influent flowrate has resolved this problem.

### 2.2.5 Operational Steps for Struvite Mitigation and Implementation of Struvite Harvesting (2013–2023)

Digestion system changes were implemented by 2013 along with struvite harvesting. Figure 16.4 presented a general process flow diagram of the upgraded facility. The plan was to produce 2.27 tonnes/day (2.5 tons/day) of struvite product for recycle. Other improvements included WAS thickening on gravity belt thickeners (GBTs) rather than the DAFs, and acid phase digestion to precede the main mesophilic digestion processes. Steam heating is used to heat raw sludge for the acid phase, and a small stream of acid-phase digested sludge is recycled and added to WAS with a 6- to 8-hour detention time prior to thickening to release phosphorus from WAS before digestion. Thermophilic digestion is used as second-stage digestion for 5% to 15% of the mesophilic digested sludge to produce a Class A biosolids product. The biosolids thermophilically digested are centrifuged to approximately 20%. The mesophilic digested sludge is thickened on a GBT to approximately 5.8%. Struvite harvesting was implemented using the Ostara Pearl process. The combination of WAS thickening filtrate and digested sludge thickening filtrate and dewatering centrate are combined and pumped to the Ostara Pearl upflow reactor for struvite harvesting. As shown in Figure 16.9, approximately 45% of the total facility phosphorus loading to the facility is fed to Ostara as soluble phosphorus.

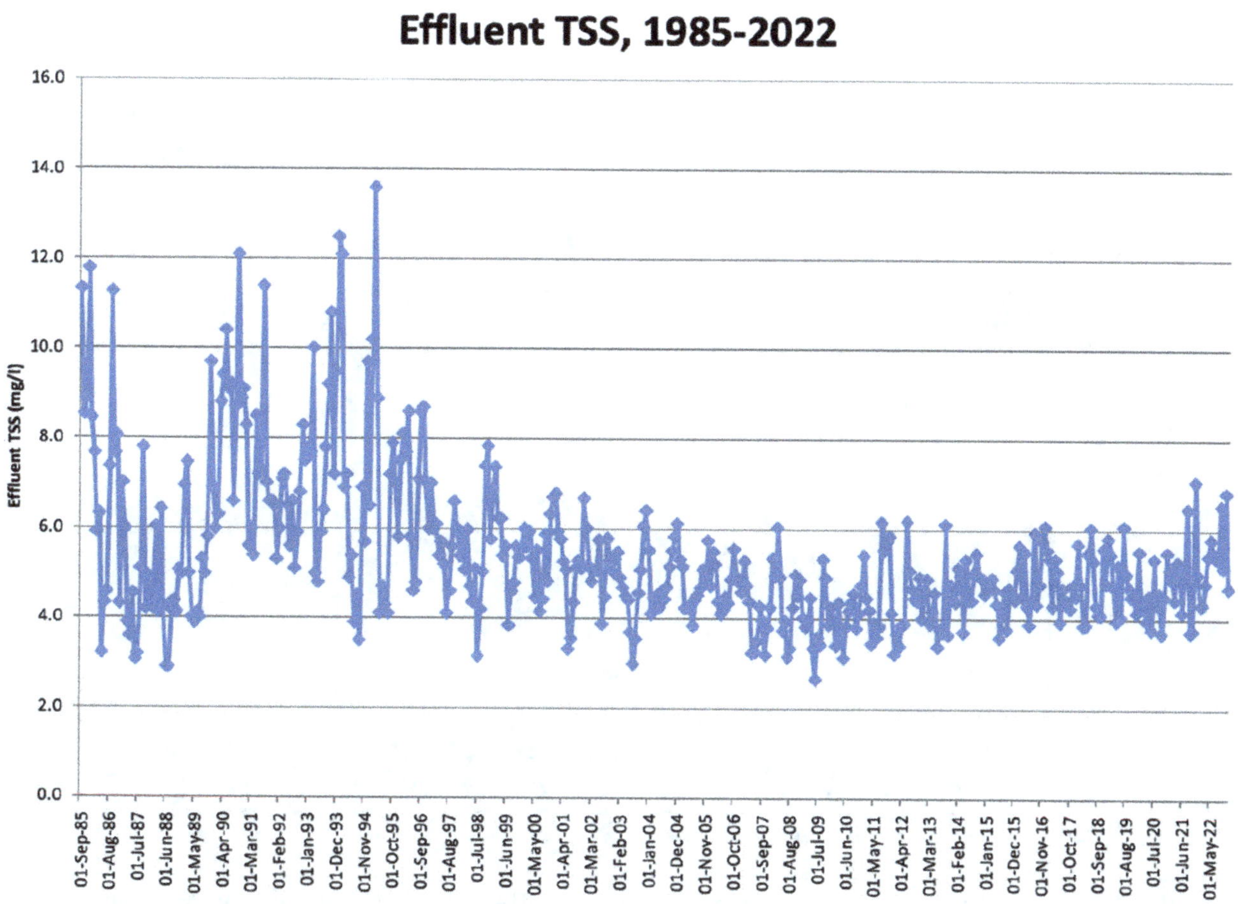


FIGURE 16.7 Nine Springs Facility Effluent TSS Concentrations From 1985 to 2023 (Courtesy of the Madison Metropolitan Sewerage District)

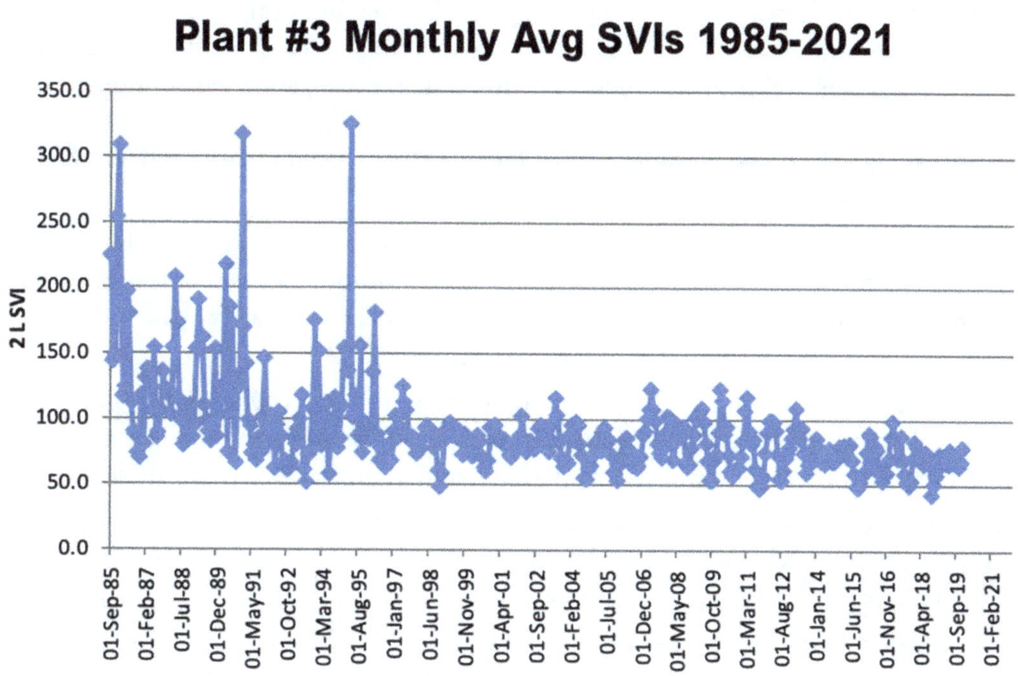


FIGURE 16.8 Nine Springs Facility Plant 3 SVI Data From 1985 to 2021 (Courtesy of the Madison Metropolitan Sewerage District)

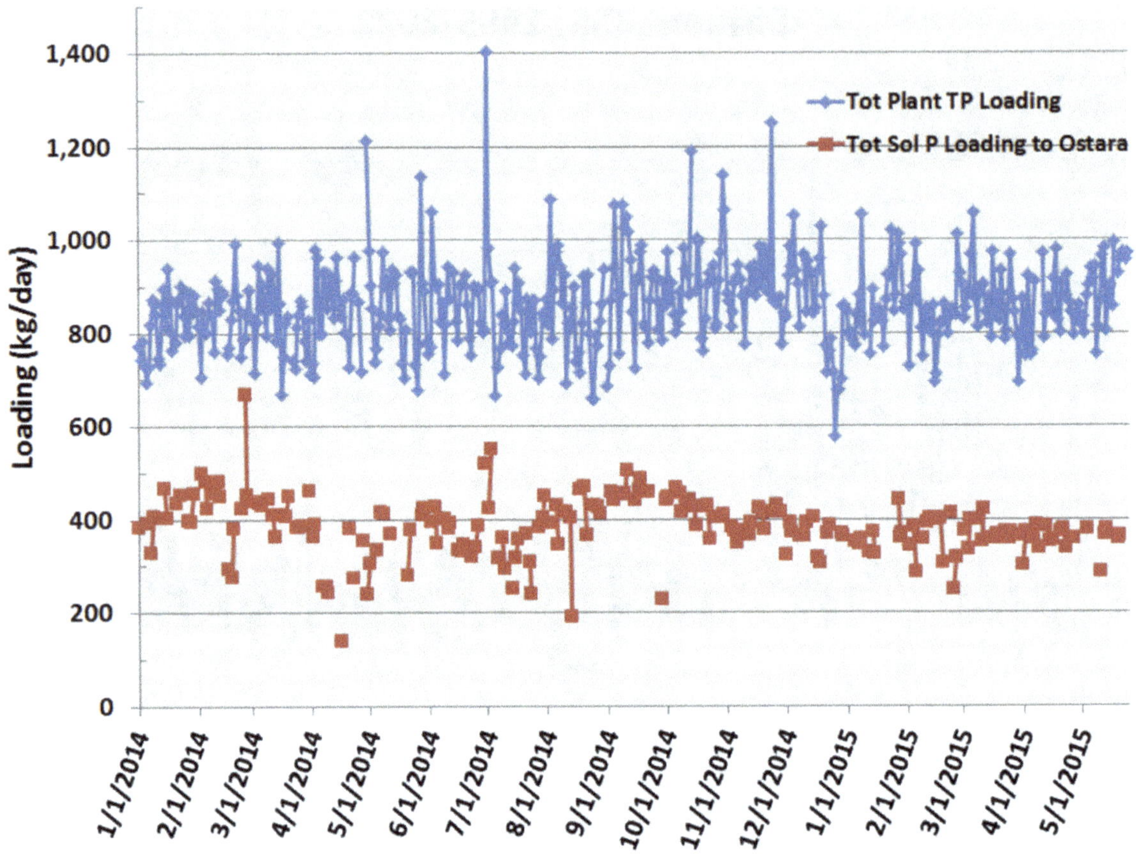


FIGURE 16.9 Nine Springs Facility Phosphorus Loadings in 2014 and 2015 (Courtesy of the Madison Metropolitan Sewerage District)

Anticipated benefits of struvite harvesting were generally met, but optimizing operation has required modifications and research to improving process performance. The following are some process details and notes on changes either implemented or for which investigation is continuing:

1. The soluble phosphorus removal in struvite harvesting has met the projected 80% to 85% removal projections.
2. Concentrations in the combined feed to the Pearl reactors are approximately 160 mg/L orthophosphate, 80 mg/L magnesium, and 220 mg/L ammonia. These concentrations were increased from design concentrations by pre-thickening approximately 30% of the WAS in a retired DAF thickening process before GBT thickening of WAS.
3. Soluble P concentrations in the digesters have decreased from ~200 mg/L to ~100 mg/L. Evidence supports that this results in significantly less struvite precipitation in the digesters and in digested sludge piping, translating to decreased operation and mainenance costs in these areas.
4. A comparable decrease in the water extractable phosphorus in the biosolids has positively affected land application limitations for phosphorus.
5. The total phosphorus removal by the struvite harvesting process was initially lower than expected due to loss of struvite fines in the reactor effluent and recycle to the facility. With operational experience, this effect has been mitigated by occasional re-seeding of the reactor with smaller size prills (the struvite beads produced). Production averaged between 1.36 and 1.82 tonnes/day (1.5 and 2.0 tons/day) from 2017 through 2023.

6. A low ferric chloride dose is still added to the sludge feed to thickening and dewatering processes. The pH elevates in the filtrate and centrate during digested sludge thickening and dewatering processes to prevent inadvertent struvite formation in the filtrate and centrate. A low ferric chloride dose added into the digested sludge feed suppresses the nuisance struvite formation and lowers hydrogen sulfite levels in the digester gas.
7. Integral in these process modifications was the new paradigm that various unit operations in the facility—such as return sludge, waste sludge, and thickening/dewatering processes—are quite intertwined with successful operation of struvite harvesting. And an efficiently operating struvite harvesting process with minimal downtime is absolutely necessary to protect the facility biological phosphorus removal process from overloading and upsets.

A detailed discussion of operational experience with the struvite harvesting process was presented in a paper by Dose and colleagues (2015), "Operating Experience With Ostara Struvite Harvesting Process."

## 2.3 Case Study 3—Geneva Wastewater Treatment Facility, Geneva, Illinois

### *2.3.1 Facility History*

The Geneva Wastewater Treatment Facility (GWWTF) has a permitted average daily flow (ADF) of 18.9 ML/d (5.0 mgd) and a peak flow of 47.3 ML/d (12.5 mgd) through the two biological treatment trains. The main liquids treatment is composed of fine screening, aerated grit removal, primary clarification, BNR targeting nitrogen and phosphorus removal, and ultraviolet disinfection. The solids train is composed of RAS/WAS pumping, WAS holding tanks, centrifuge thickening, anaerobic digestion, and dewatering via centrifuges. Centrate, digester supernate, and WAS holding tank decant and overflow are handled by aerated sidestream treatment tanks that get routed back to the screens. Biosolids are stored in a solids building and are then land applied as Class B sludge.

The GWWTF is required by NPDES permit to comply with effluent limits, including for total phosphorus (1 mg/L annual average). The GWWTF recently completed a capital improvement project that included upgrades to the biological process to implement biological nutrient removal in the anaerobic–anoxic–aerobic (A2O) process configuration. Secondary clarifier and RAS pumping system upgrades were also completed.

### *2.3.2 Process Description*

Raw sewage enters the facility through three interceptors and flows to a common influent channel to the mechanically cleaned bar screen. The screen has 10-mm (3/8-in.) clear openings. Screened flow discharges to two aerated grit removal tanks. Captured grit is dewatered using one grit washer and then disposed of at the landfill. Screened and de-gritted flow is conveyed to one of four primary clarifiers where settleable solids and associated carbonaceous biological oxygen demand (CBOD) are removed.

There are two trains of bioreactors in the activated sludge process. Each train is comprised of three passes (nos. 1–6), and the total available bioreactor volume is approximately 7.6 ML (2 MG). Figure 16.10 presents the layout of one bioreactor train. The following upgrades were completed to convert the process to A2O:

- Created an anaerobic zone (14% of each train's total volume) by using fiberglass (FRP) baffle walls and weir.
- Installed a 0.95 ML/d (0.25 mgd) internal recycle pumps for each anaerobic zone to improve EBPR.

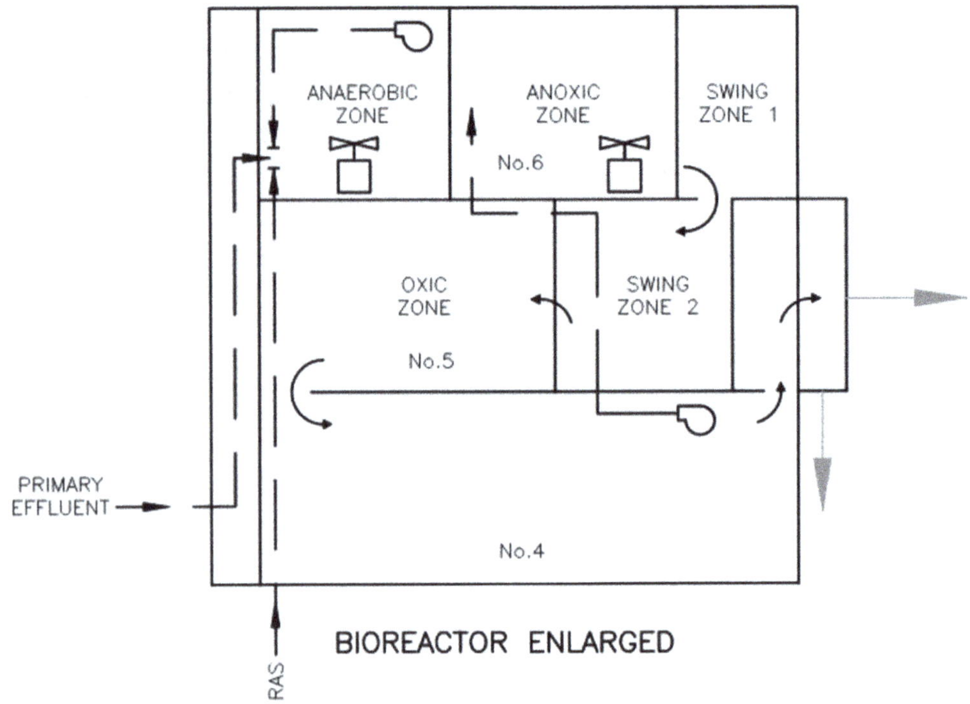


FIGURE 16.10 Geneva WWTF A2O Bioreactor Train Layout

- Created an anoxic zone (28% of each train's total volume) by using FRP baffle walls and a concrete anoxic/oxic zone weir.
- Installed two duty internal recycle pumps in each train, each with 18.9 ML/d (5 mgd) pumping capacity to recycle nitrified mixed liquor from the end of the aeration basin to the head end of anoxic basin for each bioreactor train.
- Installed mixers in each pass of the new anaerobic and anoxic zones.
- Implemented aeration system improvements consisting of new fine-bubble diffusers and turbo blowers to provide sufficient aeration system capacity at 18.9 ML/d (5 mgd) permitted treatment capacity.
- Partitioned anoxic zone into two zones. Install fine-bubble diffuser in the second part of the anoxic zone to create a swing anoxic/aerobic zone.
- Installed online probes for dissolved oxygen, ammonia, and oxidation-reduction potential for the operation and control of the new nutrient removal process.

Mixed liquor from the aeration basins can be settled in one of three 29-m (95-ft) diameter circular clarifiers. Clarifiers no. 1 and 2 have side water depths of 3 m (10 ft), while clarifier no. 3 has a side water depth of 3.7 m (12 ft).

The GWWTF has two ultraviolet disinfection reactors capable of treating average and peak secondary treated flow. To treat flows above the capacity of the biological process and ultraviolet disinfection system, the GWWTF directs excess primary effluent to a chlorine contact tank for disinfection using chlorine gas and sodium bisulfite dechlorination before blending with disinfected secondary effluent discharge to the Fox River.

Activated sludge is wasted from the secondary clarifiers and pumped to the unmixed, unaerated WAS holding tanks for storage before thickening. The WAS storage tanks have a total volume of approximately 0.44 ML (116 000 gal). Waste activated sludge is thickened using one thickening centrifuge. Primary sludge and WAS are stabilized through anaerobic digestion. The GWWTF has two digesters; each can hold up to about 2.5 ML (0.66 MG) of sludge. After secondary digestion, stabilized sludge is pumped to one dewatering centrifuge.

### 2.3.3 Effects of Unmixed In-line Fermentation

The GWWTF participated in the Water Research Foundation's Project No. 4975, *Practical Considerations for the Incorporation of Biomass Fermentation into Enhanced Biological Phosphorus Removal* (Water Research Foundation, 2023), to evaluate the approaches and benefits to Sidestream Enhanced Biological Phosphorus Removal. In addition to the A2O process improvements listed above, the final anaerobic zone was modified to allow for operation as an unmixed in-line fermenter (UMIF). The mechanical mixers in the zone are typically turned off, allowing sludge to accumulate at the bottom of the zone and ferment to produce VFAs. The mixers are only turned on for brief periods several times a day to prevent plugging. The study results show that operating in UMIF resulted in improved EBPR performance.

### 2.3.4 Operating Data

Figure 16.11 presents effluent total phosphorus concentrations for 2020 to 2022. The facility's effluent total phosphorous concentration over this period averaged about 0.45 mg/L.

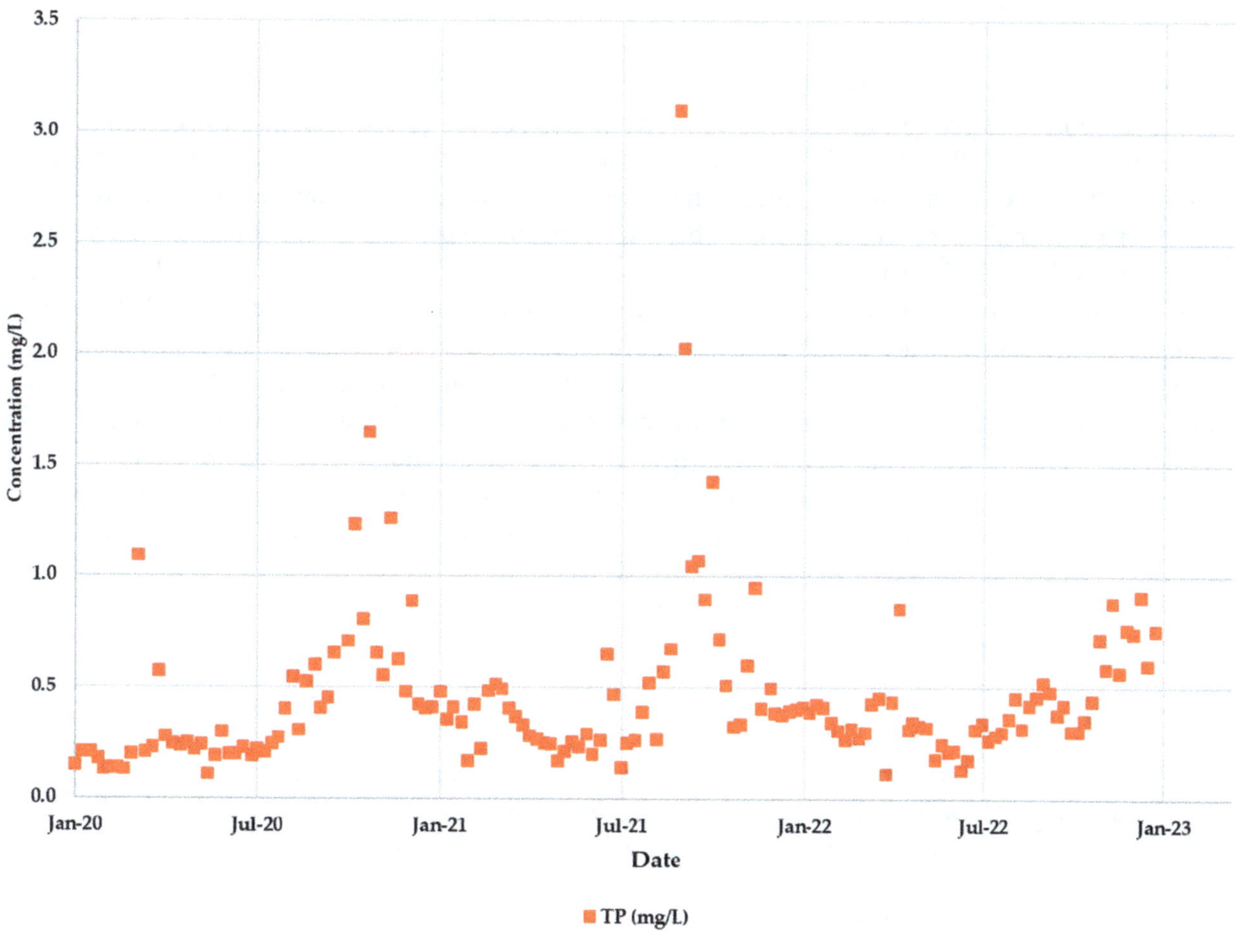


FIGURE 16.11 Geneva WWTF Effluent Total Phosphorus Concentrations From 2020 to 2022

## 2.4 Case Study 4—Goldsboro Water Reclamation Facility, Goldsboro, North Carolina

### 2.4.1 Facility History

The Goldsboro Water Reclamation Facility (WRF) is a 53.7 ML/d (14.2 mgd) advanced treatment facility. The City of Goldsboro is located in the Neuse River basin, which begins in the middle of the state and runs to the estuarine waters of the Pamlico Sound. This river basin was declared "nutrient sensitive" by the state of North Carolina in 1995.

The effluent criteria for the WRF are as follows:

- Five-day BOD, 4 and 8 mg/L (during summer and winter, respectively);
- Total suspended solids, 30 mg/L;
- Total phosphorus, 2.0 mg/L (quarterly average);
- Ammonia-nitrogen, 1.0 and 2.0 mg/L (summer and winter, respectively); and
- *Fecal Coliform*, 200 per 100 mL.

The WRF's effluent is also subject to a total nitrogen allocation in accordance with the Neuse River Nutrient Management Rule. The WRF's allocation amount of 90 828 kg/yr (199 822 lb/yr) correlates with an average effluent total nitrogen concentration of 4.6 mg/L and the facility's design flow.

### 2.4.2 Process Description

The WRF is fed influent wastewater by the Westbrook Road Pump Station. Pumped influent flows through an influent screen and two aerated grit chambers and then is split among the WRF's biological process trains. The WRF has four 8.3 ML (2.1 mil. gal.) secondary treatment trains, each with an independent A2O process, configured as an oxidation-ditch bioreactor circular loop surrounding a dedicated secondary clarifier. Figure 16.12 presents the bioreactor arrangement. The bioreactor includes unaerated zones to provide anaerobic and anoxic conditions and a main aeration channel. Fine-bubble diffusion is used for oxygen transfer in the aeration channels. Influent to the bioreactors can be directed to diffuser channels if operators have decided to change from the A2O configuration.

Mixed liquor from each bioreactor flows to the center secondary clarifier, with RAS returned to the bioreactor train. Clarified effluent flows to traveling-bridge-type effluent sand filters for polishing before disinfection using ultraviolet. Finally, the effluent passes over an aeration cascade before discharge to the Neuse River. The WRF is equipped with a flood-pumping system that can activate during flood conditions. Effluent from the WRF can also be reused at city-owned farmland and golf courses or can send up to 15.1 ML/d (4 mgd) to a constructed wetlands facility.

Waste activated sludge removed from the clarifiers is held in a storage tank before being dewatered by belt filter press and composted.

### 2.4.3 Operating Data

Figure 16.13 shows the monthly average effluent total nitrogen and total phosphorus concentrations from January 2020 to December 2022. Data records also indicate that the facility readily meets its effluent ammonia-nitrogen limit during the summer nutrient removal season.

### 2.4.4 Operating Experience

Operations staff have keyed in on the importance of checking the process RAS settings at least once or twice a year. During the initial process study and commissioning, it had been determined that the

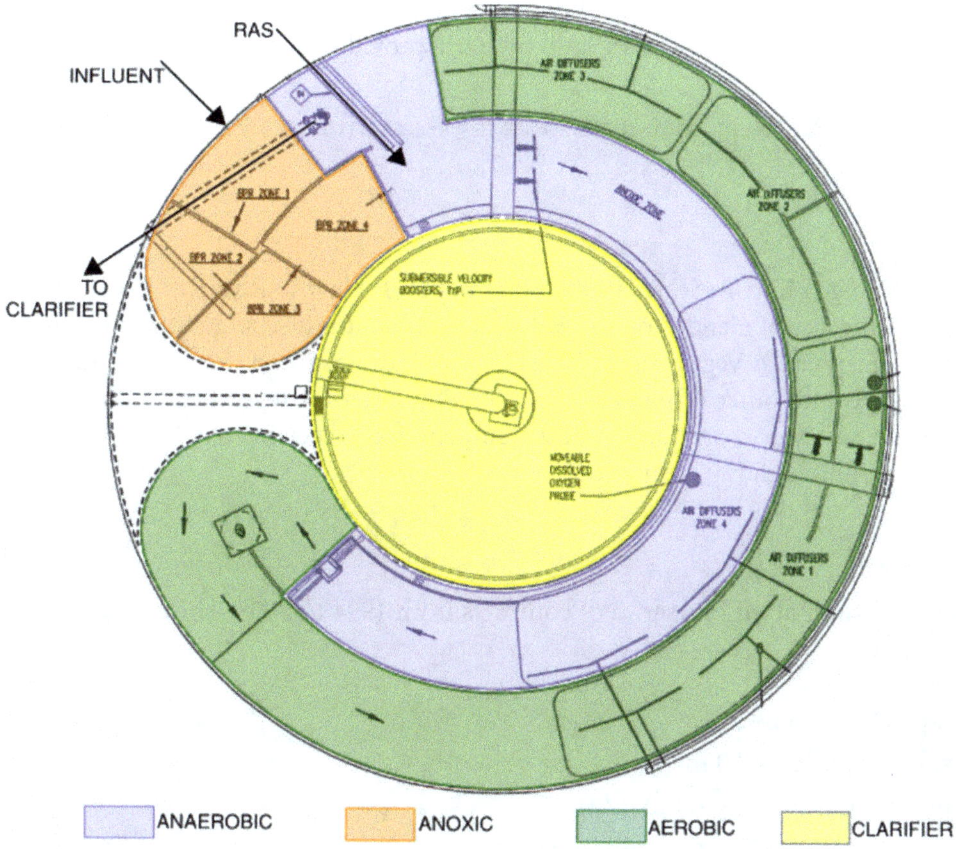


**FIGURE 16.12** Goldsboro WRF Bioreactor Layout

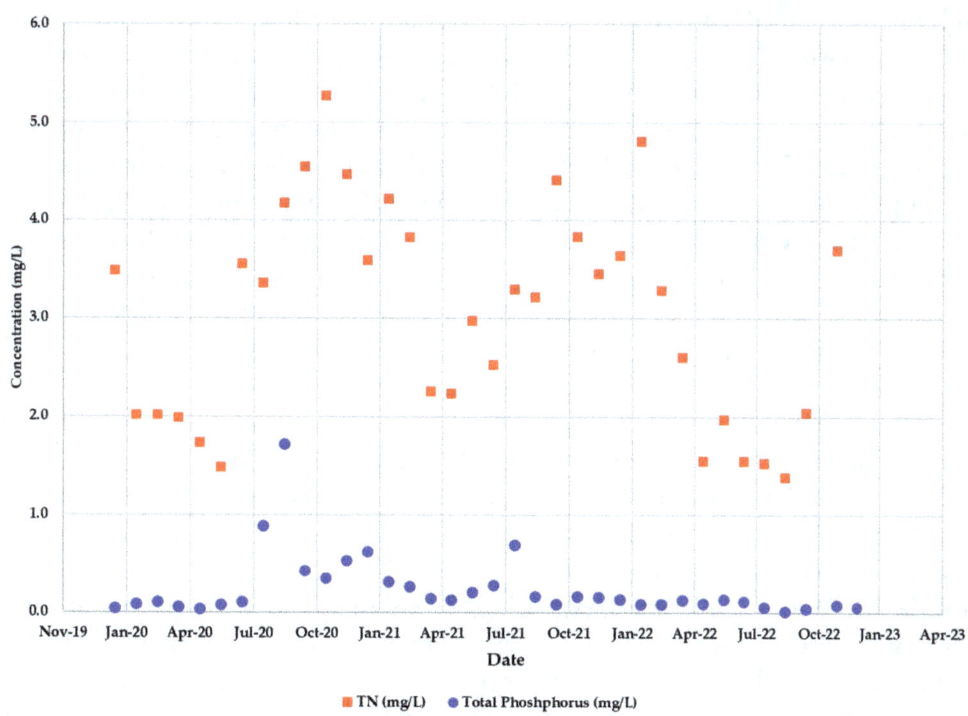


**FIGURE 16.13** Goldsboro WRF Effluent Total Nitrogen and Total Phosphorus Concentrations From 2020 to 2022

optimum RAS ratio was 125%, and the facility had operated with that setting for years. However, over time, operators found that a reduced RAS ratio of 100% provided better results.

## 2.5 Case Study 5—Kurt R. Segler Water Reclamation Facility, Henderson, Nevada

### 2.5.1 *Facility History*

The 120-ML/d (32-mgd) Kurt R. Segler Water Reclamation Facility (KRSWRF) is a state-of-the-art BNR facility that serves the city of Henderson, Nevada. Final effluent is discharged through an outfall to the environmentally sensitive Las Vegas Wash ("Wash"), which drains into Lake Mead. The lake is an important recreational lake and source of potable and irrigation water for several communities in Arizona, California, Nevada, and Mexico. A portion of the effluent is also reclaimed and used to irrigate local golf courses, a cemetery, and highway green areas.

### 2.5.2 *Process Description*

The KRSWRF was constructed in a series of contracts from 1994 to 2008. The effluent criteria for the facility are as follows:

- Five-day BOD, 30 mg/L;
- Total suspended solids, 30 mg/L;
- Total phosphorus, <19 kg/d (43 lb/d);
- Ammonia-nitrogen, <57 kg/d (126 lb/d); and
- Fecal coliforms, 200 per 100 mL.

Effluent total phosphorus limits are applied from March 1 to October 31, while effluent ammonia-nitrogen limits are applied from April 1 to September 30. The KRSWRF limits are part of a total limit set for the Wash, which receives effluent from North Las Vegas, Las Vegas, and the CCWRD. In 2001, there was an algae bloom in Lake Mead, and the dischargers volunteered to meet effluent total phosphorus and ammonia-nitrogen limits during each month. The Nevada Department of Environmental Protection accepted this offer and kept the seasonal limits in the NPDES permits.

Figure 16.14 presents a process flow diagram showing the key unit processes. The liquid treatment stream includes the headworks comprised of fine screening, grit removal, and influent pumping. Three flow equalization basins are used to shave off the peak flows and allow a steady flow to each complex. The flow is then divided into the east and west complexes. The facility does not have primary treatment. The East Complex is a single-train, 30-ML/d (8-mgd), completely mixed process, and the West Complex is a two-train, 90-ML/d (24-mgd), plug flow process. Each complex includes supplementary phosphorus removal (SPR) using alum and polymer in separate coagulation processes followed by effluent filtration. Supplementary phosphorus removal is needed to meet the low total phosphorus total maximum daily load limits of less than 18 kg/d (40 lb/d). Ultraviolet irradiation is used for effluent disinfection, and chlorine is available as a backup to ultraviolet. The WAS and chemical sludge are dewatered on belt filter presses (BFPs) and hauled to a landfill.

Currently, the KRSWRF is undergoing expansion (Phase 4 Expansion). Phase 4 will add 45 ML/d (12 mgd) of secondary treatment capacity to the East Complex. Activated sludge with anaerobic, anoxic, and aerobic zones will provide the same BNR phosphorus removal as current processes but with design enhancements for more reliable and lower nutrient levels in the effluent. The other major process change is the replacement of continuous backwash tertiary filters with mono-media backwash filters for the east side process, with extra capacity for the new secondary process as well.

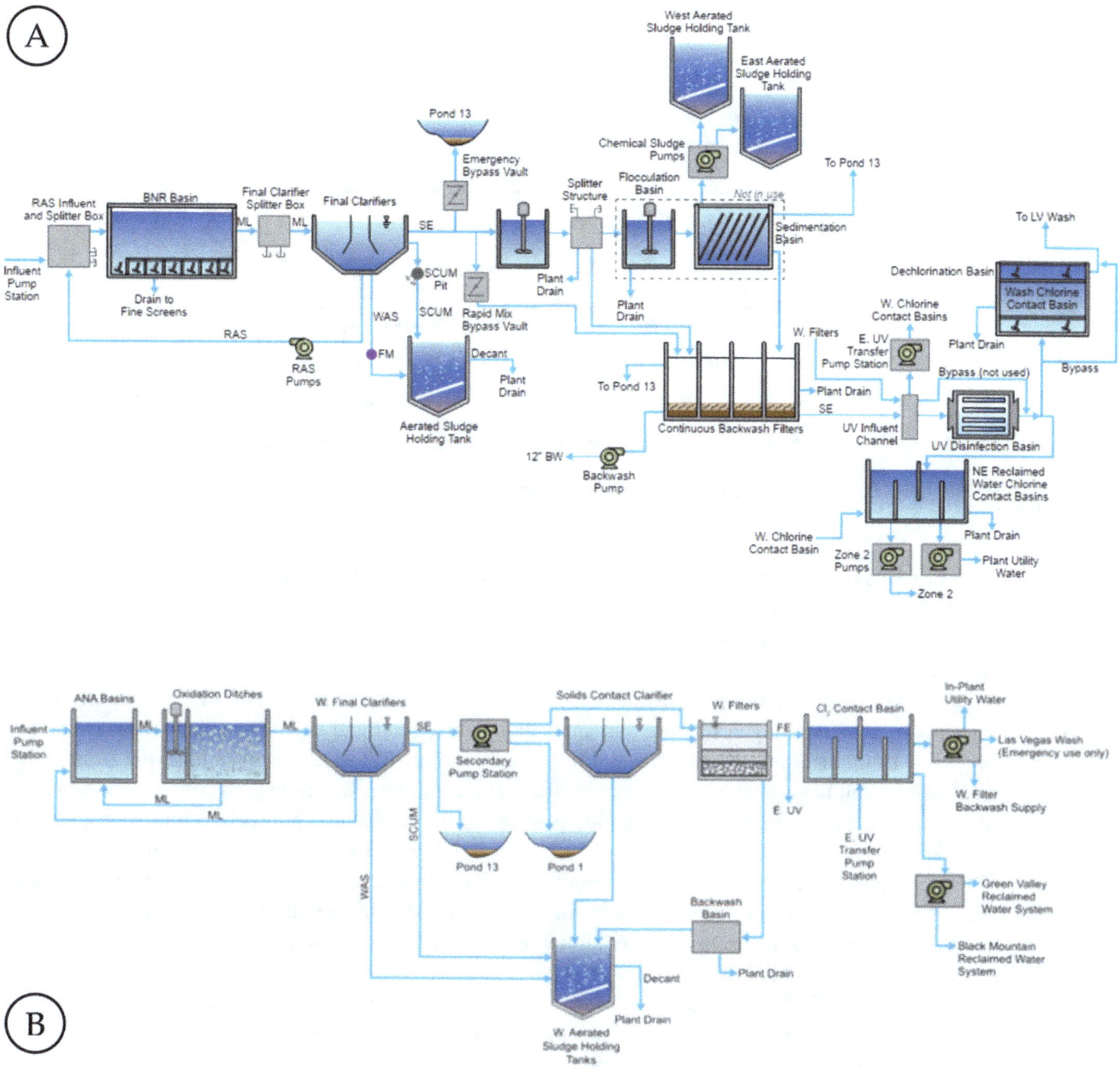


FIGURE 16.14 Kurt R. Segler Facility Process Flow Diagram. (A) East WRF liquid flow schematic; (B) West WRF liquid flow schematic (Courtesy of the City of Henderson)

The bioreactor design is based on the modified Johannesburg (JHB) process (Figure 16.15). Both complexes use pre-anoxic, anaerobic, anoxic, and aerobic zones for deoxygenation, phosphorus release, denitrification, and nitrification and phosphorus uptake, respectively. The East Complex has one pre-anoxic zone that receives RAS for deoxygenation. The anaerobic zone that receives the influent wastewater can be comprised of two or three cells, and the anoxic zone can be comprised of three or four cells. Nitrified mixed liquor is returned to the anoxic zone for denitrification. The aerobic zone is a completely mixed basin equipped with diffused air aeration and is followed by two secondary clarifiers. Return activated sludge is pumped back to the pre-anoxic zone. Alum and polymer are added to the secondary effluent in a two-cell, rapid-mix chamber. The effluent can flow to a three-cell flocculation

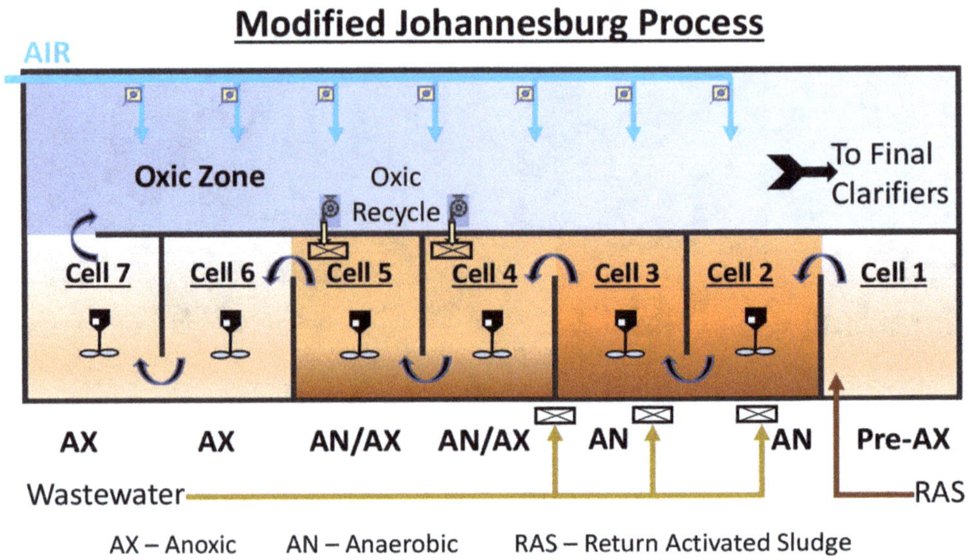


FIGURE 16.15 Kurt R. Segler Facility Bioreactor Layout (Courtesy of City of Henderson)

chamber followed by a plate settler and four continuously backwashed filters (CBF), or it can be directly filtered by CBFs.

The West Complex has two 45-ML/d (12-mgd) trains. Each train has one pre-anoxic zone that deoxygenates RAS. The anaerobic zone can be comprised of three or four cells, and the anoxic zone can be comprised of three or four cells. Nitrified mixed liquor is returned to the anoxic zone for denitrification. The aerobic zone is a plug flow process in an oxidation-ditch configuration. Oxygen is added to the aerobic zone using mechanical aerators and diffused-air aeration. Mixed liquor from the two aerobic zones is combined and flows to four secondary clarifiers. The combined RAS is divided between the two trains. Alum and polymer are added to the secondary effluent, and the water is pumped to four solids contact clarifiers followed by 16 mono-media rapid sand filters. Filter backwash is returned to the headworks. Filtered effluents from the east and west complexes are combined and flow through two ultraviolet disinfection channels with two banks of lamps in each channel. Following ultraviolet disinfection, the water can flow via gravity or be pumped to reclaimed water pumping stations for distribution. Water that is not reclaimed flows to the Wash.

The modified JHB process configuration has several unique design features, including multiple wastewater feed points to the pre-anoxic and anaerobic zones. The mixer in Anaerobic Zone 2 is operated intermittently to allow mixed liquor settlement and fermentation to occur in this cell (Barnard et al., 2010). The WAS and waste chemical sludge (WCS) are pumped to aerated sludge holding basins (ASHB). The WAS and WCS are held in separate ASHBs. The two sludges flow into a blend tank, and the blended sludge is pumped to two solids process buildings. Four BFPs in each building are used to dewater the sludge. Polymer is added to the sludge before the BFPs. The dewatered sludge drops into trailers and is hauled to a landfill for disposal.

### *2.5.3 Facility Commissioning*

The BNR process for the east and west complexes was commissioned in phases because of modifications related to the construction project. Activated sludge was transferred from ASHBs or from complex to complex to seed the phase being commissioned. Initially, half of the expected MLSS concentration was added to the activated sludge process. Wastewater was introduced at 50% of the train's capacity, and process monitoring tests were run each day. As the mixed liquor concentration increased and monitoring tests showed positive results, the wastewater flowrate was gradually increased. Typically, the wastewater flowrate was increased every 2 days until the wastewater flow being received by the KRSWRF was properly distributed between the two complexes. Enhanced biological phosphorus removal was not initially achieved.

Before commissioning, it was discovered that the influent VFA concentration ranged from 10 to 15 mg/L. During commissioning of the East Complex, the mixer in the second anaerobic cell was shut off and operated for 15 minutes each day. This adaptation produced enough VFAs to allow phosphorus to be released in the anaerobic zone and phosphorus uptake to occur in the aerobic zone. During the first commissioning exercise, the initial secondary effluent dissolved phosphorus concentration ranged between 0.15 and 0.3 mg/L. After both complexes were fully commissioned, the alum dosage was decreased to approximately 10% of the original value. Biochemical oxygen demand, TSS, and ammonia-nitrogen limits were met without any problems during commissioning; in addition, all effluent limits are currently being met.

### 2.5.4 Process Monitoring and Control

Key parameters that are monitored and used to control the BNR process include the following:

- Wastewater flow and characteristics (BOD, TSS, VFA, $NH_3$, TKN, total phosphorus);
- Pre-anoxic, anaerobic, anoxic, aerobic zone, and supernatant characteristics ($PO_4$, $NH_3$, and $NO_x$);
- Secondary effluent, SPR, and filter effluent characteristics ($PO_4$, $NH_3$, $NO_x$, and pH);
- Aerobic zone dissolved oxygen, MLSS, and settleability;
- Activated sludge SRT; and
- Chemical dose (alum and polymer).

### 2.5.5 Operating Data

Since the initial startup of the facility, effluent total phosphorus and ammonia-nitrogen concentrations have consistently been well below effluent limits. Figure 16.16 illustrates the current treatment efficiency of the KRSWRF and shows monthly average effluent total phosphorus concentrations for 2021, 2022, and 2023.

### 2.5.6 Operating Experience

Whereas the KRSWRF has consistently met or exceeded its effluent quality requirements since it was commissioned, several operating problems have emerged that relate to the EBPR performance. These include,

- Low influent VFA concentration: During the design phase of the BNR upgrade, the wastewater VFA concentration was tested and proved to be sufficient to support stable EBPR. The ratio of VFA to total phosphorus was greater than 5 to 1. Before commissioning, the ratio was found to be less than 3 to 1. As noted previously, the mixer in the second anaerobic zone was operated 15 minutes a day, and enough VFA was produced to allow the EBPR to stabilize (Barnard et al., 2010). The mixer was started and stopped manually. Eventually, timers were added to the mixer controls to allow operators to select the time of day and the duration of the intermittent mixer operation.
- Enhanced biological phosphorus removal optimization: The submersible mixers in the pre-anoxic, anaerobic, and anoxic zones (23 in all) began to fail. Working with the CCWRD, facility staff

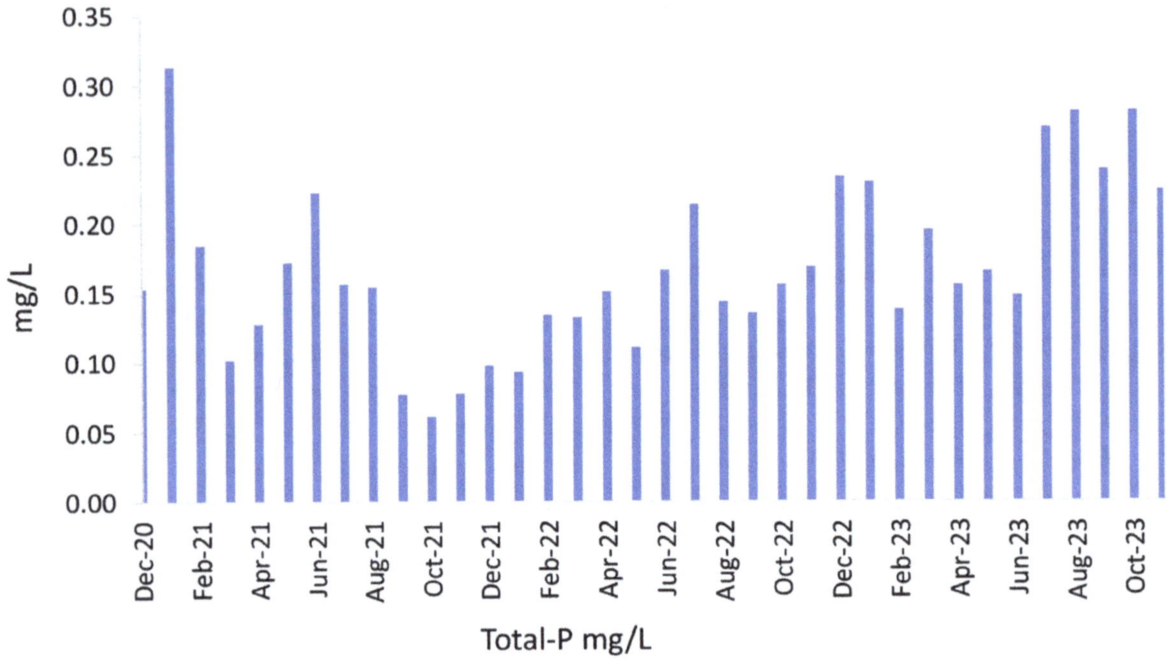


FIGURE 16.16 Kurt R. Segler Facility Effluent Total Phosphorus Concentrations From 2021 to 2023 (Courtesy of the city of Henderson)

were able to source a reliable mixer and found that the motor horsepower could be reduced from 9 to 4.5 kW (12 to 6 hp). The reduced power requirements of the new mixers resulted in considerable power savings for the facility. The secondary effluent orthophosphorus concentration is now consistently below 0.1 mg/L and, at times, averages 0.05 mg/L. The lower secondary effluent orthophosphorus concentrations have resulted in lower alum doses, which are, at times, less than 10 mg/L. Before making the change to the lower horsepower mixers, alum doses ranged from 25 to 35 mg/L. The reduced mixing power and alum dosage requirements resulted in operating cost savings of more than $175,000 in 2010.

- Heat-related EBPR problems: Henderson is in a desert climate. During the summer months, the ambient temperature can reach 46 °C (115°F) and higher. Starting in June and lasting through August or September, ambient temperatures typically range between 38 and 43 °C (100 and 110 °F). When the temperature exceeds 38 °C (100 °F), the mixed liquor temperature will climb to greater than 28 °C and, at times, be 30 °C. When the mixed liquor temperature exceeds 28 °C, the competition between the phosphorus-accumulating organism (PAO) and glycogen-accumulating organism (GAO) populations is greatly increased. During the first year of operation, the GAOs outcompeted the PAOs, the effluent dissolved phosphorus concentration hovered around 1 mg/L, and the MLSS concentration was approximately 2000 mg/L for about 6 weeks. The next year, the MLSS concentration was lowered to 1800 mg/L, and the problem lasted about 3 weeks. The following year, the MLSS was lowered to 1500 mg/L, and the problem lasted about 1 week. This same phenomenon occurred again in 2011. When operating at an MLSS concentration of 1500 mg/L, the bioreactor SRT is around 3.5 days. To date, operating in this range has not overloaded solids processing, and the operating values have been on target. The primary issue at the lower MLSS is over-wasting from the bioreactors. The facility is manned for 10 hours a day, 7 days a week. If the WAS control system fails, there could be a problem with too much sludge being wasted. When facility staff see an increase in the effluent orthophosphorus concentration from around the 0.05- to 0.1-mg/L range to the 0.2- to 0.3-mg/L range in a 4-day period, it becomes apparent that GAOs are taking over. During these times, the alum dosage is increased so that the facility meets its effluent total phosphorus limit.

## 2.6 Case Study 6—Douglas L. Smith Middle Basin Wastewater Facility, Overland Park, Kansas

### 2.6.1 Facility History

The Middle Basin facility is a 55-ML/d (14.5-mgd) facility that is owned and operated by Johnson County Wastewater (JCW) and was originally built in the early 1980s and then later expanded in phases. Over the past 40 years, the facility has been modified a total of 34 times with upgrades ranging from minor mechanical upgrades to new treatment processes and expansions. The most extensive recent modification to the treatment processes was placed into operation in 2008. The primary objective of this expansion/upgrade project was to construct the fourth BNR train in response to regulatory requirements to address wet weather discharges and ammonia limits. However, Johnson County Wastewater also proactively modified the activated sludge process to convert to a modified Johannesburg layout despite there being no regulatory limits on nitrogen and phosphorus at the time. Given the voluntary nature of the BNR conversion, efforts were made to convert to the modified Johannesburg process using textbook and design rules-of-thumb for denitrification and biological phosphorus removal from the early 2000s.

Since that time, some nutrient goals have been placed into the facility's permit, but they remain goals, not enforceable limits, at this time. The monthly average effluent requirements as of 2023 for the Middle Basin facility are as follows:

- Carbonaceous biological oxygen demand (CBOD), 10 to 15 mg/L depending on the time of year;
- Total suspended solids (TSS), 30 mg/L;
- Ammonia-N, 0.8 to 3.1 mg/L depending on the time of year; and
- *E. coli.*, 262 per 100 mL.

The following nutrient requirements are also reported monthly as annual averages:

- Limit for combined nitrate-N and nitrite-N of 10 mg/L;
- Goal for total nitrogen, 10 mg/L; and
- Goal for total phosphorus, 1.0 mg/L;

Johnson County Wastewater has operated the Middle Basin facility to maintain compliance with the limits, and it also strives to operate the facility to achieve the lowest possible total nitrogen and total phosphorus effluent concentrations with the tools available at the facility at reasonable cost.

### 2.6.2 Process Description

Currently, the liquid treatment process at Middle Basin includes coarse mechanical screens, no grit removal, followed by four modified Johannesburg BNR trains, each with its own dedicated primary and final clarifiers. Final treatment includes an ultraviolet disinfection system and a cascade aerator before discharge to Indian Creek in Overland Park, Kansas. Flowrates above the capacity of the BNR trains are diverted to lagoons for equalization. Solids are managed by separate thickening of primary sludge via a gravity thickener and WAS by mechanical thickening. The primary sludge is also fermented in a fermenter tank before the gravity thickener, and the VFA-rich supernatant is pumped back to the four BNR trains. Sludge thickening is followed by staged anaerobic digestion with co-digestion of fats, oil, and grease, and then the sludge is dewatered using belt filter presses. Figure 16.17 presents a process flow diagram of the liquid treatment facilities.

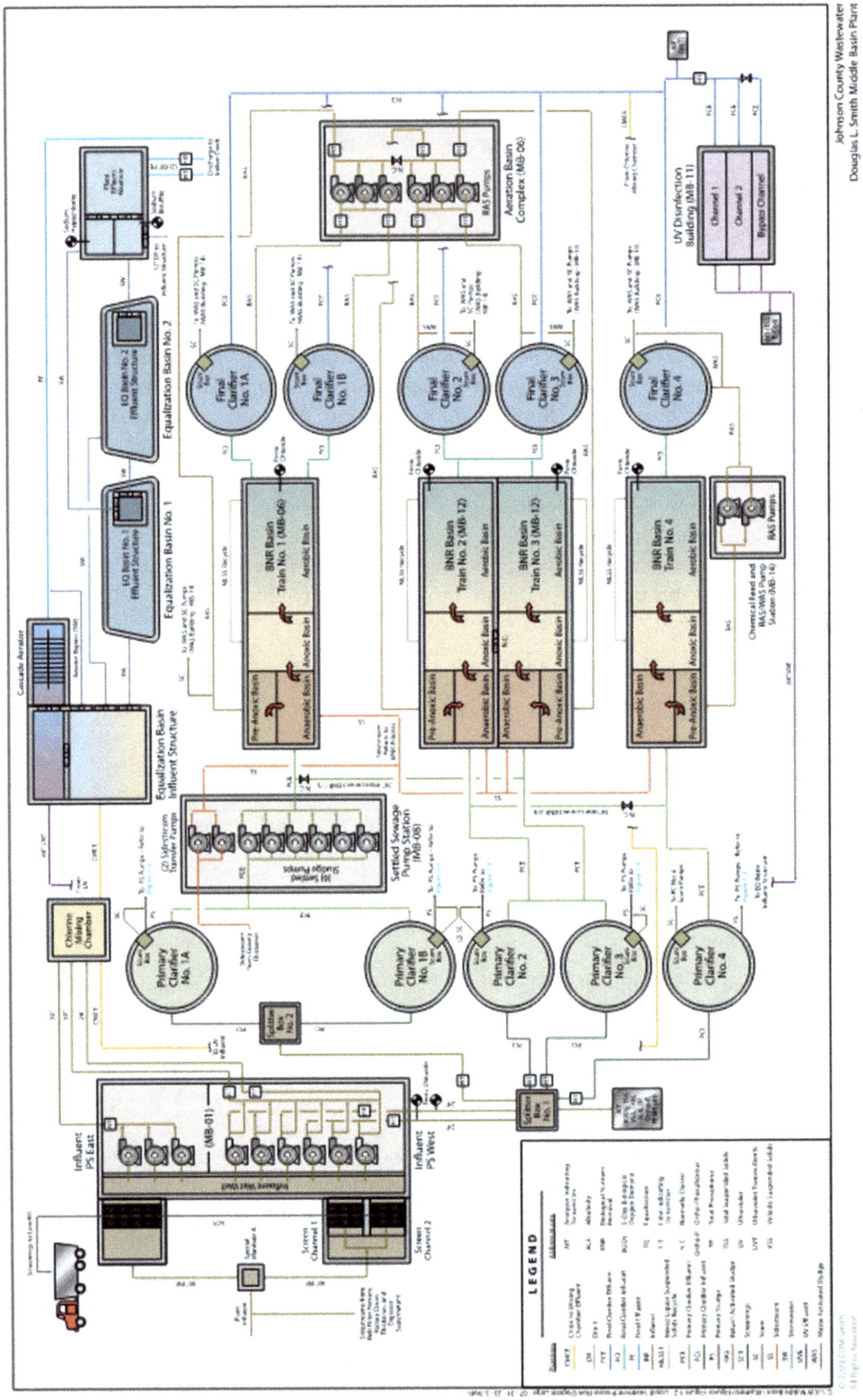


**FIGURE 16.17** Overview of Middle Basin Liquid Treatment (Courtesy of Johnson County Wastewater)

### 2.6.3 Process Monitoring and Control

The facility uses dissolved oxygen control for biological process control. One dissolved oxygen probe is installed for each diffuser aerobic zone, thus three dissolved oxygen probes per train in all four trains, for a total of 12 probes. This is one of the key operational control tools used for the process. The operations team performs regular colorimetric tests for effluent orthophosphorus and nitrate for a quick check on EBPR and nitrate-removal process performance, while the 24-hour composite samples are analyzed by JCW's off-site laboratory for regulatory reporting requirements.

Key operational parameters that are monitored and used to control the process include the following:

- Influent plus recycle characteristics (total phosphorus, TKN, $BOD_5$, and TSS) (twice per week);
- Primary sludge gravity thickener supernatant characteristics ($BOD_5$, TSS, volatile suspended solids (VSS), TKN, total phosphorus, and a variety of VFAs);
- Anaerobic, anoxic, and aerobic zone mixed liquor supernatant characteristics ($PO_4$, $NO_x$, etc.) sampled twice per week;
- Aerobic zone dissolved oxygen and MLSS concentrations using online probes;
- Aerobic zone 30-minute settleability and SVI; and
- Anoxic plus aerobic SRT.

Of note is that the Middle Basin facility used to monitor a variety of primary effluent characteristics (including VFAs and TSS) to each of the four BNR trains, but they ultimately eliminated most of the sampling regime except for $PO_4$-P to avoid "over-operating" the flow split to the four trains.

A portion of RAS is wasted using automated controls with optical TSS probes in the MLSS and in the RAS to maintain a target bioreactor SRT. Wasting rates are calculated at supervisory control and data acquisition in real time using a 3-hour rolling average of the TSS readings from the optical probes. The 24-hour composite samples of mixed liquor and RAS are analyzed for TSS concentration two times a week to verify that the process is being maintained at the target SRT. The facility targets an anoxic plus aerobic SRT at a constant value year-round of approximately 12 to 14 days.

### 2.6.4 Operating Data

Figure 16.18 presents the final effluent total phosphorus concentrations from the Middle Basin facility for the past 20 years, and Figure 16.19 presents the final effluent NOx-N concentrations for the same time period. The reduction in phosphorus and NOx-N was evident starting in 2008 when the voluntary upgrade to the modified Johannesburg process, including a primary sludge fermenter, was placed into service. The majority of the reduction in phosphorus over the next 5 years was achieved primarily using EBPR, while starting in the mid-2010s JCW began dosing ferric chloride to sludge dewatering to mitigate struvite buildup, and this also somewhat helped reduce the occasional spikes of effluent phosphorus.

### 2.6.5 Operating Experience

As noted above, the Middle Basin facility was upgraded to a modified Johannesburg process voluntarily in 2008. Since that time, the facility's operations have not had to be optimized for phosphorus removal due to lack of regulatory pressure to do so. A recent facility plan was completed in 2023 to identify improvements necessary to help in BNR operations. Following is a summary of operational challenges and experience at the facility:

- Primary sludge fermentation: A lack of headworks fine screening results in rag build up and mixer failures. Mixer failure results in stratification and crusty foam build up on top of the fermenter.

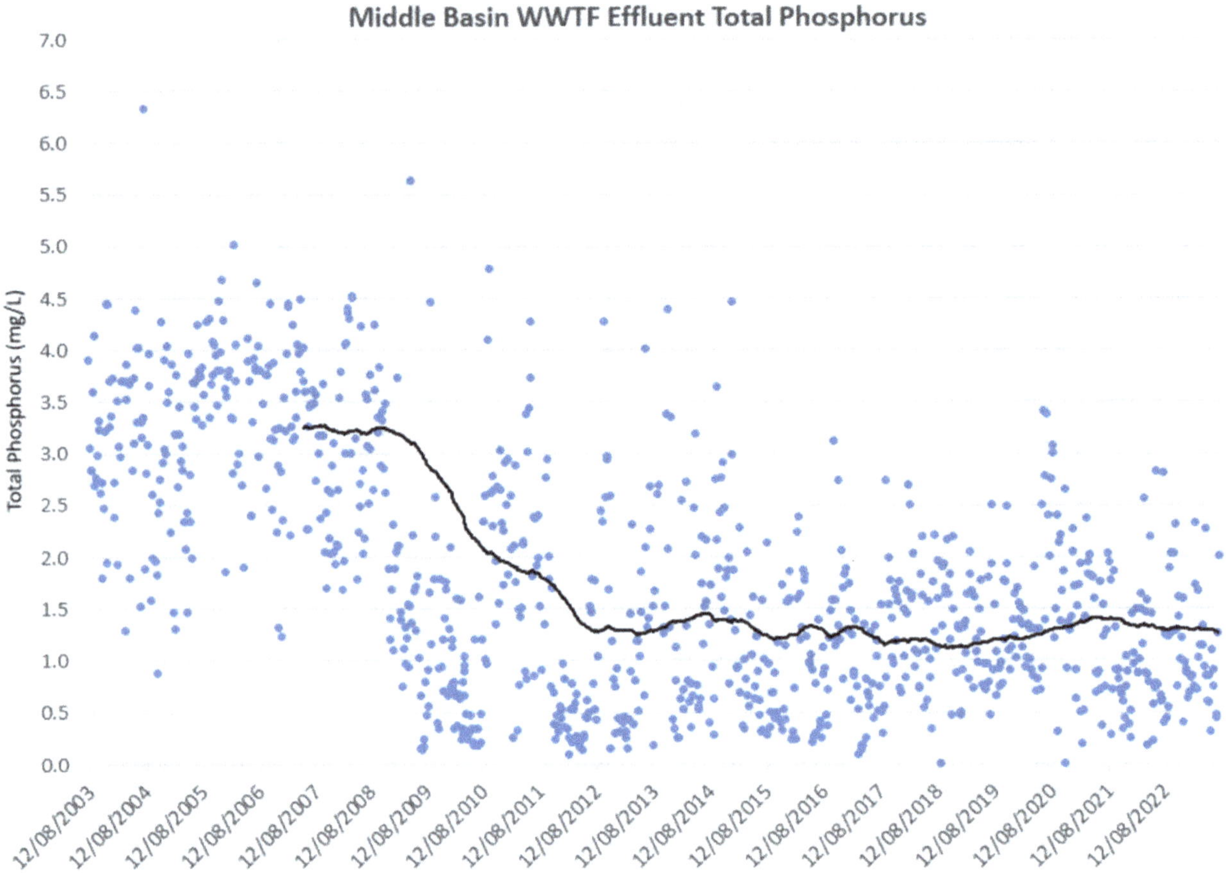


FIGURE 16.18 Middle Basin Effluent Total Phosphorus Concentrations (Courtesy of Johnson County Water)

Thick foam impedes gravity return of VFA-laden sludge and creates odor control headspace isolation. Whereas primary sludge fermentation has been shown to be beneficial to EBPR, the facility continues to experience a carbon deficiency.

- Aeration: During construction, it was learned that the aeration system was substantially oversized. Modifications to the aeration control systems were implemented, but the work was split between two control parties, and this led to an inflexible aeration control system that is difficult to optimize. Given the facility's tight ammonia permit requirements, the aeration system is currently operated with typical dissolved oxygen set points of 2.5 mg/L and 1.0 mg/L in the first and last zones, respectively. Running at low dissolved oxygen has shown promise at lowering effluent nitrogen and phosphorus, but this did result in some negative side effects, including ammonia bleed-through and foam build up.
- RAS rates: For decades, the facility has been challenged with final clarifier blankets, especially during wet weather. With the benefit of online optical TSS probes in the MLSS and RAS, and flow-pacing controls scheme installed during the expansion/upgrade project, JCW has determined the facility runs optimally with the RAS operated at 80% of the forward flow. This is higher than typical design standards and likely results in shortened HRT in the anaerobic and anoxic zones.
- Solids processing recycle: Unfortunately, the influent sampling at this facility includes recycle from the thickening and dewatering processes. Thus, the BOD, TSS, TKN, and total phosphorus are all elevated relative to "typical influent." Efforts have been made to improve the quality of the recycles

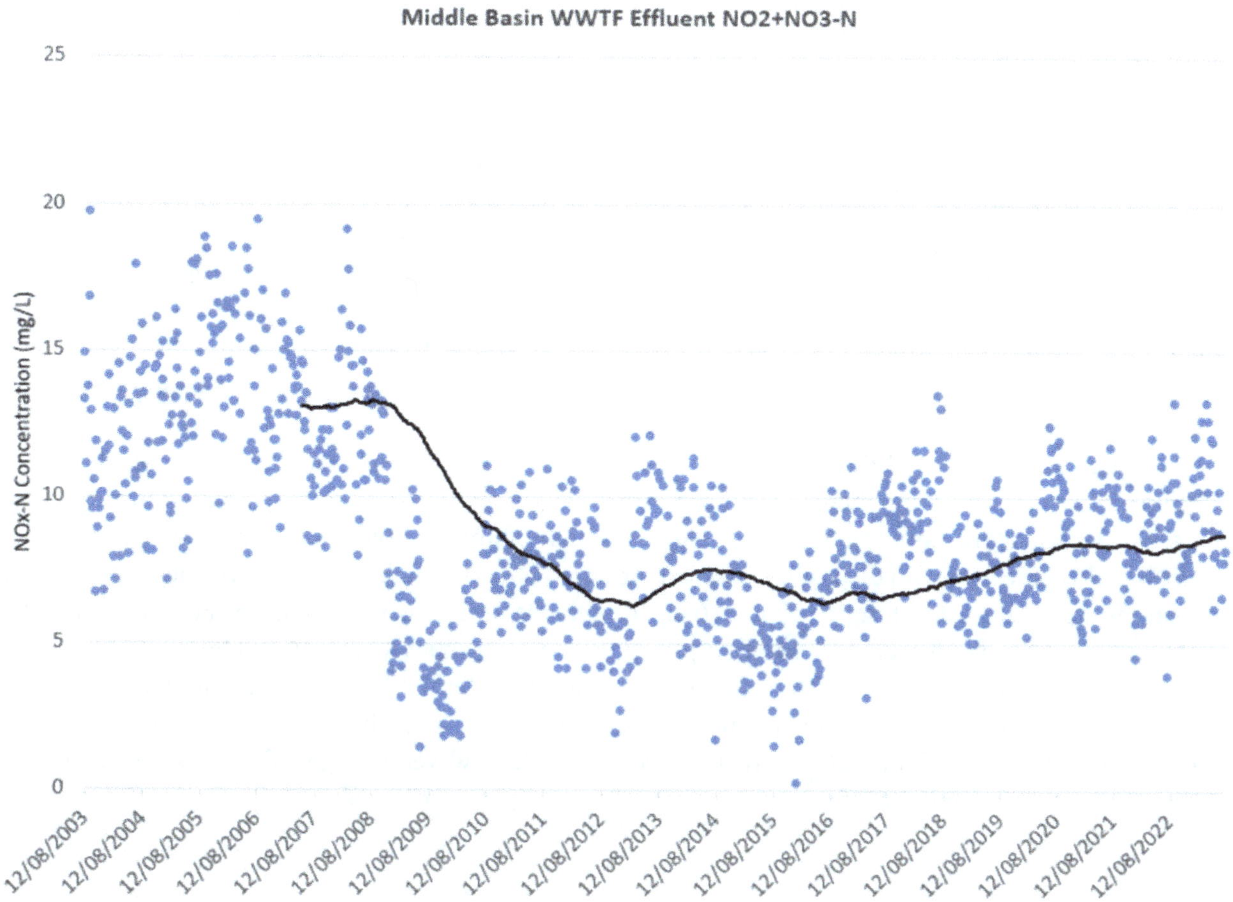


FIGURE 16.19 Middle Basin Effluent Total Nitrite Plus Nitrate Concentrations (Courtesy of Johnson County Water)

from sludge thickening with the use of polymer and to improve the quality of the recycles from sludge dewatering with the use of ferric chloride. Nominal ferric chloride addition is used on the belt filter press feed to lessen struvite build up. When ferric chloride use to the dewatering feed has not been functioning, the return of $PO_4$-P from anaerobic digestion has been noticeable at the BNR.

- WAS: Due to historical variability with operator-determined wasting and resultant ammonia challenges, JCW automated the wasting of the activated sludge (WAS) using the optical TSS probes to maintain a constant SRT year-round. Whereas this automation has proven to provide award-winning performance for conventional permitted pollutants, there has been recent discussion of possibly too long SRT in the summer, when EBPR appears less effective at some JCW facilities.
- Mixed liquor recycle pump (MLRP) rates: Initial automated controls were for flow pacing, with the typical operator-adjustable range from 1Q to 4Q. As expected, this is inversely related during dilute wet weather. Experimentation has shown a fixed MLRP speed is better.

## 3.0 SUMMARY

Since the introduction of full-scale EBPR processes to North America in the early 1980s, researchers, designers, and operators have worked hand-in-hand to optimize the performance of these processes.

The important role of VFAs in the EBPR mechanism cannot be overstated, and operators have endeavored to manage the available readily biodegradable carbon to stabilize system phosphorus removal. To this end, carbon management for optimum biological phosphorus and nitrogen removal in single sludge systems has become important because the PAOs and denitrifying heterotrophs compete for the same carbon sources. Furthermore, EBPR bioreactors often have the flexibility to vary the relative sizes of the anaerobic, anoxic, and aerobic zones so that phosphorus and nitrogen removal can be optimized under changing operating conditions. Use of primary sludge fermentation remains popular in temperate climates, in which VFAs naturally present in influent wastewater are insufficient to ensure reliable EBPR in the facility. Treatment of sludge handling return streams for phosphorus removal through intentional struvite formation and harvesting has served to reduce the recycled phosphorus loading to EBPR facilities, thus improving their performance and potentially generating an operating revenue stream through the production of a slow-release fertilizer product.

Facilities designed for EBPR now reliably achieve effluent total phosphorus concentrations below 0.1 mg/L. However, achieving these low concentrations is somewhat dependent on tertiary filtration, metal salt addition, or a combination of the two. Although membrane bioreactor processes designed for EBPR have become increasingly popular to meet stringent effluent quality parameters, these systems still rely on in-facility alum addition to achieve stringent effluent total phosphorus limits. Full-scale operating experience has shown that incorporating phosphorus removal has facility-wide effects. For example, implementing EBPR increases both the recycle loads and the potential for struvite formation in the sludge handling system. Inclusion of chemical phosphorus removal can increase sludge production and affect ultraviolet disinfection, particularly if iron salt is used. However, thoughtful facility design and operation can minimize these effects.

## 4.0 REFERENCES

Barnard, J. L., Houweling, D., Analla, H., & Steichen, M. (2010, October 2–6). Saving phosphorus removal at the Henderson NV plant. *Proceedings of the 83rd Annual Water Environment Federation Technical Exhibition and Conference* [CD-ROM], New Orleans, LA, United States. Water Environment Federation.

Dose, A., Grooms, A., Reusser, S., Britton, A., & Prasad, R. (2015, September 26–30). Operating experience with Ostara struvite harvesting process. *Proceedings of the 88th Annual Water Environment Federation Technical Exhibition and Conference* [CD-ROM], Chicago, IL, United States. Water Environment Federation.

Water Research Foundation. (2023). *Practical considerations for the incorporation of biomass fermentation into enhanced biological phosphorus removal* (Project No. 4975).

# Appendix A

# Optimization and Troubleshooting Guides



Nutrient removal processes require continuous monitoring and adjustment to maintain optimal treatment efficiency. Unusual or upset conditions can occur and are identified when indicators show changes from normal operating conditions. The upset condition may lead to effluent quality that does not meet permit limits for one or more of the following parameters: 5-day biochemical oxygen demand ($BOD_5$) (or 5-day carbonaceous biochemical oxygen demand [$CBOD_5$]), total suspended solids (TSS), ammonia-nitrogen ($NH_3$-N), total nitrogen, total inorganic nitrogen, and total phosphorus. The first step in resolving an upset condition is to find the problem causing the upset, whether it is loading, aeration, biomass inventory, clarifier operation, internal recycle, pH and alkalinity, toxicity, or a combination of factors. Similarly, if the process is performing at less than desired efficiency or reliability, the same procedures can be used to optimize the process.

When the problem is identified, the next step is to go through an appropriate troubleshooting exercise to resolve the issue. The optimization and troubleshooting guides included in this appendix are intended to assist in identifying the problem and in determining the appropriate response.

## 1.0 OPTIMIZATION AND TROUBLESHOOTING GUIDE FORMAT

Tables A.1 through A.11 each address a different operational aspect of nutrient removal processes, as follows:

Table A.1—nitrification,

Table A.2—denitrification,

Table A.3—enhanced biological phosphorus removal (EBPR),

Table A.4—chemical phosphorus removal,

Table A.5—aeration and mixing systems (diffused aeration),

Table A.6—aeration and mixing systems (mechanical aeration),

Table A.7—biomass inventory,
Table A.8—clarifier operation,
Table A.9—internal recycle,
Table A.10—pH and alkalinity, and
Table A.11—toxicity.

Tables A.1 through A.11 are arranged in columns with the following headers:

- "Indicator or observations": This shows the condition(s) observed or reported by the operator.
- "Probable cause": This shows possible causes of the upset or inefficiency.
- "Check or monitor": The operator should check the system for specific information to document the current condition and aid in diagnosing the problem; after the solution has been implemented, the operator should perform the listed monitoring step to verify process performance improvement until the process is optimized or recovered from upset.
- "References": The numbers listed in this column show where additional information is located in either another table or one of the chapters of these standards.

## 2.0 OPTIMIZATION AND TROUBLESHOOTING GUIDES

Tables A.1 through A.11 present the optimization and troubleshooting guides. Reading across the page of each table, follow the numbers and letters. For example, solution 1a in the "Solutions" column refers to the corresponding item 1a in the "Probable cause" and "Check or monitor" columns. Information from a decision-tree-type troubleshooting guide specific to EBPR, developed by Benisch et al. (2004), was incorporated into these tables.

## 3.0 REFERENCES

Benisch, M., Baur, R., & Neethling, J. B. (2004, October 2–6). Decision tree for troubleshooting biological phosphorus removal. *Proceedings of the 77th Annual Water Environment Federation Technical Exhibition and Conference* [CD-ROM], New Orleans, LA, United States. Water Environment Federation.

## 4.0 SUGGESTED READINGS

Bernardelli, A., Marsili-Libelli, S., Manzini, A., Stancari, S., Tardini, G., Montanari, D., Anceschi, G., Gelli, P., & Venier, S. (2020). Real-time model predictive control of a wastewater treatment plant based on machine learning. *Water Science and Technology, 81*(11), 2391.

Jenkins, D., Richard, M., & Daigger, G. (2003). *Manual on the causes and control of activated sludge bulking, foaming and other solids separation problems* (3rd ed.). CRC Press.

U.S. Environmental Protection Agency. (2015). *Case studies on implementing low-cost modifications to improve nutrient reduction at wastewater treatment plants* (EPA-841/R-15-004).

Wahlberg, M. (2018). *Activated sludge wastewater treatment: Process control & optimization for the operations professional.* Destech Publications.

Water Environment Federation & Association of Boards of Certification. (2018). *Wastewater treatment fundamentals I, liquid treatment.* Water Environment Federation.

Water Environment Federation & Association of Boards of Certification. (2022). *Wastewater treatment fundamentals III, advanced treatment.* Water Environment Federation.

Water Research Foundation. (2023). *Guidelines for optimizing nutrient removal plant performance, project no. 4973.*

TABLE A.1 Process Upset—Nitrification

| Indicator or observations | Probable cause | Check or monitor | Solutions | References |
|---|---|---|---|---|
| (1) Higher effluent $NH_3$-N than normal, accompanied by raw wastewater odor and/or dark color in the aerobic zone. | (1a) Excessive loading of BOD and/or total Kjeldahl nitrogen (TKN). | (1a1) Check $NH_3$-N at the end of the process before clarification for nitrification performance.<br>(1a2) Check BOD and TKN concentrations in influent to the facility. | (1a1,2) Check for and discourage discharges to the collection system that are causing unusually high strength BOD and/or TKN in the influent wastewater.<br>(1a3) See if sidestreams from solids handling are causing periodically high loads and adjust operations to even out BOD and TKN loadings to the process. If the facility has primary clarifiers, check removal efficiency, and consider adjustments. | Chapter 3<br>Chapter 4 |
| | | (1a3) Check BOD and TKN concentrations in the influent to the biological nutrient removal (BNR) process. | (1a3) If the water resource recovery facility has influent equalization, adjust release if possible to improve loading profile. | |
| | | (1a4) If activated sludge, check dissolved oxygen profile across the aeration zone; if rotating biological contactor, check dissolved oxygen in the first or second compartment; if trickling filter, check dissolved oxygen in the trickling filter effluent. If moving bed biofilm reactor (MBBR) or integrated fixed-film activated sludge (IFAS), check bulk liquid dissolved oxygen concentration. Dissolved oxygen checks should consider diurnal variation as daily, or once per shift dissolved oxygen readings may miss low dissolved oxygen periods. | (1a4) If activated sludge with diffused aeration and mixing, see Table A.5; if activated sludge with mechanical aeration and mixing, see Table A.6. If trickling filter, try to increase recirculation/aeration as a short-term solution; if MBBR or IFAS, increase the aeration rate to increase dissolved oxygen concentration, if possible. Also, see 1a2. | |
| | (1b) Effective capacity of process has been reduced. | (1b1) Determine the ratio of volatile suspended solids to TSS of the mixed liquor. | (1b1) If there is a lower ratio of volatile suspended solids to TSS than is historically seen, check the performance of upstream grit removal and chemical addition systems. Operate based on mixed liquor volatile suspended solids (i.e., volatile suspended solids) on solids retention time (SRT) until the ratio of volatile suspended solids to TSS returns to normal. | Chapter 3<br>Chapter 4 |
| | | (1b2) Observe mixing and aeration patterns. | (1b2) If activated sludge with diffused aeration and mixing, see Table A.2. If activated sludge with mechanical aeration and mixing, see Table A.3. | |
| | | (1b3) If a biofilm system, look for excessive growth or debris that hinders aeration or airflow and causes short-circuiting. | (1b3) Increase fluid velocity to decrease biofilm thickness by flooding, increase recirculation rate, flush with hose, and increase aeration rate. | |

| | | | | |
|---|---|---|---|---|
| (2) Higher effluent $NH_3$-N than normal, accompanied by appearance of uneven loading (odors, mixed liquor suspended solids [MLSS] color, and dissolved oxygen); MLSS concentrations or dissolved oxygen different in different trains. | (2a) Unequal flow distribution. | (2a) Check flow of process influent to each train. Check $NH_3$-N and $NO_2 + NO_3$-N at the end of each process train before clarification. | (2a) Adjust flow-splitting devices to provide equal flow to same-sized trains or to provide equal loadings to different-sized trains. | |
| | (2b) Unequal nitrate-nitrogen (RAS) distribution. | (2b) If activated sludge and RAS enters separately, check RAS flow to each train. | (2b) Adjust flow-splitting devices on RAS to provide equal flow to same-sized trains or to provide equal loadings to different-sized trains. | |
| | (2c) Unequal MLSS distribution. | (2c) If process influent and RAS mix and then are split, check for good mixing, and check flow to each train. | (2c) Increase mixing if poor mixing is observed before flow distribution. Adjust MLSS flow-splitting devices to provide equal flow to same-sized trains or to provide equal loadings to different-sized trains. | |
| (3) Higher effluent $NH_3$-N than normal, accompanied by dark diluted color of MLSS and sometimes white, sudsy foam in aerobic zone. | (3) MLSS (and SRT) is lower than desired because of excessive wasting. | (3) Check MLSS and dissolved oxygen, and check $NH_3$-N at end of process before clarification. | (3) If MLSS is low, dissolved oxygen in aeration zones is high, and $NH_3$-N is high, decrease wasting to allow MLSS to increase for a SRT adequate for nitrification to occur. See Chapter 3 and Table A.7. | Chapter 3 |
| (4) Higher effluent $NH_3$-N than normal, accompanied by low influent pH or low alkalinity. | (4a) Low alkalinity in potable water combined with decomposition of wastewater in collection system during warm weather. | (4a) Check influent pH, alkalinity, and TKN. | (4a1) Decrease SRT by reducing MLSS through increased wasting if nitrification is occurring adequately or if nitrification is not desired.<br>(4a2) Add alkalinity through chemical addition. See Chapters 3 and 7. | Chapter 3<br>Chapter 7 |
| | (4b) Acidic discharge(s) to collection system. | (4b) Check influent pH, alkalinity, and other parameters that could identify the source of acidic discharge(s). | (4b) Identify and eliminate source of acidic discharge(s). For short-term solution, see (4a2) above. | |
| (5) Higher effluent $NH_3$-N concentration, accompanied by high dissolved oxygen. | (5a) Inhibition of nitrification. | (5a1) Check pH profile through process, air flow volume to diffused aeration, dissolved oxygen in aeration zones, and toxicity.<br>(5a2) Check for changes in oxygen utilization rate.<br>(5a3) Conduct nitrification inhibition laboratory testing. | (5a1) Check SRT and dissolved oxygen in process; adjust higher, if possible. Refer to Tables A.10 and A.11.<br>(5a2.3) If nitrification is not occurring/recovering, reseed with active biomass once inhibitory substance has passed through. | |

TABLE A.2 Process Upset—Denitrification

| Indicator or observations | Probable cause | Check or monitor | Solutions | References |
|---|---|---|---|---|
| (1) Higher effluent $NO_2 + NO_3$ concentration than normal. | (1a) Excessive loading of TKN. | (1a1) Check $NO_2 + NO_3$ at end of aeration zone for nitrification performance. | (1a1, 2) Check for and discourage discharges to collection system that are causing unusually high TKN in the influent wastewater. | |
| | | (1a2) Check TKN load in influent to the facility. | | |
| | | (1a3) Check TKN load in influent to the BNR process. | (1a3) See if sidestreams from solids handling are causing periodically high loads and adjust operations to even out TKN loadings to process. | |
| | (1b) Effective capacity of anoxic zone has been reduced. | (1b1) Check for grit deposits at bottom of bioreactors | (1b1) If there is a lower ratio of volatile suspended solids to TSS than historically seen, check the performance of upstream grit removal and chemical addition systems. Operate based on mixed liquor volatile suspended solids (i.e., volatile suspended solids) on SRT until the ratio of volatile suspended solids to TSS returns to normal. | Chapter 7 |
| | | (1b2) Check for MLSS recirculation returning excessive dissolved oxygen. | (1b2) Reduce dissolved oxygen at MLSS recirculation pump intake. | |
| | | (1b3) Observe mixing patterns. | (1b3) If available, adjust mixing intensity for ensuring complete mixing. Install and adjust baffling to improve mixing. Refer to Chapter 6. | |

| | | | | |
|---|---|---|---|---|
| | (1c) The ratio of BOD to TKN has changed. | (1c1) Check BOD and TKN concentrations in influent to BNR process. | (1c1) Look at facility operations to see if sidestreams from solids handling are causing periodically high loads, and adjust operations to reduce or even out BOD and TKN loadings to process. See Chapter 12. | Chapter 5<br>Chapter 7<br>Chapter 12 |
| | | (1c2) Check BOD and TKN concentrations in influent to facility. | (1c2) If BOD to TKN ratio is low, there may be insufficient BOD for denitrification in anoxic zones. | |
| | | (1c2a) High TKN in influent. | (1c2a) Check for and discourage discharges to collection system that are causing unusually high TKN. | |
| | | (1c2b) Low BOD in influent. | (1c2b) Add carbon source to increase the ratio of BOD to TKN and improve denitrification. | |
| | | (1c3) Check for adequate supplemental carbon feed to anoxic zones. | | |
| | (1d) Inhibition of denitrification | (1d) Check $NO_2 + NO_3$, dissolved oxygen, and/or oxidation-reduction potential profiles through the process. | (1d) Optimize the anoxic conditions in anoxic zones; see Tables A.5 and A.6 for controlling aeration and Table A.9 for controlling internal recycle. Verify supplemental carbon addition flow. | Chapter 5<br>Chapter 7 |
| (2) Odors in anoxic zones. | (2) Under-loading of process is causing excessive detention time. | (2) Check detention time and oxidation-reduction potential in anoxic zones. | (2a) If possible, decrease volume of anoxic zone.<br>(2b) Add some dissolved oxygen to the beginning of anoxic zone.<br>(2c) Increase internal recycle to return more $NO_2$ + $NO_3$-N and dissolved oxygen. | |

TABLE A.3 Process Upset—Enhanced Biological Phosphorus Removal

| Indicator or observations | Probable cause | Check or monitor | Solutions | References |
|---|---|---|---|---|
| (1) Higher effluent total phosphorus concentration than normal. | (1a) Ratio of BOD to total phosphorus has changed. | (1a1) Check BOD, soluble BOD, total phosphorus, and orthophosphorus in influent to the biological nutrient removal process. | (1a1) Review facility operations to see if sidestreams from solids handling are causing periodically high loads, and adjust operations to reduce or even out total phosphorus loadings to process. | Chapter 6<br>Chapter 8 |
| | | (1a2) Check BOD, soluble BOD, total phosphorus, and orthophosphorus in influent to facility. | (1a2) If BOD to total phosphorus or soluble BOD to orthophosphorus is low, there may be insufficient VFAs for EBPR. | |
| | | (1a2a) High total phosphorus in influent. | (1a2a) Check for and discourage discharges to collection system that are causing unusually high total phosphorus. | |
| | | (1a2b) Low soluble BOD (low volatile fatty acids [VFAs]) in anaerobic zone. | (1a2b) Chemicals added to the collection system for odor control may inhibit fermentation and, therefore, formation of VFAs. Consider using different chemicals if this is the case. If possible, raise the sludge blanket in the primary clarifiers 0.31 to 0.61 m (1 to 2 ft). This will turn the primaries into a first-stage fermenter and produce some VFAs that will help recover the process. | |
| | | | (1a2c) Add VFAs, such as acetic acid, to anaerobic zone to increase EBPR. See Chapter 7. | |
| | | (1a3) Insufficient soluble BOD (VFAs) and additional VFAs unavailable. | (1a3) Use chemical phosphorus removal on short-term basis, or if this is a continuing situation, use chemicals and recalculate SRT and wasting based on higher inorganic content of MLSS resulting from chemical solids production. See Chapter 8. | |

| | | | | |
|---|---|---|---|---|
| | (1b) Inhibition of phosphorus release in anaerobic zone. | (1b) Check orthophosphorus, $NO_2 + NO_3$, $NH_3$-N, dissolved oxygen, and/or oxidation-reduction potential profiles through process. Observe mixing in anaerobic zone. | (1b) Optimize the anaerobic conditions in the anaerobic zone and ensure that there is no short circuiting; see Tables A.5 and A.6 for controlling aeration. If possible, reduce RAS rates to reduce $NO_3$ return to anaerobic zone. Increase anaerobic zone volume, if possible. | |
| | (1c) Inhibition of biological phosphorus uptake in aerobic zone. | (1c1) Check orthophosphorus, dissolved oxygen profiles through process. Observe mixing in aerobic zone(s). Check aerobic hydraulic retention time. Check dissolved oxygen level at beginning of aerobic zone. | (1c1) Ensure that there is sufficient hydraulic retention time and no short circuiting; see Tables A.5 and A.6 for controlling aeration. Install baffling, if needed, to block strong currents from inlet to outlet of aeration zones. Ensure ample dissolved oxygen at beginning of aeration zone; maintain minimum dissolved oxygen concentration of 2 mg/L if possible. | |
| | | (1c2) Check influent VFA and VFA orthophosphorus profiles through process. | (1c2) If shock load of influent VFA occurs, excessive phosphorus release in the anaerobic zone may exceed biological phosphorus uptake in the aerobic zone(s). Eliminate excess VFA discharge or equalize VFA load. Excessive VFA loading can promote filamentous bulking and foaming. | |
| | | (1c3) Check pH, temperature, and microorganism population for glycogen-accumulating organisms. | (1c3) If other causes of low biological phosphorus uptake have been eliminated from consideration, the treatment facility conditions (low pH, high temperature, etc.) may favor glycogen-accumulating organisms over phosphorus-accumulating organisms. Consider pH adjustment or chemical phosphorus removal. | |
| | | (1c4) Check the $PO_4$-P/TP (ratio of orthophosphorus to total phosphorus) to determine if increase is from particulate phosphorus in TSS or soluble phosphorus ($PO_4$-P). If chemical addition has been on an increasing trend, a microscopic analysis should be performed to check if chemical addition has cut off polyphosphate-accumulating organisms. | (1c4) Adjust chemical coagulant dosage if orthophosphorus has increased. Improve effluent suspended solids quality if particulate. Refer to Table A.8. | |
| (2) Odors in anaerobic zone. | (2) Under-loading of process is causing excessive detention time. | (2) Check detention time and oxidation-reduction potential in anaerobic zone. | (2a) If possible, decrease volume of anaerobic zone (e.g., decrease number of basins in service or lower operating water level).<br>(2b) Increase RAS flowrate to return more $NO_2$ + $NO_3$-N. | Chapter 6<br>Chapter 8 |

**TABLE A.4** Process Upset—Chemical Phosphorus Removal

| Indicator or observations | Probable cause | Check or monitor | Solutions | References |
|---|---|---|---|---|
| (1) Sudden increase in effluent total phosphorus concentration. | (1a1) Chemical feed pump failure. | (1a1) Verify that the pump has power.<br>(1a2) Check the manufacturer's troubleshooting information. | (1a1) Provide power.<br>(1a2) Perform system checks, and conduct maintenance as directed by the manufacturer. | |
| | (1b) Chemical precipitates have formed in the piping, plugging chemical feed. | (1b) Verify that chemical is reaching the application point, and potentially break the piping in search of the suspected restriction. | (1b) If the chemical flow is restricted, attempt to remove the restriction or restricted piping. Muriatic acid has been successful in dissolving chemical buildup. Contact the chemical distributor for additional advice. Consider discontinuing carrier water that may be promoting precipitate formation within the pipe. | |
| | (1c) High slug loading from an unknown source. | (1c1) Monitor key industrial contributors.<br>(1c2) Check sidestream contributions. | (1c1) Work with industries to control loads.<br>(1c2) Sidestreams should be treated carefully. Contingencies may include adding chemical to the sidestreams, equalizing the flow, and recycling sidestreams during periods of low loadings. | |
| | (1d) Solids carryover and breakthrough from solids separation process. | (1d) Monitor effluent suspended solids. | (1d) Improve effluent suspended solids quality. Refer to Table A.8. | |
| | (1e) Analytical error. | (1e) Laboratory quality assurance and quality control. | (1e) If laboratory data are in error, the data should be excluded from the data set for process control purposes and should be noted, as required, if used for regulatory reporting. | |
| | (1f) Change in ratio of orthophosphorus to total phosphorus. | (1f) Check ratio of orthophosphorus to total phosphorus to determine if increase in total phosphorus is attributable to phosphorus in soluble or particulate form. | (1f) Adjust chemical coagulant dosage if orthophosphorus has increased. Improve effluent suspended solids quality if particulate. Refer to Table A.8. | |

| | | | |
|---|---|---|---|
| (2) Gradual increase in effluent total phosphorus concentration. | (2a) Additional process loading because of industrial haracterist. | (2a1) Monitor key industrial contributors. | (2a1) Work with industries to control loads. |
| | | (2a2) Check sidestream loads. | (2a2) Sidestreams should be treated carefully. Contingencies may include adding chemical to the sidestreams, equalizing the flow, and recycling sidestreams during periods of low loadings. |
| | (2b) Inconsistent chemical strength. | (2b) Monitor chemical strength. | (2b1) Increase the dose rate when using weaker chemicals; decrease the dose rate when using stronger chemicals. |
| | | | (2b2) Require more consistent product from supplier. |
| | (2c) pH has shifted, resulting in less efficient phosphorus removal. | (2c) Check pH and alkalinity. | (2c) Adjust pH with appropriate chemical addition. |
| | (2d) Excessive SRT, increasing temperature. | (2d) Check SRT and temperature trends. | (2d) Polyphosphate-accumulating organisms (PAOs) perform best at SRTs around 1–2 days. Avoid excessive SRT. Target SRT to levels that will deliver needed nitrification performance at season temperatures. |

TABLE A.5 Aeration and Mixing Systems—Diffused Aeration

| Indicator or observations | Probable cause | Check or monitor | Solutions | References |
|---|---|---|---|---|
| (1) Low dissolved oxygen in aeration zones. | (1a) Poor oxygen transfer in aeration zones. | (1a) Check for diffuser problems. | (1a) Repair diffusers if needed; resolve fouling if noted. Consider scheduling routine diffuser cleaning. | |
| | (1b) Insufficient aeration. | (1b) Check airflow versus calculated airflow for loadings. Check dissolved oxygen profile through process. | (1b) Increase airflow to aeration zones until dissolved oxygen concentration is approximately 2.0 mg/L. If airflow per diffuser is too high and additional aeration zones are available, put additional aeration zones in service. | |
| | (1c) Changes in influent characteristic, decreasing field oxygen transfer efficiency. | (1c) Check airflow versus calculated airflow for loadings. Check dissolved oxygen profile through the process. | (1c) Increase airflow to aeration zones if possible. | |
| (2) High dissolved oxygen in aeration zones. | (2) Flow, BOD, and/or TKN are lower than design. | (2a) Check airflow versus calculated airflow for loadings. Check dissolved oxygen profile through process. | (2a) Decrease airflow to aeration zones until target dissolved oxygen is reached. If airflow per diffuser is too low, take aeration zone(s) out of service to maintain at least the minimum recommended airflow per diffuser. | |
| | | (2b) Compare actual flow, BOD, and TKN to design values. Calculate required airflow. | (2b) If diffusers and blowers allow, reduce the amount of air added by periodically reducing or shutting off air (not more than 1.5 to 2 hours off to prevent odors). Use a combination of on-time and either reduced airflow time or off-time throughout the day to better match aeration to the oxygen demand.<br>(2b) If multiple trains are available, reduce number of trains in service. | |

| | | | |
|---|---|---|---|
| (3) Significant dissolved oxygen into anoxic zones. | (3a) Too much dissolved oxygen at internal recycle suction. | (3a) Check dissolved oxygen in aeration zone where internal recycle originates. | (3a) For activated sludge process with internal recycle, taper dissolved oxygen profile for low dissolved oxygen at internal recycle suction. Low oxygen uptake at end of basin may require supplemental mixing to allow reduced airflow below that required for mixing to reduce dissolved oxygen. |
| | (3b) Too much turbulence at influent. | (3b) Observe turbulence or splashing at influent. | (3b) Adjust basin levels, use baffles, and/or modify inlet ports to minimize turbulence and introduction of dissolved oxygen to anoxic zones. |
| (4) Significant dissolved oxygen and/or nitrite/nitrate into anaerobic zone. | (4a) Too much dissolved oxygen in RAS. | (4a) Check dissolved oxygen in RAS. Check RAS concentration and flowrate. | (4a) Reduce dissolved oxygen in aeration zone immediately upstream of clarifiers. If clarifier operation allows, reduce RAS flow. |
| | (4b) Too much nitrite/nitrate returning with RAS. | (4b) Check nitrite/nitrate concentration in RAS. Check RAS concentration and flowrate. | (4b) If nitrification is not needed, reduce SRT to reduce nitrite/nitrate concentration going to clarifiers. If clarifier operation allows, reduce RAS flow. Optimize denitrification. |
| | (4c) Too much turbulence at influent. | (4c) Observe turbulence or splashing at influent. | (4c) Adjust basin levels, use baffles, and/or modify inlet ports to minimize turbulence and introduction of dissolved oxygen to anaerobic zones. |
| (5) Poor mixing pattern. | (5a) Diffusers need cleaning or repair. | (5a) Visual observation of mixing pattern. | (5a) Bump or chemically clean in-place diffusers; take basin out of service, and manually clean diffusers, replace broken diffusers, or periodically replace all diffusers if beyond expected service life. |
| | (5b) If blow-offs are present, air is releasing through blow-offs. | (5b) Check for excessive diffuser headloss or blow-off malfunction. | (5b) Clean or replace diffusers as in 5a and/or repair blow-off system. |
| | (5c) Diffusers are not installed at same elevation. | (5c) Check that diffusers are installed level and at the same elevation. | (5c) Take basin out of service, and adjust diffuser support systems and/or laterals so that they are level and so diffusers are at the same elevation. |

(*continued*)

TABLE A.5 Aeration and Mixing Systems—Diffused Aeration (*continued*)

| Indicator or observations | Probable cause | Check or monitor | Solutions | References |
|---|---|---|---|---|
| (6) No or too little turbulence. | (6a) Too few blowers in operation. | (6a) Check number of blowers in operation and control system. | (6a) Increase number of blowers in operation, or adjust controls to bring on more blowers if automatically controlled. | |
| | (6b) Blower malfunction. | (6b) Compare airflow at each blower to its performance curve(s). | (6b) Diagnose and repair or replace malfunctioning parts. Check setting of inlet air vanes and valves. | |
| | (6c) Dirty inlet filter. | (6c) Check inlet air pressure between filter and blower. | (6c) Replace inlet filters; clean filters if washable. | |
| | (6d) Valves need adjustment. | (6d) Check airflow to basin. | (6d) Adjust manual valves to better distribute air. | |
| | (6e) Aeration control system needs adjustment. | (6e) Compare control parameter values to set point(s). | (6e) If insufficient air is provided under automatic operation, adjust controls to increase airflow when setpoints are maintained. | |
| | (6f) Air rate is too low for proper operation of diffusers. | (6f) Compare airflow divided by number of diffusers to acceptable low value. | (6f) Increase airflow to provide at least minimum recommended airflow per diffuser. | |
| (7) Excessive turbulence over entire basin. | (7a) Too many blowers in operation. | (7a) Check dissolved oxygen in aeration zones and number of blowers in operation. | (7a) If dissolved oxygen is above desired levels, decrease airflow. Stay within range of recommended airflow per diffuser as in 7c. | |
| | (7b) Aeration control system needs adjustment. | (7b) Compare control parameter values to setpoint(s). | (7b) If excess air is provided under automatic operation, adjust controls to decrease airflow when setpoints are maintained. | |
| | (7c) Air rate is too high for proper operation of diffusers. | (7c) Compare airflow divided by number of diffusers to acceptable high value. | (7c) Decrease airflow to provide, at most, the maximum recommended airflow per diffuser. | |
| | (7d) Leak(s) in air-distribution piping system. | (7d) Check for leaks in air-distribution piping system. | (7d) Repair leaks. | |

TABLE A.6 Aeration and Mixing Systems—Mechanical Aeration

| Indicator or observations | Probable cause | Check or monitor | Solutions | References |
|---|---|---|---|---|
| (1) Low dissolved oxygen in aeration zones. | (1a) Poor oxygen transfer in aeration zones. | (1a) Check for aerator problems. | (1a) See solutions 5, 6a, and 7. | |
| | (1b) Insufficient aerators in service. | (1b) Check total horsepower in service versus calculated horsepower based on design oxygen transfer. Check dissolved oxygen profile through process. | (1b) Increase number of aerators in service until dissolved oxygen concentration is approximately 2.0 mg/L. If additional aeration zones are available, put them in service. | |
| (2) High dissolved oxygen in aeration zones. | (2) Flow, BOD, and/or TKN are lower than design. | (2a) Check total horsepower in service versus calculated horsepower. Check dissolved oxygen profile through process. | (2a) Decrease level in basin to lower impeller submergence until dissolved oxygen is approximately 2.0 mg/L. If dissolved oxygen is still too high, take aerator out of service, use low speed if two-speed aerators, or take an aeration zone out of service to maintain minimum impeller submergence. | |
| | | (2b) Compare actual flow, BOD, and TKN to design values. Calculate required aeration horsepower. | (2b) If aerators allow, periodically reduce the speed of the aerator to reduce oxygen transfer, but maintain mixing. If aerator is able to resuspend solids, consider shutting off aerator(s) (not more than 1.5 to 2 hours off to prevent odors). Use a combination of on-time and either reduced speed time or off-time throughout the day to better match aeration to the oxygen demand. For large aerators, on-off operation may be too hard on gear reducers; consider soft-start of motors if on-off operation is used. | |

(*continued*)

**TABLE A.6** Aeration and Mixing Systems—Mechanical Aeration (*continued*)

| Indicator or observations | Probable cause | Check or monitor | Solutions | References |
|---|---|---|---|---|
| (3) Significant dissolved oxygen into anoxic zones. | (3a) Too much dissolved oxygen at internal recycle suction. | (3a) Check dissolved oxygen in aeration zone where internal recycle suction originates. | (3a) For activated sludge process with internal recycle, locate internal recycle suction where dissolved oxygen is lowest and adjust submergence, as in 2a. If nitrite/nitrate concentration in effluent is low, reduce internal recycle rate and monitor response. | |
| | (3b) Too much turbulence at influent. | (3b) Observe turbulence or splashing at influent. | (3b) Adjust basin levels, use baffles, and/or modify inlet ports to minimize turbulence and introduction of dissolved oxygen to anoxic zones. | |
| (4) Significant dissolved oxygen and/or $NO_2 + NO_3$-N into anaerobic zone. | (4a) Too much dissolved oxygen returning with RAS. | (4a) Check dissolved oxygen in RAS. Check RAS concentration and flow. | (4a) Reduce dissolved oxygen in aeration zone immediately upstream of clarifiers. If clarifier operation allows, reduce RAS flow. | |
| | (4b) Too much nitrite/nitrate returning with RAS. | (4b) Check nitrite/nitrate returning with RAS. Check RAS concentration and flowrate. | (4b) If nitrification is not needed, reduce SRT to reduce nitrite/nitrate concentration going to clarifiers. If clarifier operation allows, reduce RAS flow. | |
| | (4c) Too much turbulence at influent. | (4c) Observe turbulence or splashing at influent. | (4c) Adjust basin levels, use baffles, and/or modify inlet ports to minimize turbulence and introduction of dissolved oxygen to anaerobic zones. | |
| (5) Surging noise and waves in aeration zone. | (5a) Water level is too low for mechanical aerator impeller. | (5a) Visual observation of waves in basin and surging noise; check impeller submergence level. | (5a) Increase level in basin to provide recommended minimum impeller submergence; repair leaks in weirs or gates that allow water level to drop below desired level at low flows. | |
| | (5b) Basin design is prone to wave formation at aerator. | (5b) Visual observation of waves reflecting off surfaces and creating standing wave. | (5b) Adjust aerator speed or position, if possible, or modify basin using baffles to reduce the standing wave effect. | |

| | | | |
|---|---|---|---|
| (6) Motor overload. | (6a) Water level is too low for mechanical aerator impeller. | (6a) Visual observation of waves in basin, surging noise; check impeller submergence level. | (6a) See 5a in this guide. |
| | (6b) Water level is too high for mechanical aerator impeller. | (6b) Check impeller submergence level, especially at peak flow. | (6b) Decrease level in basin to limit impeller submergence to recommended maximum at peak flow. |
| (7) Vibration, reduced splashing, low dissolved oxygen. | (7a) Impeller is fouled with debris. | (7a) Check for vibration, visual observation of impeller, and visual observation of splash pattern. | (7a) Take aerator out of service and remove debris from impeller. |
| | (7b) Impeller is fouled with ice. | | (7b) Take aerator out of service, and remove ice from impeller, install manufacturer-approved shields that keep splash confined to basin, and minimize ice formation. |

**TABLE A.7** Biomass Inventory

| Indicator or observations | Probable cause | Check or monitor | Solutions | References |
|---|---|---|---|---|
| (1) Wastewater temperature is low, dissolved oxygen is higher; higher effluent TKN than normal. | (1) Cold weather is slowing down biomass activity. | (1) Check wastewater temperature in process. | (1) Increase SRT to continue nitrification, if required. If nitrification is not required during cold weather, then adjust process to discourage nitrification (lower SRT, less air, and lower MLSS and RAS flowrate). | |
| (2) MLSS is low, dark diluted color of MLSS, accompanied by | | | | |
| (2a) Higher effluent TKN than normal. | (2a1) MLSS is lower than desired because of excessive wasting. | (2a1) Check MLSS, dissolved oxygen, ammonia, and nitrite/nitrate at end of process before clarification. | (2a1) Decrease wasting to allow MLSS to increase for an SRT adequate for nitrification to occur. Select a better time to obtain more consistent MLSS concentration and/or use moving average of 7 or more days on which to base wasting. Institute operational limits on the amount of wasting that can occur in a single day. | Table A.8 |
| | (2a2) High effluent TSS. | (2a2) Check effluent TSS. | (2a2) Refer to Table A.8. | |
| (2b) If using an EBPR removal process without nitrification, higher effluent total phosphorus than normal. | (2b) SRT, MLSS is low to inhibit nitrification, but is too low for the temperature to EBPR PAOs. | (2b) Check SRT, MLSS, and dissolved oxygen, and check orthophosphorus in anaerobic zone and at end of process before clarification. | (2b) Decrease wasting as in 2a1. If nitrification occurs and there is insufficient aeration capacity to maintain stable operation, operate at lower SRT, MLSS, and implement chemical phosphorus removal until temperature increases to where PAOs are restored; see Chapter 8. | Chapter 8 |
| (3) MLSS and SRT are high, pin floc in secondary clarifier effluent, high sludge volume index (SVI), sometimes dark tan foam on aeration basin. Higher effluent TKN or total phosphorus than normal. | (3a) High SRT, MLSS resulting from reduced wasting or seasonally higher temperatures. | (3a1) If multiple trains are in service, calculate loadings if fewer trains are in service.<br>(3a2) Check MLSS and waste activated sludge (WAS) solids concentrations and WAS flowrate, and calculate SRT. | (3a1) If calculated loadings are within design values with fewer trains in service, decrease number of trains in service.<br>(3a2) Lower MLSS concentration to lower SRT, and consider lowering RAS rate. | |

| | | | |
|---|---|---|---|
| (4) Secondary release of phosphorus in the second anoxic basin yielding high total phosphorus concentration in the effluent. | (4a) Detention time in the second anoxic is too high. | (4a1) Monitor RAS recycle rate.<br>(4a2) Check for excessive SRT. | (4a1) Increase RAS recycle rate.<br>(4a2) Adjust SRT for securing nitrification and no higher. |
| | (4b) Lack of carbon in the second anoxic basin preventing denitrification. | (4b) BOD or chemical oxygen demand into the second anoxic zone. | (4b) Supplement carbon to the second anoxic basin. |
| | (4c) First anoxic basin completing denitrification. | (4c1) Monitor the MLSS recycle rate to the first anoxic basin.<br>(4c2) Monitor the oxygen supply to the aerobic zone. | (4c1) Decrease the MLSS recycle to the first anoxic basin to shift more nitrate load to the second zone.<br>(4c2) Increase the oxygen supply to the last stage before the second anoxic zone. |
| (5) High chemical demand for phosphorus removal affecting the nitrification performance of the system. | (5) Too much chemical sludge produced by metal salts used for phosphorus precipitation. | (5) Verify where the metals salts are being fed. Verify the ratio of mixed liquor volatile suspended solids to MLSS in the system. | (5) Implement staged chemical addition to reduce the dosage fed to the secondary clarifiers and trim at a tertiary treatment process. |

TABLE A.8 Clarifier Operation

| Indicator or observations | Probable cause | Check or monitor | Solutions | References |
|---|---|---|---|---|
| (1) Floating solids at surface, sometimes thick solids layer, possibly bulking solids. Higher effluent TSS or total phosphorus than normal. | (1a) Denitrification in clarifier is causing rising sludge. | (1a) Check nitrite/nitrate and dissolved oxygen concentrations entering and leaving clarifiers. | (1a) Optimize denitrification process (increase internal recycle, lower dissolved oxygen in internal recycle, add supplemental carbon, etc.) to achieve <10 mg/L nitrate in final effluent. Reduce sludge blanket in clarifier. Increase dissolved oxygen in clarifier feed. Decrease number of clarifiers in service if loading allows. | |
| | (1b) RAS is not being removed quickly enough. | (1b) Monitor blanket height, and check RAS rate and detention time in clarifiers. Check orthophosphate in RAS. | (1b) Increase RAS return rate and/or increase dissolved oxygen in clarifier influent. Verify that RAS is flowing normally from each clarifier. Reduce solids detention time in clarifiers (lower blanket height, increase RAS rate, and reduce number of clarifiers in service). | |
| | (1c) Excessive turbulence is causing air bubbles in floc. | (1c) Observe turbulence and presence of bubbles on floc from mixed liquor effluent. | (1c) Reduce turbulence in aeration tank immediately upstream of clarifiers, and reduce turbulence between aeration tank and clarifier inlet. | |
| | (1d) Excessive filamentous organisms in MLSS are causing bulking sludge. | (1d) Check settleability and sludge volume index. | (1d) Minimize growth of filamentous organisms by adjusting causative conditions in process basins. Filamentous bacteria identification will support process data for identifying causative conditions. | |
| (2) Solids overflowing clarifier weir. Higher effluent TSS and BOD than normal. | (2a) RAS is not being removed quickly enough. | (2a) Monitor blanket height, and check RAS rate. | (2a) See Solution 1b. | |
| | (2b) Unequal flow distribution. | (2b) Check flow to each clarifier. | (2b) Adjust flow-splitting devices to provide equal flow to same-sized clarifiers, or provide equal loadings to different-sized clarifiers. | |
| | (2c) Hydraulic and/or solids overloading. | (2c) Check surface overflow rates at peak flow. Check solids loading rate. | (2c) If available, place another clarifier in service. If equipped, switch to step-feed operation to reduce solids loading. Reduce surges to the clarifiers by controlling pumping stations in collection system and recycle streams from solids handling to equalize flows. Decrease solids inventory if process allows. Alter mode of operation to improve settleability. | |
| | (2d) Turbulence from collection rake. | (2d) Observe solids carryover when rake passes by. | (2d) If possible, reduce rake travel speed. | |

| | | | | |
|---|---|---|---|---|
| (3) Phosphorus in floc. Higher effluent TSS than normal. | (3a) Excessive turbulence upstream of clarifiers. | (3a) Observe turbulence upstream, and check dissolved oxygen in last aeration zone. | (3a) Reduce aeration in aeration zone upstream of clarifiers. Adjust flow-splitting devices or water levels to reduce turbulence upstream of clarifiers. Add chemical flocculant (polymer) on an interim basis. | Chapter 12 |
| | (3b) SRT is too long. | (3b1) Calculate SRT (check MLSS and WAS solids concentration and WAS flowrate).<br>(3b2) Check SVI and filamentous population. | (3b1) Reduce SRT by increasing sludge wasting. Limit wasting increase to 10% higher for 2 times SRT to minimize unstable operation. Consider polymer addition to secondary clarifier influent to enhance TSS capture. | |
| | (3c) Short-circuiting in clarifier. | (3c) Look for areas where floc carryover is heaviest. Observe water level along effluent weir. | (3c) Level weirs. If possible, add or adjust baffling to reduce excessive velocities and density currents. | |
| | (3d) Upset resulting from loading or colloidal solids in sidestreams from solids handling. | (3d) Measure TSS and turbidity in sidestreams from solids handling. Check dissolved oxygen profile through process. | (3d) Look at facility operations to see if sidestreams from solids handling are causing periodically high loads, and adjust operations to reduce colloidal solids and even out BOD and TKN loadings to the process. See Chapter 12. | |
| | (3e) Toxicity. | (3e) Refer to Table A.11. | (3e) Refer to Table A.11. | |
| (4) Thin RAS. Higher effluent total phosphorus than normal. | (4a) Plugging of sludge withdrawal. | (4a) Check RAS concentration; observe sludge withdrawal equipment. | (4a) Backflush RAS collection system. | |
| | (4b) RAS return rate too high. | (4b) Check RAS concentration. | (4b) Reduce RAS return rate. | |
| | (4c) Poor sludge settling. | (4c) Check SVI and compare to historical values. | (4c) Minimize formation of filamentous organisms by adjusting causative conditions in process basins. | |
| (5) Turbid effluent. Higher effluent TSS or TKN than normal. | (5) Toxic or acid constituents in wastewater. | (5) See Table A.11. | (5) See Table A.11. | |

TABLE A.9 Internal Recycle

| Indicator or observations | Probable cause | Check or monitor | Solutions | References |
|---|---|---|---|---|
| (1) If denitrifying through internal recycle, high nitrite/nitrate in effluent upstream of clarifiers. Higher effluent nitrite/nitrate than normal. | (1a) Internal recycle flowrate is insufficient. | (1a1) Verify that internal recycle pumps are operating. | (1a1) If pumps are not operating, fix the problem and restore internal recycle flow. | |
| | | (1a2) Compare internal recycle flowrate versus design rate. Look for clog in valve, line, or pump. | (1a2) If pumps are operating and flowrate is lower than normal, unclog internal recycle system and restore internal recycle flow. | |
| | (1b) Internal recycle is returning too much. | (1b) Check dissolved oxygen concentration in internal recycle. | (1b) Reduce dissolved oxygen concentration in internal recycle. See Tables A.5 and A.6. | |
| | (1c) Insufficient carbonaceous BOD in anoxic zones. | (1c) See Table A.2. | (1c) See Table A.2. | |
| (2) Dissolved oxygen in anoxic zone. Higher effluent nitrite/nitrate than normal. | (2) Internal recycle rate is too high for the loading to the process. | (2) Calculate internal recycle flowrate needed to achieve nitrogen removal and compare to actual rate. | (2) Lower internal recycle flowrate. | |
| (3) Low dissolved oxygen in aeration zone with aeration system at maximum; average or high ammonia and low nitrite/nitrate in effluent upstream of clarifiers. | (3) Oxygen-limited condition in aeration zone is causing denitrification to occur simultaneously in aeration zone. | (3) Check profiles of dissolved oxygen, ammonia, and nitrite/nitrate through the process. | (3) For short-term solution, consider discontinuing internal recycle flow if aeration zone is large enough to accomplish simultaneous nitrification and denitrification. Increase aeration in aeration zone to reduce effluent ammonia, raise dissolved oxygen in aeration zone, and allow resumption of internal recycle. | |

**TABLE A.10** pH and Alkalinity

| Indicator or observations | Probable cause | Check or monitor | Solutions | References |
|---|---|---|---|---|
| (1) pH is average in influent, but there is low pH in process effluent. Higher effluent TKN than normal. | (1a) Solids retention time is too low. | (1a) Check pH profile through process and effluent ammonia. | (1a) If SRT is low, increase SRT by reducing MLSS through decreased wasting to ensure adequate nitrification achieved. | Chapter 3<br>Chapter 7 |
| | (1b) Insufficient alkalinity to maintain pH. | (1b1) Check MLSS, WAS flowrate, and WAS concentration, and calculate SRT. | (1b1) If alkalinity is low, add alkalinity through chemical addition. See Chapters 3 and 7. | |
| | | (1b2) Check effluent alkalinity. | (1b2) Consider implementing anoxic zone if none exists to recover alkalinity through denitrification. The anoxic zone may be created by shutting off aeration/isolating an aeration zone and ensuring good mixing through other means. Internal recycle may be used to maximize alkalinity recovery. See Chapter 7. | |
| (2) Low pH or alkalinity in influent to BNR process. | (2a) Low pH or alkalinity entering the facility. | (2a) Check influent pH and alkalinity relative to TKN load. | (2a) See Table A.1. | Chapter 3<br>Chapter 7<br>Chapter 12 |
| | (2b) Sidestreams in facility or acidic discharge in facility. | (2b) Check pH and alkalinity in sidestreams. | (2b) Look at facility operations to see if sidestreams are causing low pH or alkalinity, and adjust operations to even out pH and alkalinity entering the process. If acid is used in the facility, consider neutralizing before returning to facility. See Chapter 12. | |
| | (2c) Insufficient alkalinity. | (2c) Check pH and alkalinity in process influent and effluent. | (2c) Add alkalinity through chemical addition. See Chapters 3 and 7. | |

TABLE A.11 Toxicity

| Indicator or observations | Probable cause | Check or monitor | Solutions | References |
|---|---|---|---|---|
| (1) Light color and/or high dissolved oxygen in aeration zones. Higher effluent TSS, TKN, or total phosphorus than normal. | (1) Toxic load inhibited nitrification, leading to over-aeration. | (1a) Check pH profile through process, dissolved oxygen in aeration zones, ammonia, and nitrite/nitrate through process. Check oxygen utilization rate of MLSS.<br>(1b) Check raw wastewater samples and return streams from solids handling for toxic components. | (1a) Maintain as much biomass in areas unaffected by the toxic load through rerouting of RAS, reduction in aeration or mixing, or isolation of basins to preserve biomass for reseeding. Consider an increase in return sludge flowrate for more dilution to the influent that has toxic substance.<br>(1b) Identify and eliminate the source of toxic load. | |
| (2) Raw or chemical odor in process, pin floc in clarifier effluent, or cloudy effluent. Higher effluent TSS or TKN than normal. | (2) Toxic load has affected the biomass. | (2a) Take sample and look for dead (inactive) microorganisms under the microscope to confirm.<br>(2b) See 1b above. | (2a) See solution 1a.<br>(2b) See solution 1b. | |
| (3) For attached growth process, excessive sloughing with little or no biomass on media. Higher effluent TSS or TKN than normal. | (3) Toxic load has affected the biomass. | (3) See 1b and 2a. | (3) See Solution 1b. If recirculation is being used, turn off recirculation until toxic load has passed; restart recirculation to dilute any toxic residue after the load has passed. | |

# Index

www.ingramcontent.com/pod-product-compliance
Lightning Source LLC
LaVergne TN
LVHW050557240526
839643LV00006B/79
* 9 7 8 1 5 7 2 7 8 4 6 6 6 *